Fundamentals of
REINFORCED CONCRETE DESIGN

Fundamentals of REINFORCED CONCRETE DESIGN

M.L. Gambhir
Formerly Professor and Head
Department of Civil Engineering
Dean, Planning and Resource Generation
Thapar Institute of Engineering and Technology
Patiala, Punjab

PHI Learning Private Limited
Delhi - 110092
2026

*In fond memory of **Shri Asoke K. Ghosh** (October 1942 – February 2024), Founder Chairman and Managing Director of PHI Learning, whose vision endlessly inspires.*

The Legacy Continues. . . .

Published by Pushpita Ghosh, PHI Learning Private Limited, Rimjhim House, 111, Patparganj Industrial Estate, Delhi-110092 and Printed by Syndicate Binders, A-20, Hosiery Complex, Noida, Phase-II Extension, Noida-201305 (N.C.R. Delhi).

₹995.00

FUNDAMENTALS OF REINFORCED CONCRETE DESIGN
M.L. Gambhir

ISBN-978-81-203-3048-1 (Print Book)
ISBN-978-93-5443-075-6 (e-Book)

To
The Society
To whom we owe a lot

Contents

Preface

The book is primarily designed to be an applied text to cater to the class-room or self-study needs of students at undergraduate level in Civil Engineering. It covers all the basic topics of reinforced concrete design generally taught in first course in Civil Engineering curriculum in Indian Universities. It presents, in simple terms, the basic principles of reinforced concrete design, a thorough knowledge of which is essential for proper understanding of current design practices and code provisions. It conforms to the limit states design method as given in the latest revision of IS:456. The traditional working stress design method is given in the appendix for the use in investigation of limit states of serviceability and for design in the situations where the use of limit states design approach is not convenient.

In writing this text I have mainly drawn on my lecture notes developed while teaching the subject and on the experience accumulated over the years as a result of both research and consultancy. Considerable effort has been devoted to the detailed discussion of basic concepts, behaviour of various structural components under loads, and development of fundamental expressions for analysis and design. The emphasis is on clarity of concept and development of structural sense needed for proper detailing. The text presents efficient and systematic procedures for solving design problems. In addition to the discussion of basis for design calculations, a large number of worked-out practical design examples based on the appropriate codes and current design practices have been included to illustrate salient features of reinforced concrete design. A wide variety of well-labelled diagrams are provided throughout the text to help the reader to develop sound judgement in practical design. Review questions and tutorial problems are included at the end of each chapter for class-room or self-study to facilitate thorough comprehension of the fundamentals.

In Chapter 1, a concise discussion on the properties of concrete and steel has been included to give the reader a feel for constituent materials of total structural system. Chapter 2 compares various design approaches to reinforced concrete design and discusses the codal recommendations. Chapters 3 to 6 deal with limit states of collapse in flexure; shear, bond, and torsion; compression, and limit state of serviceability. In Chapter 7, the application of basic principles discussed in preceding chapters to the design of key building components has been given to enable the reader to undertake the practical design of a common range of structures. Chapter 8 on detailing the structures describes good detailing and construction practices. The detailing of steel is considered to be an art and is carried out according to the stipulations given in approved manuals for integrated action in various parts of the structure. The durability and serviceability aspects have been given due consideration for an efficient reinforced concrete design.

Time saving analysis and design aids in the form of tables and charts have been developed for use in the design office. The relevant algorithms used in development of design aids are explained in details. The design aids would prove extremely useful to the practicing engineers engaged in actual practice. There has been conscious effort to present results in non-dimensional form to facilitate the application to different materials and cross-sectional dimensions. A large amount of practical data in tabular form is given in the appendix for the use in design office.

Extensive reference is made to IS code provisions, but care is taken to avoid overdependence on the code to enable the reader to rationally assess the design situation rather than blindly follow the code provisions. However, to maximize the benefits, the readers are advised to use IS:456, SP:16 and SP:34 along with the text keeping in mind that the code stipulations should be used as a guide only. A structural engineer must use his/her judgement in addition to calculations in interpretation of various provisions of the code to obtain an efficient and economical structure.

The subject matter, its format and presentation sequence has been class tested. It is hoped that the text will prove to be a dependable companion for teachers and practicing engineers.

I am thankful to the Bureau of Indian Standards for their published material to which references are made at numerous places in the text. I thank all those who have assisted in various ways in preparation of the text. Particularly I wish to acknowledge the assistance rendered by Dr. Puneet Gambhir in preparation of the manuscript. I am extremely grateful to my wife, Ms. Saroj Gambhir for the patience she has shown while I was busy completing the job. I express my gratitude to Ms. Neha Gambhir who has developed design aids and contributed in making the text possible.

I welcome suggestions from the readers for improvement in the subject matter in any manner.

M.L. GAMBHIR

CHAPTER

1

Introduction to Reinforced Concrete

1.1 INTRODUCTION

Concrete is the most widely used material for construction. It consists of a binding medium of cement and water called **cement paste**, and particles of relatively inert filler called **aggregate** (and sometimes *admixture*). The mixture, when placed in forms and allowed to cure, becomes hard like stone. The hardening is caused by chemical reaction between water and the cement, which continues for a long time, and consequently the concrete grows stronger with age.

The popularity of the concrete is due to the fact that from the common ingredients, it is possible to tailor the properties of concrete to meet the demands of any particular situation. The advances in concrete technology have paved the way to make the best use of locally available materials by judicious mix proportioning and proper workmanship, so as to produce concrete satisfying performance requirements.

The finished product (hardened concrete) has high compressive strength, but its tensile strength is very low—approximately one-tenth of its compressive strength. In situations where tensile stresses are developed the concrete is strengthened by steel bars forming a composite construction called **reinforced cement concrete**. The concrete without reinforcement is called **plain concrete** or simply **concrete**.

Thus, concrete making is not just a matter of mixing ingredients to produce a plastic mass, but good concrete has to satisfy the performance requirements in plastic or green state and also in the hardened state. In the plastic state the concrete should be workable and free from *segregation* and *bleeding*. Segregation is the separation of coarse aggregate, and bleeding is the separation of cement paste from the main mass. In its hardened state concrete should be strong, durable and impermeable; and it should have minimum dimensional changes. In reinforced concrete, the concrete and steel components are to be so arranged and proportioned that the optimal use is made of the materials involved.

In order to obtain quality concrete for structural use, knowledge of concrete-producing materials and their proportioning becomes essential. This chapter describes briefly the concrete-making materials, namely, cement, aggregate, water, admixture, and the resulting product. In addition, types and properties of the reinforcement are described, and a general overview of basic concrete structural systems is given.

1.2 CONCRETE-MAKING MATERIALS

1.2.1 Cement

Depending upon the intended use the cement may be any of the: Ordinary Portland Cement (OPC) of grades 33, 43 and 53; rapid hardening Portland cement, Portland slag cement, Portland pozzolana cement, low heat Portland cement and sulfate resisting cement. However, the cement should comply with the requirements of the relevant code.

Concrete mixes having high cement content give rise to increased shrinkage, creep and cracking. For high strength concrete increase in cement content beyond a certain value of the order of 550 kg/m^3 of concrete may not increase the concrete strength. IS:3370 (Part-1) stipulates a maximum limit of cement content to 530 kg/m^3. The minimum cement content is fixed by durability considerations.

A designer should be conscious of the following about the grade of cement:

1. The use of high grade cement does not necessarily yield high grade concrete as the strength of concrete depends on the mix proportions of ingredients. It is possible to produce concrete of wide ranging strengths using a particular grade of cement as it is done in USA.
2. The use of higher grade of cement does not produce better quality concrete. Two cements of different grades can produce concrete of same quality. Any grade or brand of cement that satisfies the minimum requirements prescribed by the code should generally produce concrete of desired properties with regard to strength and durability if proper mix proportions are chosen.
3. For satisfying the strength requirements proper cement content, limited water content, adequate compaction and appropriate curing of hardened concrete should be ensured. Durability of concrete on the other hand, mainly depends on the characteristics of both the materials and environment in which it is to be placed.
4. Beyond a certain age all the grades of cement give same strength, and the only advantage in using higher grade cement is the faster rate of gain in strength during the initial two or three weeks. But the faster rate of hydration results in higher heat of hydration, increased shrinkage, and lower later-age strength.

5. From durability considerations, the code has specified a minimum quantity of cement, independent of grade of cement. A lesser amount of cement produces harsh mix with poor workability resulting in honey-combing. In such a situation, if the quantity of water is based on workability, it will lead to higher water-cement ratio resulting in higher void ratio and permeability.

1.2.2 Mineral Additives

The pozzolanic material such as fly ash, silica fume, rice husk ash, metakaoline; and ground granulated blast furnace slag conforming to relevant standards may be used provided uniform blending with cement is ensured.

1.2.3 Aggregates

The aggregate is used primarily for the purpose of providing bulk to the concrete, and constitute 60 to 80 per cent of the finished product. To increase the density of the resulting mix, the aggregate is frequently used in two or more sizes. The most important function of smaller size particles (fine aggregate) is to assist in producing workability and uniformity in the mixture. The fine aggregate also assists the cement paste to hold the larger particles (coarse aggregate) in suspension. This action promotes plasticity in the mixture and prevents the possible segregation of paste and coarse aggregate. Aggregates shall comply with the requirements of IS:383. As far as possible the natural aggregates shall be preferred.

Coarse aggregate: An aggregate such as crushed stone, crushed gravel or uncrushed gravel most of which is retained on 4.75 mm IS sieve is termed as **coarse aggregate**. The graded coarse aggregate is described by its nominal size, e.g. a graded aggregate of nominal size of 10 mm means an aggregate most of which passes through 10 mm sieve. The particle size distribution of an aggregate as determined by sieve analysis is termed as **grading of an aggregate**. The strength of fully compacted concrete with a given water-cement ratio is independent of the grading of an aggregate but it affects the workability of the concrete. The aggregate of coarser grading require more fines to produce cohesive concrete where as finer grading requires more water for the given workability.

Size of aggregate: The nominal maximum size of coarse aggregate should be as large as possible within the limits specified but in no case greater than one-fourth of the minimum thickness of the member, provided that the concrete can be placed without difficulty so as to surround all reinforcement thoroughly and fill the corners of the form. For most reinforced cement concrete works, 20 mm aggregate is suitable. In concrete elements with thin sections, closely spaced reinforcement or small cover, use of 10 mm nominal maximum size should be considered. For heavily reinforced concrete members as in the case of ribs or webs of main beams, the nominal maximum size of the aggregate should usually be restricted to 5 mm less than the minimum clear distance between the main bars or 5 mm less than the minimum cover to the reinforcement whichever is smaller.

Fine aggregate: An aggregate such as natural sand or crushed stone or crushed gravel sand, which consists of particles most of which pass through 4.75 mm IS sieve is termed **fine aggregate**. The sand is generally considered to have a lower size limit of about 0.07 mm. According to size, the fine aggregate may be described as coarse, medium and fine sands. Depending upon the particle size distribution IS: 383 have divided the fine aggregate in four grading zones primarily based on percentage passing the 600 μm IS sieve. The grading zones become progressively finer from grading zone I to grading zone IV. Sand of grading zone IV produces nearly gap-graded concrete (low sand content), and zone I produces harsh mix and high cement content may be necessary for a higher workability.

1.2.4 Water

Water used for mixing and curing shall be clean and free from injurious amounts of oils, acids, alkalis, salts, sugar, organic materials or other substances that may be deleterious to concrete or steel. Potable water is generally considered satisfactory for mixing concrete.

Average 28 days compressive strength of concrete cubes prepared with water proposed to be used shall not be less than 90 per cent of the average of strength of similar concrete cubes prepared with distilled water. The process of preparation, curing and testing of both types of cubes shall be in accordance with IS: 516.

The initial setting time of cement using the proposed water shall not be less than 30 minutes and shall not differ by ±30 minutes from the initial setting time of the same cement using distilled water.

The pH value of water shall not be less than 6. Mixing or curing of concrete with seawater is not recommended because of presence of harmful salts in the seawater. Water found satisfactory for mixing is also suitable for curing concrete. However, water used for curing should not produce an objectionable strain or bad looking deposit on the concrete surface.

1.2.5 Admixtures

Admixtures are introduced in a concrete mix to modify the properties of concrete in its fresh and hardened states. The properties commonly modified are: the rate of hydration or setting time, workability, dispersion and air-entrainment. The commonly used admixtures are: accelerators, retarders, air-entraining agents, plasticizers (water-reducers), super plasticizers and bonding admixtures. A degree of control must be exercised to ensure proper quantity, as an excess quantity may be detrimental to the properties of concrete.

An admixture, if used shall comply with IS: 9103. It should neither impair the durability of concrete nor combine with the constituent of cement to form harmful compounds nor increase the risk of corrosion of reinforcement.

1.3 WORKABILITY OF CONCRETE

The diverse requirements of mixability, stability, transportability, placeability, mobility, compactability and finishability of fresh concrete are collectively referred to as **workability**. The workability of fresh concrete is thus a composite property. It is difficult to define precisely all the aspects of the workability of concrete in a single definition. IS:6461(Part VII) defines workability as *that property of freshly mixed concrete or mortar which determines the ease and homogeneity with which it can be mixed, placed, compacted and finished.* The optimum workability of fresh concrete varies from situation to situation, e.g. the concrete which can be termed as **workable** for pouring into large sections with minimum reinforcement may not be equally workable for pouring into heavily reinforced thin sections. A concrete may not be workable when compacted by hand but may be satisfactory when vibration is used.

Measurement of workability: A number of different empirical tests are available for measuring the workability of fresh concrete, but none of them is wholly satisfactory. Each test measures only a particular aspect of it and there is really no unique method, which measures the workability of concrete in its totality. However, by checking and controlling the uniformity of the workability, it is easier to ensure a uniform quality of concrete and hence uniform strength for a particular job. The empirical tests widely used are: (i) the slump test, (ii) the compacting factor test, (iii) the Vee-Bee consistency test and (iv) the flow test.

1.4 PROPORTIONING OF CONCRETE MIXES

Several procedures of concrete mix proportioning have been evolved and codified over the years in different countries. The commonly used methods in India are: ACI method, DoE (British) method and Indian Standard Recommended Guidelines. It should be noted that these are merely the recommendations; in practice, any proven method of proportioning may be adopted which produces concrete that meets the performance requirements in the fresh and hardened states. The general step-by-step procedure for concrete mix design can be summarized as follows:

1. The mean target strength is estimated from the specified characteristic strength and the level of quality control.
2. The water-cement ratio is selected for the mean target strength and is checked for the requirements of durability.
3. The degree of workability required in terms of slump, compacting factor or Vee-Bee time is selected.
4. The percentage of fine aggregate in the total aggregate is determined from the characteristics of coarse and fine aggregates. Alternatively, the aggregate-cement ratio may be determined.
5. The water content for the required workability is arrived at.
6. The cement content is calculated and its quantity is checked for the requirements of durability.

7. The concrete mix proportions for the first trial mix are computed and concrete cubes are cast in the laboratory following the standard procedure. After the required period of curing, the cubes are tested for the compressive strength of the mix.
8. The trial batches obtained by making suitable adjustment in water-cement ratio or aggregate-cement ratio are tested till the final mix composition is arrived at.
9. The final proportions are expressed either on mass or volume basis.

Most of the available mix design methods are essentially based on the above procedure and due consideration should be given for the moisture content of aggregate and the entrained air. The desirable amount of water in a concrete mix is such that the sum of volumes of entrapped air and water is minimum, i.e., density achievable is maximum.

1.4.1 Quantities of Materials to Produce Specified Volume of Concrete

When the mix proportions have been determined, the quantities of materials required to produce a specified quantity of concrete can be calculated by *absolute volume method.* The method is based on the principle that the volume of fully compacted concrete is equal to the absolute volume of all the ingredients. If W, C, F_a and C_a are the mass of water, cement, fine aggregate and coarse aggregate, respectively, used in making the concrete; S_c, S_{fa} and S_{ca} are the specific gravities of cement, fine aggregate and coarse aggregate, respectively; and v is the percentage of entrained air in the concrete then the absolute volume of fully compacted fresh concrete (ignoring air content) is given by

$$V_c = \frac{W}{1000} + \frac{C}{1000S_c} + \frac{F_a}{1000S_{fa}} + \frac{C_a}{1000S_{ca}} \tag{1.1}$$

The method is illustrated in Example 1.1.

Example 1.1 Calculate the quantities of ingredients required to produce one cubic metre of structural concrete. The mix is to have proportions of 1 part of cement to 1.40 parts of sand to 2.79 parts of 20 mm nominal size crushed coarse aggregate (by dry-volumes) with a water-cement ratio of 0.48 (by mass). Assume the bulk densities of cement, sand and coarse aggregate to be 1500, 1700 and 1650 kg/m^3, respectively. The specific gravities of cement, sand and coarse aggregates are 3.15, 2.6 and 2.6, respectively. The percentage of entrained air is 2.

Solution The mix proportions of 1:1.40: 2.79 by dry-volume to be used in the production of structural concrete can be expressed in terms of masses as follows:

Water	*Cement*	*Sand*	*Coarse aggregate*
–	1 × 1500	1.40 × 1700	2.79 × 1650 (kg)
0.48	1.0	1.59	3.07

The absolute volume of concrete produced by one bag of cement of 50 kg is

$$V_c = \frac{0.48 \times 50}{1000} + \frac{1 \times 50}{1000 \times 3.15} + \frac{1.59 \times 50}{1000 \times 2.6} + \frac{3.07 \times 50}{1000 \times 2.6} = 0.1295 \text{ m}^3$$

With an entrained air of 2 per cent, the absolute volume of ingredients in one cubic metre of fully compacted fresh concrete is 1.0 – 0.02 = 0.98 m^3. Therefore, cement content per cubic metre of concrete is expressed as

$$C = \frac{0.98}{0.1295} = 7.57 \text{ bags or } C = 379 \text{ kg}$$

Therefore, ingredients requirement (kg/m^3) are:

Cement	379
Sand	602
Coarse aggregate	1162
Water	182

1.4.2 Acceptance Criteria for Concrete

In order to ensure proper quality of concrete produced, it is required to cast a number of specimens from random samples and test them at suitable intervals to obtain results as quickly as possible to enable the level of control to be established with reasonable accuracy in a short time. IS:456 stipulates that random samples from fresh concrete shall be taken as specified in IS:1199 and the cubes shall be made, cured and tested at 28 days as described in IS:516. The test result of a sample shall be the average of the strength of three specimens (constituting the sample). The individual variation should not be more than ±15 per cent of the average. If more, the test result of the sample is invalid. The random sampling procedure is adopted to ensure that each concrete batch shall have a reasonable chance of being tested, i.e., the sampling should be spread over the entire period of concreting and cover all mixing units. The code prescribes minimum frequency of sampling of 1, 2, 3 and 4 numbers of samples, respectively, for 1-5, 6-15, 16-30 and 31-50 m^3 of concrete being used in the job. For concrete quantity of 51 m^3 and above, the number of samples shall be 4 plus one additional sample for every 50 m^3 of concrete or part thereof. At least one sample should be taken from each shift. In case of continuous production unit, e.g. ready mixed concrete plant, the frequency of sampling may be as per agreement. Additional samples may be required for various purposes, e.g. for determination of 7 days strength, accelerated strength, time of striking the formwork etc. As far as the requirements of specifications with regard to the acceptance criteria for concrete is concerned, IS:456 stipulates that the concrete shall be deemed to satisfy the strength requirements provided the mean strength of any group of four non-overlapping consecutive test samples satisfies the following:

For M15 grade concrete

$\geq f_{ck}$ + 0.825 times the standard deviation

or

f_{ck} + 3 MPa; whichever is greater

with strength of individual test sample being $\geq f_{ck} - 3$ MPa.

For M20 and higher grade concretes

$\geq f_{ck}$ + 0.825 times the standard deviation

or

f_{ck} + 4 MPa whichever is greater

with strength of individual test sample being $\geq f_{ck} - 4$ MPa.

1.5 PROPERTIES OF HARDENED CONCRETE

1.5.1 Compressive Strength

It is the most important property of concrete and is often taken as the index of the overall quality of concrete. The strength increases with the age as hydration of cement takes considerable time. The age factors defined as the ratios of strength of concrete at 7 days, 1, 2, 3, 6 and 12 months to the strength of concrete at 28 days are: 0.665, 1.0, 1.1, 1.15, 1.2 and 1.24, respectively.

Since there is significant variability in the compressive strength of concrete, the concept of characteristic strength has been introduced to provide the designer a reasonable assurance of a certain minimum strength of the concrete. The characteristic strength is defined as the strength of the material below which not more than 5 per cent of test results are expected to fall.

Based on its compressive strength the concrete is designated by different grades. The grade of concrete is designated by letter **M** followed by a number, where the letter **M** refers to mix and the number to the specified characteristic strength of 150 mm size cubes at 28 days, expressed in MPa. The concrete of grades M10, M15 and M20 is termed as **ordinary concrete** and that of grades M25 to M55 as **standard concrete** and the concrete of grade 60 and above is termed as **high strength concrete**. Concrete of grade lower than M10 may be used only for plain concrete constructions, lean concrete, simple foundations, foundation for masonry walls and other simple or temporary reinforced concrete construction.

Nominal mix concrete may be used for concrete of grades M20 or lower. The proportions of materials for nominal mix concrete shall be in accordance with the Table 9 of clause 9.3 of IS:456.

The design parameters given in the following sections are applicable to the concrete of grades M15 to M55.

1.5.2 Grades of Concrete

The selection of minimum grade of concrete is dictated by durability considerations which are based on kind of environment to which the structure is exposed. Though, minimum grade of concrete for reinforced concrete structures is specified as M20 under mild exposure conditions, it is advisable to adopt a higher grade. Under more adverse environmental exposure conditions higher grades of concrete are recommended. For moderate, severe, very severe, and extreme exposure conditions, the grades prescribed are: M25, M30, M35 and M40, respectively. In some specific applications the grade of concrete is primarily selected on the basis of strength rather than durability, e.g. tall buildings, reactors and so on.

1.5.3 Tensile Strength

The tensile strength of concrete is relatively low, usually varying from 10 to 20 per cent of the compressive strength. It may be measured indirectly in terms of tensile splitting strength of cylindrical specimens. For members subjected to flexure, the value of modulus of rupture f_{cr} rather than tensile splitting strength is used in design. The flexural and splitting tensile strengths of concrete are obtained as described in IS:516 and IS:5816, respectively. The flexural strength is on an average 50 per cent greater than the strength of concrete in direct tension. IS:456 specify a value of $0.7\sqrt{f_{ck}}$ MPa for flexural strength for normal weight concrete, where f_{ck} is the characteristic compressive strength of concrete.

1.5.4 Bond Strength

It is shear stress at the interface of reinforcement bar and concrete developed to resist the slippage of bar relative to its surrounding concrete.

1.5.5 Stress-Strain Characteristics

The knowledge of the stress-strain curve of concrete is essential in developing analysis and design procedure for concrete structures. Figure 1.1 presents typical stress-strain curves obtained from concrete cylinders having height-to-lateral dimension ratio of 2, loaded in axial compression in a test conducted under uniform rate of strain. The curves arc almost linear up to about one-half the compressive strength. The strain at the maximum stress is approximately 0.002. *The higher the grade of concrete, the steeper is the initial portion of stress-strain curve, the sharper the peak and lesser the failure strain.* If a uniform rate of stress is adopted, it will not be possible to obtain the descending portion of the stress-strain curve beyond the maximum stress.

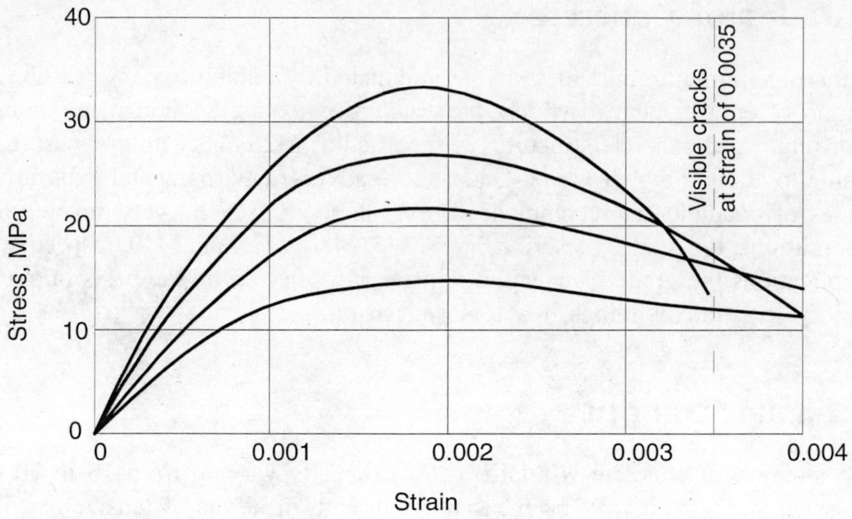

Fig. 1.1 *Stress-strain curves for concrete (cylinder specimen).*

An equation representing the stress-strain curve completely should satisfy the following conditions:

(i) at $f = 0$, $\varepsilon = 0$ and $\dfrac{df}{d\varepsilon} = E_t$

(ii) at $f = f_o$, $\varepsilon = \varepsilon_o$ and $\dfrac{df}{d\varepsilon} = 0$

(iii) at $f = f_f$, $\varepsilon = \varepsilon_u$

The equation satisfying all these conditions is used in the limit state design method. In another case, some simplifying assumptions are made. One of the major assumptions is made in approximating the stress-strain curve to a straight line, i.e. treating the concrete as linearly elastic material. This approximation is used in working stress method of design of structural concrete without much loss of accuracy up to about 50 per cent of f_o. Concrete is not strictly elastic in the sense that if it is unloaded after being stressed at $0.5f_o$ or less, a permanent set is noticed. However, the magnitude of the permanent set gradually decreases with more cycles of loading and unloading (within $0.5f_o$) and the stress-strain curve tend to become a straight line. The creep deformation of concrete also varies linearly with the sustained stress up to a value of $0.5f_o$. Hence, for all practical purposes, the concrete could be considered as a linear elastic material when stress does not exceed $0.5f_o$.

1.5.6 Modulus of Elasticity

The modulus of elasticity determined from the stress-strain curve of the first cycle could be defined as initial tangent modulus, secant modulus, tangent modulus or chord modulus, as shown in Fig. 1.2. In the laboratory determination of modulus of elasticity of concrete, a cylinder is loaded and unloaded (stress not exceeding one-third of f_o) for three or four cycles, the stress-strain curve is plotted after residual strain has become almost negligible and the average slope of stress-strain curve is taken.

The modulus of elasticity obtained by vibration tests on concrete prisms or cylinders is

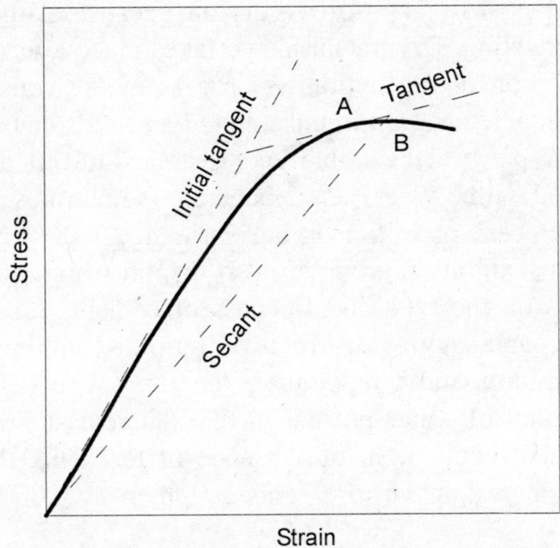

Fig. 1.2 Different modulii of elasticity.

termed as **dynamic modulus of elasticity**. The dynamic modulus is approximately equal to the initial tangent modulus and hence greater than the static or secant modulus. The short-term static modulus of elasticity for structural concrete defining the slope of the tangent to the stress-strain diagram may be estimated from $E_c = 5000\sqrt{f_{ck}}$ MPa. However, the actual measured values may differ by ±20 per cent from the values obtained from this expression.

1.5.7 Poisson's Ratio

The Poisson's ratio is determined as the ratio of lateral to longitudinal strains in a concrete specimen under axial compression. Its value ranges from 0.15 to 0.20 and remains constant up to 80 per cent of ultimate strength. If the dynamic tests are performed its value is slightly higher and is about 0.24. For standard structural concrete its value may be taken as 0.175.

1.5.8 Shrinkage

The contraction in volume of concrete as it hardens and dries is called **shrinkage**. The shrinkage strains are independent of the stress conditions in the concrete. If restrained, shrinkage strains can cause cracking of concrete and will generally result in an increase of deflection of the structural member with time. The calculations of deflection due to shrinkage are given in Chapter 6. Two type of shrinkage strains are recognized, namely, the plastic and the drying shrinkage. Plastic shrinkage occurs during the first few hours after placing the fresh concrete in forms. The hydration of cement causes a reduction in the volume of system of cement plus water to an extent of about 1 per cent of the volume of dry cement. Exposed surfaces are more easily affected by exposure to dry air because of their large contact surface. This can result in surface cracking. Drying shrinkage takes place after the concrete has attained its full set and a major portion of the hydration process in the cement gel is accomplished. Withdrawal of water from concrete stored in unsaturated air voids causes drying shrinkage. The shrinkage on first drying is partly irreversible and is called **initial drying shrinkage**. If the dry concrete is re-saturated with water, an expansion (sometimes referred to as moisture movement) of about 60 per cent of initial drying shrinkage will occur.

As a rule, concrete that exhibits high creep also displays high shrinkage. The magnitude of shrinkage strain depends on the type and fineness of cement, cement paste content, water content, ratio of fine to coarse aggregate, relative humidity, temperature and duration of exposure. For a given humidity and temperature, the total shrinkage of concrete is highly influenced by the total amount of water present in the concrete at the time of mixing and, to a lesser extent, by the cement content. In the absence of test data, the approximate value of the total shrinkage strain for design purposes may be taken as 0.0003.

1.5.9 Temperature Variation

Effects of temperature variation on concrete could be considered similar to those of shrinkage and the two can be usually combined in the analysis of concrete structures. A drop in temperature causes contraction and a rise in temperature an expansion. The thermal coefficient of expansion of concrete is of the order of $10 \times 10^{-6}/^{\circ}C$. The effect of temperature variation in slabs is controlled by the provision of temperature reinforcement. Expected movements due to temperature variation are again accommodated in the expansion and contraction joints.

1.5.10 Creep of Concrete

The concrete under constant compressive stress shows an *immediate strain* followed by a further deformation which progresses at a diminishing rate, as illustrated in Fig. 1.3. The immediate strain is often referred to as the elastic strain and the subsequent time-dependent continuous strain under sustained load is referred to as creep strain, or simply the creep. If the load is removed, the part of the strain, which is immediately recovered, is called the **elastic recovery** and the delayed recovery is termed as the **creep recovery**. The elastic recovery is less than the initial elastic strain because the elastic modulus increases with age; the creep recovery is a small portion of the total creep strain.

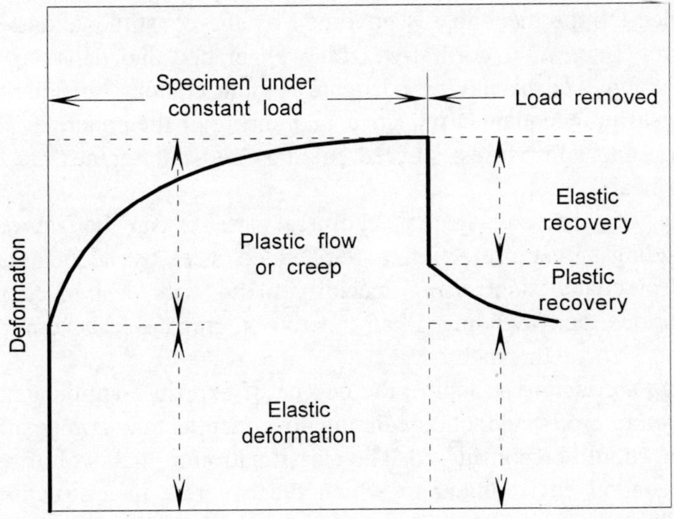

Fig. 1.3 Deformation of hardened concrete under load.

The magnitude of creep strain depends upon concrete grade, environment and stress-time history (i.e., stress level, duration of loading and age at loading). The creep generally increase with *high* cement content, water-cement ratio, air-entraiment and ambient temperature; *low* aggregate content and relative humidity; *loading* at early age and sustained over long periods. Generally, the creep has little effect on the strength of a structure, but it will cause redistribution of stress in reinforced concrete members at the service loads and leads to an increase in the service load deflections. As long as the stress in concrete does not exceed one-third of its characteristic compressive strength, creep may be assumed to be proportional to the stress.

The creep strain can be determined by deducting the elastic strain and shrinkage strain from the total strain. According to IS:456, in the absence of experimental data, the ultimate creep strain may be estimated from the values of creep coefficient which is defined as the ratio of creep strain to elastic strain at the age of loading. The values of creep coefficients may be taken to be 2.2, 1.6 and 1.1 for the duration of loading of 7, 28 and 365 days, respectively.

1.5.11 Durability of Concrete

A durable concrete is one that performs satisfactorily in the working environment with its anticipated exposure conditions during its service life. Thus, the term durability encompasses resistance to weather, chemical attack, corrosion of reinforcement and fire. The materials and mix proportions should be such as to maintain its integrity and protect embedded steel from corrosion.

One of the major characteristics influencing the durability of concrete is its permeability to the ingress of water, oxygen, carbon dioxide, chlorides, sulfate and other potentially deleterious substances. Impermeability is governed by the constituents and workmanship used in producing the concrete. With normal-weight, well-graded and dense aggregates a suitably low permeability is achieved by having adequate cement content, sufficiently low free water-cement ratio, by ensuring adequate compaction and curing of the concrete. Appropriate mineral additives (pozolana) and admixtures; surface coatings and membranes are commonly used to enhance the impermeability.

The durability of concrete against chemical attack can be increased by reducing permeability, providing direct protection to embedded steel by adequate clear cover, using corrosion resistant or coated steel, using special cements with desired chemical resistance to sulfate and/or chlorides, and avoiding alkali-reactive aggregates. The durability also depends upon the shape and size of the member.

While designing a concrete structure, the degree of exposure anticipated during its service life together with other relevant factors relating to concrete mix composition, workmanship, design and detailing should be considered. The classification of the level of severity of exposure conditions of the general environment to which the concrete in a structure will be exposed during its service life is given in Table 3 of Clause 8.22 of IS:456.

Requirement of concrete cover: The cover to embedded steel is identified as one of the major factors influencing durability of concrete. Cover is important as it provides

1. protection against degrading due to corrosive materials like chlorides, sulfates, moisture, oxygen and other pollutants.
2. protection to steel against corrosion.

Thus, the protection of the steel in concrete against corrosion depends upon an adequate thickness of good quality concrete over the reinforcement. The nominal or clear cover to the reinforcement shall be provided as per Table 1.1.

TABLE 1.1 Nominal Cover to Meet Durability Requirements

Exposure	*Mild*	***Moderate***	*Severe*	*Very severe*	*Extreme*
Minimum nominal concrete cover, mm	20	**30**	45	50	75

Notes:

1. For main reinforcement up to 12 mm diameter bar for mild exposure the nominal cover may be reduced by 5 mm.
2. Unless specified otherwise, actual concrete cover should not deviate from the required nominal cover by +10 mm.
3. For the 'severe' and 'very severe', exposure conditions a reduction of 5 mm may be made for the concrete grade of M35 and above.

The code does not prescribe maximum clear cover to reinforcement. The provision of very large covers of 100 mm or more is undesirable, as it may cause increased crack-widths particularly in flexural members, thereby increasing the risk of corrosion of reinforcement and deterioration of concrete. The conventional practice of providing maximum clear cover of the order of 75 mm (as in the case of foundations) should be followed for all the constructions. It is necessary to check the limit state of cracking in case of large covers.

1.6 REINFORCEMENT

As concrete is strong in compression but weak in tension, steel bars are used to reinforce the concrete to resist tensile stresses resulting from the applied loads. Additional reinforcement is occasionally provided to reinforce the compression zone of the concrete beam section and also in columns. Such reinforcement is necessary for heavy loads in order to reduce long-term deflection. The material to be used for reinforcing the concrete must possess the following properties: (i) a high yield strength f_y, (ii) a high modulus of elasticity E_s, (iii) an ability to develop a good bond with concrete, (iv) a coefficient of expansion which is nearly equal to that for concrete, and (v) easy availability. Steel as reinforcement satisfies all the above requirements. The resulting composite material is ideal because as the concrete sets, it contracts and thus grips the reinforcement. This adhesion makes concrete and steel to work together as a single material.

1.6.1 Types of Reinforcement

Types of steel commonly used for reinforcement are: mild steel, medium tensile steel, hot rolled deformed bars, cold twisted high yield strength deformed (HYSD) bars, and hard drawn steel wire fabric. In all cases, the modulus of elasticity of steel shall be taken as 2×10^5 MPa. Typical actual stress-strain curves are shown in Fig. 1.4(a) and idealized curves in Fig. 1.4(b).

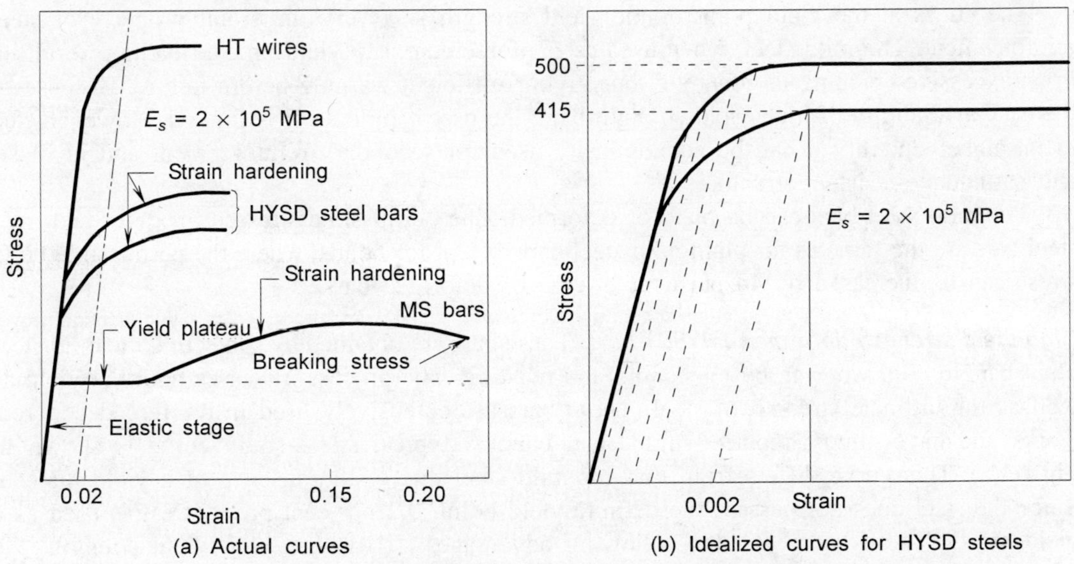

Fig. 1.4 Stress-strain curve for reinforcement bars.

Steel reinforcing bars are generally round in cross-section since they possess the maximum possible perimeter to provide a better bond with concrete. To minimize the slippage of bars relative to surrounding concrete, lugs or protrusions (projections) called **deformations** are rolled on to the bar surface. The lugs generally have an average spacing not exceeding 0.7ϕ (nominal diameter of bar), and a height at least 0.04 to 0.05 of the nominal bar diameter; also they must be present around at least 75 per cent of the nominal bar perimeter. The deformations are placed so that the angle to the axis of the bar is not less than 45°. Generally, longitudinal ribs are also present on the surface of the bar. The nominal diameter of a deformed bar is equivalent to that of a plain bar having the same mass per unit length as the deformed bar. The steel reinforcement is normally designated as grade Fe250, Fe415 and Fe500. The corresponding yield strengths are 250, 415 and 500 MPa, respectively, and generally have well-defined yield points.

The important tests conducted on reinforcing bars are the tension test and cold bent test. The tension test is performed to determine the stress-strain curve, yield stress or 0.2 per cent proof stress and ultimate strength of the reinforcing bars whereas cold bent test is conducted to determine whether steel bars can be bent without fracturing. The test specimen is bent until two sides of the specimen are parallel. The internal radius of the bent must not be greater than the diameter of the bar for diameter up to 25 mm and 1.5 times the diameter of the bar for diameter greater than 25 mm.

Mild steel bars: The minimum specified yield strength of mild steel bars conforming to IS:432 (Part I) is 250 MPa. Among all kinds of steel, it is the most ductile. A typical stress-strain curve for a mild steel bar used in reinforced concrete construction loaded monotonically in tension is shown in Fig.1.4(a). The curve exhibits an initial linear elastic portion, a yield plateau, strain hardening range in which stress again increases with strain, and, finally, a range in which the stress drops off until fracture occurs.

The stress at the yield point or the yield strength is a very important property of steel reinforcement. The mild steel, when used as reinforcement, can yield on overloading, resulting in an excessive cracking in concrete, thus giving sufficient warning before failure. Because of the easy availability of high yield strength bars, the use of mild steel is limited. However, due to its higher ductility it can be economically used for secondary reinforcement and in blast- and earthquake-resistant structures.

The mild steel bars can be plain or deformed. The permissible stresses for deformed mild steel bars are the same as for plain mild steel bars except for bonds, where the permissible bond stress can be increased by 40 per cent in case of deformed bars.

High yield strength deformed (HYSD) bars: These bars are obtained by subjecting the mild steel plain bars to cold working by tensioning and twisting. The process increases both the yield as well as the ultimate strength of steel. These bars are extensively used in reinforced concrete works and have almost replaced mild steel bars. A typical stress-strain curve is shown in Fig.1.4(a). This curve differs from that for mild steel due to the absence of a yield plateau. Since the steel does not possess the definite yield point, 0.2 per cent proof stress is used as a yield stress. These bars have the following advantages; (i) due to high yield strength, the amount of steel required is considerably reduced and (ii) the twisted or deformed bars provide a better bond with concrete than mild steel ones and thus end hooks in bars can be omitted, resulting in further saving of steel.

There are some disadvantages associated with the use of high yield strength steel, namely, reduced ductility; the lower percentage of longitudinal steel in the beam reduces the capacity of the section in shear; with a higher stress level, the deflection and cracking of the member are large.

High yield strength bars are available under trade names such as TOR steel, Tistrong, and Tiscon. These bars are available in two grades, namely, Fe415 and Fe500. Recently IS:1786 has introduced a higher grade Fe550. In the grade designation Fe denotes ferrous material and the number, the yield strength of steel in MPa. These steels lack a well-defined yield point; the yield strength is taken as the stress corresponding to a strain of 0.002. The specified yield strength normally refers to a guaranteed minimum. The actual yield strength is usually somewhat higher than this specified value.

1.6.2 Fabrication and Placement of Bars

The fabrication and fixing of reinforcement bars are the other factors of major importance. The reinforcement bars shall be free from loose scales, loose rust and coats of paints, oil, mud or other coating, which may reduce or destroy the bond. Before fixing the bars in position, the forms are usually oiled so as to facilitate stripping and to get proper finishing.

Placing of reinforcement: Bar bending schedule should be prepared for all reinforcement work. Reinforcement shall be bent and fixed in accordance with procedure specified in IS:2502. The reinforcement shall be placed and maintained in the position shown in the structural drawings. Reinforcement should be secured against displacement beyond the specified limits. Crossing bars should not be tack-welded for assembly of reinforcement unless permitted by engineer-in-charge. The concrete cover to the reinforcement can be obtained by using pre-cast support blocks. In order to maintain the bars in position during concreting, the bars are tied with each other using either binding wire or tack welded. For general work, soft iron wire of 16 gauge (1.6 mm ϕ) is suitable. After tying the bars, the surplus wire shall be cut off so that it does not touch the formwork as this leads to rusting of reinforcement. About 7 to 8 kg of wire is required for one tonne of steel bars.

Tolerances on placing of reinforcement: Unless otherwise specified by engineer-in-charge, the reinforcement shall be placed within the following tolerances:

1. for effective depth up to 200 mm ±10 mm
2. for effective depth more than 200 mm ±15 mm

Cover to reinforcement: Nominal cover as given in Table 1.1 should be provided to all steel reinforcement including links. Spacers, chairs and other supports detailed on drawings, together with such other supports as may be necessary, should be used to maintain the specified nominal cover to the steel reinforcement. Spacers or chairs should be placed at a maximum spacing of 1000 mm and closer spacing may sometimes be necessary. Spacers, cover blocks should be of concrete of same strength or PVC. Unless specified otherwise, actual concrete cover to reinforcement should not deviate from the required nominal cover by 0 to +10 mm

1.7 FORMWORK

The formwork or shuttering may be defined as the moulds of timber, plywood or steel or some other material into which the freshly mixed concrete is poured at the site and which holds the concrete till it sets. The formwork includes the total system of support of freshly placed concrete. The formwork should be designed, built accurately, substantially and efficiently.

1.7.1 Stripping of Forms

The removal of form after the concrete has set is termed as **stripping of forms**. The stripping or striking of forms should proceed in a definite order. The formwork should be so designed and constructed as to allow them to be stripped in the desired order. The period up to which the forms must be left in place before they are stripped is called **stripping time**. The factors affecting the stripping time are the position of the forms, the loads coming on the elements immediately after stripping, temperature of the atmosphere and the subsequent loads coming on the element. Using ordinary Portland cement with temperatures above 20°C, the stripping time normally required are given in Table 1.2.

TABLE 1.2 Stripping Times for Different Conditions

Element and supporting conditions	*Stripping time, days*
Walls, columns, vertical sides of beams	1
Slabs with props left in position	3
Beam sofit with props (supports) left in position	7
Slabs: removal of props	
1. Span up to 4.5 m	7
2. Span over 4.5 m	14
Beam and arches: removal of props	
1. Span up to 6 m	14
2. Span over 6 m	21

For rapid hardening Portland cement, the stripping period can be reduced to three-sevenths of that given in Table 1.2 except for the vertical sides of walls, beams and columns where the forms are to be retained for 24 hours.

1.8 CONCRETE STRUCTURAL SYSTEMS

A structure should basically fulfill four requirements, namely, structural, functional, aesthetic and economic, i.e., it should serve its functional purpose while being safe, serviceable, attractive and economically cost efficient. Although most of the structures are designed for a life span of 50 years, a properly proportioned concrete structure may have a much longer useful life.

A majority of the concrete structures can be considered to consist of some basic structural elements, broadly classified into beams, slabs, columns, walls and foundations. These elements shown in Fig. 1.5 are briefly described in this section.

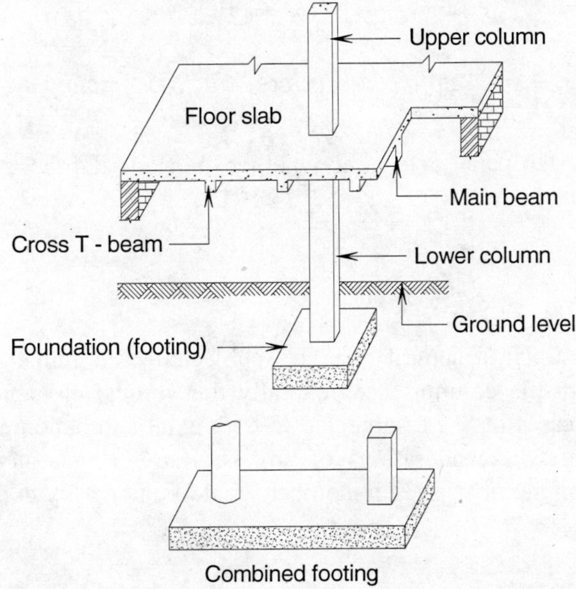

Fig. 1.5 Typical elements of a structural system.

1.8.1 Slabs

Slabs are the structural elements in flexure that are subjected to loads predominantly acting perpendicular to their surfaces. The commonly encountered floor slabs are horizontal, two-dimensional elements that transmit gravity (live and dead) loads to the vertical frames of a structure. They can be a slab on *beams,* as shown in Fig. 1.6, a slab without beams (flat slab or plates) resting directly on columns. Depending upon the plan dimensions, a slab may transfer the loads in one direction while essentially acting as a *one-way slab* or in two perpendicular directions with a *two-way slab* action.

1.8.2 Beams

These are one-dimensional structural elements in flexure that usually transmit the load from slabs to vertical supporting columns. They are normally cast monolithically with slabs. They may form a T-beam section for interior beams or L-beam at the exterior of the building. The

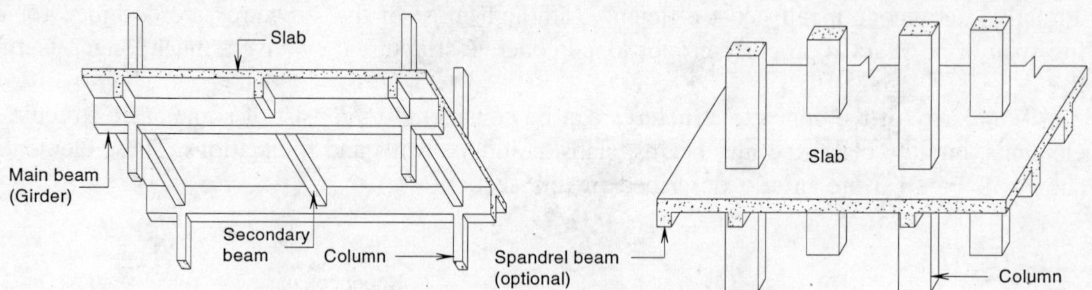

Fig. 1.6 Reinforced concrete structural framing systems.

plan dimensions of a slab panel determine whether the floor slab behaves essentially as a one-way or a two-way slab.

1.8.3 Columns

These are one-dimensional structural elements subjected predominantly to compression. In a structural floor system the columns are generally the vertical elements supporting beams. In most of the practical cases they are subjected to both axial and bending loads, and are of major importance in the safety considerations of any structure. If a structural system includes a horizontal compression member, such a member would be normally considered a beam-column.

1.8.4 Foundations

These are the structural concrete elements transmitting the weight of the superstructure and superimposed loads to the soil. They could be of many forms, the simplest being the isolated footing. It can be viewed as an inverted slab transmitting a distributed load from the soil to the column. Other forms of foundations are combined footings supporting more than one column, mat foundations, and piles driven to hard strata.

1.9 DESIGN APPROACHES

As discussed earlier in the chapter, in reinforced concrete, the steel provides the tensile strength and concrete the compressive strength. The tensile force in concrete is transferred to the steel through the bond between the interfaces of the two materials. From the point of view of economy, steel is cheaper than concrete for every unit of load carried in tension, and concrete is cheaper when compressive load is considered. Thus, the reinforced concrete member should be proportioned to make optimal use of both the materials. This proportioning of reinforced concrete members is termed as **reinforced concrete member design**. The commonly used methods for the design of the structural section are: (i) the working stress design method and (ii) the limit states design method.

Working stress design method: It is based on the behaviour of a section under the loads expected to be encountered by it during its service life. It ensures satisfactory behaviour under service load and is assumed to possess adequate safety against collapse. Both the concrete and steel are assumed to behave perfectly elastically, i.e., the stress is proportional to the strain. The distribution of strain across the section is assumed to be linear, i.e., the plane sections that are plane before bending remain plane after bending. Thus the strain and hence the stress at any point are proportional to the distance of the point from the neutral axis. With this a triangular stress distribution in concrete is obtained, varying from zero at the neutral axis to maximum at the extreme fibre of the section. This ensures that the stresses in concrete and steel do not exceed the allowable stresses at service loads. The allowable stresses are taken as the fixed proportions of the ultimate strength of the materials. The factor of safety for the concrete in compression due to bending, and that for steel in tension are 3.0 and 1.78, respectively. This elastic method fails to give the maximum load that the structure can support and an idea of reserve strength of the member. The actual stresses developed in the structure at the working loads often differ considerably from theoretical values. The reason for this difference is the assumption that the concrete is a perfectly elastic material, which is not correct; on the contrary it is a highly complex heterogeneous material.

Limit states design method: The method is based on statistical probability. A structure at failure may reach a limit state, due to the coincidental occurrence of both overload and excessive weakening of materials at the critical section. The strength of materials and the loads are the two factors, which vary considerably. The statistical probability of strength and load can be estimated, which will provide a limiting state of material strength and load. The statistical base is covered by characteristic strength and characteristic load. Partial factors of safety are introduced to reduce the probability of failure. The limit state is defined as the acceptable limit for the safety and serviceability of the structure before failure occurs. Thus, the concept of limit states design helps to achieve acceptable probabilities so that the structure will not become unfit for use and will not reach the limit states of collapse and of serviceability. The method thus is based on the behaviour of the structure at the ultimate load and ensures its safety against collapse. The design of the section is based on the non-linear stress diagram taking inelastic strain into consideration.

The vast majority of reinforced concrete structures throughout the world have been proportioned based on the working stress design method. Over the last two decades or so, the limit states design techniques are fast replacing the traditional working stress design procedures. In India also we are still in a period of transition from the working stress design procedure to the limit states design. IS:456 have permitted the use of both the procedures for proportioning the concrete members. However, it has emphasized the use of limit states design method.

In the following chapters the design and detailing of the structural elements described in Section 1.8 are discussed in detail. It is to be realized that a sound structure can be obtained by using quality concrete and following sound construction practices. The quality concrete can be produced if adequate quality control is exercised in all stages of production, and in the selection of all constituent materials. The placing of concrete in forms and subsequent curing, and the sequence of stripping the formwork should be well planned and correctly executed. Control tests should be frequently carried out as per the code stipulations. Temporary loads during construction should be strictly controlled, since, they can reach levels higher than the actual design loads when green concrete can least sustain them. The reinforcement should be accurately fabricated and placed in the forms, and well anchored to perform its function efficiently. For the designed structure to be safe and enduring, all these phases must be well monitored in order to comply with the structural design requirements discussed in the following chapters.

Review Questions

1.1 Name the principal compounds in Portland cement, their relative rates of reaction with water and their approximate proportions.

1.2 List the products of hydration and their influence on the properties of cement.

1.3 Explain the difference between various grades of ordinary Portland cement.

1.4 Explain how the Portland-pozzolana cement and super-sulfate cement differ from ordinary Portland cement and specific circumstances in which these cements could be used.

1.5 What are the types of cement suitable for: (a) mass concreting, (b) a concrete resistant to sulfate attack?

1.6 Explain: (i) the grading zones of sand and (ii) the grading of coarse aggregate.

1.7 What is the basis for selecting the maximum size of coarse aggregate in structural concrete?

1.8 What is meant by the term workability? List the methods used for its measurement. Clearly indicate the aspect of workability measured by each method.

1.9 What is meant by segregation and bleeding of concrete? Under what circumstances, do they take place?

1.10 Why is the cube strength different from the cylinder strength for the same grade of concrete?

1.11 Define characteristic strength. Determine the mean target strength required for mix proportioning for a concrete of grade M30, assuming good quality control.

1.12 Distinguish between static and dynamic modulii of elasticity of concrete.

1.13 Briefly explain the general steps involved in a mix design procedure.

1.14 Using a mix design procedure, mix proportions of the desired grade of concrete have been obtained as 1: 2.2: 3.6 (by mass) with water-cement ratio of 0.49 and air content of 3 per cent. Calculate weights of individual ingredients required to produce 0.60 m^3 of concrete. The specific gravities of cement, sand and aggregate were 3.15, 2.65 and 2.72, respectively.

1.15 Differentiate between modulus of rupture and split tensile strength of concrete.

1.16 What is meant by creep of concrete? What structural parameters does it affect?

1.17 How is the deflection of a simply supported reinforced concrete beam affected by the shrinkage of concrete?

1.18 Describe the major factors influencing the permeability of concrete.

1.19 List the factors that affect durability of reinforced concrete structures.

1.20 What is corrosion of steel in reinforced concrete members and enumerate the steps normally taken for reducing the possibility of corrosion of steel?

1.21 What is meant by sulfate attack on concrete? Which of the sulfate salts are most aggressive? How does one provide for conditions where the subsoil water has high sulfate content?

1.22 Why is it necessary to reduce the permeability of concrete in reinforced concrete construction?

1.23 Give reasons why the minimum and maximum cement contents are specified in IS:456. What are the values specified?

1.24 Explain the terms mild, moderate, severe and very severe as applied to exposure condition.

1.25 Define the term clear or nominal cover for steel in reinforced concrete construction. Can cover be made up by plastering after the removal of the shuttering of the reinforced concrete members?

1.26 Distinguish the terms nominal cover (for normal conditions) and additional cover (under special conditions) as given in the IS Code.

1.27 Discuss advantages and disadvantages of providing large nominal cover to reinforcement in flexural members.

1.28 What is meant by cold working of mild steel? How does it affect the structural properties of steel?

1.29 Differentiate between mild steel and HYSD steel bars.

1.30 What is meant by the strain hardening of steel? How is it related to the grade of reinforcing steel?

1.31 Distinguish between beam-supported slab system and flat slab system.

1.32 How are the beam, column and foundation systems distinguished?

CHAPTER

2

Design Principles

2.1 INTRODUCTION

The design of a structure is aimed at ensuring its safety under critical loading conditions, and to limit the deformations so as not to affect its appearance, performance and durability during service conditions. In other words the aim of structural design is to evolve an economical structure so that it fulfils its functional requirements during its service life with adequate safety and serviceability.

Safety: It implies that the likelihood of partial or total collapse of the structure is acceptably low not only under service load but also under probable overloads due to earthquake or wind. Collapse may occur due to various possibilities such as exceeding the load-bearing capacity, overturning, sliding, buckling, fatigue and fracture.

Serviceability: It implies satisfactory performance of the structure under service loads, without discomfort to the user due to excessive deflection, cracking, vibration and so on. Other considerations that come under the purview of serviceability are durability, acoustic and thermal insulation, etc. A design, which adequately addresses to the safety requirement, need not necessarily satisfy the serviceability requirements.

Safety and serviceability can be enhanced by increasing the design safety margins; but this increases the cost of the structure. While considering overall economy, the increased cost associated with increased safety margins should be weighed against the potential losses that could result from any damage.

2.2 REINFORCED CONCRETE DESIGN PHILOSOPHIES

The earliest codified design philosophies of the working stress design based on linear elastic theory and ultimate load method which was introduced as an alternative to working stress design method in the ACI Code in 1956 and the British Code in 1957, and subsequently in the Indian Code (IS:456) in 1964 have since been superseded by the limit states design philosophy.

The present day approach is based on the premises that the various uncertainties in a design can be handled in the mathematical framework of probability theory. The risk involved in the design is quantified in terms of a probability of failure. Such probabilistic methods are reliability based methods. The original complicated probabilistic 'reliability-based' design approach has since been simplified and reduced to a deterministic format involving multiple safety factors rather than probability of failure.

Thus, the past several decades have witnessed an evolution in design philosophy—from the traditional 'working stress method', through the ultimate load method, to the modern limit states method of design.

2.2.1 Working Stress Design

The traditional working stress design approach basically assumes that the structural material behaves in a linearly elastic manner and the members are proportioned so as to sustain the anticipated service or working loads without the stress in the concrete or reinforcement exceeding the permissible values for the individual materials. In this approach the entire margin of safety is provided for by the fact that the calculated stresses in the members under the action of working loads are such that they are considerably below the ultimate or yield stress of constituent materials. As the specified allowable (permissible) values of stresses are restricted to a fraction of their ultimate or yield stress the assumption of linear elastic behaviour is considered justifiable. The ratio of strength of material to the permissible stress is often referred to as the factor of safety. The factors of safety used in getting the permissible stresses are as follows:

Material	Concrete	Steel
Factor of safety	3.0	1.78

In the case where the stresses due to seismic or wind forces are combined with those due to dead, superimposed and impact loads, IS:456 permits the members to be designed for permissible stresses in concrete and steel one-third greater than the normal values. The wind and seismic forces need not be considered as acting simultaneously.

Effect of reinforcement: As stated earlier, the concrete members are reinforced at the places where they are subjected to tensile stresses. This strengthening is accomplished by the embedment of steel bars, which must then resist almost 100 per cent of tensile forces. Figure 2.1(a) shows a schematic elevation of a reinforced concrete simply supported beam subjected to vertical loads. Under the action of these loads the bottom most fibres of the beam are elongated and hence subjected to tensile stress. As the magnitude of loads is increased, the

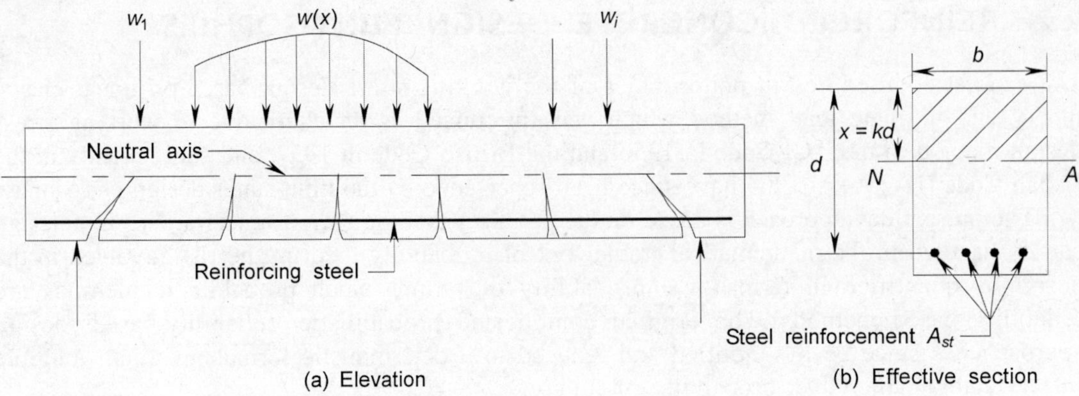

***Fig. 2.1** Schematic elevation of simply supported beam illustrating effective section.*

elongation of the bottom most fibre exceeds the ultimate tensile strain of concrete and it cracks. Since the strain in a beam fibre is proportional to its distance from the neutral axis, a further increase in load will increase the tensile stress causing cracks to increase in number and spread upward towards the neutral axis. These cracks will be perpendicular to the direction of maximum principal tensile stress in concrete.

Once the concrete is cracked it is no longer capable of resisting tensile forces and tensile forces in the bottom of the beam must be resisted by the reinforcement, and the compressive forces at the top are resisted by the concrete. The effective cross-section resisting flexure is shown in Fig. 2.1(b).The shaded area above the neutral axis is the compression zone, and the reinforcement area, A_{st}, is all that resists the tensile stress below the neutral axis since the tensile strength of concrete in tension zone is negligible.

The analysis of stresses under the applied loads assumes strain compatibility whereby the strain in the reinforcing steel is assumed to be equal to that in the adjoining concrete to which it is bonded. Furthermore, as the stresses in concrete and steel are assumed to be linearly related to their respective strains, it follows that the stress in steel is linearly related to that in the adjoining concrete by a constant factor called the **modular ratio**, which is defined as the ratio of the modulus of elasticity of steel to that of concrete.

However, the assumption of linear elastic behaviour and the tacit assumption that the stresses under service loads can be kept within the permissible stresses are not realistic due to the long-term effects of creep and shrinkage, the effects of stress concentration, and other secondary effects. All such effects result in a significant redistribution of the calculated stresses. Moreover, WSM does not provide a realistic measure of the actual factor of safety underlying a design and does not differentiate between different types of loads that act simultaneously, but have different degrees of uncertainty.

The modular ratio defined earlier is an imaginary quantity because due to creep and nonlinear stress-strain behaviour, concrete does not have definite modulus of elasticity unlike the case of steel. The modular ratio results in a larger percentage of compression steel (if required) than that obtained by using limit states design method, thus leads to an uneconomical design. However, the relatively large sections obtained for the structural members result in better serviceability performance, i.e., lesser deflections, smaller crack widths, etc. under the normal service loads. The method is notable for its simplicity in concept, as well as application. It should be noted that WSM with due allowance for change of values of modulus of concrete to allow for creep and shrinkage is only method available for investigating the service stresses and serviceability states of deflection and cracking. The working stress design method is explained in detail in Appendix A.

2.2.2 Ultimate Load Design

In this method, sometimes referred to as the load factor or the ultimate strength method, the stress condition near collapse of the structure is analyzed, and the nonlinear stress-strain curves of concrete and steel are used. The safety measure in the design is introduced by an appropriate choice of the load factor, defined as the ratio of the ultimate load the section can carry to the working load it has to support. The ultimate load method makes it possible for different types of loads to be assigned different load factors under combined loading conditions, thereby overcoming the related shortcomings of working stress method.

The method generally results in more economical designs, i.e., slender sections of beams and columns as compared to WSM, particularly when HYSD steel bars and high strength concrete are used. However, the satisfactory strength performance at ultimate loads does not guarantee satisfactory serviceability performance at the service loads. The designs with the slender sections sometimes result in excessive deflections and crack widths under service load. Moreover, the use of the nonlinear stress-strain behaviour for the design of sections without appropriate nonlinear limit analysis of the structure is wrong, because significant redistribution of stress resultants takes place, as the loading is increased from service loads to ultimate loads. This method has also been superseded by limit states method.

2.2.3 Probabilistic Design

As discussed earlier in the chapter the safety margins are provided in design to safeguard against the risk of failure, i.e., collapse or unserviceability. In the traditional working stress and ultimate load methods of design, the safety margins are assigned in terms of permissible stresses and load factors, respectively, based primarily on the engineering judgement. Structures designed according to these traditional methods are, in general, free from failure. However, the scientific basis underlying the provision of safety margins in design has been questioned again and again. The reliability-based design provides a rational solution to the problem of adequate safety.

The main variables in design calculations that are subject to varying degrees of uncertainty and randomness are the loads, material properties and dimensions. Further, there are idealizations and simplifying assumptions used in the theories of structural analysis and design. Other factors that influence the prediction of strength and serviceability are construction methods, workmanship and quality control, intended service life of the structure, possible future change of use and frequency of loading.

These random variations necessitate a realistic, rational and quantitative representation of safety based on statistical and probabilistic analysis. A rational and quantitative solution to the problem of adequate safety can be obtained by quantifying the acceptable risk in terms of target probability of failure or target reliability. Reliability is generally expressed as the complement of the probability of failure, i.e., equal to $(1-P_f)$.

2.2.4 Limit States Design

The working stress design method fails to give the load-carrying capacity of the structure and also fails to ensure its reliability under service loads since the actual stresses developed in the structure at service loads often differ considerably from the theoretical values. Moreover, the design parameters involving the strength of materials, and the loads to which a structure is subjected, vary considerably. The influence of variability of loads and strength characteristics of materials can be taken into account by using the concept of statistical probability. The statistical probability technique provides limiting states of material strength and loads. These limiting values are represented by the characteristic strength and characteristic load. A design approach embodying these principles aims at achieving acceptable probabilities that the structure shall not become unfit for its intended use, i.e., it shall withstand safely all loads liable to act on it throughout its life; it shall also satisfy the serviceability requirements such as limitations on excessive deflections, cracking, local damage and vibrations. The acceptable limit for the safety and serviceability requirements, before failure occurs, is called the **limit state**. Thus, a structure in all probability shall not be allowed to reach a limit state, i.e., probability of a limit state being attained is kept at an acceptable low value. However, the exact probabilistic design is not possible at present because of lack of statistical data concerning design variables. The design approach recommended by IS:456 is, hence, semi-probabilistic in which the design variables are treated in a probabilistic manner and other uncertainties in imposed loads, inadequacies in the method of analysis, and design and construction are covered through partial safety factors.

All relevant limit states shall be considered in design to ensure an adequate degree of safety and fulfilment of serviceability requirements. In general, the structure shall be designed on the basis of the most critical state, i.e., ultimate limit state and shall be checked for the other limit states.

Thus, limit state design philosophy uses a *multiple safety factor format, which* attempts to provide adequate safety at ultimate loads as well as adequate serviceability at service loads, by considering all possible limit states. The selection of the various multiple safety factors is based on probabilistic analysis involving separate consideration for different kinds of failure, type of materials and types of loads.

2.2.5 Limit States

The acceptable limit for safety and serviceability requirements before failure or collapse is termed as a **limit state**. In other words, a limit state is a state of impending failure, beyond which a structure ceases to perform its intended function satisfactorily, i.e., it either collapses or becomes unserviceable. Two principal types of limit state are the limit state of collapse and limit state of serviceability.

Limit states of collapse: This limit state requires that the structure shall be able to withstand the design loads with an adequate factor of safety against collapse. The limit state of collapse of the structure or a part thereof may arise from: (i) the rupture of one or more critical sections, (ii) the buckling due to elastic or plastic instability, (iii) the loss of static equilibrium due to overturning or sliding, and (iv) the fatigue failure. The limit state of collapse, which deals with the strength, includes any one or more of: (i) flexure, (ii) shear, (iii) torsion and (iv) compression.

The resistance to flexure, shear, torsion and axial loads at every section shall not be less than the corresponding value of action at that section produced by the probable most unfavourable combination of loads on the structure using appropriate partial safety factors.

Limit state of serviceability:

Deflection. The most important serviceability limit states are of deflection and cracking. Excessive deflections or deformations can impair the appearance and the efficiency of a structure and cause discomfort or alarm to the users. The effects are visible in the form of local damage to finishes and to non-load-bearing members even though the limit state of local damage is not exceeded for the structure itself. For example, excessive deflections can cause cracking and possible separation of plaster finishes, crushing of partition walls or cracking of gazing units. The overall final deflection including all effects shall not normally exceed 1/250th of span. After erection of partition such a deflection shall not exceed 1/350th of span.

Cracking. The limit state of cracking or local damage is concerned with the damage to the structure, which could occur due to the application of normal service loads. This damage due to cracking generally results from the volumetric changes including that due to drying shrinkage, creep under sustained load and thermal stresses, and can appear in the form of spalling of concrete. The local damage can occur due to either excessive tension or compression in either concrete or steel. Tension in concrete can result in the limit state of crack width. The stresses under service loads in both concrete and steel must be restricted if compression spalling or micro-cracking of concrete, and excessive tension in steel are to be avoided. Cracking of concrete should not adversely affect the appearance or durability of the structure; the acceptable limits of cracking would vary with the type of structure and environment. The actual width of cracks varies over a wide range and prediction of the absolute maximum width is not possible. In general, the surface width of cracks should not exceed 0.30 mm. The compliance to the spacing requirements of reinforcement stipulated by code should be adequate to control flexural cracking. In case of beam-column member carrying compressive load less than $0.2f_{ck}A_c$, it may be treated as a flexural member for the purpose of crack control. However, for the member with design axial load greater than $0.2f_{ck}A_c$, the cracks due to bending action are not critical.

The other limit states are:

Limit state of vibration: Excessive vibration causes discomfort, alarm or actual damage, or interferes with the proper functioning of the structure.

Limit state of fatigue: Effects of fatigue should be considered; thereby deflections or stresses may have to be limited.

Limit state of impact resistance: The structure or structural elements, which may be subjected to impact, explosion or earthquake must always be considered for structural collapse.

Limit state of durability: The durability is an important factor influencing the long-term performance of a concrete structure. The codes stipulate the checking of structure for the durability of concrete and steel as an integral part of the design. Special precautions may be necessary for exceptionally severe environmental exposure. The durability of reinforced concrete structure can be enhanced by providing adequate cover to reinforcement bars, specifying the minimum cement content and maximum water-cement ratio for the given exposure conditions, using ingredients of concrete which do not contain chlorides and sulfates beyond permissible limits and following good construction practices.

Limit state of fire resistance: For a structural element, which may be subjected to fire, the conditions considered are: (i) resistance to structural collapse, (ii) resistance to penetration of flames, and (iii) resistance to heat penetration.

The relative importance of each limit state will depend upon the nature of the structure and its intended purpose. Each structure is designed for the relevant critical (principal) limit state and checked for its adequacy with respect to other (secondary) limit states. The limit states of collapse dealing with the strength include any one or more of: (1) flexure, (ii) shear, (iii) torsion, (iv) bond, and (v) compression.

2.2.6 Multiple Safety Factors

The objective of limit states design is to ensure that the probability of any limit state being reached is acceptably low. This is accomplished by specifying appropriate factors of safety both to the loads and material strengths. The multiple safety factors for each limit state are primarily based on experience, tradition and engineering judgement. With every revision of Code, conscious attempts are made to specify more rational reliability-based safety factors, in order to achieve practical designs that are satisfactory and consistent in terms of the functional requirements, degree of safety, reliability and economy. A classical reliability model requires the following condition to be satisfied for adequate safety:

$$\text{Design resistance } (\phi R_n) \geq \text{design load effect } (\gamma S_n) \tag{2.1}$$

where R_n and S_n denote the *nominal* or *characteristic* values of *resistance R* and *load effect S*, respectively; ϕ and γ denote the *resistance factor* and *load factor*, respectively. The *resistance factor* ϕ accounts for under-strength, i.e., possible shortfall in the computed nominal resistance, owing to uncertainties related to material strengths, dimensions and inaccuracies in assumptions, and accordingly, it is less than unity. On the contrary, the load factor γ which accounts for overloading and the uncertainties associated with S_n, is generally greater than unity. Equation (2.1) may be written as:

$$S_n < \frac{R_n}{(\gamma/\phi)} \tag{2.2}$$

which is a representation of the safety concept underlying working stress method and (γ/ϕ) denotes the factor of safety applied to the material strength, in order to arrive at the permissible stress for design.

Alternatively, Eq. (2.2) may be rearranged as:

$$R_n \geq (\gamma/\phi)S_n \tag{2.3}$$

Equation (2.3) represents safety concept underlying ultimate load method with (γ/ϕ) denoting the so-called **load factor** applied to the load in order to arrive at the *ultimate load for design.*

2.2.7 Partial Safety Factors

The partial safety factor format adopted by IS:456 is also a multiple safety factor format, which may be expressed as follows:

$$R_d \geq S_d \tag{2.4}$$

where R_d is the design resistance computed using the reduced material strengths $0.67f_{ck}/\gamma_c$ and f_y/γ_s , thus involving two separate partial safety factors γ_c and γ_s for concrete and steel, respectively, in lieu of a single overall resistance factor ϕ and S_d is the design load effect computed for the enhanced loads ($\gamma_d DL$; $\gamma_1 LL$; $\gamma_s SL/WL$) involving separate partial load factors γ_d, γ_1, and γ_s for dead, live and seismic or wind loads, respectively. The other parameters involved are the nominal compressive strength of concrete $0.67f_{ck}$ and the nominal yield strength of steel f_y on the side of the resistance, and the nominal or characteristic load effects *DL, LL* and *SL/WL* representing dead, live and seismic/wind loads, respectively, on the side of design load effects.

It may be noted that, whereas the multiplication factor ϕ is generally less than unity, the dividing factors γ_c and γ_s are greater than unity giving the same effect. All the load factors are generally greater than unity, because overestimation usually results in improved safety. However, one notable exception to this rule is the dead load factor γ_d which is taken as 0.8 or 0.9 while considering stability against overturning or sliding, or while considering reversal of stress when dead loads are combined with seismic/wind loads; in such cases, underestimating the counteracting effects of dead load results in greater safety.

One other effect to be considered in the selection of load factors is the reduced probability of different types of loads (*DL, LL, SL/WL*) acting simultaneously at their peak values. Thus, it is usual to reduce the load factors when three or more types of loads are considered acting concurrently; this is referred to sometimes as the **load combination effect**.

2.3 CODAL RECOMMENDATIONS FOR LIMIT STATES DESIGN

2.3.1 Characteristic Values

The Code regulations assist the designer in evolving satisfactory designs and are conservative to certain extent to minimize the risks. The current regulations have recognized that concrete is inherently much more variable than steel and hence average stress in concrete at failure has been limited to $0.67f_{ck}$. Secondly as primary compression failure occurs without warning, the codes specify a maximum depth of stress block (neutral axis) or limit the percentage of steel to some fraction of that which would produce a balanced condition.

The strengths of both concrete and steel although determined by the laboratory tests conducted on the specimens obtained from the site, they do not represent the strength of materials in construction. Consequently, provision is made in the design for such deficiencies. Similarly, the load conditions imposed on the structure cannot be defined with precision. For example, the live or superimposed loads on the buildings and other constructions depend not only on the random positioning of the loads but on the type of occupancy that generates the general level of loading. The dead loads are generally less likely to vary. However, sometimes the construction errors can be substantial. Metrological conditions produce loading due to wind, snow, ice or earthquakes, all of which are highly variable. Thus for the majority of structures, it would be uneconomical to consider the maximum loads that could be applied and it is therefore, reasonable to design only for those loads that are likely to be applied in the life of the structure.

The above mentioned variability of the strength of materials and of the loads which are considered to have random variations is accounted for through the concept of characteristic values which are defined by IS:456 as outlined in the following sections. However, these values are based on the assumption that the variations of the values have normal distribution.

Characteristic strength of materials: The general definition of characteristic strength of a material corresponds to the 5 percentile strength value, i.e., the characteristic strength of a material is a value of strength below which not more than 5 per cent of test results of the samples are expected to fall. The characteristic strength of concrete is determined from the crushing strength of 150 mm concrete cubes.

The characteristic strength for reinforcing steel is the yield strength in tension for hot rolled bars (mild steel) and 0.2 per cent proof stress in tension for cold worked or high yield strength deformed (HYSD) bars. The specified yield stress i.e. minimum value guaranteed by the manufacturer is usually taken as the characteristic strength.

Characteristic load: The characteristic load is a value of the load that has a 95 per cent probability of not being exceeded during the life span of the structure. However, in the absence of statistical data regarding loads, the nominal values specified for dead, live and wind loads are to be taken from IS:875 (Parts 1-3) and the values for seismic loads are to be taken from IS:1893.

2.3.2 Design Values

As explained earlier, the characteristic values take care of inherent statistical variations in the loads and material strengths. However, a number of random and subjective influences, not amenable to statistical study, modify the characteristic values. For example, accidental overloads, inaccuracies of design assumptions, modification of structural behaviour due to constructional errors could adversely affect the probability of failure through modifying the load effects. Similarly, defects in workmanship in the production of concrete, i.e., in mixing, placing, compacting and curing may result in actual concrete strength being lower than the intended values given by test specimens. In case of reinforcement the strength may be affected by slight bends, presence of rust at the surface caused by storage conditions, etc. To account for such effects, the characteristic values are modified through the use of suitable partial safety factors for each limit state to give design values appropriate to that limit state.

Partial safety factors for materials: It is necessary to reduce the *characteristic strength* values by suitable factors to obtain *design strength values*. The factors by which characteristic strength values are reduced are termed as **partial safety factor** γ_m. The partial safety factors vary depending upon the material and limit state considered. The values for γ_m specified in IS:456 are given in Table 2.1.

TABLE 2.1 Values of Partial Safety Factors

Material	*Limit state*		
	Collapse	*Deflection*	*Cracking*
Concrete	1.50	1.0	1.3
Steel	1.15	1.0	1.0

In the calculation of relative stiffness values for the analysis of continuous beams, frames etc, a uniform value 1.0 is used for γ_m irrespective of the limit state. Design strength f_m of the material is obtained by dividing the characteristic strength f by the partial safety factor γ_m. Thus,

$$f_m = \frac{f}{\gamma_m} \tag{2.5}$$

For example, in the case of concrete while f_{ck} is the cube characteristic strength, the characteristic strength of concrete in actual structure is taken to be $0.67f_{ck}$ $(= f_{ck}/\gamma)$. All other properties such as bond strength, modulus of elasticity and stress-strain relationship, are derived from the design strength.

It should be noted that for ultimate limit states, the Code specifies $\gamma_c = 1.5$ and $\gamma_s = 1.15$. A higher partial safety factor has been assigned to concrete, compared to reinforcing steel, evidently because of the higher variability associated with it.

For serviceability limit states, $\gamma_c = \gamma_s = 1.0$. A safety factor of unity is appropriate here, because the interest is in estimating the actual deflections and crack widths under the service loads, and not safe (conservative) values.

Partial safety factors for loads: The multiplying factor used to modify (increase) the characteristic load F to obtain the design load F_d is termed as the **partial safety factor of load** γ_f.

The value of the load factor depends upon the load combination and the limit state considered. Thus,

$$F_d = \gamma_f F \quad (2.6)$$

The values of γ_f for different load combinations and limit states are given in Table 2.2.

TABLE 2.2 Values of γ_f for Different Limit States and Load Combinations

Load combination	*Ultimate limit state*	*Serviceability limit state*
Dead and live loads	1.5 (*DL* + *LL*)	*DL* + *LL*
Dead and seismic/wind loads		
Case: (i) Dead load contributes to stability	0.9DL + 1.5(*SL*/*WL*)	*DL* + *SL*/*WL*
Case: (ii) Dead load assists overturning	1.5(*DL* + *SL*/*WL*)	*DL* + *SL*/*WL*
Dead, live and seismic/wind loads	1.2(*DL* + *LL* + *SL*/*WL*)	(1.0*DL* + 0.8*LL* + 0.8*SL*/*WL*)

Note: The value of γ_f = 0.9 for dead load in case (i) is applicable when *stability against overturning* or *stress reversal* is critical. The unfavourable case for overturning is the one with minimum dead load.

Total safety factors: Total safety factors may be obtained by multiplying γ_f and γ_m. Thus, for steel, the safety factor is 1.725, compared with the usual average of 1.80 used in working stress design approach.

Limit states of collapse: The code recommends three different weighted combinations involving the combined effects of dead loads (*DL*), live loads (*LL*) and seismic/wind loads (*SL*/*WL*) for estimating the *ultimate* and the *serviceability load effects*.

The reduced load factor of 1.2 in the third load combinations of (*DL* + *LL* + *SL* or *WL*) recognizes the reduced probability of all the three imposed loads reaching their characteristic or peak values simultaneously.

For the purpose of structural design, the design resistance using the partial safety factors should be greater than or equal to the maximum load effect that arises from the above load combinations.

IS:456 have recommended an average load factor of 1.5 for the combined gravity loads, (*DL* + *LL*), for convenience. This is, likely to have about the same effect as (1.4*DL* + 1.6*LL*), adopted by BS: 8110.

Serviceability limit states: The use of a partial load factor of unity in general is appropriate because it implies the service load conditions required for serviceability design. However, when live loads and wind loads are combined, it is improbable that both will reach their characteristic values simultaneously; hence a lower load factor is assigned to *LL* and *SL* in the third combination, to account for the reduced probability of simultaneous occurrence.

2.3.3 Factored Loads

The loads obtained by multiplying the characteristic loads by the appropriate partial safety factors are termed as **factored loads**. Factored bending moment, factored shear force, factored thrust and factored torsion can be computed from the factored load values by the usual procedures.

2.3.4 Design Stress-Strain Curves

Concrete: The characteristic and design stress-strain curves adopted by the IS:456 for concrete in flexural compression are shown in Fig. 2.2. The maximum stress in the characteristic curve is restricted to $0.67f_{ck}$. The curve consists of a parabola in the initial region up to a strain of 0.002 (where the slope becomes zero), and a straight line thereafter, at a constant stress level of $0.67f_{ck}$ up to an ultimate strain of 0.0035.

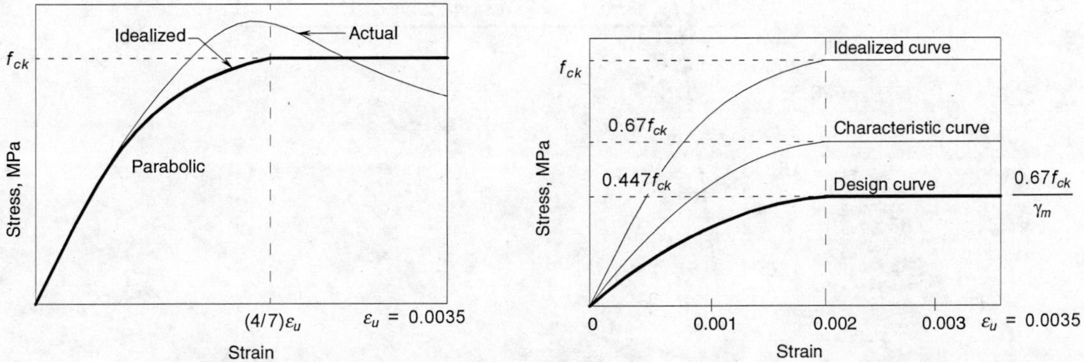

Fig. 2.2 *Stress-strain curves for concrete in flexural compression.*

For the purpose of limit states design, γ_c is equal to 1.5. Thus, the design curve is obtained by simply scaling down the ordinates of the characteristic curve dividing by γ_c. Accordingly, the maximum design stress becomes equal to $0.447f_{ck}$, and the relationship for the design compressive stress f_c corresponding to a strain $\varepsilon \leq 0.0035$ is given by:

$$f_c = 0.447 f_{ck}\left[2\left(\frac{\varepsilon}{0.002}\right) - \left(\frac{\varepsilon}{0.002}\right)^2\right] \quad \text{for } \varepsilon < 0.002 \tag{2.7}$$

$$= 0.447 f_{ck} \quad \text{for } 0.002 \leq \varepsilon \leq 0.0035 \tag{2.8}$$

The stress-strain curve for compression zone of reinforced concrete flexural member is approximately identical to that obtained for a member under uniaxial compression. Thus, when concrete is subjected to uniform compression, as in the case of a concentrically loaded short column, the ultimate strain is limited to 0.002, and the corresponding maximum design stress is $0.447f_{ck}$. The stress-strain curve has no relevance in the limit state of collapse by direct compression of concrete, and hence is not given by the Code.

When concrete is subject to axial compression combined with flexure, the ultimate strain is limited to a value between 0.002 and 0.0035, depending on the location of the *neutral axis*. However, the maximum design stress level remains unchanged at $0.447f_{ck}$.

Reinforcing steel: The characteristic and design stress-strain curves specified by the IS:456 for various grades of reinforcing steel are shown in Figs. 2.3 and 2.4. The design yield strength f_{yd} is obtained by dividing the specified yield strength f_y by the partial safety factors $\gamma_s = 1.15$; thus, $f_{yd} = 0.87f_y$. In the case of mild steel of grade Fe250, which has a well-defined yield point, the behaviour is assumed to be perfectly linearly elastic up to a design stress level of $0.87f_y$ and a corresponding design yield strain = $0.87f_y/E_s$; for larger strains, the design stress level remains constant at $0.87f_y$ i.e. stress-strain curve is bi-linear as shown in Figs. 2.3.

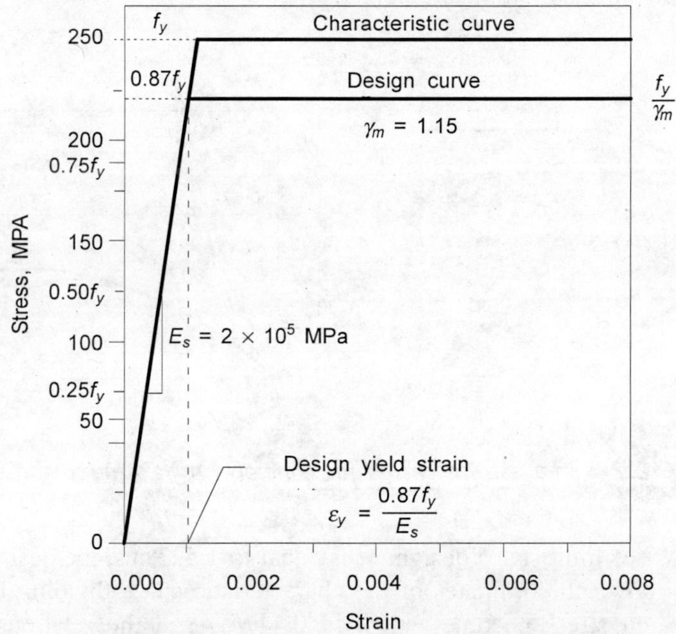

Fig. 2.3 *Characteristic and design stress-strain curves for mild steel.*

The failure strain on the design stress-strain curve is:

$$\varepsilon_{yd} = \frac{f_{yd}}{E_s} = \frac{0.87 \times 250}{2 \times 10^5} = 0.00109$$

In the case of HYSD bars of grades Fe415 and Fe500, which have no specific yield point, the transition from linear elastic behaviour to nonlinear behaviour is assumed to occur at a stress level equal to 0.8 times f_y in the characteristic curve and 0.8 times f_{yd} in the design curve. The full design yield strength $0.87f_y$ is assumed to correspond to a proof strain of 0.002, i.e., the residual strain on unloading is 0.002 with unloading curve having same slope as the initial loading curve.

The design yield strain corresponding to stress level [$= 0.80 \times (0.87f_y)$] beyond which the design stress-strain curve becomes nonlinear is taken as $(0.80 \times 0.87f_y/E_s)$, as shown in the Fig. 2.4.

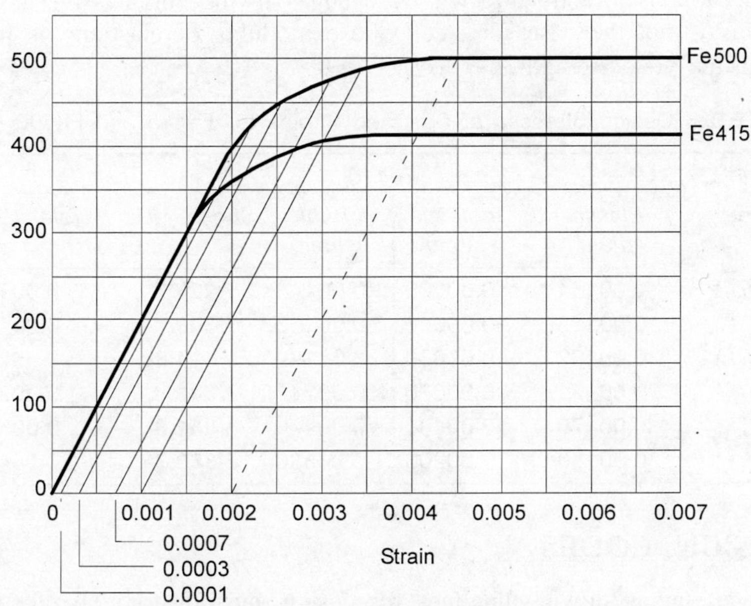

Fig. 2.4 *Characteristic stress-strain curves for HYSD steels.*

For Fe415 steel,

$$\varepsilon_{yd} = \frac{0.80 \times 0.87 f_y}{E_s} = \frac{0.80 \times 0.87 \times 415}{2 \times 10^5} = 0.00144$$

As there is no residual strain at this point, the total strain is 0.00144. At the end of non-linear portion i.e. at proof stress or yield strain level, residual strain is 0.002, hence total strain at this point is:

$$\varepsilon_{yd} = 0.002 + \frac{0.87 f_y}{E_s} = 0.002 + \frac{0.87 \times 415}{2 \times 10^5} = 0.00380$$

The nonlinear region between above two points is a combination of linear and nonlinear strains, and approximated piecewise by linear segments. At the stress level of 0.95 times the yield stress, the residual strain is 0.0007 and corresponding total strain is:

$$\varepsilon_{yd} = 0.00070 + \frac{0.95 \times (0.87 f_y)}{E_s} = 00070 + \frac{0.95 \times 0.87 \times 415}{2 \times 10^5}$$

$$= 0.00070 + 0.00171 = 0.00241$$

The coordinates of the salient points of the design stress-strain curve for the steel of grades Fe415 and Fe500 are listed in Table 2.3 and the design stress, corresponding to any given strain, can be obtained by linear interpolation from Table 2.3. The accurate values of stress in steel f_s can be obtained from Table 5.3.

It may be noted that whereas in the case of concrete the partial safety factor γ_c is applicable at all stress levels, in the case of reinforcing steel, the partial safety factor γ_s is applicable only for the inelastic stress levels. This may be ascribed to the fact that in the case of concrete, the stress-strain curve is directly affected by changes in the compressive strength of concrete; however, this is not the case for steel whose modulus of elasticity is independent of the variations in the yield strength.

TABLE 2.3 Design Stresses at Specified Strains for Fe415 and Fe500 Grade Steels

Salient points	*Fe415*				*Fe500*	
Stress level	*Elastic strain*	*Inelastic strain*	*Total strain*	*Stress, MPa*	*Total Strain*	*Stress, MPa*
0.800 ($0.87f_y$)	0.00144	0.00000	0.00144	**288.7**	0.00174	**347.8**
0.850 ($0.87f_y$)	0.00153	0.00010	0.00163	**306.7**	0.00195	**369.6**
0.900 ($0.87f_y$)	0.00162	0.00030	0.00192	**324.8**	0.00226	**391.3**
0.950 ($0.87f_y$)	0.00171	0.00070	0.00241	**342.8**	0.00277	**413.0**
0.975 ($0.87f_y$)	0.00176	0.00100	0.00276	**351.8**	0.00312	**423.9**
1.000 ($0.87f_y$)	0.00180	0.00200	≥0.00380	**360.9**	≥0.00417	**434.8**

2.4 DESIGN CODES

National codes laying down guidelines for design and construction of structure have been formulated. These codes are based on collective wisdom of experts in the field gained over the years and are periodically updated to incorporate the latest research results. The aim of the codes is: (i) to ensure adequate structural safety by specifying minimum requirements for design, (ii) to make available the results of sophisticated analyses in the form of simple formulae or charts for day to day use, (iii) to ensure a measure of consistency among designers, and (iv) to provide some legal validity for helping the structural engineers and constructors to solve their differences regarding inadequacies of design and supervision and thus help in fixing the liability.

It should be realized that the code provisions are not meant to serve as a substitute for basic understanding and engineering judgement, but to serve as guidelines. The design procedures discussed in this book generally conform to IS:456—Plain and reinforced concrete; IS:875 (Part 1–5)—Code of practice for design loads for building and structures; IS:1893-Criteria for earthquake resistant design of structures and IS:13920—Ductile detailing of reinforced concrete structures subjected to seismic forces. In addition to these basic codes, the book has frequently referred to the handbooks published by Bureau of Indian Standards (BIS) which serve as useful supplement to the basic codes. The handbooks referred to are:

1. SP:16—Design Aids (for reinforced concrete) to IS:456;
2. SP:24—Explanatory Handbook on IS:456, and
3. SP:34—Handbook on Concrete Reinforcement and Detailing.

Review Questions

2.1 List the different methods of design of reinforced concrete members which are accepted in practice.

2.2 Briefly describe the following design procedures for design of reinforced concrete structures:

(a) Working stress design
(b) Ultimate load design
(c) Probabilistic design
(d) Limit states method

State the differences between the probabilistic design and the limit states design.

2.3 Discuss the relative merits and demerits of the traditional methods of working stress and ultimate load designs.

2.4 What is meant by modular ratio? Why is this quantity considered to be unreliable?

2.5 How is the modular ratio affected by creep of concrete?

2.6 What is the basic difference between deterministic and probabilistic design approaches?

2.7 What is the impediment in using a fully probabilistic design approach to a structure in practice?

2.8 What is meant by a limit state? Enumerate the five limit states commonly used in the limit states design.

2.9 Why the limit states design method is considered superior to the working stress design method?

2.10 What is a limit state of collapse?

2.11 What are various serviceability requirements recommended by IS:456?

2.12 Define the term partial safety factors as used in limit states design. List the various factors and state the values stipulated in IS:456.

2.13 Differentiate between the terms factor of safety and partial safety factors for material strength.

2.14 Draw the shape of the experimental stress-strain curve for concrete as obtained from compression tests on cylinder specimen.

2.15 Draw the stress-strain curve assumed in bending compression along with the curve used for limit states design of reinforced concrete section in bending, using a partial safety factor of concrete equal to 1.5.

2.16 Draw the experimental and design stress-strain curves using a partial safety factor of 1.15 for: (a) mild steel bars of grade Fe250 and (b) HYSD bars of grade Fe415.

2.17 With regard to the design stress-strain curves of concrete and steels a reduction by γ_m is applied all along the curve in concrete, whereas in steel the reduction is applied only from the yield point ($0.87f_y$). Give reasons for this difference.

2.18 The maximum moments at a section in a reinforced concrete beam for different independent service load conditions are obtained from structural analysis, as −40 kNm, −85 kNm, ±105 kNm and ±55 kNm under dead, live, wind and earthquake loads, respectively. Determine the ultimate design negative as well as positive moments to be considered for the limit state of collapse in flexure.

2.19 What is meant by characteristic strength of material as specified by IS:. 456?

2.20 Explain the terms: characteristic and factored loads.

2.21 Explain the basis for the selection of partial safety and load factors recommended by the IS: 456 for serviceability limit states.

2.22 Why does IS:456 limit the design compressive strength in structural concrete to 0.67 f_{ck} and not f_{ck}?

2.23 Why is the partial safety factor for concrete γ_c, greater than that for reinforcing steel γ_s in the consideration of ultimate limit states?

2.24 Why does IS:456 specify same partial safety factors for dead and live loads? What are the corresponding values recommended by BS Code?

2.25 Why is γ_c applicable at all the stress levels, whereas γ_s is applicable only near the yield stress level?

2.26 Determine the design stress levels at ultimate limit states in (a) Fe250, (b) Fe415 and (c) Fe500 grades of steel, corresponding to tensile strains of (i) 0.001, (ii) 0.002, (iii) 0.003 and (iv) 0.0035.

2.27 Name the special publications by the Bureau of Indian Standards to supplement IS:456.

CHAPTER

3

Limit State of Collapse—Flexure

3.1 INTRODUCTION

This chapter deals with the proportioning of beams which are primarily subjected to flexural action. The design principles that apply to a simple beam can be extended for designing relatively complicated structures like continuous beams, slabs, staircases, footings, and retaining walls. The beam can be of any shape. However, rectangular-shaped beams, being most commonly used, will be studied in detail.

The flexural members may fail due to the crushing of concrete in compression or yielding of steel in tension. The failure will occur when either steel or concrete reaches its limiting strain. As compared to limiting strain in concrete which is in the range 0.003–0.008, the limiting strain in steel is very high: of the order of 0.25–0.35. Thus, the actual collapse of the flexural member is caused always by the crushing of concrete. If the strain in the steel at failure is above the yield point strain the section is termed as **under-reinforced** and the failure is **tension failure**. On the other hand, if the strain in steel at failure is below the yield point strain, the section is **over-reinforced** and failure is the **compression failure**. If the concrete reaches its full compressive strength simultaneously with the steel reaching its yield point stress, the section is termed as **balanced section**.

3.2 ANALYSIS AND DESIGN FOR FLEXURE

In assessing the strength of any section at failure by limit state analysis, the following basic assumptions are made:

1. Plane sections normal to the axis remain plane after bending, i.e., the strain is proportional to the distance from the neutral axis. The maximum strain in concrete at the outermost compression fibre is 0.0035 in bending.
2. The tensile strength of concrete is neglected.
3. The steel area is assumed to be concentrated at the centroid of the steel.
4. The stress-strain curve for the concrete in compression may be assumed to be rectangle, trapezium, parabola or any other shape which results in prediction of strength in substantial agreement with the test results.

   An acceptable stress-strain curve for design purposes is shown in Fig. 2.2. The stress distribution for strain varying from 0 to 0.002 is parabolic and, thereafter, for strain from 0.002 to 0.0035, it is constant, i.e., the curve has a rectangular shape. The compressive strength of concrete in a structure shall be assumed to be 0.67 times the characteristic strength. In addition to this, a partial safety factor γ_m =1.5 shall be applied. Thus, the maximum stress on the *design curve* which corresponds to design compressive strength of concrete is:

$$f_{cd} = \frac{0.67 f_{ck}}{\gamma_m} = \frac{0.67 f_{ck}}{1.5} = 0.447 f_{ck} \tag{3.1}$$

5. The stress in reinforcement is derived from the stress-strain curve for the type of steel used. Typical stress-strain curves for reinforcements of different grades are shown in Figs. 2.3 and 2.4. For design purposes, a partial safety factor γ_m = 1.15 shall be applied.
6. In order to ensure ductile failure, i.e., the tension reinforcement undergoes a certain degree of inelastic deformation before the concrete fails in compression, the maximum strain in tension reinforcement in the section at failure shall not be less than $[(f_y/1.15)/E_s] + 0.002 \approx (0.87f_y/E_s) + 0.002$, where f_y and E_s are characteristic strength of steel and its modulus of elasticity, respectively.
   The modulus of elasticity for all types of steel is $E_s = 2 \times 10^5$ MPa. The design yield stress or 0.2 per cent proof stress f_{yd}, is equal to $f_y/\gamma_m = f_y/1.15 = 0.87fy$. The stress-strain relations shown in Figs. 2.3 and 2.4 are assumed to be applicable to steel both in tension and compression. In the relation for mild steel, the stress is proportional to the strain up to the yield point, and thereafter the strain increases at a constant stress as shown in Fig. 2.3. On the other hand, in cold deformed steels (Fe415 and Fe500), the stress is proportional to the strain up to a stress level of $0.8f_y$ and thereafter the stress-strain curve has inelastic strains as shown in Fig. 2.4. The inelastic strain values at different stress levels are given in Table 2.3.
7. There exists a perfect bond or adhesion between the concrete and reinforcing steel up to the point of failure, so that there is no slippage between the two materials.

3.3 ANALYSIS OF SINGLY REINFORCED RECTANGULAR SECTIONS

Various symbols designating the cross-sectional dimensions, the stresses and strains developed at the limit state of collapse are shown in Fig. 3.1. The aim of analysis is to determine design parameters, to locate the depth of the neutral axis, and to compute the limiting or ultimate moment of resistance of the section.

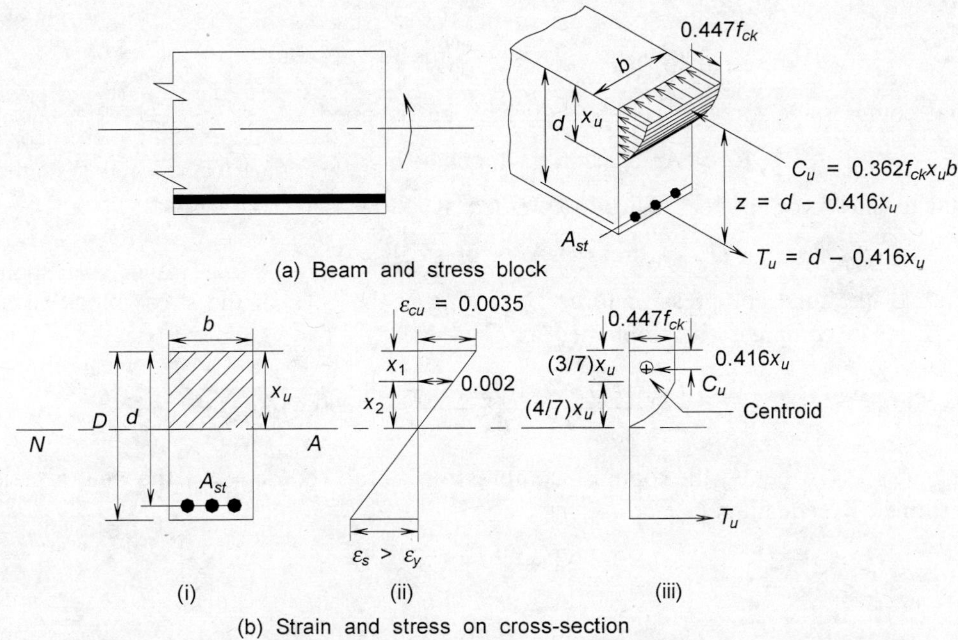

Fig. 3.1 *The parameters of stress block for limit state calculations.*

The external forces are assumed to be resisted by internal compressive forces C_u developed in concrete, and tensile forces T_u in steel. The external moment at a section is resisted by the internal couple formed due to the above two equal and opposite forces in concrete acting through the centroid of compressive stress block and tension acting at the centroid of steel. The distance between the lines of action of these resistive forces is known as **lever arm** denoted by $z(= jd)$, as shown in Fig. 3.1(a).

3.3.1 Depth of the Neutral Axis

Consider the neutral axis be at a depth x_u from the top compression fibre. The distance x_2 from the neutral axis to the point up to which the stress variation is parabolic can be obtained from the strain diagram:

$$\frac{\varepsilon_{cu}}{x_u} = \frac{0.002}{x_2}$$

or

$$x_2 = \frac{0.002x_u}{0.0035} = \frac{4x_u}{7} \tag{3.3}$$

Thus, the depth of the rectangular portion of the stress block is:

$$x_1 = x_u - x_2 = (3/7)x_u \tag{3.4}$$

Area of the stress block = Area of rectangle + Area of parabola

$$= C_1 + C_2 = \left(\frac{3x_u}{7}\right)(0.447 f_{ck}) + \left(\frac{2}{3}\right)(0.447 f_{ck})\left(\frac{4x_u}{7}\right)$$

$$= 0.3619 f_{ck} x_u \approx 0.362 f_{ck} x_u \tag{3.5}$$

The total compressive force in concrete C_u may be obtained as:

$$C_u = \text{Area of stress block} \times b = 0.362 f_{ck} x_u b \tag{3.6}$$

The total tensile force in steel can be expressed as:

$$T_u = \text{Stress} \times \text{Area of steel} = 0.87 f_y A_{st} \tag{3.7}$$

The depth of the total compressive force C_u (acting at the C.G. of the stress block) from the extreme fibre in compression is obtained as:

$$\bar{x} = \frac{C_1(x_1/2) + C_2[(3x_2/8) + x_1]}{C_1 + C_2} = 0.416 x_u \tag{3.8}$$

Thus, C_u acts 0.416 below the topmost compression fibre. The depth of the neutral axis can be determined by equating C_u and T_u, i.e.,

$$0.362 f_{ck} x_u b = 0.87 f_y A_{st}$$

or

$$x_u = \frac{0.87 f_y A_{st}}{0.362 b f_{ck}} = 2.403 \frac{f_y A_{st}}{f_{ck} b}$$

or

$$\frac{x_u}{d} = 2.403 \left(\frac{f_y}{f_{ck}}\right)\left(\frac{A_{st}}{bd}\right) \tag{3.9}$$

The lever arm, i.e., the distance between compressive and tensile forces is:

$$z = jd = d - \bar{x} = d - 0.416 x_u \tag{3.10}$$

3.3.2 Maximum Depth of the Neutral Axis, $x_{u,\max}$

The compression failure in a singly reinforced beam takes place with sudden release of energy. A compression failure without yielding of steel means a collapse without warning. On the other hand if steel undergoes large deformation there is some increase in strength due to strain hardening before it fails. The failure by yielding of steel is gradual giving sufficient warning before total collapse. Therefore, a good design ensures ultimate failure by first yielding of steel in tension followed by compression failure of concrete. This can be achieved by ensuring $x_u \not> x_{u,\max}$. However, in case of doubly reinforced beams due to reduction of brittleness by provision of steel in compression zone, this rule can be relaxed.

The assumptions (2) and (6) govern the maximum depth of the neutral axis in flexural members. The limiting or maximum depth of the neutral axis, $x_{u,\max}$, as shown in Fig. 3.2(a), is obtained directly from the strain diagram corresponding to the limiting conditions of strain in two materials:

$$\frac{x_{u,\max}}{d - x_{u,\max}} = \frac{0.0035}{(0.87 f_y / E_s + 0.002)}$$

or

$$\frac{x_{u,\max}}{d} = \frac{0.0035}{(0.87 f_y / E_s + 0.0055)} \tag{3.11}$$

The limiting values of the depth of the neutral axis for different grades of steel, based on the above expression, are given in Table 3.1.

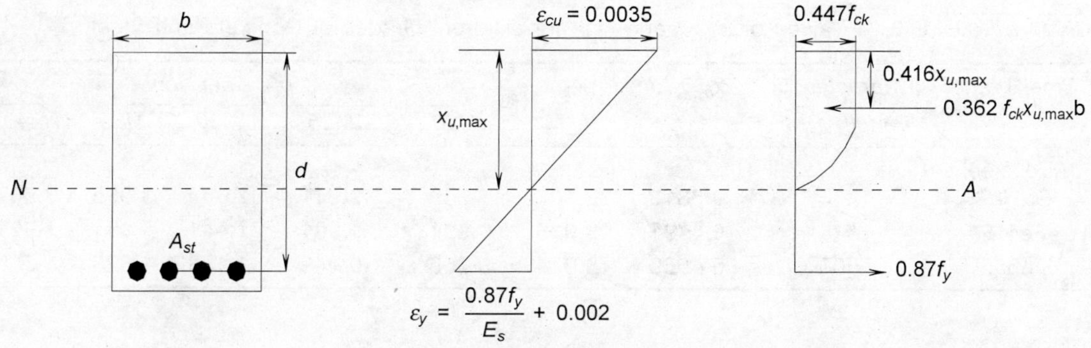

(a) Strain and stress distributions for balanced condition

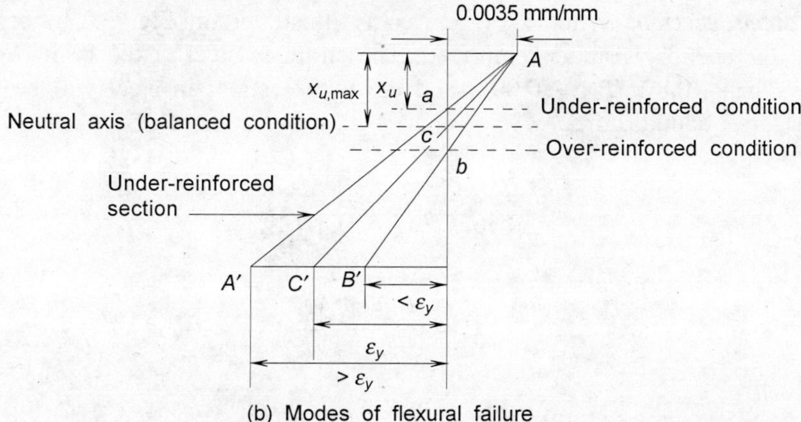

(b) Modes of flexural failure

Fig. 3.2 *Strain distribution for various modes of flexural failure.*

3.3.3 Maximum Percentage of Steel, $p_{t,lim}$

The maximum value of the percentage of steel permitted in a singly reinforced beam can be obtained by equating the actual depth of the neutral axis x_u to its maximum value $x_{u,\max}$.

$$x_{u,\max} \equiv x_u = \frac{0.87\, f_y A_{st}}{0.362\; b\, f_{ck}}$$

$$\frac{A_{st}}{bd} = \frac{p_{t,\lim}}{100} = \frac{x_{u,\max} \times 0.362 f_{ck}}{0.87 f_y d} = 0.4161\left(\frac{f_{ck}}{f_y}\right)\left(\frac{x_{u,\max}}{d}\right) \tag{3.12}$$

or

$$p_{t,\lim} = 41.61\left(\frac{f_{ck}}{f_y}\right)\left(\frac{x_{u,\max}}{d}\right)$$

The values of $p_{t,lim}$ for different grades of steel and concrete are tabulated in Table 3.1.

TABLE 3.1 Values of $x_{u,\max}$ and $p_{t,\lim}$ for Different Grades of Concrete and Steel

Steel grade	*Characteristic strength, f_y, MPa*	*$x_{u,max}/d$*	*$p_{t,lim} \times (f_y/f_{ck})$*	*$p_{t,lim}$, per cent*				
				M20	*M25*	*M30*	*M35*	*M40*
Mild steel								
Grade-I	250	0.5313	22.107	1.769	2.211	2.653	3.095	3.537
Fe415	**415**	**0.4791**	**19.935**	**0.961**	**1.201**	**1.441**	**1.681**	**1.921**
Fe500	500	0.4560	18.974	0.759	0.949	1.138	1.328	1.518

3.3.4 Limiting or Ultimate Moment of Resistance

Under-reinforced section: When $x_u < x_{u,\max}$, as illustrated in Fig. 3.2(b) or $p_t < p_{t,\lim}$ or $M_u < M_{u,\lim}$, the section is under-reinforced. The strain in steel at the limit state of collapse will be more than $[(0.87 f_y/E_s) + 0.002]$ and the tensile stress in steel will be $0.87 f_y$. For internal force equilibrium,

$$C_u = T_u$$

or

$$0.362 f_{ck} b x_u = 0.87 f_y A_{st}$$

Thus,

$$x_u = \frac{0.87\, f_y A_{st}}{0.362\, f_{ck} b}$$

From Eq. (3.10),

$$\text{Lever arm } z = d - 0.416 x_u$$

The ultimate moment of resistance of the section can be determined by taking moment about the compressive force or the tensile force.

M_u in terms of the tensile force is given as:

$$M_u = T_u Z = 0.87 f_y A_{st}(d - 0.416 x_u)$$

Substituting the value of x_u,

$$M_u = 0.87 f_y A_{st}\left[1 - 0.9998\left(\frac{f_y A_{st}}{bdf_{ck}}\right)\right] d \tag{3.13}$$

Equation (3.13) has been approximated in IS:456 as:
In terms of percentage of steel the equation can be written as:

$$M_u = 0.87 f_y\left(\frac{p_t}{100}\right)\left[1 - \left(\frac{f_y}{f_{ck}}\right)\left(\frac{p_t}{100}\right)\right] bd^2$$

or

$$\frac{M_u}{0.87 f_{ck} bd^2} = \left(\frac{f_y}{f_{ck}}\frac{p_t}{100}\right)\left[1 - \left(\frac{f_y}{f_{ck}}\frac{p_t}{100}\right)\right]$$

or

$$\left(\frac{f_y}{f_{ck}}\frac{p_t}{100}\right)^2 - \left(\frac{f_y}{f_{ck}}\frac{p_t}{100}\right) + \frac{M_u}{0.87 f_{ck} bd^2} = 0$$

or

$$\left(\frac{f_y}{f_{ck}}\frac{p_t}{100}\right) = \frac{1}{2}\left[1 - \sqrt{1 - \frac{4}{0.87}\frac{M_u}{f_{ck} bd^2}}\right]$$

or

$$p_t = 50\left(\frac{f_{ck}}{f_y}\right)\left[1 - \sqrt{1 - \left(\frac{4.6 M_u}{f_{ck} bd^2}\right)}\right] \tag{3.14a}$$

$$A_{st} = \left(\frac{f_{ck} bd}{2 f_y}\right)\left[1 - \sqrt{1 - \left(\frac{4.6 M_u}{f_{ck} bd^2}\right)}\right]$$

$$= \left(\frac{f_{ck} bd}{2 f_y}\right)\left[1 - \sqrt{1 - R_u}\right] \tag{3.14b}$$

where

$$R_u = \left(\frac{4.6M_u}{f_{ck}bd^2}\right)$$

Now, consider the M_u in terms of compressive force:

$$M_u = C_u z = 0.362 f_{ck} b x_u (d - 0.416 x_u) = 0.362 f_{ck}\left(\frac{x_u}{d}\right)\left[1 - 0.416\left(\frac{x_u}{d}\right)\right] bd^2$$

$$= Q_u bd^2 \tag{3.15a}$$

where

$$Q_u = 0.362 f_{ck}\left(\frac{x_u}{d}\right)\left[1 - 0.416\left(\frac{x_u}{d}\right)\right] \tag{3.15b}$$

An alternative procedure for determinations of tension steel similar to that of working stress format can be easily formulated. Equation 3.15(b) can be rearranged as:

$$\frac{Q_u}{0.362 f_{ck}} = \left(\frac{x_u}{d}\right) - 0.416\left(\frac{x_u}{d}\right)^2$$

$$\left(\frac{x_u}{d}\right) = \left[\frac{1}{2}\left(\frac{1}{0.416}\right) - \sqrt{\frac{1}{4}\left(\frac{1}{0.416}\right)^2 - \frac{Q_u}{0.416 \times 0.362 f_{ck}}}\right]$$

$$= 1.202\,[1 - \sqrt{1 - R_u}\,] \tag{3.15c}$$

This expression can conveniently be used to compute x_u/d for the given values of M_u, b, d and f_{ck}.

The lever arm coefficient is given by

$$j = 1 - 0.416\left(\frac{x_u}{d}\right)$$

$$= 1 - 0.5[1 - \sqrt{1 - R_u}\,] = 0.5[1 + \sqrt{1 - R_u}\,] \tag{3.16a}$$

With this value of j, A_{st} can be computed from:

$$A_{st} = \frac{M_u}{(0.87 f_y)(jd)} \tag{3.16b}$$

Balanced section: When $x_u = x_{u,\max}$, as shown in Fig. 3.2(b) or $p_t = p_{t,\lim}$ or $M_u = M_{u,\lim}$ the section is a balanced section, and the strain in the concrete at the limit state of collapse is

0.0035 and that in steel is $[(0.87f_y/E_s) + 0.002]$. The limiting moment of resistance of the section can be expressed as:

$$M_{u,\text{lim}} = M_{u,\text{max}} = Q_{u,\text{max}} f_{ck} bd^2$$

where the constant $Q_{u,\text{max}}$ is given by

$$Q_{u,\text{max}} = \frac{M_{u,\text{lim}}}{f_{ck} bd^2} = 0.362\left(\frac{x_{u,\text{lim}}}{d}\right)\left[1 - 0.416\left(\frac{x_{u,\text{lim}}}{d}\right)\right] \tag{3.17}$$

(a) For mild steel: $f_y = 250$ MPa

$$\frac{x_{u,\text{max}}}{d} = 0.5313, \quad M_{u,\text{lim}} = 0.1498 f_{ck} bd^2 \tag{3.18a}$$

(b) For Fe415 steel: $f_y = 415$ MPa

$$\frac{x_{u,\text{max}}}{d} = 0.4791, \quad M_{u,\text{lim}} = 0.1389 f_{ck} bd^2 \tag{3.18b}$$

(c) For Fe500 steel: $f_y = 500$ MPa

$$\frac{x_{u,\text{max}}}{d} = 0.4560, \quad M_{u,\text{lim}} = 0.1338 f_{ck} bd^2 \tag{3.18c}$$

The $M_{u,\text{lim}}$ values for different grades of steel are given in Table 3.2. In proportioning a singly reinforced rectangular section the dimension can be obtained from the relation:

$$d \geq \sqrt{\left(\frac{M_u}{Q_u b}\right)}$$

with

$$Q_u \leq 0.362 f_{ck}\left(\frac{x_{u,\text{max}}}{d}\right)\left[1 - 0.416\left(\frac{x_{u,\text{max}}}{d}\right)\right] \tag{3.19}$$

The difference between Q_u and $Q_{u,\text{max}}$ will indicate the magnitude of under reinforcement.

TABLE 3.2 $M_{u,\text{lim}}$ Values for Different Grades of Steel

Steel	*Grade*
Mild steel: Grade-I, Fe250	$0.1498 f_{ck} bd^2$
Fe415	**$0.1388 f_{ck} bd^2$**
Fe500	$0.1338 f_{ck} bd^2$

Over-reinforced section: When $x_u > x_{u,\max}$, i.e., $M_u > M_{u,\lim}$, the section is over-reinforced. As the over-reinforced section is not permitted it should be redesigned.

3.4 ANALYSIS AND DESIGN OF RECTANGULAR BEAMS

The analysis of a beam of known cross-sectional dimensions and reinforcement details consists in calculating the maximum allowable or limiting moment of resistance. On the other hand, design of a beam for flexure consists in proportioning the dimensions of cross-section and detailing the reinforcement to resist the given moment safely and economically.

3.4.1 Types of Problems

Two types of problems are generally encountered in practice, namely, the analysis problems and the design problems. The former deals with the analysis of the beam section of given dimensions and reinforcement details wherein the limiting moment of resistance of the section is determined. It is based on the requirements of strain compatibility and equilibrium of forces. The later deals with the design of a beam section wherein cross-sectional dimensions and the area of steel for resisting the given factored or design moment are determined. The dimensions of the section can be proportioned such that limiting stresses/strains in concrete and steel are induced simultaneously. Such a section, known as a **balanced section**, though efficient in the use of materials is not necessarily the most economical section. There may be other considerations in fixing the dimensions of the section.

Analysis of rectangular beams: In this type of problems the dimensions of the beam section (b, d), area of reinforcement (A_{st}), and grades/characteristic strengths of materials (f_{ck},f_y) are given. The following types of problems are generally encountered in the analysis of concrete beams reinforced in tension only.

Type I: Determination of Limiting or Ultimate Moment Carrying Capacity of a Beam Section. The steps involved are:

1. Find the position of actual neutral axis x_u from the known values of b, d, A_{st}, f_{ck}, and f_y.
2. Find the position of critical neutral axis $x_{u,\max}$.
3. Compare x_u with $x_{u,\max}$ to determine the type of beam section:
   (a) if $x_u > x_{u,\max}$, the section is over-reinforced.
   (b) if $x_u < x_{u,\max}$, the section is under-reinforced.
4. Calculate the moment carrying capacity for the appropriate type of beam section. For the over-reinforced section

$$M_{u,\lim} = (0.362 f_{ck} x_{u,\max} b)(d - 0.416 x_{u,\max})$$

For the under-reinforced section:

$$M_{u,\lim} = T_u \times z = 0.87 f_y A_{st}(d - 0.416 x_u)$$

Example 3.1 Determine the limiting moment-carrying capacity of a reinforced concrete rectangular section of size 250 × 550 mm deep (effective) reinforced on the tension side with four 20 mm ϕ bars. The concrete used is of grade M20 and reinforcing steel is mild steel of grade Fe250.

Solution For the given section and materials:

$$b = 250 \text{ mm}, d = 550 \text{ mm}, A_{st} = 4 \times \left(\frac{\pi}{4}\right) \times 20^2 = 1257 \text{ mm}^2, f_{ck} = 20 \text{ MPa and } f_y = 250 \text{ MPa}$$

Equating tensile and compressive forces,

$$0.362 f_{ck} b x_u = (0.87 f_y) A_{st}$$

$$x_u = \frac{0.87 f_y A_{st}}{0.362 f_{ck} b} = \frac{0.87 \times 250 \times 1257}{0.362 \times 20 \times 250} = 151.05 \text{ mm}$$

The maximum permitted value $x_{u,\max}$ is given by

$$x_{u,\max} = 0.531d \text{ (for } f_y = 250 \text{ MPa)}$$

$$= 0.531 \times 550 = 292.05 \text{ mm.}$$

Since $x_u < x_{u,\max}$, the section is under-reinforced, and hence

$$M_{u,\lim} = T_u \times z = 0.87 f_y A_{st}(d - 0.416 x_u)$$

$$= 0.87 \times 250 \times 1257 \times (550 - 0.416 \times 151.05)$$

$$= 133.19 \times 10^6 \text{ Nmm} = 133.19 \text{ kNm}$$

Type II: Determination of load carrying capacity of a beam section. If in Type-I problems the effective span and support conditions of the beam are known, load-carrying capacity can be computed.

Example 3.2 A reinforced concrete rectangular section of size 300 × 600 mm (effective) is reinforced by 3 bars of 20 mm ϕ. Determine the safe uniformly distributed load that the beam can carry over a simply supported effective span of 6 m. The concrete mix and steel used are of M20 and Fe415 grades, respectively. For the given environmental exposure the effective cover to reinforcement is 50 mm.

Solution For the given section and materials:

$$b = 300 \text{ mm}, d = 600 \text{ mm}, D = 600 + 50 = 650 \text{ mm},$$

$$f_{ck} = 20 \text{ MPa}, f_y = 415 \text{ MPa, and } A_{st} = 3 \times \frac{\pi \times 20^2}{4} = 942 \text{ mm}^2$$

For computation of x_u, equate tensile and compressive forces:

$$0.362 f_{ck} b x_u = 0.87 f_y A_{st}$$

or

$$x_u = \frac{0.87 f_y A_{st}}{0.36 f_{ck} b} = \frac{0.87 \times 415 \times 942}{0.362 \times 20 \times 300} = 156.59 \text{ mm}^2$$

Maximum permitted value, $x_{u,\max} = 0.479\ d = 0.479 \times 600 = 287.4$ mm
Since $x_u < x_{u,\max}$, the beam is under-reinforced

$$M_{u,\lim} = T_u \times z = 0.87 f_y A_{st}(d - 0.416 x_u)$$

$$= 0.87 \times 415 \times 942 \times (600 - 0.416 \times 156.59) = 181910{,}260 \text{ Nmm}$$

$$= 181.91 \text{kNm}$$

Let w_d be the factored design load in kN/m then the factored moment is:

$$M_d = \frac{w_d L^2}{8} = \frac{w_d (6)^2}{8} = 4.5 \text{ w}_\text{d} \text{ kNm}$$

Equating factored moment to the limiting moment capacity,

$$4.5\ w_d = 181.91, \text{ giving } w_d = 40.4 \text{ kN/m}$$

Total service load on the beam $w = \dfrac{w_d}{\gamma_d} = \dfrac{40.4}{1.5} = 26.93$ kN/m.

Self-weight of the beam $w_g = 0.30 \times 0.65 \times 1 \times 25 = 4.875$ kN/m

Safe imposed load $w_s = 26.93 - 4.875 = 22.055$ kN/m

Design of rectangular beams: The following types of problems are generally encountered:

Type-I: Determination of tensile steel. In this type of problems the cross-sectional dimensions are fixed from architectural or other considerations. The materials to be used and the design moment to be carried are known. The steps involved are:

1. Compute the design constants $p_{t,\lim}$ and $Q_{u,\max}$ for the given materials. The design constants for some of the materials are given in Tables 3.1 and 3.2.
2. Determine the limiting moment capacity of the section as follows:

$$M_{u,\lim} = (0.362 f_{ck} x_{u,\max} b)(d - 0.416 x_{u,\max})$$

3. Compute the area of reinforcement to be provided.

Example 3.3 Find the limiting moment-carrying capacity of a reinforced concrete rectangular section of size 200 × 425 mm effective, and the area of mild steel required. The concrete used is of grade M20.

Solution For the given section and materials:

$$b = 200 \text{ mm}, d = 425 \text{ mm}, f_{ck} = 20 \text{ MPa and } f_y = 250 \text{ MPa}$$

Limiting depth of the neutral axis is:

$$x_{u,\max} = 0.531\ d \text{ (for } f_y = 250 \text{ MPa)}$$
$$= 0.531 \times 425 = 225.675 \text{ mm.}$$

Hence the limiting moment capacity of the section is expressed as:

$$M_{u,\lim} = (0.362 f_{ck} x_{u,\max} b)(d - 0.416 x_{u,\max})$$
$$= (0.362 \times 20 \times 225.675 \times 200)(425 - 0.416 \times 225.675) \times 10^{-6}$$
$$= 108.20 \text{ kNm}$$

Alternatively,

$$M_{u,\lim} = 0.1498 f_{ck} bd^2 = 0.1498 \times 20 \times 200 \times (425)^2 = 108.23 \text{ kNm}$$

With M20 grade concrete and mild steel reinforcement:

$$p_{t,\lim} = 1.767 \text{ or } \frac{100 A_{st}}{bd} = 1.767$$

Thus,

$$A_{st} = \frac{1.767 bd}{100} = \frac{1.767 \times 200 \times 425}{100}$$
$$= 1501.95 \text{ mm}^2 \text{ (say, } 1502 \text{ mm}^2\text{).}$$

Alternatively, A_{st} can be obtained by equating tensile and compressive forces:

$$0.362 f_{ck} b x_{u,\max} = 0.87 f_y A_{st}$$

Thus,

$$A_{st} = \frac{0.362 f_{ck} b x_{u,\max}}{0.87 f_y}$$
$$= \frac{0.362 \times 20 \times 200 \times 225.675}{0.87 \times 250}$$
$$= 1502.42 \text{ mm}^2$$

The difference in two values is due to the fact that figures in the expression for $p_{t,\lim}$ have been rounded off.

Example 3.4 Determine the area of tensile reinforcement required for a singly reinforced beam section of size 300 × 550 mm effective to carry a factored moment of 175 kNm. The concrete mix used is M20 and tensile steel reinforcement is of grade Fe415.

Solution For the given section and materials:

$$b = 300 \text{ mm},\ d = 550 \text{ mm},\ f_{ck} = 20 \text{ MPa and } f_y = 415 \text{ MPa}$$

Limiting moment of resistance of the section is obtained as:

$$M_{u,\lim} = 0.1388\ f_{ck} bd^2$$
$$= 0.1388 \times 20 \times 300 \times 550^2 \times 10^{-6} = 251.92 \text{ kNm}$$

The actual factored moment M_u = 175 kNm. Since M_u is less than the limiting moment of resistance, the section must be designed as an under-reinforced section. Let the depth of the neutral axis be x_u, therefore,

$$0.362\ f_{ck}bx_u\ (d - 0.416\ x_u) = 175 \times 10^6$$

or

$$0.362 \times 20 \times 300\ x_u(550 - 0.416\ x_u) = 175 \times 10^6$$

$$x_u^2 - 1322.12x_u + 193680.05 = 0$$

Thus,

$$x_u = 169.78 \text{ mm}$$

Equating the total compression to total tension, we have

$$0.362f_{ck}bx_u = 0.87f_yA_{st}$$

$$A_{st} = \frac{0.362f_{ck}bx_u}{0.87f_y} = \frac{0.362 \times 20 \times 300 \times 169.78}{0.87 \times 415} = 1021.36 \text{ mm}^2$$

Alternatively from Eq. 3.14(a),

$$p_t = \frac{(50 \times 20)}{415}\left[1 - \sqrt{1 - \frac{4.6 \times 175 \times 10^6}{20 \times 300 \times (550)^2}}\right] = 0.61212 \text{ per cent}$$

$$A_{st} = \frac{p_t bd}{100} = \frac{0.61212 \times 300 \times 550}{100} = 1010 \text{ mm}^2$$

The difference between two values is due to the approximation in derivation of later expression. Provide 3 bars of 20 mm ϕ and one bar of 10 mm ϕ(A_{st} = 1020.5 mm^2).

Type-II: Design or Proportioning of a section to support the given loads. In this type of problems, cross-sectional dimensions and amount of tension reinforcement to be provided are computed by assuming the section to be a balanced section. The steps involved are:

1. Calculate the maximum factored bending moment M_u due to applied loads.
2. Determine the design constants $M_{u,\lim}$ and $Q_{u,\max}$ for balanced design. The design constants for some of the materials are given in Tables (3.1) and (3.2).
3. Choose a depth-to-breadth ratio of the beam section preferably between 2 and 3.
4. Determine the sectional dimensions b and d from the relation $M_u = Q_{u,\max}\ bd^2$.
5. Obtain the overall depth D by adding concrete cover to steel to the effective depth d.
6. Compute the area of tension reinforcement using the appropriate expression.

Example 3.5 Design a reinforced concrete beam having an effective simply supported span of 5.5 m. The beam is required to support live and superimposed loads of 15 kN/m and 10 kN/m, respectively. The materials to be used are M20 grade concrete and HYSD steel of grade Fe415.

Solution For M20 grade concrete and Fe415 grade steel:

$$f_{ck} = 20 \text{ MPa}, f_y = 415 \text{ MPa}, w_l = 15 \text{ kN/m and } w_s = 10 \text{ kN/m}$$

For estimation of the self-weight of the beam, consider the depth of the beam to be in the range of $L/12$ i.e., $D = 500$ mm and $b = 250$ mm ($b = D/2$).

Self-weight of the beam $w_g = 0.25 \times 0.50 \times 1 \times 25 = 3.125$ kN/m

Total dead load $w_d = w_g + w_s = 3.125 + 10 = 13.125$ kN/m

The partial safety factors for loads are: $\gamma_d = \gamma_l = 1.5$. Factored design moment due to factored loads will occur at the mid-span of the beam;

$$M_u = \frac{(\gamma_d w_d)L^2}{8} + \frac{(\gamma_1 w_1)L^2}{8}$$

$$= \frac{1.5(13.125 + 15) \times 5.5^2}{8} = 159.52 \text{ kNm}$$

For the balanced failure condition, i.e. $M_u = M_{u,\lim}$

$$M_{u,\lim} = 0.1388bd^2 = 159.52 \times 10^6$$

Assuming $b = d/2$, the preceding relation reduces to:

$$0.1388(d/2)d^2 = 159.52 \times 10^6$$

$$d = \left[\frac{159.52 \times 10^6}{0.1388 \times 20 \times (1/2)}\right]^{1/3} = 486.19 \text{ mm (say 490 mm).}$$

Adopt 20 mm ϕ bars.

Overall depth $D = d$ + (half diameter of bar) + clear cover

$= 486.19 + 10 + 20 = 516.19$ mm (say 520 mm)

Effective depth $d = 520 - 10 - 20 = 490$ mm.

Thus, provide $b = 250$ mm, $D = 520$ mm and $d = 490$ mm with an effective cover of 30 mm. The dimensions assumed for calculating the self-weight are close to the actual values. Hence there is no necessity for repeating the calculations.

For M20 grade concrete and Fe415 grade steel:

$$p_{t,\lim} = 0.960$$

or

$$A_{st} = \frac{0.960 \times 250 \times 490}{100} = 1176 \text{ mm}^2$$

Alternatively, $A_{st,\text{bal}} = p_{t,\lim}\ bd\left(\frac{f_{ck}}{f_y}\right)$

$$= 0.19926 \times 250 \times 490 \times \frac{20}{415} = 1176.35 \text{ mm}^2$$

Provide 4 bars of 20 mm ϕ ($A_{st} = 1257$ mm^2). The reinforcement details are shown in Fig. 3.3.

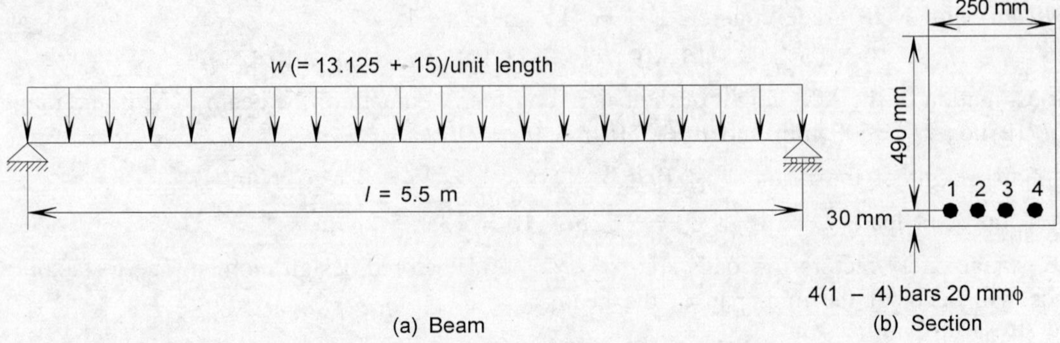

Fig. 3.3 Design of simply supported singly reinforced beam of Example 3.5.

3.5 BEAMS REINFORCED IN TENSION AND COMPRESSION

In practice, very frequently, it is desirable or even mandatory to have a section of restricted depth in order to comply with some architectural or structural requirements wherein the section has to carry a moment more than it can resist as a balanced section. If a section is arbitrarily made shallower than the balanced design, it results in over-reinforced section and the concrete is overstressed, while steel is stressed to its permissible value. As mentioned earlier IS:456 do not permit the use of the over-reinforced section. In such cases it is preferable to design it as a doubly reinforced beam where the reinforcement is also provided in compression to give additional strength to the concrete. When this condition exists, the allowable compressive force can be increased by providing reinforcing steel in the compression zone, as shown in Fig. 3.4. Such a section reinforced both in tension and compression is known as a **doubly reinforced section**. They are also provided in the cases of load where reversal of stresses may take place, e.g., wind and earthquake loads. The doubly reinforced sections are recommended only when they are absolutely necessary.

When the cross-sectional dimensions of beam (b and d) have been predetermined and design or factored moment M_d exceeds the limiting moment of resistance of the singly reinforced section, $M_{u,\lim}$, the additional moment of resistance required is obtained by providing compression reinforcement in addition to tension reinforcement. The moment of resistance of a doubly reinforced section is thus the sum of the limiting moment of resistance $M_{u,\lim}$ of the singly reinforced section and the additional moment of resistance $M_{da}(= M_d - M_{u,\lim})$. The lever arm for additional moment of resistance is equal to the distance between the centroids of tension and compression reinforcements $(= d - d')$, where d' is the effective cover to the compression reinforcement. The additional tensile force is balanced by the additional compressive force, i.e.,

$$A_{st2}(0.87f_y) = A_{sc}(f_{sc} - f_{cc}) \tag{3.18}$$

Thus,

$$M_{da} = A_{st2}(0.87f_y)\ (d - d')$$

$$= A_{sc}(f_{sc} - f_{cc})(d - d') \tag{3.19}$$

where

A_{st2} = Area of additional tension reinforcement

A_{sc} = Area of compression reinforcement

f_{sc} = Design stress in compression steel

f_{cc} = Design stress in concrete at the level of compression steel at a distance d' below the top edge.

The stress in concrete is constant from the top to a depth $0.43x_{u,\max}$. Generally,

$$d' \le 0.43x_{u,\max} \text{ hence } f_{cc} = 0.447f_{ck}.$$

The stress in compression steel f_{sc} can be determined from its strain. The strains and stresses developed in a doubly reinforced section are shown in Fig. 3.4. The strain in concrete at a distance d' from the extreme compression fibre and hence in compression steel can be determined as:

$$\frac{\varepsilon_{cu}}{x_{u,\max}} = \frac{\varepsilon_{cc}}{(x_{u,\max} - d')}$$

where $\varepsilon_{cu} = 0.0035$

Thus,

$$\varepsilon_{sc} = \varepsilon_{cc} = \frac{0.0035(x_{u,\max} - d')}{x_{u,\max}} \tag{3.20}$$

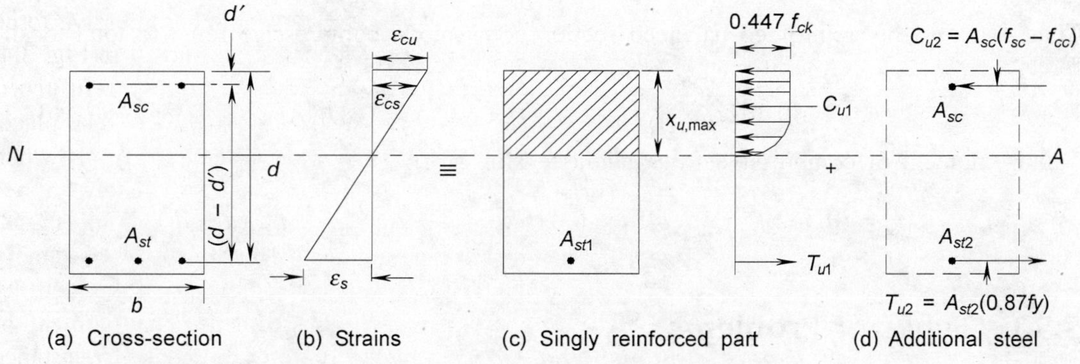

Fig. 3.4 Design of beam reinforced both in tension and compression.

For mild steel, f_y = 250 MPa, the stress-strain curve shown in Fig. 2.3 indicates that stress remains constant with the increase in strain beyond the yield point. For the values of (d'/d) up to 0.2, f_{cc} is equal to $0.447f_{ck}$ and f_{sc} is equal to $0.87f_y$, i.e.,

$$f_{sc} = 0.87f_y = 0.87 \times 250 = 217.5 \text{ MPa}.$$

When the reinforcement to be used is HYSD steel of grades Fe415 and Fe500, the stress in compression steel f_{sc} can be obtained from Table 3.3 for the given values of d'/d.

TABLE 3.3 Design Stress f_{sc} for Different Values of (d'/d)

d'/d	Stress f_{sc}, *MPa*			*d'/d*	Stress f_{sc}, *MPa*		
	Fe250	*Fe415*	*Fe500*		*Fe250*	*Fe415*	*Fe500*
0.040	217.5	358.73	425.86	0.125	217.5	347.61	404.45
0.045	217.5	358.37	424.85	0.130	217.5	346.66	402.87
0.050	**217.5**	**357.97**	**423.81**	0.135	217.5	345.68	401.24
0.055	217.5	357.54	422.73	0.140	217.5	344.67	399.54
0.060	217.5	357.07	421.62	0.145	217.5	343.64	397.77
0.065	217.5	356.56	420.47	**0.150**	**217.5**	**342.59**	**395.92**
0.070	217.5	356.01	419.30	0.155	217.5	341.50	393.98
0.075	217.5	355.42	418.10	0.160	217.5	340.39	391.95
0.080	217.5	354.79	416.87	0.165	217.5	339.24	389.80
0.085	217.5	354.13	415.62	0.170	217.5	338.06	387.55
0.090	217.5	353.43	414.34	0.175	217.5	336.85	385.16
0.095	217.5	352.70	413.03	0.180	217.5	335.59	382.63
0.100	**217.5**	**351.93**	**411.69**	0.185	217.5	334.30	379.95
0.105	217.5	351.12	410.31	0.190	217.5	332.95	377.11
0.110	217.5	350.29	408.91	0.195	217.5	331.55	374.10
0.115	217.5	349.42	407.46	**0.200**	**217.5**	**330.08**	**370.89**
0.120	217.5	348.53	405.98	0.205	217.5	328.55	367.49

Thus, for a doubly reinforced balanced section equate total compressive force to total tensile force, i.e.,

$$0.362 f_{ck} x_{u,\max} b + A_{sc}(f_{sc} - f_{cc}) = 0.87 f_y A_{st} \tag{3.21}$$

Taking moment of compressive force about tension steel:

$$M_u = (0.362\, f_{ck} x_{u,\max} b)(d - 0.416 x_{u,\max}) + A_{sc}(f_{sc} - f_{cc})(d - d') \tag{3.22}$$

3.5.1 Types of Problems

As in case of singly reinforced beams two types of problems are generally encountered in practice, namely, the analysis problems and the design problems.

Type-I: Determination of limiting moment of resistance or load-carrying capacity. In this case, cross-sectional dimensions, area of reinforcements in tension and compression, and grades of materials used are known. The various steps involved are:

1. Determine the depth of the neutral axis of the section, $x_{u,\max}$, by considering it to be a balanced section:

$$x_{u,\max} = \frac{0.0035d}{0.0055 + \dfrac{0.87 f_y}{E_s}}$$

2. Determine the total compressive and tensile forces:

$$C_u = 0.362 f_{ck} b x_{u,\max} + A_{sc}(f_{sc} - f_{cc})$$

where the stresses f_{sc} and f_{cc} correspond to the strain ε_{sc} at the level of compression steel which is given by

$$\varepsilon_{sc} = \frac{0.0035(x_{u,\max} - d')}{x_{u,\max}}$$

With mild steel reinforcement, for $d'/d \le 0.2$,

$$f_{cc} = 0.447 f_{ck} \text{ and } f_{sc} = 0.87 f_y$$

Tensile force, $T_u = 0.87 f_y A_{st}$

Whereas, in the case of Fe415 or Fe500 grade steel reinforcement, the stresses f_{sc} and f_{cc} are obtained from design stress-strain curves of the steel and the concrete, respectively.

3. Compare C_u with T_u, to ascertain whether the section is balanced, under-reinforced or over-reinforced.
   (i) If $C_u = T_u$ it is a balanced section. The limiting moment of resistance with respect to compressive force is given by:

$$M_{u,\text{lim}} = 0.362 f_{ck} b x_{u,\max}(d - 0.416 x_{u,\max}) + (f_{sc} - f_{cc}) A_{sc}(d - d')$$

   (ii) If $C_u > T_u$, it is an under-reinforced section. In this case, tension steel reaches its yield strength and the extreme compression fibre reaches its ultimate strain.
   (iii) If $C_u < T_u$, it is an over-reinforced section.

4. Obtain the depth of the actual neutral axis from the internal force equilibrium relation:

$$C_u = T_u$$

   (i) For the under-reinforced section, the internal force equilibrium equation is:

$$0.362 f_{ck} b x_u + (f_{sc} - f_{cc}) A_{sc} = 0.87 f_y A_{st}$$

   or

$$x_u = \frac{0.87 f_y A_{st} - A_{sc}(f_{sc} - f_{cc})}{0.362 f_{ck} b}$$

   where f_{sc} and f_{cc} correspond to strain ε_{sc} which is given by

$$\varepsilon_{sc} = \frac{0.0035(x_u - d')}{x_u}$$

The value x_u satisfying the above two equations can be obtained by an iterative procedure, starting with $x_{u,\max}$.

(ii) For the over-reinforced section, the internal force equilibrium equation is:

$$0.362 f_{ck} b x_u + (f_{sc} - f_{cc}) A_{sc} = f_s A_{st}$$

or

$$x_u = \frac{A_{st} f_s - A_{sc}(f_{sc} - f_{cc})}{0.362 f_{ck} b}$$

where stresses f_{sc} and f_{cc} correspond to the strain ε_{sc} at the level of compression steel which is given by

$$\varepsilon_{sc} = \frac{0.0035(x_u - d')}{x_u}$$

and the stress f_s corresponds to the strain in tension steel ε_s given by

$$\varepsilon_s = \frac{0.0035(d - x_u)}{x_u}$$

The values of x_u, f_{sc}, f_{cc} and f_s satisfying the above three equations can be obtained by the iterative procedure, starting with $x_u = x_{u,\max}$.

5. From the converged values of x_u, f_{sc}, f_{cc} and f_s, determine the limiting moment of resistance of the section:

$$M_u = 0.362 f_{ck} b x_u (d - 0.416 x_u) + A_{sc}(f_{sc} - f_{cc})(d - d')$$

6. If the effective span and the support conditions of the beam are known, compute the load-carrying capacity.

Example 3.6 A reinforced concrete beam of rectangular section has a width of 300 mm and an effective depth of 600 mm. The effective covers to tension and compression reinforcements are 50 mm and 30 mm, respectively. The beam is reinforced with 4 bars of 25 mm ϕ mild steel in tension and 3 bars of 20 mm ϕ mild steel in compression. Find the limiting moment-carrying capacity of the beam. The concrete mix used is M20.

Solution For this analysis problem:

$$b = 300 \text{ mm},\ d = 600 \text{ mm},\ d' = 30 \text{ mm and } \left(\frac{d'}{d}\right) = 0.05$$

The areas of tension and compression steels are:

$$A_{st}(4 \times 25 \text{ mm } \phi) = 1963 \text{ mm}^2$$

$$A_{sc}(3 \times 20 \text{ mm } \phi) = 942 \text{ mm}^2$$

For M20 grade concrete and mild steel reinforcement:

$$f_{ck} = 20 \text{ MPa and } f_y = 250 \text{ MPa}$$

$$f_{cc} = 0.447 f_{ck} = 0.447 \times 20 = 8.94 \text{ MPa}$$

Let the compression steel also yields along with tension steel at the collapse, i.e.,

$$f_{sc} = 0.87 f_y = 0.87 \times 250 = 217.5 \text{ MPa}$$

The limiting strain in mild steel at the yield point is:

$$\varepsilon_{su} = \varepsilon_{scu} = \frac{0.87 f_y}{E_s}$$

$$= \frac{217.5}{200,000} = 0.00109$$

To ascertain whether the section is balanced, under-reinforced or over-reinforced, consider it to be a balanced section. The limiting depth of the neutral axis is:

$$x_{u,\max} = 0.531d = 0.531 \times 600 = 318.6 \text{ mm}$$

Let x_u be the depth of the actual neutral axis. For the equilibrium of internal forces on the section,

$$C_u = T_u$$

$$0.362 f_{ck} x_u b + A_{sc}(f_{sc} - f_{cc}) = 0.87 f_y A_{st}$$

or

$$0.362 \times 20 \times 300 x_u + 942 \times (217.5 - 8.92) = 217.5 \times 1963$$

or

$$x_u = \frac{217.5 \times 1963 - 942 \times (217.5 - 8.94)}{0.362 \times 20 \times 300} = 106.12 \text{ mm}$$

Since $x_u < x_{u,\max}$, the section is under-reinforced. Check whether tension and compression steels have yielded as assumed.

Strain in tension steel:

$$\varepsilon_s = \frac{0.0035(d - x_u)}{x_u}$$

$$= \frac{0.0035 \times (600 - 106.12)}{106.12} = 0.0163 > \varepsilon_{su}(= 0.00109)$$

Hence tension steel yields. The actual strain in compression steel is given as:

$$\varepsilon_{sc} = \frac{0.0035(x_u - d')}{x_u}$$

$$= \frac{0.0035 \times (106.12 - 30)}{106.12} = 0.00251 > \varepsilon_{scu}(= 0.00109)$$

Hence compression steel also yields. The limiting moment of resistance or limiting moment capacity can be obtained by taking moment about tension steel:

$$M_{u,\lim} = 0.362 f_{ck} x_u b(d - 0.416x_u) + A_{sc}(f_{sc} - f_{cc})(d - d')$$

$$= 0.362 \times 20 \times 106.12 \times 300 \times (600 - 0.416 \times 106.12) + 942 \times (217.5 - 8.94) \times (600 - 30)$$

$$= 240.10 \times 10^6 \text{ Nmm} = 240.10 \text{ kNm}$$

Example 3.7 Analyze a rectangular beam section of 300 mm width and 600 mm effective depth, reinforced with 4 bars of 25 mm ϕ on the tension side and 3 bars of 20 mm ϕ on the compression side at an effective cover d' of 30 mm. The grade of reinforcement steel used is Fe415 and that of the concrete mix is M20. Determine: (i) the limiting moment of resistance of the section and (ii) the safe uniformly distributed load the beam can support in addition to its self-weight over an effective span of 9 m.

Solution For the given beam cross-section and materials:

$$b = 300 \text{ mm}, \; d = 600 \text{ mm}, \; d' = 30 \text{ mm and } \frac{d}{d'} = 0.05$$

$$A_{st} = 1963 \text{ mm}^2 \text{ and } A_{sc} = 942 \text{ mm}^2, \; f_{ck} = 20 \text{ MPa} \quad \text{and} \quad f_y = 415 \text{ MPa}$$

To ascertain whether the section is balanced, under-reinforced or over-reinforced, consider it to be a balanced section. Then

$$x_{u,\max} = 0.0035 \; d/(0.0055 + 0.87 f_y/E_s)$$

$$= 0.0035 \times 600/[0.0055 + (0.87 \times 415)/(2 \times 10^5)]$$

$$= 0.479 \times 600 = 287.4 \text{ mm}$$

The permissible value of strain in Fe415 grade steel at the yield point is

$$\varepsilon_{su} = 0.002 + \frac{0.87 f_y}{E_s}$$

$$= 0.002 + \frac{0.87 \times 415}{2 \times 10^5} = 0.0038$$

The design stresses f_{sc} and f_{cc} in compression reinforcement and concrete at the level of the centroid of compression reinforcement, respectively, correspond to ε_{sc} given by

$$\varepsilon_{sc} = \frac{0.0035(x_{u,\max} - d')}{x_{u,\max}} = \frac{0.0035 \times 287.4 - 30}{287.4} = 0.00313$$

The value of f_{sc} and f_{cc} are obtained from the design stress-strain curves of steel and concrete, respectively.

$$f_{cc} = 0.447 f_{ck} = 0.447 \times 20 = 8.94 \text{ MPa}$$

From Table 3.3,

$$f_{sc} = 357.97 \text{ MPa}$$

and

$$f_{st} = 0.87f_y = 0.87 \times 415 = 361.05 \text{ MPa}$$

Let x_u be the depth of the actual neutral axis; then for the equilibrium of internal forces:

$$0.362f_{ck}bx_u + A_{sc}(f_{sc} - f_{cc}) = 0.87f_yA_{st}$$

or

$$x_u = \frac{0.87 f_y A_{st} - A_{sc}(f_{sc} - f_{cc})}{0.362 f_{ck} b}$$

$$= \frac{361.05 \times 1963 - 942 \times (357.97 - 8.94)}{(0.362 \times 20 \times 300)} = 174.93 \text{ mm}$$

Since, $x_u < x_{u,\max}$ the section is under-reinforced. Assuming $x_u = 174.93$ mm for the next trial, the corresponding strain in compression steel is obtained as:

$$\varepsilon_{sc} = \frac{0.0035(x_u - d')}{x_u} = \frac{0.0035 \times (174.93 - 30)}{174.93} = 0.00290$$

The design stresses f_{cc} and f_{sc} corresponding to ε_{sc} are:

$$f_{cc} = 8.94 \text{ MPa}, f_{sc} = 354.5 \text{ MPa and } f_{st} = 361.05 \text{ MPa}$$

The modified value of x_u can be computed as:

$$x_u = [\ 361.05 \times 1963 - 942 \times (354.5 - 8.94)]/(0.362 \times 20 \times 300) = 176.44 \text{ mm}$$

The above procedure is repeated till the solution converges. Taking the converged values as:

$$x_u = 176 \text{ mm}, f_{cc} = 8.94 \text{ MPa}, f_{sc} = 354.5 \text{ MPa, and } f_{st} = 361.05 \text{ MPa}$$

The limiting moment of resistance of the section is:

$$M_{u,\text{lim}} = 0.362f_{ck}bx_u(d - 0.416x_u) + A_{sc}(f_{sc} - f_{cc})(d - d')$$

$$= 0.362 \times 20 \times 300 \times 176 \times (600 - 0.416 \times 176) + 942 \times (354.5 - 8.94) \times (600 - 30)$$

$$= 386.74 \times 10^6 \text{ Nmm} = 386.92 \text{ kNm}$$

Consider the factored load on the beam to be w_d kN/m, then the maximum moment is given by

$$M_d = \frac{w_d L^2}{8} = \frac{w_d (9)^2}{8} = 10.125\ w_d \text{ kNm}$$

Equating factored moment to the limiting moment capacity:

$$10.125w_d = 386.92,$$

or

$$w_d = 38.21 \text{ kN/m}$$

Total service load on the beam $w = \dfrac{w_d}{\gamma_d}$ =38.21/1.5 = 25.48 kN/m.

Self-weight of the beam $w_g = 0.30 \times 0.65 \times 1 \times 25 = 4.875$ kN/m.

Superimposed uniformly distributed load w_s the beam can support safely is:

$$w_s = w - w_g = 25.48 - 4.875 = 20.605 \text{ kN/m}$$

Type II: Design of a Section. In this case cross-sectional dimensions, and grades of materials used are predetermined. The area of reinforcements in tension and compression are to be computed. The various steps involved are:

1. Determine the limiting moment of resistance of the section $M_{u,\lim}$ considering the section to be singly reinforced.
2. Compare $M_{u,\lim}$ with the factored design moment M_d.
   If $M_{u,\lim} > M_d$, a singly reinforced section will be adequate, and if $M_{u,\lim} < M_d$, the beam shall be designed as a doubly reinforced section.
3. Determine the area of tension reinforcement A_{st1} required for singly reinforced beam of Step (1):
$$A_{st1} = \frac{M_{u,\lim}}{0.87 f_y (d - 0.416 x_{u,\max})}$$
4. Calculate the balance or additional factored moment to be resisted by the beam:
$$M_{da} = M_d - M_{u,\lim}$$
   This moment is to be balanced by the additional tension steel A_{st2} and the compression steel A_{sc}, the forces in these steels form an additional internal couple of resistance.
5. Determine A_{st2} and A_{sc} from the relations:
$$A_{st2} = \frac{M_{da}}{0.87 f_y (d - d')}$$
   and
$$A_{sc} = \frac{M_{da}}{(f_{sc} - f_{cc})(d - d')}$$
   where f_{sc} and f_{cc} are design stresses in compression steel and concrete at the level of the centroid of compression steel.
6. Determine total tension steel

Example 3.8 A reinforced concrete beam of rectangular section of size 250 × 550 mm overall is to be designed for a factored moment of 225 kNm. Compute the reinforcement required at the effective cover of 50 mm. The concrete mix to be used is M20 and the grade of steel is Fe415.

Solution The cross-sectional dimensions and material properties are:

$$b = 250 \text{ mm},\ d = 550 - 50 = 500 \text{ mm and } d' = 50 \text{ mm}$$

$$f_{ck} = 20 \text{ MPa},\ f_y = 415 \text{ MPa and } M_d = 225 \text{ kNm.}$$

The limiting depth of the neutral axis:

$$x_{u,\max} = 0.479 \times 500 = 239.5 \text{ mm}$$

$M_{u,\lim}$ for a singly reinforced section

$$M_{u,\lim} = 0.1388\ f_{ck} b d^2 = 0.1388 \times 20 \times 250 \times 500^2$$
$$= 173.50 \times 10^6 \text{ Nmm} = 173.50 \text{ kNm}$$

For M20 grade concrete and Fe415 grade steel:

$$p_{t,\text{lim}} = \frac{100A_{st}}{bd} = 0.96$$

Therefore,

$$A_{st1} = \frac{0.96 \times 250 \times 500}{100} = 1200 \text{ mm}^2.$$

Since $M_d > M_{u,\text{lim}}$, additional moment of resistance required is:

$$M_{da} = 225 - 173.5 = 51.5 \text{ kNm}$$

Additional tension reinforcement is obtained as follows:

$$A_{st2} = \frac{M_{da}}{(0.87f_y)(d - d')}$$

$$= \frac{51.5 \times 10^6}{0.87 \times 415 \times (500 - 50)} = 316.98 \text{ mm}^2$$

Total tensile steel $A_{st} = A_{st1} + A_{st2} = 1200 + 316.98 \approx 1517 \text{ mm}^2$
The compression steel is given by the relation:

$$A_{sc} = \frac{M_{da}}{(f_{sc} - f_{cc})(d - d')}$$

For $d'/d = 0.10$ and Fe415 grade steel, from Table 3.3, $f_{sc} = 351.93$ MPa, and

$$f_{cc} = 0.447f_{ck} = 0.447 \times 20 = 8.94 \text{ MPa}$$

Therefore,

$$A_{sc} = \frac{51.5 \times 10^6}{(351.93 - 8.94) \times (500 - 50)} = 333.67 \text{ mm}^2.$$

3.6 ANALYSIS OF THE FLANGED BEAM SECTION

In many reinforced concrete structures and, particularly in floor systems, a concrete slab is cast monolithically with and, connected to, rectangular beams. In such a construction a portion of the slab above the beam behaves structurally as a part of the beam in compression. The slab portion is called the **flange** and beam the **web**. If the flange projections are on either side of the rectangular *web* or *rib*, the resulting cross-section resembles the T-shape and hence is called a **T-beam section**. On the other hand, if the flange projects on one side, the resulting cross-section resembles an inverted L and hence is termed as **L-beam**. The flanged beams are shown in Fig. 3.5. The terms b_f, b_w, D_f and d represent the width of flange, breadth of the web, thickness of the flange (i.e., of the slab), and effective depth of the beam, respectively. In the

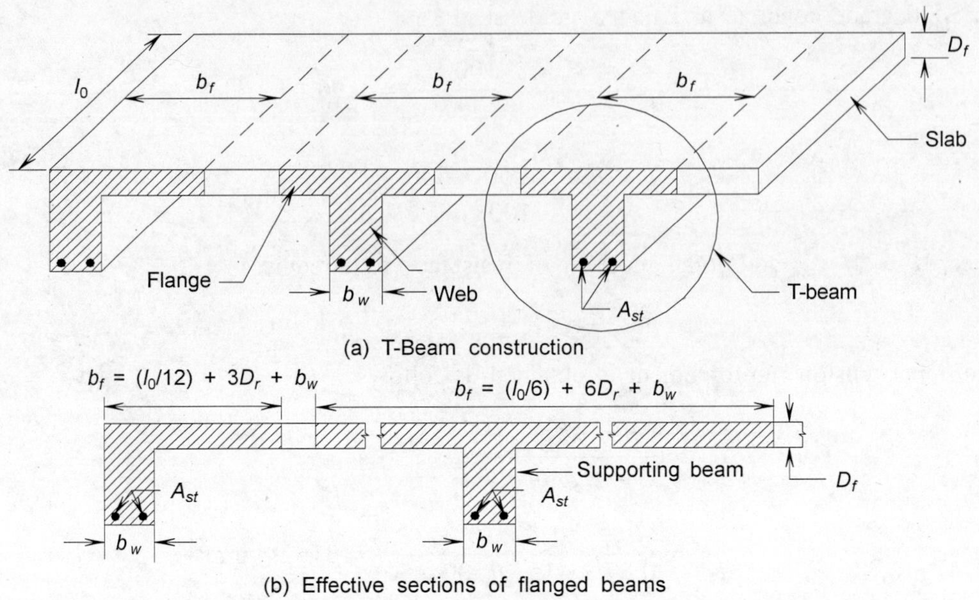

Fig. 3.5 Idealization of a floor system using flanged beams.

absence of more accurate determination the effective width of the flange, b_f that acts along with the rectangular rib, may be taken as stipulated by IS:456. The values are:

$$\text{For T-beam: } b_f = \left(\frac{l_0}{6}\right) + b_w + 6D_f$$

$$\text{For L-beam: } b_f = \left(\frac{l_0}{12}\right) + b_w + 3D_f \tag{3.23a}$$

where l_o is the distance between zero moment (inflection) points in the beams. This is equal to the effective span for simply supported beams and is 0.7 times the effective span for continuous beams. However, the effective flange width in no case should exceed the breadth of web plus half the sum of the clear distances to the adjacent beams on either side. In case of isolated beams, the effective flange width obtained below shall not exceed the actual width of the flange.

$$\text{For T-beam: } b_f = b_w + \frac{l_o}{(l_o/b) + 4}$$

$$\text{For L-beam: } b_f = b_w + \frac{0.5l_o}{(l_o/b) + 4} \tag{3.23b}$$

As in rectangular beams, the flanged sections may be either singly or doubly reinforced.

3.6.1 Analysis of Flanged Section with Tension Reinforcement

Type-I: Determination of limiting moment of resistance or load-carrying capacity. The analysis of reinforced concrete T- and L-beam sections requires the determination of the limiting moment capacity of the section of given dimensions and reinforcement details. The flexural strength of the section is reached when the compression strain in the extreme compression fibre of concrete reaches its ultimate value. Depending on the cross-sectional dimensions and area of reinforcement, the section may be balanced, under-reinforced or over-reinforced. In addition, the neutral axis may lie in the flange or in the web. The beam section is analyzed accordingly. To ascertain the type of section consider it to be a balanced section. The limiting depth of the neutral axis for this case, $x_{u,\max}$, can be obtained as:

$$\frac{0.0035}{x_{u,\max}} = \frac{0.002 + 0.87 f_y / E_s}{(d - x_{u,\max})}$$

or

$$x_{u,\max} = \frac{0.0035d}{0.0055 + (0.87 f_y / E_s)} \tag{3.24}$$

As in the case of rectangular beams T- and L-beam sections are generally, under-reinforced ($x_u < x_{u,\max}$) and hence tension steel will reach its limiting strength.

CASE I $x_u < D_f$. The actual neutral axis falls within the flange as shown in Fig. 3.6(a) and the beam is analyzed as a rectangular beam of width b_f instead of a beam of width b_w and depth

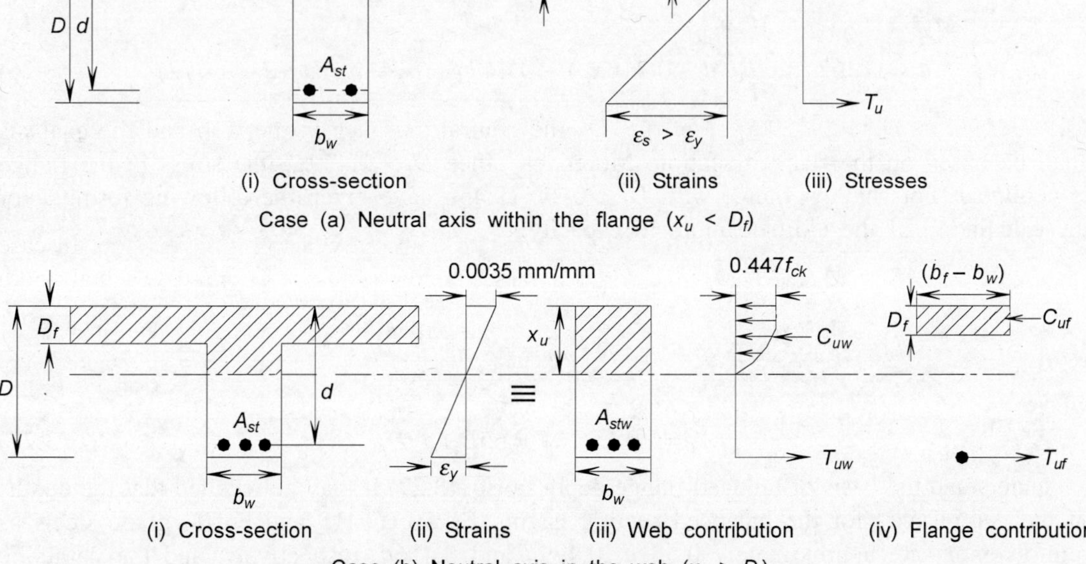

Fig. 3.6 *Normal stress distribution in a reinforced concrete T-beam under flexure.*

d. Such an idealization is possible since the concrete in tension can be ignored. In this case the expressions developed earlier for rectangular beam can be utilized. The depth of the actual neutral axis x_u must be evaluated by using the internal force equilibrium condition:

$$C_u = T_u \text{ or } 0.362\, f_{ck} x_u b_f = 0.87\, f_y A_{st}$$

Thus,

$$x_u = \frac{0.87\, f_y A_{st}}{0.362\, f_{ck} b_f}$$

The limiting or ultimate moment capacity of the beam can be obtained as:

$$M_u = T_u \times z = 0.87 f_y A_{st}(d - 0.416 x_u)$$

CASE *II* $x_u > D_f \leq 0.429 x_u$. For the case $x_u > D_f$, the neutral axis lies in the web as shown in Fig. 3.6(b) and the section will be analyzed as a flanged section. However, for the condition $D_f < 0.429 x_u$, the stress in the flange is uniform. The T-beam can be considered as a rectangular beam of width b_w and depth *d*, and the remaining portion of flange as a beam of width $(b_f - b_w)$ and depth D_f as shown in Fig. 3.6(b).
The depth of the neutral axis x_u can be determined by using the force equilibrium equation:

$$\text{Compressive force in web + Compressive force in flange = Total tension}$$

$$C_{uw} + C_{uf} = T_u$$

or

$$0.362\, f_{ck} x_u b_w + 0.447 f_{ck}(b_f - b_w) D_f = 0.87 f_y A_{st} \tag{3.25}$$

The limiting or ultimate moment capacity of beam can be obtained by taking moment about tensile steel:

$$\begin{aligned} M_u &= M_{u,\text{web}} + M_{u,\text{flange}} \\ &= 0.362 f_{ck} x_u b_w (d - 0.416 x_u) + 0.447 f_{ck}(b_f - b_w) D_f (d - D_f/2) \end{aligned} \tag{3.26}$$

CASE *III* $x_u > D_f > 0.429\, x_u$. For $x_u > D_f$, the neutral axis lies in the web and the analysis shall be based on the flanged section. Moreover, since $D_f > 0.429 x_u$, the stress in the flange is nonlinear. For shallow flange with $D_f < 0.2d$, IS:456 have given the following formula for the calculation of the ultimate moment capacity:

$$M_u = M_{u,\text{web}} + M_{u,\text{ flange}}$$

$$M_u = 0.362\, f_{ck} x_u b_w (d - 0.416 x_u) + 0.447 f_{ck}(b_f - b_w) Y_f \left(d - \frac{Y_f}{2}\right)$$

where

$$Y_f = (0.15 x_u + 0.65 D_f) \not> D_f \tag{3.27}$$

To understand the basis of reduced flange depth in Eq. (3.27) it may be recalled that the depths of the neutral axis for the balanced section having Fe250, Fe415 and Fe500 grade steels as reinforcement are approximately 0.531*d*, 0.479*d* and 0.456*d*, respectively, and the parabolic portion of the parabolic-rectangular stress-strain curve extends to a height of $0.002x_u/0.0035 = 0.571x_u$ from the neutral axis beyond which rectangular stress distribution extends to a height of $0.429x_u$. Considering the worst case, the rectangular stress distribution may extend to a depth $0.429 \times 0.456d = 0.196d$. Hence a depth of $0.2d$ has been chosen as the limiting depth for the shallow flanges.

The limiting value of Y_f is equal to D_f for the case when the bottom fibre of the flange is subjected to a compressive strain equal to or greater than 0.002 corresponding to the design stress equal to $0.447f_{ck}$. When the strain in the bottom fibre of the flange is less than 0.002, the design stress distribution in the flange is nonlinear. This makes it necessary to use the reduced flange depth D_f, if the uniform design stress of $0.447f_{ck}$ is to be used over the entire flange depth.

The value of x_u in the preceding expressions must be evaluated by using the internal force equilibrium equation, for which it is essential to assume for thc first trial that either $x_u < D_f$ or $x_u > D_f$. To reduce the calculations for the first trial, assume that the neutral axis coincides with the underneath of the flange slab, and calculate total compression in concrete flange and total tension in steel, C_u and T_u, respectively. Then

1. If $C_u > T_u$, the neutral axis lies in the flange.
2. If $C_u = T_u$, the neutral axis coincides with the bottom of the flange.
3. If $C_u < T_u$, the neutral axis lies in the web.

The doubly reinforced flanged beams may also be encountered in practice, in that case, the expressions developed earlier for rectangular beams of width b_f should be utilized. The following examples will illustrate the nature of analysis involved in the case of flanged beams.

Example 3.9 Analyze a beam of T-shaped cross-section having an effective flange width of 1500 mm, flange thickness of 100 mm, web width of 300 mm and an effective depth of 600 mm, to determine the limiting or ultimate moment of resistance of the beam for the cases of tension reinforcement of: (i) 5 × 22 mm ϕ bars, and (ii) 5 × 28 mm ϕ bars. The materials used are concrete mix of grade M20 and HYSD steel of grade Fe415.

Solution For the given cross-section and materials:

$$b_f = 1500 \text{ mm}, \; b_w = 300 \text{ mm}, \; d = 600 \text{ mm}, \; D_f = 100 \text{ mm}$$

$$f_{ck} = 20 \text{ MPa and } f_y = 415 \text{ MPa.}$$

CASE I Area of tension steel A_{st} (5 × 22 mm ϕ) = 1900 mm^2

Consider that the neutral axis coincides with bottom fibre of the flange, i.e., $x_u = D_f$.

$$\text{Total compression in flange} = 0.362 f_{ck} b_f D_f$$

$$= 0.362 \times 20 \times 1500 \times 100 = 1.086 \times 10^3 \text{ kN}$$

$$\text{Total tension in steel} = 0.87 f_y A_{st} = 0.87 \times 415 \times 1900$$

$$= 0.686 \times 10^3 \text{ kN}$$

Since the total compression in the flange is greater than the total tension in steel, the neutral axis is located in the flange, i.e., $x_u < D_f$.

The actual value of x_u may be obtained as:

$$0.362 f_{ck} b_f x_u = 0.87 f_y A_{st}$$

$$x_u = \frac{0.87 f_y A_{st}}{0.362 f_{ck} b_f}$$

$$= \frac{0.87 \times 415 \times 1900}{0.362 \times 20 \times 1500} = 63.17 \text{ mm}$$

Thus, $x_u < D_f$ confirms that the neutral axis lies in the flange. Now, the section may be balanced, under-reinforced or over-reinforced. To ascertain this consider the section to be a balanced one:

$$x_{u,\max} = 0.479d = 287.4 \text{ mm} > x_u$$

Thus, the section is under-reinforced. Hence

$$M_u = 0.87 f_y A_{st}(d - 0.416 x_u)$$

$$= 0.87 \times 415 \times 1900 \times (600 - 0.416 \times 63.17)$$

$$= 393.57 \times 10^6 \text{ Nmm} = 393.57 \text{ kNm}$$

CASE II Area of tension steel $A_{st}(5 \times 28 \text{ mm } \phi) = 3078 \text{ mm}^2$

Consider that the neutral axis coincides with bottom fibre of the flange, i.e., $x_u = D_f$.

$$\text{Total compression in flange } C_u = 0.362 f_{ck} b_f D_f$$

$$= 0.362 \times 20 \times 1500 \times 100 = 1.086 \times 10^3 \text{ kN}$$

$$\text{Total tension in steel } T_u = 0.87\, f_y A_{st}$$

$$= 0.87 \times 415 \times 3078 = 1.111 \times 10^3 \text{ kN}$$

Since $C_u < T_u$, the neutral axis lies in the web. To evaluate x_u, the approximate internal force equilibrium condition based on the assumption that $D_f < 0.429x_u$, i.e., stress is uniform in the flange, can be used:

$$C_{uw} + C_{uf} = T_u$$

$$0.362 f_{ck} x_u b_w + 0.447 f_{ck} D_f (b_f - b_w) = 0.87 f_y A_{st}$$

or

$$x_u = \frac{0.87 f_y A_{st} - 0.447 f_{ck} D_f \left(b_f - b_w\right)}{0.362 f_{ck} b_w}$$

$$= \frac{0.87 \times 415 \times 3078 - 0.447 \times 20 \times (1500 - 300) \times 100}{0.362 \times 20 \times 300} = 17.73 \text{ mm} < D_f$$

Hence, the computed value of x_u is not valid. This points to the fact that the stress in the flange is non-uniform. Thus D_f is to be replaced by Y_f.

$$Y_f = 0.15 x_u + 0.65 D_f$$

Therefore,

$$0.362 \times 20 \times 300x_u + 0.447 \times 20 \times (0.15x_u + 0.65 \times 100) \times (1500 - 300) = 0.87 \times 415 \times 3078$$

or

$$2172x_u + 1609.2x_u + 697320 = 1111311.9$$

Giving $x_u = 109.49$ mm > D_f and $0.429x_u = 0.429 \times 109.49 = 46.97$ mm < D_f.
Hence,

$$Y_f = 0.15x_u + 0.65\ D_f = 0.15 \times 109.49 + 0.65 \times 100 = 81.42 \text{ mm}$$

The depth of the balanced neutral axis is:

$$x_{u,\max} = 0.479d = 0.479 \times 600 = 287.4 \text{ mm}$$

Since $x_{u,\max} > x_u$, the section is under-reinforced. The ultimate moment of resistance of the section is:

$$M_u = M_{u,\text{web}} + M_{u,\text{flange}}$$

$$= 0.362 f_{ck} x_u b_w (d - 0.416x_u) + 0.447 f_{ck} Y_f (b_f - b_w)\left(d - \frac{Y_f}{2}\right)$$

$$= 0.362 \times 20 \times 109.49 \times 300 \times (600 - 0.416 \times 109.49)$$

$$+ 0.447 \times 20 \times 81.42 \times (1500 - 300) \times \left(600 - \frac{81.42}{2}\right)$$

$$= 131.86 \times 10^6 + 488.53 \times 10^6 = 620.39 \times 10^6 \text{ Nmm} = 620.39 \text{ kNm}$$

Example 3.10 Determine the limiting moment of resistance of beam of T-shaped cross-section having a flange width of 1250 mm, a web width of 300 mm, a flange thickness of 125 mm and an effective depth of 550 mm. The beam is reinforced with 8 bars of 25 mm ϕ on the tension side. The concrete mix used is of grade M20 and HYSD steel is of grade Fe415.

Solution For the given cross-section and materials:

$$b_f = 1250 \text{ mm},\ b_w = 300 \text{ mm},\ D_f = 125 \text{ mm, and } d = 550 \text{ mm}$$

Area of tension steel $A_{st}(8 \times 25 \text{ mm } \phi) = 3927 \text{ mm}^2$

$$f_{ck} = 20 \text{ MPa and } f_y = 415 \text{ MPa}$$

Then

$$x_{u,\max} = 0.479\ d = 0.479 \times 550 = 263.45 \text{ mm}$$

Consider that the neutral axis coincides with bottom of the flange, i.e., $x_u = D_f$

Total compression $C_u = 0.362 f_{ck} b_f D_f = 0.362 \times 20 \times 1250 \times 125 = 1.131 \times 10^6$ N

Total tension $T_u = 0.87 f_y A_{st} = 0.87 \times 415 \times 3927 = 1.418 \times 10^6$ N

Since $C_u < T_u$, the neutral axis lies in the web. Assuming that the stress in the flange is constant, the internal force equilibrium condition, $C_{uw} + C_{uf} = T_w$ gives:

$$0.362 f_{ck} x_u b_w + 0.447 f_{ck} D_f (b_f - b_w) = 0.87 f_y A_{st}$$

$$0.362 \times 20 \times x_u \times 300 + 0.447 \times 20 \times 125 \times (1250 - 300) = 0.87 \times 415 \times 3927$$

$$x_u = 164.00 \text{ mm}$$

$$0.429x_u = 0.429 \times 164.00 = 70.36 \text{ mm} < D_f$$

Therefore, stress in the flange is non-uniform and D_f is to be replaced by its reduced value:

$$Y_f = (0.15x_u + 0.65D_f)$$

Thus,

$$0.362 \times 20x_u \times 300 + 0.447 \times 20 \times (0.15x_u + 0.65 \times 125) \times (1250 - 300)$$

$$= 0.87 \times 415 \times 3927$$

$$x_u = 211.20 \text{ mm} < x_{u,\max}$$

Hence the section is under-reinforced. The reduced depth of the flange is given by

$$Y_f = 0.15 \times 211.20 + 0.65 \times 125 = 112.93 \text{ mm} < D_f$$

The limiting or ultimate moment of resistance is expressed as:

$$M_u = M_{uw} + M_{uf} = 0.362x_u b_w(d - 0.416x_u) + 0.447f_{ck}Y_f(b_f - b_w)\left(d - \frac{Y_f}{2}\right)$$

$$= 0.362 \times 20 \times 300 \times 211.20 \times (550 - 0.416 \times 211.20)$$

$$+ 0.447 \times 20 \times 112.93 \times (1250 - 300)\left(550 - \frac{112.93}{2}\right)$$

$$= 211.996 \times 10^6 + 473.357 \times 10^6 = 685.35 \times 10^6 \text{ Nmm}$$

$$= 685.35 \text{ kNm}$$

It should be noted that the contribution from the web is not really negligible.
Alternatively, the problem can also be analyzed as follows:
Since the flange thickness 125 mm > $0.2d$ (= 0.2 × 550 = 110 mm), the equivalent flange depth, $Y_f = (0.15x_u + 0.65D_f)$ must be utilized. For this purpose the value of x_u should be evaluated or estimated. It can conveniently be estimated by trial and modification.

For Fe415 grade steel: $x_{u,\max} = 0.479d = 0.479 \times 550 = 263.45$ mm

First trial: Consider the trial depth of the neutral axis x_u to be 240 mm and beam to be under-reinforced ($x_u < x_{u,\max}$).

$$Y_f = 0.15 \times 240 + 0.65 \times 125 = 117.25 \text{ mm} < 125 \text{ mm}$$

Use the internal force equilibrium condition to obtain fresh estimate of the neutral axis depth,

$$0.362f_{ck}b_w x_u + 0.447f_{ck}Y_f(b_f - b_w) = 0.87\, f_y A_{st}$$

or

$$0.362 \times 20 \times 300x_u + 0.447 \times 20 \times 117.25 \times (1250 - 300) = 0.87 \times 415 \times 3927$$

$$x_u = \frac{1417843.35 - 995804.25}{2172} = 194.31 \text{ mm}$$

The new value is much less than assumed x_u = 240 mm.

Second trial: Let the trial depth of the neutral axis be $\frac{1}{2}(240 + 194) = 217$ mm (say 215 mm):

$$Y_f = 0.15 \times 215 + 0.65 \times 125 = 113.5 \text{ mm} < 125 \text{ mm}$$

Using the internal force equilibrium condition,

$$0.362 \times 20 \times 300x_u + 0.447 \times 20 \times 113.5 \times (1250 - 300) = 0.87 \times 415 \times 3927$$

or

$$x_u = \frac{1417843.35 - 963955.5}{2172} = 208.97 \text{ mm}$$

The new value is closer to the assumed value of 215 mm

Third trial: Let the depth of the neutral axis be 212 mm

$$Y_f = 0.15 \times 212 + 0.65 \times 125 = 113.05 \text{ mm}$$

From the internal force equilibrium condition,

$$x_u = \frac{1417843.35 - 961407.6}{2172} = 210.73 \text{ mm}$$

The new value agrees closely with the assumed value of 212 mm.
Thus, the depth of the neutral axis may be taken as 211.36 mm [= $\frac{1}{2}(212 + 210.73)$] and

$$Y_f = 0.15 \times 211.36 + 0.65 \times 125 = 112.95 \text{ mm}$$

Check as per clause of IS:456;

$$\frac{D_f}{x_u} = \frac{125}{211.36} = 0.591 > 0.429$$

Hence satisfactory,

$$M_{u,\lim} = 0.362 \times 20 \times 300 \times 211.36 \times (550 - 0.416 \times 211.58)$$

$$+ 0.447 \times 20 \times (1250 - 300) \times 112.99 \times 550 - \frac{112.99}{2}$$

$$= 212.30 \times 10^6 + 473.58 \times 10^6$$

$$= 685.88 \times 10^6 \text{ Nmm} = 685.88 \text{ kNm}$$

It should be noticed that contribution from the web is significant.

Type II: Checking the Adequacy of the Section. The cross-sectional dimensions b_f, D_f, b_w, d, and area of tension reinforcement are given. It is required to compute the limiting moment of resistance and compare it with the factored moment due to the action of given loading. If the effective span and the support conditions of the beam are known, the load-carrying capacity can be computed.

3.7 DESIGN OF THE FLANGED BEAM SECTION

3.7.1 Flanged Beam with Tension Reinforcement

The design of the flanged section requires determination of cross-sectional dimensions and area of steel for resisting the given factored moment. Normally, the depth of the flange is fixed by the design of the slab. The width of the flange is governed by the part of the slab which deflects monolithically with the web and is governed by the criteria given in IS:456. The width of the web is governed by shear considerations as well as by the minimum width required for providing reinforcement. Thus, the design of the T- or L-beam section requires the determination of the effective depth of the beam and the area of reinforcement. The steps involved are:

1. Estimate the depth of the flanged beam for preliminary computations as follows:
   (a) 1/12th of the span for heavy loads,
   (b) 1/12th to 1/15th of the span for medium loads, and
   (c) 1/15th to 1/20th for light loads.

   Since the width of the flange of a T-beam is generally large and the neutral axis falls within the flange, the depth of the beam may be selected such that the neutral axis coincides with the bottom of the flange so that the concrete in flange is fully utilized. Thus,

   $$M_u = 0.362 f_{ck} b_f D_f (d - 0.416 D_f)$$

   or

   $$d = 0.416 D_f + M_u/(0.362 f_{ck} b_f D_f)$$

   Revise the design moment M_u, if required, to be carried by the beam.
2. Estimate the width of rib b_w which is approximately one-half of the rib projection below the flange.
3. Ascertain the type of section, i.e., whether it is balanced, under-reinforced or over-reinforced.
4. Calculate the sectional area of reinforcement.
5. Based on the sectional area of reinforcement obtained in Step 4, determine the actual position of the neutral axis and hence the actual value of the lever arm.
6. Recalculate the sectional area of reinforcement.

Example 3.11 Design a T-beam section having a flange width of 1250 mm, flange thickness of 100 mm and web width of 250 mm, which is subjected to a factored moment of 400 kNm. Use concrete mix of grade M20 and mild steel of grade Fe250.

Solution The design of section requires determination of the depth of beam and area of reinforcement. Consider the depth of beam to be such that the neutral axis coincides with the bottom of the flange, with the beam being either balanced or under-reinforced. Thus,

$$M_u = 0.362 f_{ck} b_f D_f (d - 0.416 D_f) \quad \text{(as } x_u = D_f\text{)}$$

and hence

$$d = 0.416D_f + \frac{M_u}{0.362 f_{ck} b_f D_f}$$

$$= \frac{0.416 \times 100 + 400 \times 10^6}{0.362 \times 20 \times 1250 \times 100} = 483.59 \text{ mm}$$

Adopt an effective depth of 500 mm. Therefore, the limiting depth of the neutral axis is

$$x_{u,\max} = 0.531 \times 500 = 265.5 \text{ mm}$$

Consider $x_u (< D_f)$ to be the depth of the actual neutral axis; then from the moment equation:

$$M_u = 0.362 f_{ck} b_f x_u (d - 0.416 x_u)$$

$$400 \times 10^6 = 0.362 \times 20 \times 1250 x_u (500 - 0.416 x_u)$$

or

$$x_u^2 + 1201.92 x_u - 106247.34 = 0$$

or

$$x_u = 82.71 \text{ mm}$$

$$\text{Lever arm } z = jd = d - 0.416 x_u$$

$$= 500 - 0.416 \times 82.71 = 466.09 \text{ mm}$$

The area of steel required A_{st} can be determined from:

$$M_u = (0.87 f_y A_{st}) z$$

or

$$A_{st} = \frac{M_u}{0.87 f_y z} = \frac{400 \times 10^6}{0.87 \times 250 \times 466.09} = 3945.76 \text{ mm}^2$$

Provide 7 bars of 28 mm ϕ.

Example 3.12 Design a T-beam section with a flange width of 1250 mm, a flange depth of 100 mm, a web width of 250 mm, and an effective depth of 500 mm, which is subjected to a factored moment of 560 kNm. The concrete mix to be used is of grade M20 and steel is of grade Fe415.

Solution For the given cross-section and materials:

$$b_f = 1250 \text{ mm}, \; D_f = 100 \text{ mm}, \; b_w = 250 \text{ mm and } d = 500 \text{ mm}$$

$$f_{ck} = 20 \text{ MPa and } f_y = 415 \text{ MPa}$$

Consider it to be a balanced section. Then,

$$x_{u,\max} = 0.479d = 0.479 \times 500 = 239.5 \text{ mm}$$

Limiting moment of resistance of the balanced section is given by

$$M_{u,\lim} = 0.362 f_{ck} b_w x_{u,\max}(d - 0.416 x_{u,\max}) + 0.447 f_{ck} Y_f (b_f - b_w)(d - 0.5 Y_f)$$

where

$$Y_f = (0.15 x_{u,\max} + 0.65 D_f) \text{ or } D_f \text{ whichever is less}$$
$$= 0.15 \times 239.5 + 0.65 \times 100 = 100.925 \text{ mm} > D_f$$

Hence,

$$Y_f = 100 \text{ mm}$$

Therefore,

$$M_{u,\lim} = 0.362 \times 20 \times 250 \times 239.5 \times (500 - 0.416 \times 239.5)$$
$$+ 0.447 \times 20 \times 100 \times (1250 - 250) \times (500 - 0.5 \times 100)$$
$$= 575.86 \times 10^6 \text{ Nmm} = 575.86 \text{ kNm}$$

Since $M_{ud} < M_{u,\lim}$, it is an under-reinforced section. The neutral axis may lie either in the flange or in the web. To ascertain this consider that the neutral axis coincides with the bottom of the flange, i.e., $x_u = D_f$ then the moment of resistance of the section is given by

$$M_{uf} = 0.362 f_{ck} b_f D_f (d - 0.416 D_f)$$
$$= 0.362 \times 20 \times 1250 \times 100 \times (500 - 0.416 \times 100) \times 10^{-6} = 414.85 \text{ kNm}$$

Since $M_{uf} < M_{ud}$, the actual neutral axis lies in the web. The value of x_u can be determined from the moment equation.

$$M_{ud} = 0.362 f_{ck} b_w x_u (d - 0.416 x_u) + 0.447 f_{ck} (b_f - b_w) Y_f (d - 0.5 Y_f)$$

where

$$Y_f = 0.15 x_u + 0.65 D_f$$

Consider $Y_f = D_f = 100$ mm then

$$560 \times 10^6 = 0.362 \times 20 \times 250 \times x_u(500 - 0.416 x_u) + 0.447 \times 20$$
$$\times (1250 - 250) \times 100 \times (500 - 50)$$

or

$$x_u(500 - 0.416 x_u) = 87127.07$$
$$(x_u)^2 - 1201.92 x_u + 209440.08 = 0$$

Therefore, $x_u = 211.46$ mm and $Y_f = 0.15 x_u + 0.65 \times 100 = 96.72$ mm $< D_f$.

Since $Y_f < D_f$, the value of Y_f considered above is not correct. Now consider $Y_f = 98.35$.

$$560 \times 10^6 = 0.362 \times 20 \times 250 x_u (500 - 0.416 x_u)$$
$$+ 0.447 \times 20 \times (1250 - 250) \times 98.35 \times (500 - 0.5 \times 98.35)$$

or

$$(x_u)^2 - 1201.92 x_u + 217292.51 = 0$$

Therefore, $x_u = 221.67$ mm, and $Y_f = 0.15\ x_u + 0.65 D_f = 98.25$ mm.

Hence, x_u may be taken as 222 mm and Y_f = 98.30 mm.
The tension reinforcement A_{st} can be determined from the internal force equilibrium condition:

$$0.87 f_y A_{st} = 0.362 f_{ck} b_w x_u + 0.447 f_{ck}(b_f - b_w) Y_f$$

or

$$361.05 A_{st} = 0.362 \times 20 \times 250 \times 222 + 0.447 \times 20 \times (1250 - 250) \times 98.3$$
$$= 1280622$$

or

$$A_{st} = 3546.94 \text{ mm}^2$$

Alternatively, the tension reinforcement may be considered as consisting of two parts; one corresponding to the force carried by the flange and the other to that carried by the web. Thus,

$$A_{st} = A_{sf} + A_{sw}$$

Since $D_f / D \le 0.02$, the stress in the flange is uniform:

$$A_{sf} = \frac{0.447 f_{ck} D_f (b_f - b_w)}{0.87 f_y}$$

$$= \frac{0.447 \times 20 \times 100 \times (1250 - 250)}{0.87 \times 415} = 2476.11 \text{ mm}^2$$

The corresponding limiting or ultimate moment carried by the flange is:

$$M_{u,\text{flange}} = 0.447 \times 20 \times 100 \times (1250 - 250) \times \left(500 - \frac{100}{2}\right)$$
$$= 402.3 \times 10^6 \text{ Nmm} = 402.3 \text{ kNm}$$

The moment carried by the web portion is given as:

$$M_{u,\text{web}} = M_{ud} - M_{u,\text{flange}} = 560 - 402.3 = 157.7 \text{ kNm}$$

The reinforcement corresponding to web component of moment can be determined as:

$$R_u = \left(\frac{4.6 M_u}{f_{ck} b d^2}\right) = \left(\frac{4.6 \times 157.7 \times 10^6}{20 \times 250 \times 500^2}\right) = 0.58034$$

and

$$A_{st} = \left(\frac{f_{ck} b d}{2 f_y}\right)[1 - \sqrt{1 - R_u}]$$

$$A_{stw} = \left(\frac{20 \times 250 \times 500}{2 \times 415}\right)[1 - \sqrt{1 - 0.58034}] = 1060.81 \text{ mm}^2$$

Therefore,

$$A_{st} = A_{sf} + A_{sw} = 2476.11 + 1060.81 = 3536.92 \text{ mm}^2$$

3.7.2 Flanged Beam with Tension and Compression Reinforcements

When a flanged section is reinforced on the compression side also, it is known as a **doubly reinforced flanged section**. It can be designed by using a procedure similar to that for a doubly reinforced rectangular beam. The neutral axis may lie either in the flange or in the web. To ascertain this consider that the neutral axis coincides with the underneath of the slab flange, and compute forces C_u and T_u, and compare.

3.8 SLABS

The slab is a two-dimensional plane element with its depth *D* being much smaller than the other two dimensions, namely, the width and the span. It may be supported on masonry walls or on beams or directly on columns. It carries gravity loads acting normal to its surface and transfers the same to the supports primarily by flexure. Slabs supported on two parallel sides carry loads by flexure in the direction perpendicular to the supports; such slabs are known as **one-way slabs** and virtually act as wide, shallow rectangular beams. The slab supported on all four sides and with a ratio of longer and shorter spans being greater than 2 is also considered as a one-way slab.

Analysis and design of a one-way slab are normally carried out by considering 1.0 m wide strip acting as a shallow rectangular beam spanning in the direction of bending. Since the loads on slabs are frequently specified in terms of load per square metre, the unit load on a 1.0 m wide strip becomes the load per linear metre. The reinforcing steel is usually spaced uniformly over its width as illustrated in Fig. 3.7, and effective A_{st} may correspond to a fractional number of bars in the 1.0 m wide strip. In this chapter, the one-way slab is being introduced simply to highlight its similarity with a rectangular beam. They will be treated separately in Chapter 7 in detail.

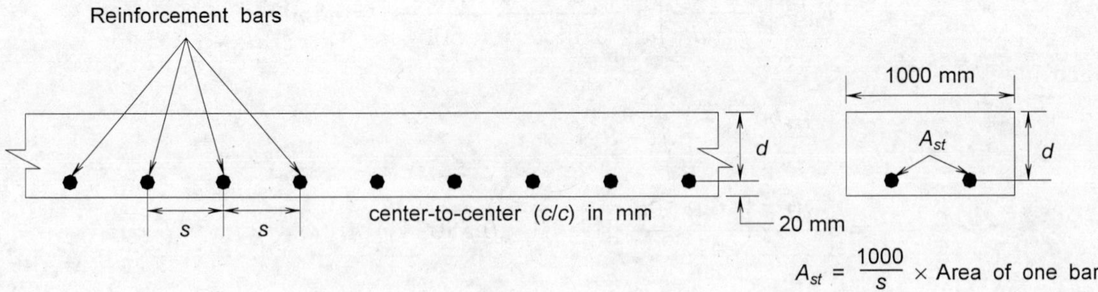

Fig. 3.7 Idealization of a one-way slab as rectangular beams.

Example 3.13 Design a one-way continuous slab to support a maximum factored sagging (positive) moment at mid-span of 10.5 kNm and a maximum factored hogging (negative) moment at a support of 12.4 kNm. The materials used are M20 grade concrete and HYSD steel bars of grade Fe415.

Solution For the given materials:

$$f_{ck} = 20 \text{ MPa, and } f_y = 415 \text{ MPa}$$

For a 1.0 m wide strip, b = 1000 mm. The minimum depth may be based on the maximum design moment, i.e., the negative moment at the support, M_u = 12.4 kNm.
Equate limiting moment of resistance of the section to the design moment M_u:

$$M_{u,\lim} = 0.1388 f_{ck} bd^2 = 0.1388 \times 20 \times 1000 \times d^2$$
$$= 12.4 \times 10^6 \text{ kNm}$$

Therefore,

$$d = 66.83 \text{ mm}$$

Provide 100 mm thick slab with d = 75 mm.
The reinforcements are:

1. For negative moment:

$$R_u = \left(\frac{4.6M_u}{f_{ck}bd^2}\right) = \left(\frac{4.6 \times 12.4 \times 10^6}{20 \times 1000 \times 75^2}\right) = 0.50702$$

$$A_{st} = \left(\frac{f_{ck}bd}{2f_y}\right)[1 - \sqrt{1 - R_u}\,]$$

$$A_{stw} = \left(\frac{20 \times 1000 \times 75}{2 \times 415}\right)[1 - \sqrt{1 - 0.50702}] = 538.33 \text{ mm}^2.$$

Spacing of 10 mm ϕ (a_{st} = 78.5 mm²) bars, s = $(1000a_{st})/A_{st}$ = 145.82 mm
Maximum spacing limits:

$$3d = 3 \times 75 = 225 \text{ mm or } 300 \text{ mm (whichever is smaller)}$$

Provide 10 mm ϕ bars @ 145 mm c/c for negative reinforcement.

2. For positive moment:

$$R_u = \left(\frac{4.6M_u}{f_{ck}bd^2}\right) = \left(\frac{4.6 \times 10.5 \times 10^6}{20 \times 1000 \times 75^2}\right) = 0.42933$$

$$A_{stw} = \left(\frac{20 \times 1000 \times 75}{2 \times 415}\right)[1 - \sqrt{1 - 0.42933}] = 442.0 \text{ mm}^2.$$

Spacing of 10 mm ϕ (a_{st} = 78.5 mm²) bars, s = $(1000a_{st})/A_{st}$ = 177.6 mm

Maximum spacing limits:

$$3d = 3 \times 75 = 225 \text{ mm or } 300 \text{ mm (whichever is smaller)}$$

Provide 10 mm ϕ bars @ 175 mm c/c for positive reinforcement.

Review Questions

3.1 (a) State the basic assumptions used in theory of bending as applied to the limit states design of reinforced concrete structures. Discuss their validity near the collapse.

(b) Determine the percentage of tension reinforcement to be provided to obtain a balanced section.

3.2 Does the assumption that plane section before bending remains plane after bending true till failure of the beam?

3.3 What is meant by strain compatibility? State the fundamental assumptions which ensure strain compatibility.

3.4 Discuss the validity of the assumption that concrete resists no flexural tensile stress in reinforced concrete beams.

3.5 Explain briefly the following: (a) balanced section, (b) under-reinforced section and (c) over-reinforced section; explain why is an over-reinforced section not permitted in reinforced concrete.

3.6 A reinforced concrete beam of rectangular cross-section has a width of b mm and effective depth of d mm. Derive expressions for; (i) the position of the neutral axis, (ii) the leverarm, (iii) the moment of resistance and (iv) the percentage of balanced steel, for various combinations of M20 and M25 grade concrete mixes, with mild steel and HYSD steel of grade Fe415.

3.7 Why is $x_{u,\max}$ dependent on the grade of steel only and not on the grade of concrete?

3.8 How is the procedure for analysis of flexural strength of reinforced concrete slab similar to that for beam?

3.9 Why is it necessary to provide transverse steel in a one-way slab?

3.10 Why is it necessary to impose minimum and maximum limits on the percentage of flexural tension reinforcement? What are their values?

3.11 Why is concrete slab generally singly reinforced?

3.12 Describe how does effective span-to-effective depth ratio helps in controlling the deflections in flexural members. Explain how percentage of tension and compression steels, and their grades influence this ratio.

3.13 What is the theoretical cut-off point? Why is it necessary to extend reinforcement beyond this point?

3.14 Why is it necessary to limit the x_u/d allowed in singly reinforced beams? Can this condition be relaxed for doubly reinforced beams? Give reasons for your answer.

3.15 Explain the terms *balanced, over-reinforced* and *under-reinforced* sections in bending. Which section is generally recommended in design? How is this ensured in design of beams according to IS: 456?

3.16 In most designs the depth of a beam or slab is taken larger than those obtained from bending moment considerations only. Does this produce an under-reinforced, over-reinforced or balanced section?

3.17 Derive the expression for the area of steel required to carry a given bending moment by a beam of rectangular cross-section where depth is much larger than that required for a balanced section.

3.18 Sketch the curve showing the relationship between M_u/bd^2 and p the percentage of steel required and explain the salient points on the curve.

3.19 Explain the term $M_{u,\text{lim}}$ and give the expression for this value for Fe250 and Fe415 grade steels.

3.20 Enumerate the advantages of design charts based on IS:456 for use in the design office. Can these charts be used for non-rectangular sections?

3.21 Under what circumstances are doubly reinforced beams used in practice?

3.22 How will you decide whether a doubly reinforced rectangular section is under-reinforced or over-reinforced?

3.23 What are the assumptions in the theory of design of doubly reinforced beam regarding the role of compression steel?

3.24 What is the effect of creep and shrinkage on the stress in compression steel in a doubly reinforced concrete beam?

3.25 Explain why a minimum steel area of 0.2 per cent of the cross-section should be present as compression steel if this is to be taken into account in the calculation of the strength of doubly reinforced beams.

3.26 What is the maximum and minimum percentage of tension and compression reinforcements in doubly reinforced beams?

3.27 How will you check whether a beam of given dimensions is to be designed as a doubly reinforced?

3.28 Derive the expressions for determination of the steel areas required in a doubly reinforced section of given dimensions to carry a given moment.

3.29 What are the recommendations for preventing the compression steel from buckling in a doubly reinforced beam?

3.30 Describe with sketches T- and L-beams and indicate their principal components.

3.31 Describe the effective flange width and list the various factors that influence it and write down the expressions for the same.

3.32 What are the three possible positions of neutral axis encountered in the design of T-beam?

3.33 Describe the method for locating the position of neutral axis in a T-beam section.

3.34 Describe the procedure for the design of T-beam section.

3.35 Why is continuous T-beam at supports designed as rectangular beam?

3.36 What is the distance usually assumed between points of zero moments or inflection points in a continuous T-beam?

3.37 How is the value of $M_{u,\text{lim}}$ calculated for a T-beam?

3.38 What are the functions of transverse steel in the slab portion of flanged beams?

3.39 What are the assumptions made in the design of T- and L-beams for shear?

3.40 What is the minimum percentage of tension steel that to be provided in a T-beam? Is it based on the flange width or web width?

3.41 What is the maximum percentage of steel that is permitted in T-beams?

3.42 The edge beams of a building are to be designed as L-beams. What special considerations are to be taken in its design?

3.43 Give the approximate formula that can be used to determine the area of steel for a T-beam subjected to a given factored moment.

Tutorial Problems

T.3.1 Analyze a rectangular beam section of size 200 × 450 mm effective depth to determine its ultimate moment of resistance when reinforced with: (a) 3 × 16 mm ϕ bars, and (b) 3 × 20 mm ϕ bars. Use concrete mix of grade M20 and steel of grade Fe415.

T.3.2 A reinforced concrete beam section of size 250 × 500 mm overall is reinforced with four bars of 20 mm ϕ on the tension side at an effective cover of 30 mm. Determine the concentrated load that the beam can support at the centre on an effective simply supported span of 5.0 m. The grades of concrete and steel used are M20 and Fe415, respectively.

T.3.3 A reinforced concrete beam of 250 mm width × 550 mm overall depth is reinforced with 3 bars of 20 mm ϕ at an effective cover of 50 mm (i.e., bars are placed with their centres at 50 mm from the bottom face). Determine the moment carrying capacity, and the uniformly distributed superimposed load the beam can support over an effective span of 6.0 m. The materials used are: M20 grade concrete mix and HYSD bars of grade Fe415. The unit weight of concrete is 25 kN/m^3.

T.3.4 A reinforced concrete simply supported beam of cross-section 300 × 650 mm overall is reinforced with 4 × 22 mm ϕ bars on the tension side with an effective cover of 50 mm. Determine moment carrying capacity and the superimposed concentrated load the beam can support at its midpoint over an effective span of 6.0 m. The materials used are: M20 grade concrete and HYSD steel of Fe415 grade. The unit weight of reinforced concrete is 25 kN/m^3.

T.3.5 A reinforced concrete cantilever beam of length 2.5 m and having a cross-section of size 250 × 450 mm effective is reinforced with 4 bars of 20 mm ϕ on the tension side with a nominal cover of 50 mm. Determine the moment carrying capacity and the uniformly distributed superimposed load the beam can support. The materials used are: M20 grade concrete and mild steel of grade Fe250. The unit weight of reinforced concrete is 25 kN/m^3.

T.3.6 A reinforced concrete beam of rectangular cross-section of size 300 × 650 mm effective is reinforced with 4 bars of 20 mm ϕ on the tension side. Check the adequacy of the section, if the beam carries a uniformly distributed load of 20 kN/m (inclusive of self-weight) over an effective span of 7.0 m with simply supported ends. The concrete mix used is of grade M20 and the steel is of grade Fe415.

T.3.7 A reinforced concrete simply supported beam of rectangular cross-section carries a uniformly distributed service load of 12.5 kN/m inclusive of self-weight over an effective span of 8.0 m. From architectural considerations the width of the beam has been fixed at 300 mm. Determine the effective depth and area of tension steel. The materials used are: M20 grade concrete and HYSD steel of grade Fe415.

T.3.8 A simply supported 125 mm thick roof slab of effective span of 3.5 m is reinforced with 12 mm ϕ bars at 100 mm c/c at an effective cover of 20 mm (i.e., the reinforcement is placed with its centre at 20 mm from the bottom). Determine the moment carrying capacity of 1.0 m wide strip of the slab, and uniformly distributed superimposed load the slab can support safely. The materials used are: M25 grade concrete mix and HYSD steel reinforcement of grade Fe415.

T.3.9 A simply supported reinforced concrete rectangular beam carries a uniformly distributed service load of 12.5 kN/m inclusive of its own weight, and also a service concentrated load of 20 kN at the midpoint over an effective span of 6.5 m. Design a suitable section for the flexural action taking: (a) $d = 2b$ and (b) $d = 2.5b$. The materials used are: M25 grade concrete and HYSD steel of grade Fe415.

T.3.10 Design a rectangular beam section 300 mm wide and 500 mm deep (effective) subjected to a factored moment of: (a) 150 kNm and (b) 175 kNm. The grades of concrete and HYSD steel used in construction are M20 and Fe415, respectively.

T.3.11 Calculate the moment carrying capacity of a reinforced concrete beam of rectangular cross-section of width b and effective depth d which is reinforced with 1.15 per cent steel both on tension and compression sides, the compression steel being placed at an effective cover of $0.10d$. The materials used are: M20 grade concrete mix and HYSD steel of grade Fe415.

T.3.12 A doubly reinforced concrete beam of rectangular section of size 300 × 725 mm overall is reinforced by 4 × 25 ϕ bars each on tension and compression sides. The effective covers to compression and tension steels are 40 mm and 50 mm, respectively.

Check the adequacy of the section when the beam carries a uniformly distributed load of 40 kN/m inclusive of its dead weight over an effective span of 6.5 m.

T.3.13 Analyze a rectangular beam section of size 250 × 500 mm effective depth to determine the limiting moment of resistance when reinforced with:

(i) 4 bars of 20 mm ϕ on the tension side and 3 bars of 16 mm ϕ on the compression side, and

(ii) 4 bars of 22 mm ϕ on the tension side and 4 bars of 16 mm ϕ on the compression side.

The concrete used is of grade M20 and steel is of grade Fe415. The effective cover to compression steel is 50 mm.

T.3.14 Design a reinforced concrete rectangular beam section of size 300 × 550 mm overall, constructed with concrete mix of grade M20 and steel of grade Fe415. The beam is subjected to factored moments of: (a) 175 kNm and (b) 300 kNm. The effective covers to reinforcement may be taken as 50 mm.

T.3.15 Design a reinforced concrete beam with its size restricted to 200 × 400 mm deep overall from the architectural considerations. The beam is subjected to a maximum factored moment of 180 kNm at the mid-span. Use M20 concrete and Fe415 steel at an effective cover of 35 mm.

T.3.16 A simply supported reinforced concrete beam carries a uniformly distributed load of 20 kN/m inclusive of self-weight over an effective span of 8.25 m. The cross-section of the beam is restricted to 250 mm wide and 600 mm deep up to the centre of the tension steel and the compression steel is placed at an effective cover of 50 mm from the top of the beam. Determine: (a) the areas of tension and compression reinforcements to be provided, if the materials used are: M20 grade concrete and mild steel reinforcement, and (b) the area of reinforcement if the effective depth is reduced by 25 per cent?

T.3.17 A reinforced concrete beam of rectangular cross-section has to carry a uniformly distributed load of 8.0 kN/m over an effective span of 10.0 m. Design the section using M20 grade concrete and Fe415 grade HYSD steel, when: (a) there is no restriction on the size of the beam section and (b) the size of the section is restricted to 250 × 600 mm effective.

T.3.18 Determine the moment carrying capacity of a reinforced concrete beam of rectangular cross-section of size $b \times d$ mm effective, for the conditions given below. The materials are: M20 grade concrete mix and mild steel reinforcement. The effective cover to compression steel is 0.10d (i.e. $d'/d = 0.10$):

(a) the area of compression steel is one-third that of tension steel,

(b) the area of compression steel is limited to 2 per cent of the cross-sectional area *bd*, and

(c) the area of compression steel is limited to 1 per cent of the cross-sectional area.

T.3.19 A reinforced concrete T-beam section has an effective flange width of 1750 mm, and an effective depth of 600 mm. The depth of the flange is 120 mm and the width of the web is 300 mm. Determine the limiting moment of resistance of the section when it is reinforced with: (i) 6 × 25 mm ϕ bars and (ii) 6 × 25 mm ϕ + 2 × 28 mm ϕ bars. The grades of concrete and steel used are: M20 and Fe415, respectively.

T.3.20 Determine the ultimate moment of resistance of a T-beam section having flange width and depth of 1200 mm and 110 mm, respectively. The effective depth of the beam is 500 mm with a web width of 250 mm. It is reinforced with 6 × 25 mm ϕ tension bars. Consider concrete of grade M20 and steel of grade Fe250.

T.3.21 Design a T-shaped beam section consisting of a flange of width 1100 mm and depth of 120 mm. The effective depth of the beam is 550 mm and the width of the rib is 275 mm. The beam carries a factored moment of 500 kNm. Evaluate the area of reinforcing steel required. Consider concrete of grade M20 and steel of grade Fe500.

T.3.22 A reinforced concrete T-beam having a flange of effective width 1250 mm and thickness of 100 mm is reinforced with 4 × 22 mm ϕ tension bars provided at a depth of 500 mm below the top of the flange in a 250 mm wide rib. Determine the uniformly distributed load, inclusive of self-weight, that the beam can support safely over a simply supported effective span of 6.5 m, if the materials used are: concrete mix of grade M20 and HYSD steel of grade Fe415.

T.3.23 Analyze a flanged beam spanning an effective distance of 9.5 m, and reinforced with 6 × 25 mm ϕ HYSD bars of grade Fe415 on the tension side. The flange is of size 1500 × 120 mm and the rib of size 300 × 425 mm effective (projected below the flange). The beam carries a uniformly distributed load of 25 kN/m inclusive of self-weight. Check the adequacy of the section at the centre of the span. The concrete used is grade M20.

T.3.24 A reinforced concrete T-beam section has a flange of size 1350 × 120 mm and a rib of effective depth (below the flange) of 450 mm. The tension and compression steels provided are 5 × 22 mm ϕ bars, and 2 × 18 mm ϕ bars, respectively. The compression steel is provided at an effective depth of 35 mm from the top of flange slab. Check the adequacy of the section when the beam is subjected to a service bending moment of 200 kNm.

T.3.25 Analyze the end beam of a reinforced concrete floor system consisting of a 125 mm thick slab cast monolithically and connected with 300 mm wide beams spaced 3.25 m centre-to-centre. The effective length of the beam is 7.5 m. The beam is reinforced with 6 × 25 mm ϕ bars placed on the tension side at a depth of 550 mm from the top of the flange slab. Check the adequacy of the section when the beam is subjected to a service bending moment of 225 kNm. The concrete mix used is of grade M20.

T.3.26 A reinforced concrete flanged beam has a flange of size 1000 × 100 mm. The beam is to support a uniformly distributed load of 24 kN/m inclusive of self-weight over an effective span of 10 m. Determine the area of medium tensile steel to be placed in the rib at a depth of 600 mm below the top of the flange slab. The concrete mix used is of grade M20.

T.3.27 An isolated L-beam has an effective width of 750 mm, and slab thickness of 100 mm. A rib of size 300 × 500 mm projects below the flange slab. Find the areas of steel to be provided at an effective cover of 40 mm at a section of maximum service bending moment of 200 kNm. The materials used are: M20 grade concrete and mild steel reinforcement.

T.3.28 A 120 mm thick slab of clear span of 4.0 m is cast monolithically with two simply supported edge beams having an effective span of 6 m. The service load on each of the two end beams is 20.011 kN/m inclusive of self-weight. The width of the rib is 250 mm. The materials to be used are: M20 grade concrete and HYSD reinforcement of grade Fe415. Ignoring torsion, design the beam for flexure.

T.3.29 Design a reinforced concrete T-beam section with the following dimensions:

Flange width	1250 mm
Thickness of flange	110 mm
Width of web	275 mm

The beam is subjected to a factored moment of 475 kNm. Use concrete of grade M20 and steel of grade Fe415.

T.3.30 A T-beam with flange of effective width, and thickness of 1500 mm and 25 mm, respectively, web beam of effective depth and the width of 600 mm and 300 mm, respectively, is reinforced with four bars of 25 mm ϕ on the tension side and three bars of 22 mm ϕ on the compression side. The effective cover to compression steel is 50 mm. The materials used are: M20 grade concrete and HYSD steel of grade Fe415. Determine: (i) allowable moment of resistance, and (ii) the uniformly distributed load the beam can support safely over an effective simply supported span of 8 m.

T.3.31 The floor of a rectangular hall 6 m wide to the centre of supports consists of a 100 mm thick slab cast monolithically with and connected to rectangular beams spaced 3 m centre-to-centre. The width of the web and effective depth of the beam are 250 mm and 550 mm, respectively. The beams are reinforced with 6×25 mm ϕ mild steel bars on the tension side. M20 grade concrete is used in construction. Analyze the end beam for: (a) determination of allowable moment of resistance and (b) check the adequacy of the section, if the beam supports a uniformly distributed load of 40 kN/m (inclusive of self-weight).

CHAPTER

4

Limit State of Collapse—Shear, Bond and Torsion

4.1 INTRODUCTION

The majority of structural members subjected to bending are also subjected to shear forces. In a reinforced concrete member under flexure the shear-resisting mechanism interacts with the bond between concrete and reinforcement, and is related to what is known as the **development** or **anchorage length** of reinforcement. Shear transfer is dependent on the tensile and compressive strengths of concrete. The value of shear strength lies between the two. The behaviour of reinforced concrete beams at failure in shear is distinctly different from their behaviour in flexure. The experimental investigations have shown that in the absence of adequate provision of shear reinforcement, the element fails prematurely and suddenly with diagonal cracks that develop being considerably wider than the flexural cracks. Shear failures which are brittle in nature are essentially diagonal tension failures. This necessitates a provision of reinforcement for diagonal tension.

It is desirable to have the flexural member attain its full design flexural capacity before premature failure occurs in shear or bond. The modern structural design concepts emphasize that a structure should have adequate ductility to give adequate warning of impending failure. In case of flexural members, the impending failure may be pointed out by yielding of tension steel which is preceded by gradual, excessive deflection and noticeable widening of cracks. Fortunately the presence of shear reinforcement ensures reasonable ductility and possible suppression of shear failure before the onset of more ductile flexural failure, even when the shear stresses are well within allowable limits. Thus, with the object of increasing the load-carrying capacity, the reinforced concrete beams are provided with shear or web reinforcement in the form of: (i) vertical stirrups, (ii) inclined stirrups, and/or (iii) bent-up bars as shown in

Fig. 4.1. The bent-up bars usually consist of the part of longitudinal steel bent up where it is no longer required for moment resistance.

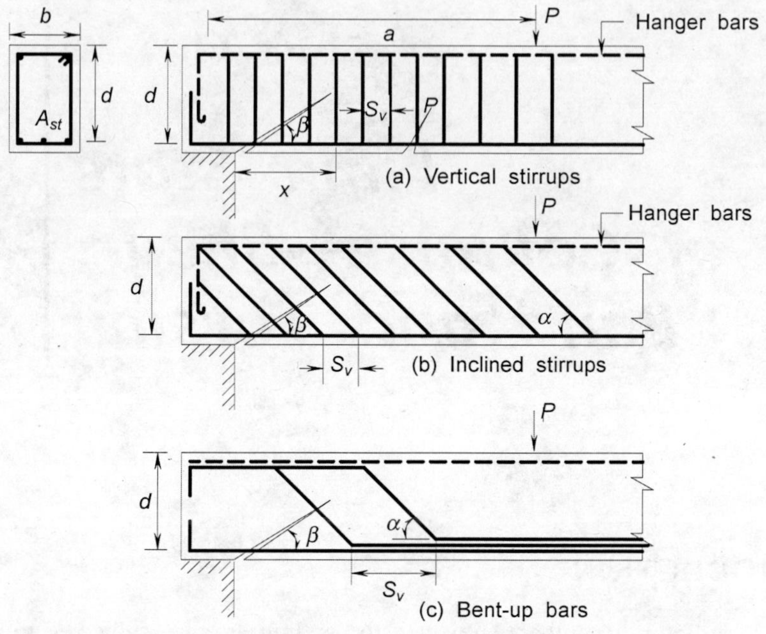

Fig. 4.1 *Various types of shear reinforcement in beam.*

Following observations based on laboratory tests on beams reported in the literature should be noted:

Shear stress, acting along with flexural stress, gives rise to principal tensile and compressive stresses. If the principal tensile stress exceeds the value of tensile strength of concrete it will result in diagonal tension cracks near the supports inclined at 45° to the axis of the beam.

The effect of shear reinforcement on the behaviour of the structural element before cracking is insignificant and does not influence the diagonal cracking load. Stirrups come into action only after the formation of diagonal cracks and they increase the ultimate load considerably. The shear reinforcement augments the shear resistance of the beam in three different ways:

1. The stirrups do not carry any shear as such; they only perform the function of reinforcing a beam transversely against diagonal cracking.
2. They serve to restrict the growth of diagonal cracks and reduce their penetration into the compression zone.
3. Stirrups tie the longitudinal reinforcement into the main bulk of the concrete.

The diagonal cracks always develop at some distance away from the supports. This may be due to the compressive stress induced in the web by concentrated support reaction.

Therefore, the code recommends the calculation of maximum nominal shear at a distance d from the face of the support except in the case of short cantilevers and other brackets. In the latter case, it should be calculated at the face of the support itself.

In order that each potential diagonal tension crack originating in the zone of excessive shear be contained by the web reinforcement, it is required that the web reinforcement be provided for a distance d beyond the point at which it is no longer theoretically required.

Bond and shear resistance are not easy to separate and hence the bond between concrete and steel has also been included in this chapter. Due to availability of deformed bars the emphasis has shifted to development length.

The complexity of the problem coupled with inadequacy of test results has resulted in many different approaches being proposed in the literature. Most of the design recommendations of various codes of practices are based on empirical relations derived from laboratory tests. The design provisions based on IS:456 have been discussed in this chapter.

4.2 LIMIT STATE OF COLLAPSE—SHEAR

Shear in the flexural member is induced due to variation of bending moment along its length. The shear force V_u at a section is related to variation of bending moment M_u, by the expression:

$$V_u = \frac{\delta M_u}{\delta x}$$

or

$$V_u \delta x = \delta M_u \tag{4.1}$$

The increment moment δM_u induces incremental compressive and tensile forces δC_u and δT_u, respectively, such that

$$\delta C_u = \delta T_u = \frac{\delta M_u}{z} \tag{4.2}$$

where z $(=jd)$ is lever arm. The incremental tensile force δT_u must be balanced by the average unit stress τ_u over a horizontal area $(b\delta x)$, as illustrated in Fig. 4.2, where $\delta T_u = \tau_u b \delta x$, and thus

$$\tau_u = \frac{\delta T_u}{\delta x b} = \frac{\delta M_u / z}{\delta x b} = \left(\frac{\delta M_u}{\delta x}\right)\left(\frac{1}{bz}\right) = \frac{V_u}{bz} \tag{4.3}$$

The variation of shear stress along the depth of the section above the neutral axis is parabolic with zero at the extreme compression fibre at a distance x_u from the neutral axis and increasing to its maximum value (V_u/bz) at the neutral axis. Below the neutral axis, the concrete is in tension and is assumed not to take any stress. Hence the value of shear stress below the neutral axis will remain constant. However, at the section passing through tension steel the incremental compressive force causing shear stress is neutralized by the incremental tensile force induced in tension steel. Hence shear stress falls to zero at the level of steel bars as shown in Fig. 4.2.

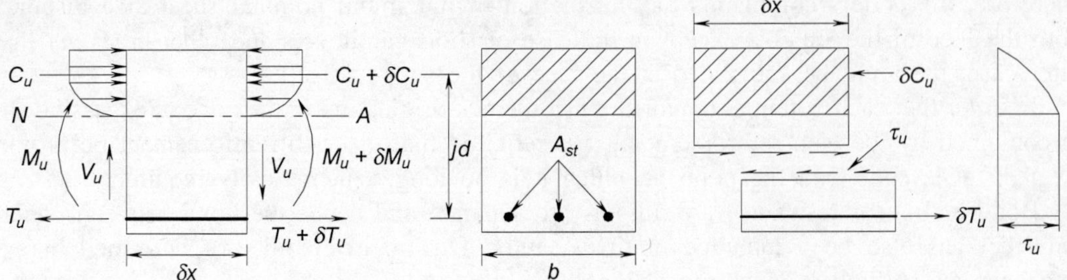

Fig. 4.2 Development and variation of shear stress in a beam.

For ease of computation, IS: 456 have adopted a simple equation:

$$\tau_v = \frac{V_u}{bd} \tag{4.4}$$

The shear stress obtained by dividing the ultimate design shear V_u by the cross-sectional area is known as **nominal shear stress** τ_v.

The average shear at which the inclined crack is first formed is taken as the shear strength of the concrete without shear reinforcement. It is based on the value of limiting nominal shear stress τ_v corresponding to the load at which first inclined crack develops using a partial safety factor of 1.2. Shear reinforcement is provided only for the part of the shear that is in excess of shear strength. The design shear strength of concrete τ_c is dependent on the crack control parameter expressed as A_{st}/bd (or $p_t = 100A_{st}/bd$, the percentage of steel) and is different for different grades of concrete. Based on numerous test result, several empirical relations have been proposed to predict the shear strength of concrete. IS:456 have recommended the shear strength values based on the following empirical formula:

$$\tau_c = \frac{0.85\sqrt{(0.80 f_{ck})}\left[\sqrt{(1+5\beta)} - 1\right]}{6\beta} \tag{4.5}$$

where

$0.80 f_{ck}$ = Cylinder strength in terms of cube strength

0.85 = Reduction factor

β is greater of:

$$\frac{0.116 f_{ck} bd}{100 A_{st}} \left(= \frac{0.116 f_{ck}}{p_t} \right) \text{ and } 1.0$$

The typical values of τ_c corresponding to p_t varying from 0.20 to 3.0 at intervals of 0.02 are listed in Table 4.1, for different grades of concrete. It may be noted that for a value of f_{ck} there is a value of p_t ($\geq 0.116 f_{ck}$) with $\beta = 1$ beyond which τ_c remains constant implying that the contribution of dowel action increases only up to this limiting value. The ultimate shear strength of concrete is obtained by $V_c = \tau_c bd$.

TABLE 4.1 Permissible Shear Stresses in Concrete with Various Values of Crack Control Parameter

$p_t = 100$ (A_{st}/bd)	Permissible shear stress, MPa — Grade of concrete					$p_t = 100$ (A_{st}/bd)	Permissible shear stress, MPa — Grade of concrete				
	M20	*M25*	*M30*	*M35*	*M40*		*M20*	*M25*	*M30*	*M35*	*M40*
≤ 0.20	0.326	0.331	0.334	0.337	0.339	0.98	0.618	0.637	0.651	0.662	0.671
0.22	0.340	0.345	0.349	0.352	0.354	**1.00**	**0.623**	**0.642**	**0.656**	**0.668**	**0.677**
0.24	0.353	0.358	0.362	0.366	0.368	1.02	0.627	0.646	0.661	0.673	0.682
0.26	0.365	0.371	0.375	0.379	0.382	1.04	0.632	0.651	0.666	0.678	0.688
0.28	0.377	0.383	0.388	0.392	0.394	1.06	0.636	0.656	0.671	0.683	0.693
0.30	0.388	0.395	0.400	0.404	0.407	1.08	0.640	0.660	0.676	0.688	0.698
0.32	0.399	0.406	0.411	0.415	0.419	1.10	0.644	0.665	0.680	0.693	0.703
0.34	0.409	0.416	0.422	0.426	0.430	1.12	0.648	0.669	0.685	0.698	0.708
0.36	0.419	0.427	0.432	0.437	0.441	1.14	0.652	0.673	0.690	0.702	0.713
0.38	0.428	0.436	0.443	0.447	0.451	1.16	0.656	0.678	0.694	0.707	0.718
0.40	0.437	0.446	0.452	0.457	0.462	1.18	0.660	0.682	0.698	0.712	0.723
0.42	0.446	0.455	0.462	0.467	0.471	1.20	0.664	0.686	0.703	0.716	0.727
0.44	0.455	0.464	0.471	0.477	0.481	1.22	0.668	0.690	0.707	0.721	0.732
0.46	0.463	0.473	0.480	0.486	0.490	1.24	0.672	0.694	0.712	0.725	0.737
0.48	0.471	0.481	0.489	0.495	0.499	1.26	0.675	0.698	0.716	0.730	0.741
0.50	**0.479**	**0.489**	**0.497**	**0.503**	**0.508**	1.28	0.679	0.702	0.720	0.734	0.746
0.52	0.486	0.497	0.505	0.512	0.517	1.30	0.683	0.706	0.724	0.738	0.750
0.54	0.493	0.505	0.513	0.520	0.525	1.32	0.686	0.710	0.728	0.743	0.755
0.56	0.501	0.512	0.521	0.528	0.533	1.34	0.690	0.714	0.732	0.747	0.759
0.58	0.508	0.519	0.528	0.535	0.541	1.36	0.693	0.718	0.736	0.751	0.763
0.60	0.514	0.527	0.536	0.543	0.549	1.38	0.697	0.721	0.740	0.755	0.768
0.62	0.521	0.533	0.543	0.550	0.557	1.40	0.700	0.725	0.744	0.759	0.772
0.64	0.527	0.540	0.550	0.558	0.564	1.42	0.703	0.729	0.748	0.763	0.776
0.66	0.534	0.547	0.557	0.565	0.571	1.44	0.707	0.732	0.752	0.767	0.780
0.68	0.540	0.553	0.564	0.572	0.579	1.46	0.710	0.736	0.755	0.771	0.784
0.70	0.546	0.560	0.570	0.579	0.586	1.48	0.713	0.739	0.759	0.775	0.788
0.72	0.552	0.566	0.577	0.585	0.592	**1.50**	**0.716**	**0.743**	**0.763**	**0.779**	**0.792**
0.74	0.557	0.572	0.583	0.592	0.599	1.52	0.719	0.746	0.766	0.783	0.796
0.76	0.563	0.578	0.589	0.598	0.606	1.54	0.723	0.749	0.770	0.786	0.800
0.78	0.568	0.584	0.595	0.605	0.612	1.56	0.726	0.753	0.774	0.790	0.804
0.80	0.574	0.590	0.601	0.611	0.619	1.58	0.729	0.756	0.777	0.794	0.808
0.82	0.579	0.595	0.607	0.617	0.625	1.60	0.732	0.759	0.780	0.798	0.812
0.84	0.584	0.601	0.613	0.623	0.631	1.62	0.735	0.762	0.784	0.801	0.815
0.86	0.589	0.606	0.619	0.629	0.637	1.64	0.738	0.766	0.787	0.805	0.819
0.88	0.595	0.611	0.624	0.635	0.643	1.66	0.740	0.769	0.791	0.808	0.823
0.90	0.599	0.617	0.630	0.640	0.649	1.68	0.743	0.772	0.794	0.812	0.826
0.92	0.604	0.622	0.635	0.646	0.655	1.70	0.746	0.775	0.797	0.815	0.830
0.94	0.609	0.627	0.641	0.651	0.660	1.72	0.749	0.778	0.801	0.819	0.834
0.96	0.614	0.632	0.646	0.657	0.666	1.74	0.752	0.781	0.804	0.822	0.837

(Contd.)

TABLE 4.1 Permissible Shear Stresses in Concrete with Various Values of Crack Control Parameter (*Contd.*)

$p_t = 100$ (A_{st}/bd)	Permissible shear stress, MPa					$p_t = 100$ (A_{st}/bd)	Permissible shear stress, MPa				
	Grade of concrete						Grade of concrete				
	M20	M25	M30	M35	M40		M20	M25	M30	M35	M40
1.76	0.755	0.784	0.807	0.826	0.841	2.40	0.821	0.867	0.896	0.920	0.939
1.78	0.757	0.787	0.810	0.829	0.844	2.42	0.821	0.869	0.899	0.922	0.942
1.80	0.760	0.790	0.813	0.832	0.848	2.44	0.821	0.871	0.901	0.925	0.945
1.82	0.763	0.793	0.817	0.835	0.851	2.46	0.821	0.874	0.903	0.927	0.947
1.84	0.765	0.796	0.820	0.839	0.855	2.48	0.821	0.876	0.906	0.930	0.950
1.86	0.768	0.799	0.823	0.842	0.858	**2.50**	**0.821**	**0.878**	**0.908**	**0.932**	**0.953**
1.88	0.770	0.802	0.826	0.845	0.861	2.52	0.821	0.880	0.910	0.935	0.955
1.90	0.773	0.804	0.829	0.848	0.865	2.54	0.821	0.882	0.913	0.937	0.958
1.92	0.775	0.807	0.832	0.851	0.868	2.56	0.821	0.884	0.915	0.940	0.960
1.94	0.778	0.810	0.835	0.855	0.871	2.58	0.821	0.887	0.917	0.942	0.963
1.96	0.780	0.813	0.838	0.858	0.874	2.60	0.821	0.889	0.920	0.945	0.966
1.98	0.783	0.815	0.841	0.861	0.878	2.62	0.821	0.891	0.922	0.947	0.968
2.00	**0.785**	**0.818**	**0.843**	**0.864**	**0.881**	2.64	0.821	0.893	0.924	0.949	0.971
2.02	0.788	0.821	0.846	0.867	0.884	2.66	0.821	0.895	0.926	0.952	0.973
2.04	0.790	0.823	0.849	0.870	0.887	2.68	0.821	0.897	0.929	0.954	0.976
2.06	0.793	0.826	0.852	0.873	0.890	2.70	0.821	0.899	0.931	0.956	0.978
2.08	0.795	0.829	0.855	0.876	0.893	2.72	0.821	0.901	0.933	0.959	0.980
2.10	0.797	0.831	0.857	0.879	0.896	2.74	0.821	0.903	0.935	0.961	0.983
2.12	0.799	0.834	0.860	0.882	0.899	2.76	0.821	0.905	0.937	0.963	0.985
2.14	0.802	0.836	0.863	0.884	0.902	2.78	0.821	0.907	0.939	0.966	0.988
2.16	0.804	0.839	0.866	0.887	0.905	2.80	0.821	0.909	0.941	0.968	0.990
2.18	0.806	0.841	0.868	0.890	0.908	2.82	0.821	0.911	0.944	0.970	0.992
2.20	0.808	0.844	0.871	0.893	0.911	2.84	0.821	0.913	0.946	0.972	0.995
2.22	0.811	0.846	0.873	0.896	0.914	2.86	0.821	0.915	0.948	0.975	0.997
2.24	0.813	0.848	0.876	0.898	0.917	2.88	0.821	0.916	0.950	0.977	0.999
2.26	0.815	0.851	0.879	0.901	0.920	2.90	0.821	0.918	0.952	0.979	1.002
2.28	0.817	0.853	0.881	0.904	0.923	2.92	0.821	0.918	0.954	0.981	1.004
2.30	0.819	0.856	0.884	0.907	0.925	2.94	0.821	0.918	0.956	0.983	1.006
2.32	0.821	0.858	0.886	0.909	0.928	2.96	0.821	0.918	0.958	0.986	1.009
2.34	0.821	0.860	0.889	0.912	0.931	2.98	0.821	0.918	0.960	0.988	1.011
2.36	0.821	0.862	0.891	0.914	0.934	**3.00**	**0.821**	**0.918**	**0.962**	**0.990**	**1.013**
2.38	0.821	0.865	0.894	0.917	0.937	3.02	0.821	0.918	0.964	0.992	1.015

Note: The term A_{st} is the area of longitudinal tension reinforcement which continues at least one effective depth beyond the section being considered except at supports where the full area of tension reinforcement may be used provided the detailing conforms to clauses 26.2.2. and 26.2.3 of IS:456.

For solid slabs (excluding ribbed slabs) the code has suggested increased design shear strength of concrete $k\tau_c$. The multiplication factor k varies from 1.0 to 1.3 depending upon the overall depth D (in mm) of the slab. For $D \leq 150$ mm and $D \geq 300$ mm the values of k are 1.3 and 1.0, respectively. For the values of D between 150 to 300 mm, k is obtained from the relation $k = 1.6 - 0.002D$.

These provisions are applicable to flexural or one-way shear only. For flat slabs and column footings punching or two-way shear considerations are involved. The presence of uniaxial compression improves shear strength. The compression delays the formation of both flexural and inclined cracks and increases their angle of inclination with longitudinal axis. On the other hand, uniaxial tension reduces the shear strength and accelerates process of cracking and decreases the angle of inclination. The shear strength in the presence of compression is taken is $\delta\tau_c$ where multiplying factor δ is given by

$$\delta = 1 + \frac{3P_u}{A_g f_{ck}} \text{ but not exceeding 1.5} \tag{4.6}$$

where P_u is factored axial compressive force in N and A_g is the gross area of the concrete section in mm^2. In case of axial tension the following relationship suggested by ACI may be used:

$$\delta = 1 + \frac{P_u}{3.45A_g} \tag{4.7}$$

where P_u is negative.

4.2.1 Maximum Shear Stress in Concrete, $\tau_{c,\max}$

When nominal shear stress τ_v exceeds the shear strength of concrete τ_c, suitable shear reinforcement is provided. However, under no circumstances, even with the shear reinforcement, shall the nominal shear stress τ_v exceed $\tau_{c,\max}$. When the shear stress τ_v is greater than $\tau_{c,\max}$ the section is redesigned by using higher grade of concrete mix or by providing greater depth. By this provision the failure of member by diagonal compression taking over, even when the diagonal tension is taken care of by the steel reinforcement, is prevented. IS:456 recommend $\tau_{c,\max}$ values which can be obtained approximately from the expression:

$$\tau_{c,\max} = \frac{0.83\sqrt{f_c}}{\gamma_m} = \frac{0.83\sqrt{0.8f_{ck}}}{1.2} = 0.62\sqrt{f_{ck}} \tag{4.8}$$

where f_c is the cylinder strength = $0.8f_{ck}$ in terms of characteristic cube strength. The values of $\tau_{c,\max}$ for different grades of concrete are given in Table 4.2.

TABLE 4.2 Maximum Shear Stress $\tau_{c,\max}$

Concrete grade	*M20*	*M25*	*M30*	*M35*	*M40*
$\tau_{c,\max}$, MPa	2.8	3.1	3.5	3.7	4.0

For solid slabs, the nominal shear stress shall not exceed $0.5\tau_{c,\max}$.

4.2.2 Shear Reinforcement

When τ_v exceeds the shear strength of concrete τ_c, the shear reinforcement shall be provided to carry the design shear force:

$$V_{us} = V_u - \tau_c bd$$

where V_u is the shear force due to externally applied design loads. As shown in Fig. 4.1 the shear reinforcement is provided in the form of (i) vertical stirrups, (ii) inclined stirrups, (iii) bent up or inclined bars, (iv) combination of (i) and (iii). *Vertical stirrup* is the most commonly used form of shear reinforcement. The stirrups are generally two-legged, which may have either open or closed form with different anchorage bends as shown in Fig. 4.3(a). However, closed stirrups are preferred. Four-legged, six-legged, and so on, stirrups can also be formed, as shown in Fig. 4.3(b). Stirrups with more than two legs are used for the region having a high shear force.

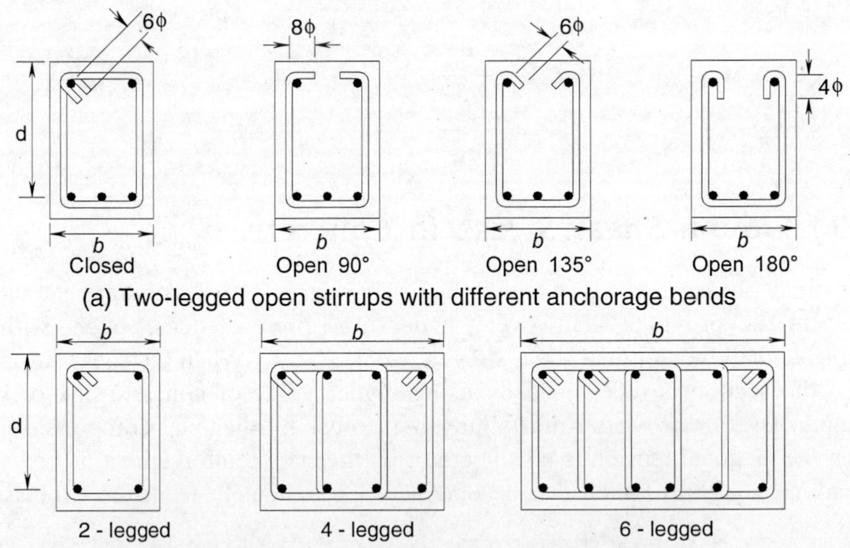

***Fig. 4.3** Different types of vertical stirrups.*

Design for Shear: The total shear carried by beam section consists of two parts, namely the one carried by concrete and other by shear reinforcement, i.e.,

$$V_u = V_{\text{concrete}} + V_{\text{stirrups}} = V_c + V_s$$

or

$$V_s = V_u - V_c = (\tau_v - \tau_c)bd \tag{4.9}$$

The design of stirrups is based on vertical component of diagonal tension, while horizontal component is resisted by the main tesion steel. Consider vertical equilibrium of shear force carried by 2-legged stirrups with total area of legs of A_{sv} spaced S_v crossing a crack line at 45°:

Number stirrups cut by the crack line $n = \dfrac{d}{S_v}$

Shear resistance of these vertical stirrups $V_s = (0.87 f_y) A_{sv} \left(\dfrac{d}{S_v} \right)$

Therefore,

$$S_v = \frac{0.87 f_y A_{sv} d}{V_s}$$

or

$$\frac{V_s}{d} = \frac{A_{sv}}{S_v}(0.87 f_y) \tag{4.10}$$

This expression can be used to develop design aids for computing spacing S_v of 2-legged stirrups directly from V_s/d. Equation (4.10) can also be written as:

$$S_v = \frac{0.87 f_y A_{sv}}{b(\tau_v - \tau_c)}$$

or

$$\frac{A_{sv}}{S_v} = \frac{(\tau_v - \tau_c)}{0.87}\left(\frac{b}{f_y}\right) \tag{4.11}$$

where

A_{sv} = Total cross-sectional area of stirrup legs effective in shear,
S_v = Spacing of stirrups along the length of the member,
b = Breadth of the beam (or the web in the flanged beam), and
f_y = Characteristic strength of stirrup steel in MPa which shall not exceed 415 MPa.

Minimum Shear Reinforcement: If $\tau_v < \tau_c$, the minimum or nominal shear reinforcement in the form of stirrups shall be provided in all the beams such that

$$\frac{A_{sv}}{S_v} \geq 0.40\left(\frac{b}{f_y}\right) \tag{4.12}$$

Comparing Eq. (4.12) with Eq. (4.11),

$$(\tau_v - \tau_c) = 0.35 \text{ MPa}$$

i.e., design is to be made for a value of $(\tau_v - \tau_c) = 0.35$ MPa.

In members of minor structural importance such as lintels or where the design shear force V_u is less than half the ultimate shear strength of concrete V_c this provision need not be complied with.

4.2.3 Maximum Spacing of Stirrups, $S_{v,max}$

It is seen that the horizontal distance between two successive diagonal cracks is approximately equal to the effective depth d. A stirrup shall be provided such that it crosses the potential crack and also no potential crack shall remain uncrossed. To ensure this, the maximum spacing $S_{v,\max}$ should not exceed 0.75d for vertical stirrups and d for stirrups inclined at 45°, where d is the effective depth of the section or 300 mm, whichever is less.

4.2.4 Design Procedure for Shear Reinforcement

1. Determine the factored or ultimate design shear force V_u.
2. Compute nominal shear stress on the section τ_v by dividing V_u by bd:

$$\tau_v = \frac{V_u}{bd}$$

3. Compute the percentage of tension reinforcement provided in the section. Based on this percentage of tension steel, determine the shear strength of concrete τ_c for the given concrete mix, and hence the shear-resisting capacity of the unreinforced concrete section, $V_c = \tau_c bd$.
4. Compare τ_v with $\tau_{c,\max}$. If $\tau_v > \tau_{c,\max}$, redesign the cross-section of the beam such that

$$\tau_v < \tau_{c,\max}$$

5. Compare V_u with V_c (or τ_v with τ_c). If $V_u \le V_c$, no shear design is required. Provide nominal reinforcement in form of vertical stirrups throughout the beam at the spacing S_v given by

$$\frac{A_{sv}}{bS_v} \ge \frac{0.4}{f_y}$$

or

$$S_v \le \frac{A_{sv} f_y}{0.4b}$$

where A_{sv} is the total cross-sectional area of stirrup legs.

6. If $V_u > V_c$, provide shear reinforcement for resisting the design shear force V_{us} given by

$$V_{us} = V_u - V_c$$

7. Choose the diameter of the stirrup bar (generally 6, 8, 10 or 12 mm) and the type of stirrup. Determine the spacing of the stirrups as follows:

For vertical stirrups:

$$S_v = \frac{0.87 f_y A_{sv} d}{V_{us}}$$

or

$$S_v = \frac{0.87 f_y A_{sv}}{(\tau_v - \tau_c) b}$$

For inclined stirrups:

$$S_v = \frac{0.87 f_y A_{sv} d}{V_{us}} (\sin \alpha + \cos \alpha) \tag{4.13}$$

For bent-up bars: If the tension steel is available for shear, bend one or more bars from tension steel at 45° at a distance d from the supports. Compute the shear force taken by bent up bars.

$$V_s = \sigma_{sv} A_{sv} \sin \alpha \ (\not> \tfrac{1}{2} V_{us})$$

Here A_{sv} is the total cross-sectional area of bent-up bars.
Design the vertical stirrups for the shear force:

$$V_d = V_{us} - V_s$$

$$V_{us} = 0.87 f_y A_{sv} \sin \alpha \tag{4.14}$$

where

A_{sv} = Total cross-sectional area of stirrup legs or bent-up bars within a distance S_v,
S_v = Spacing of stirrups or bent-up bars along the length,
b = Breadth of section (for a flanged section, $b = b_w$),
f_y = Characteristic strength of steel used for stirrups $\not>$ 415 MPa,
α = Angle between the inclined stirrups and the axis of the member and shall not be less than 45°, and
d = Effective depth of the section.

8. Check whether the spacing of stirrups obtained in steps 5 and 7 satisfies the Code design requirements:
   (i) $S_v \not> 0.75d$ (spacing governed by the depth of the beam)
   (ii) $S_v \not> (A_{sv} f_y)/(0.4b)$ (based on the minimum shear reinforcement)
   (iii) $S_v \not< 60$ mm (suggested for better compaction of concrete)
9. Find the distance from the support up to which the designed stirrups are required. For the rest of the portion provide minimum shear reinforcement.

Since the diagonal tension is resisted by combined action of its vertical component (carried by vertical stirrups) and horizontal component (carried by main reinforcement), the shear reinforcement always pass around and enclose all tension reinforcement of the beam for equilibrium of forces.

Example 4.1 A reinforced concrete beam of rectangular cross-section of 300 mm width and 550 mm overall depth is reinforced with 6 bars of 20 mm ϕ HYSD steel of grade Fe415, placed at an effective cover of 50 mm. Out of 6 bars, 3 bars have been bent up at 45°. Design the shear reinforcement if the beam is subjected to a uniformly distributed factored load of 100 kN/m over a simply supported clear span of 7 m. The concrete mix used is M20.

Solution The cross-sectional dimensions and material properties of the beam are:

$$b = 300 \text{ mm}, \ d = 550 - 50 = 500 \text{ mm}, \ f_{ck} = 20 \text{ MPa and } f_y = 415 \text{ MPa}$$

The maximum factored shear force is calculated at a distance d from the face of the support.

Thus,

$$V_u = \frac{w_u(l-2d)}{2}$$

$$= \frac{100 \times 6}{2} = 300 \text{ kN}$$

Area of inclined bars $A_{st}(3 \times 20 \text{ mm } \phi) = 942 \text{ mm}^2$.
Shear resistance of inclined bars bent up at the same cross-section is obtained as:

$$V_{ui} = 0.87 \; f_y A_{st} \sin \alpha$$

$$= 0.87 \times 415 \times 942 \times \sin 45°$$

$$= 240493 \text{ N} = 240.49 \text{ kN}$$

Area of tension reinforcement at the section of maximum shear i.e., the area of continuing bars is given as:

$$A_{st}(3 \times 20 \text{ mm } \phi) = 942 \text{ mm}^2$$

Then

$$p_t = \frac{100 A_{st}}{bd}$$

$$= \frac{100 \times 942}{100 \times 500} = 0.628 \text{ per cent}$$

Shear strength of M20 grade concrete for $p_t = 0.628$ per cent from Table 4.1 is:

$$\tau_c = 0.524 \text{ MPa}$$

Shear-resisting capacity of the un-reinforced concrete section is:

$$V_c = \tau_c bd$$

$$= 0.524 \times 300 \times 500 = 78600 \text{ N} = 78.6 \text{ kN}$$

Maximum permissible shear strength with shear reinforcement:

$$\tau_{c,\max} = 2.8 \text{ MPa}$$

Nominal shear stress on the cross-section is given by

$$\tau_v = \frac{V_u}{bd}$$

$$= \frac{300 \times 10^3}{300 \times 500} = 2.0 \text{ Mpa}$$

Since $\tau_c < \tau_v < \tau_{c,\max}$ the section is acceptable with shear reinforcement.

Design shear force $V_{us} = V_u - V_c = 300 - 78.6 = 221.4$ kN

Since $V_{us} < V_{ui}$, it is resisted by 3 inclined bars. However, IS:456 stipulates that half of the design shear force must be resisted by the vertical stirrups. Provide 8 mm ϕ 2-legged vertical stirrups with $A_{sv} = 100 \text{ mm}^2$.

$$\text{Spacing } S_v = \frac{0.87 f_y A_{sv} d}{0.5 V_{us}}$$

$$= \frac{0.87 \times 415 \times 100 \times 500}{0.5 \times 221.4 \times 10^3} = 163.08 \text{ mm}$$

Maximum spacing based on minimum shear reinforcement is:

$$S_v = \frac{A_{sc} f_y}{0.4b} = \frac{100 \times 415}{0.4 \times 300} = 345.83 \text{ mm}$$

Maximum spacing based on the depth of the beam is:

$$S_v = 0.75d$$

$$= 0.75 \times 500 = 375 \text{ mm or } 300 \text{ mm whichever is smaller}$$

Therefore, provide 8 mm ϕ 2-legged vertical stirrups at 160 mm c/c which is also satisfactory from the placement point of view. The first stirrup is usually placed at a distance $S_v/2$ from the support to prevent any possible propagation of shear crack.

Example 4.2 Determine the transverse or shear reinforcement required for a reinforced concrete beam of T-shaped cross-section with a flange width of 1500 mm, a flange depth of 100 mm, a web width of 300 mm and an effective depth of 500 mm. The beam is reinforced with 6 bars of 25 mm ϕ on the tension side and 4 bars of 14 mm ϕ on the compression side. The reinforcement continues uninterrupted into the supports. The beam supports a uniformly distributed factored load of 75 kN/m over a clear span of 9 m. The concrete used is of grade M20 and HYSD steel is of grade Fe415.

Solution The cross-sectional dimensions are:

b_f = 1500 mm, b = 300 mm, D_f = 100 mm, d = 500 mm

A_{st} (6 × 25 mm ϕ) = 2945 mm^2 and A_{sc}(4 × 14 mm ϕ) = 615 mm^2

The procedure for design of shear reinforcement for a flanged section is similar to that for a rectangular section of width equal to the width of the web of the flanged section, i.e., $b = b_w$.

The shear force is critical at a section at a distance d from the face of the support. It is given by

$$V_u = \frac{w_u(l - 2d)}{2}$$

$$= \frac{75 \times (9 - 2 \times 0.50)}{2} = 300 \text{ kN}$$

$$\text{Maximum nominal shear } \tau_v = \frac{V_u}{b_w d}$$

$$= \frac{300 \times 10^3}{300 \times 500} = 2.0 \text{ MPa}$$

Percentage of tension reinforcement is given as:

$$p_t = \frac{100A_{st}}{b_w d}$$

$$= \frac{100 \times 2945}{300 \times 500} = 1.96$$

For p_t = 1.96 per cent, the shear strength of concrete τ_c from the Table 4.1 is 0.78 MPa.

Maximum shear strength of M20 grade concrete with shear reinforcement is:

$$\tau_{c,\max} = 2.8 \text{ MPa}$$

Since $\tau_c < \tau_v < \tau_{c,\max}$, the section is acceptable with shear reinforcement.

Shear-resisting capacity of the un-reinforced concrete section is given by

$$V_c = \tau_c b_w d$$

$$= 0.78 \times 300 \times 500 = 117000 \text{ N} = 117.0 \text{ kN}$$

Consider 10 mm ϕ 2-legged stirrups(A_{sv} = 157 mm^2)

$$S_v = \frac{0.87 f_y A_{sv} d}{V_u - V_c} = \frac{0.87 \times 415 \times 157 \times 500}{(300 - 117.0) \times 10^3} = 154.88 \text{ mm}$$

Maximum spacing based on minimum shear reinforcement:

$$S_v = \frac{A_{sv} f_y}{0.4 b_w}$$

$$= \frac{157 \times 415}{0.4 \times 300} = 542.96 \text{ mm}$$

Maximum spacing based on the depth of the beam:

$$S_v = 0.75d$$

$$= 0.75 \times 500 = 375 \text{ mm or } 300 \text{ mm whichever is smaller.}$$

Consider that the shear reinforcement is required up to x m from the support on either end of the beam. Then from similar triangles, we have .

$$\frac{V_u}{L} = \frac{V_c}{L - x}$$

or

$$x = \frac{(V_u - V_c)L}{V_u}$$

where

L = Distance of the critical section from the centre of the beam

$= l/2 - d = 4.5 - 0.5 = 4$ m

Thus,

$$x = (300 - 117.0) \times \frac{4}{300} = 2.44 \text{ m}$$

Hence provide 10 mm ϕ 2-legged stirrups @ 150 mm c/c in the distance of 2.94 (= $x + d$) m from the support on either side. In the middle 3.12 m distance provide 10 mm ϕ 2-legged stirrups @ 300 mm c/c.

Example 4.3 A reinforced concrete cantilever beam of span 3 m carries a distributed factored load of 120 kN/m. The beam has a uniform width of 300 mm, with overall depth varying from 650 mm at the support to 400 mm at the free end. The beam is reinforced with 2 bars of 20 mm ϕ on the compression side and 5 bars of 28 mm ϕ on the tension side (top) at effective covers of 50 mm. Two of the tension bars are curtailed at a distance of 1.0 m from the support. Design the shear reinforcement, when concrete used is of grade M20 and HYSD steel reinforcement is of grade Fe415.

Solution For the given cross-section and materials:

$$f_{ck} = 20 \text{ MPa and } f_y = 415 \text{ MPa}$$

1. At support: b = 300 mm, d = 600 mm, A_{st} (5 × 28 mm ϕ) = 3078 mm^2
2. At free end: b = 300 mm, d = 350 mm

This example illustrates a case of beam of varying depth, for which β = angle between top and bottom edges of the beam is given by

$$\tan \beta = \frac{600 - 350}{3000} = 0.08333$$

Critical shear at the face of support V_u = 120 × 3.0 = 360 kN

Moment at the face of support M_u = 360 × 1.5 = 540 kNm

$$\tau_v = \frac{V_u \pm (M_u/d)\tan\beta}{bd}$$

As depth and moment both increase in the same direction, adopt –ve sign.

$$\text{Nominal shear stress } \tau_v = \frac{360 \times 10^3 - (540 \times 10^6/600) \times 0.08333}{300 \times 600} = 1.58 \text{ MPa}$$

Maximum shear stress with shear reinforcement $\tau_{c,\max}$ = 2.8 MPa

Effective depth at the point of curtailment is given as:

$$d_c = 350 + \frac{250}{3000} \times 2000 = 516.67 \text{ mm}$$

The curtailed bars are effective for resisting shear up to a distance of (1000 – d_c) = 483.33 mm from the support. Shear strength of concrete in this portion is:

$$p_t = \frac{100A_{st}}{bd}$$

$$= \frac{100 \times 3078}{300 \times 600} = 1.71$$

and hence τ_c = 0.7475 MPa.

Since $\tau_c < \tau_v < \tau_{c,\max}$ the section is acceptable with shear reinforcement.

For shear resistance of concrete in the portion of cantilever beyond 483.33 mm from the support, the effective depth at the distance of 483.33 mm from the face of the support is:

$$d = 350 + \frac{250}{3000}(2000 + 516.67) = 559.72 \text{ mm}$$

Area of tension steel $A_{st}(3 \times 28 \text{ mm } \phi) = 1847 \text{ mm}^2$

$$p_t = \frac{100A_{st}}{bd}$$

$$= \frac{100 \times 1847}{300 \times 559.72} = 1.10$$

Shear strength of concrete $\tau_c = 0.644$ MPa

Nominal shear stress at this point $\tau_v = \dfrac{V_u - (M_u/d)\tan\beta}{bd}$

where

$$V_u = 120 \times (2.000 + 0.51667) = 302 \text{ kN}$$

and

$$M_u = \frac{[120 \times (2.0 + 0.517)^2]}{2} = 380.02 \text{ kNm}$$

Therefore,

$$\tau_v = \frac{302 \times 10^3 - (380.02 \times 10^6/600) \times 0.08333}{300 \times 559.72} = 1.484 \text{ MPa}$$

Adopt 8 mm ϕ 2-legged vertical stirrups (A_{sv} = 100 mm^2). Spacing of stirrups in the length 483.33 mm (say 500 mm) from the support is the least of the following:

1. Maximum spacing based on shear:

$$S_v = \frac{0.87 f_y A_{sv}}{(\tau_v - \tau_c)b}$$

$$= \frac{0.87 \times 415 \times 100}{(1.58 - 0.7475) \times 300} = 144.56 \text{ mm}$$

2. Maximum spacing based on minimum shear reinforcement:

$$S_v = \frac{A_{sv} f_y}{0.4b}$$

$$= \frac{100 \times 415}{0.4 \times 300} = 345.83 \text{ mm}$$

3. Maximum spacing governed by the depth of the beam:

$$S_v = 0.75d$$

$$= 0.75 \times 559.72 = 419.79 \text{ mm}$$

4. 300 mm

Hence provide 8 mm ϕ 2-legged vertical stirrups @ 140 mm c/c.

Spacing of stirrups in the length 2516.67 mm (say 2500 mm) from the free end is the least of the following:

1. $$S_v = \frac{0.87 f_y A_{sv}}{(\tau_v - \tau_c)b}$$

$$= \frac{0.87 \times 415 \times 100}{(1.484 - 0.644) \times 300} = 143.27 \text{ mm}$$

2. Maximum spacing based on minimum shear reinforcement:

$$S_v = 345.83 \text{ mm}$$

3. Maximum spacing governed by the depth of the beam:

$$S_v = 0.75d$$

$$= 0.75 \times 350 = 262.5 \text{ mm}$$

4. 300 mm

Hence provide 8 mm ϕ 2-legged stirrups @ 140 mm c/c

4.3 LIMIT STATE OF COLLAPSE—TORSION

Most of the structures are subjected to torsion of varying magnitude in addition to flexure and shear. Torsion moment is predominant in structures like curved or ring beams at the bottom of circular water tanks, and balcony beams with cantilever slabs. Torsion moment occurs as a secondary effect in many structures, e.g. a slab built monolithic with floor beams where the torsion rigidity of the beams provides rotational edge restraint to the slab. This causes corresponding flexural restraining moment in the slab which, in turn, is balanced by torsion moment in the beams. Peripheral beams in each floor of a multistory building are subjected to significant torsion loading in addition to conventional flexure and shear. The beams supporting cantilever canopy slabs are always subjected to torsion. Other important examples are grid beams, three-dimensional building frames or space frames. Finally, the helicoidal staircases are to be designed for torsional resistance. Torsion in all these cases is induced because the centre of gravity of all the loads and reactions on one side of vertical section of beam does not lie on the axis of beam at the section. Some of the cases are illustrated in Fig. 4.4.

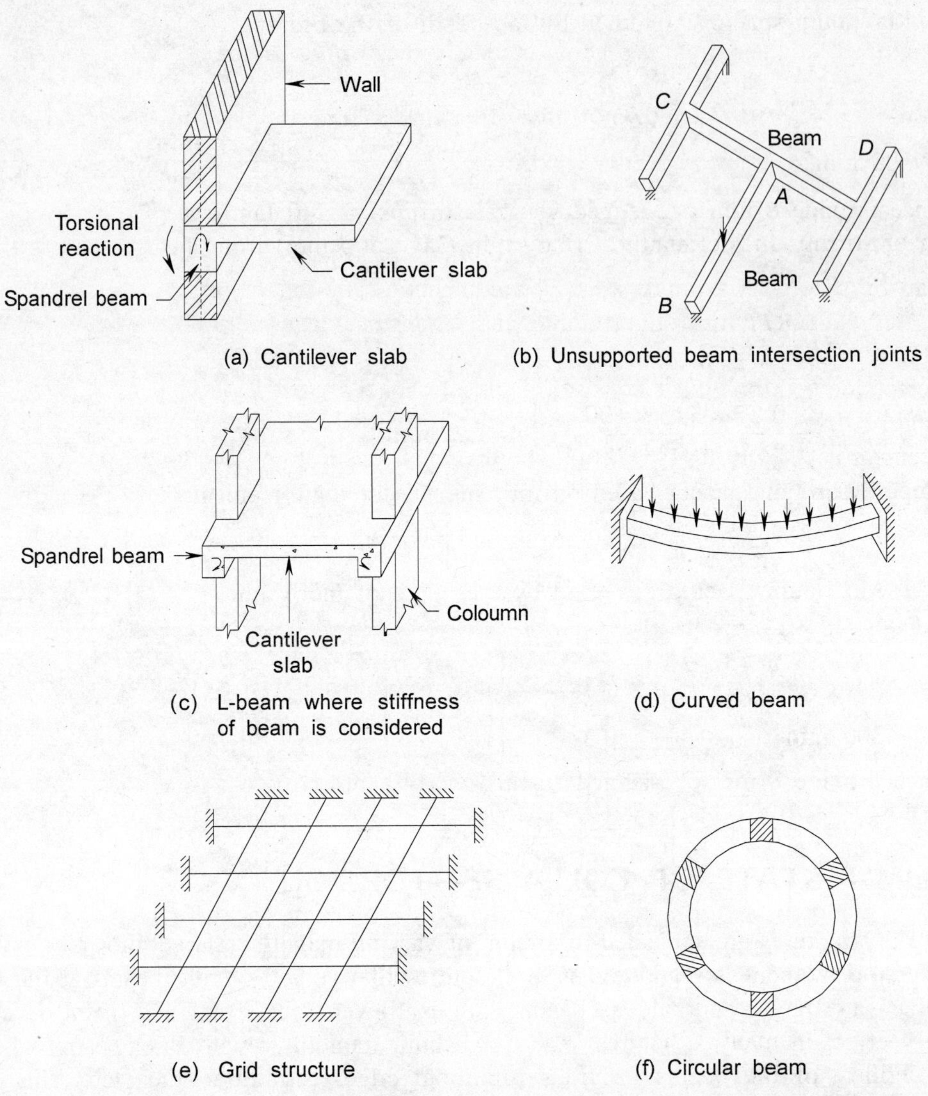

Fig. 4.4 Torsion in reinforced concrete structures.

The safety factors incorporated in ordinary design procedures are sufficient to accommodate the secondary torsional effects. In general, where the torsional resistance or stiffness of members has not been taken into account in the analysis of the structure, no specific calculations for torsion will be necessary, adequate control of any torsional cracking being provided by the required nominal shear reinforcement. On the other hand, when the torsional resistance or stiffness of members is taken into account in the analysis, the member shall be designed for torsion.

A plain concrete beam of rectangular cross-section subjected to torsion develops spiral cracks and fails suddenly. The diagonal cracks in the form of a spiral develop when diagonal tension exceeds the tensile strength of the concrete, resulting in the warping of the surface.

In reinforced concrete members subjected to flexure and torsion, the flexural cracks are the first to be formed. As soon as this stage is reached, the flexural and torsional stiffnesses are considerably affected and diagonal tension cracks are formed on the beam face in which torsional shear stresses and conventional shear stress are additive; eventually these cracks are formed on the other sides of the beam also. Finally, the tension reinforcement yields and the beam fail by secondary compression. Depending upon the relative magnitudes of flexure and torsion, this behaviour may slightly alter and compression failure may occur on the sides of the beam.

The torsion reinforcement consists of closely spaced closed stirrups or hoops with good anchorage being provided by hooking the stirrup bar ends around the longitudinal reinforcement as shown in Fig. 4.5(a). The diagonal tensile force due to torsion is resisted by the component of tensile force in the hoops and in the longitudinal steel, while diagonal compressive force is resisted by concrete.

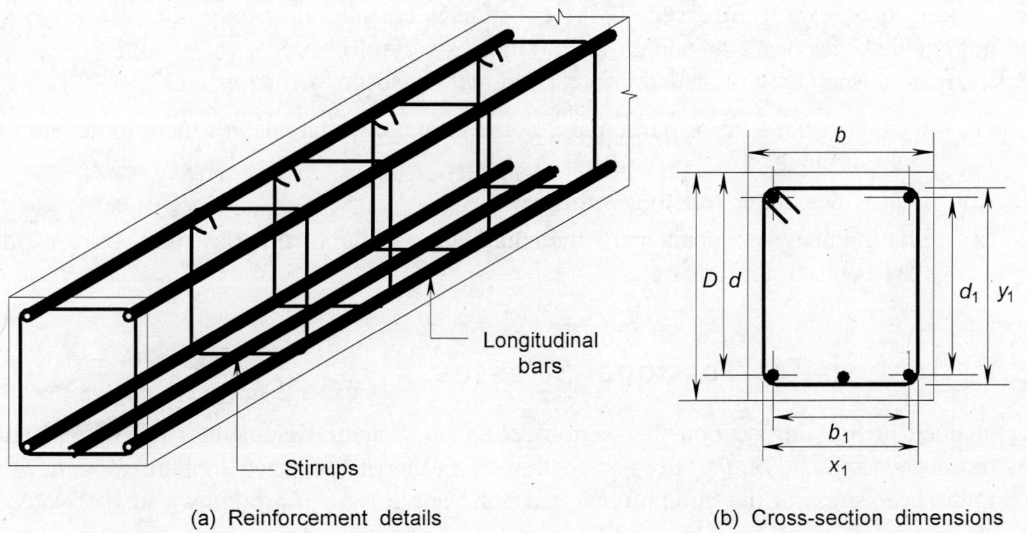

(a) Reinforcement details

(b) Cross-section dimensions

Fig. 4.5 *Detailing scheme for transverse reinforcement for torsion.*

Where torsional moment is predominant, it is necessary to weld the hoops (stirrups) to the longitudinal steel so that relative slips at the reinforcement skeleton joints are prevented. Open hooks or bends should be avoided. However, hoops and longitudinal bars placed in the webs of T- and L-beams can have good anchorage when the hoops are extended well into the flange, as shown in Fig. 4.6.

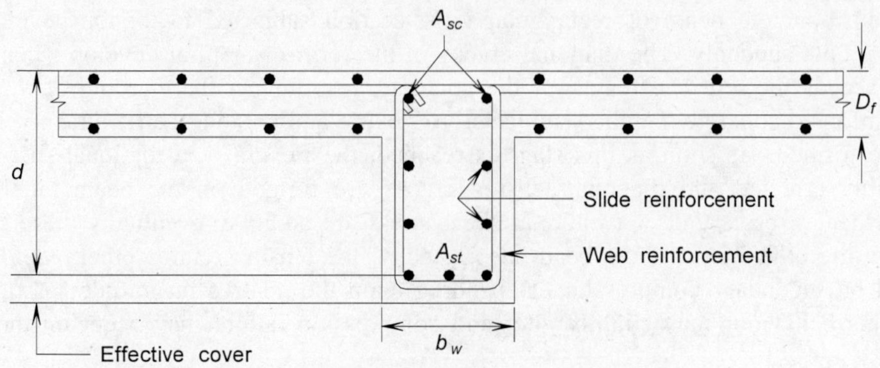

Fig. 4.6 Web reinforcement in a flange beam.

To avoid sudden collapse, it is desirable that the torsional members have ductile rather than brittle behaviour, i.e., the stirrups should yield before the compression zone of the failure surface is crushed. If both the steels yield before the crushing of concrete, the beam is said to be **under-reinforced**. If the concrete is crushed before both types of steels have yielded, the beam is said to be **over-reinforced**. If the concrete is crushed after only one of the types of steel has yielded, the beam is said to be **partially over-reinforced**.

The functions of longitudinal reinforcement with respect to torsion are:

1. It anchors the stirrups particularly at the corners, which enables them to develop full yield strength.
2. It provides some resisting torque.
3. Longitudinal steel counteracts the lengthening of the member as spiral cracks widen, thus controls the cracks.

4.3.1 Analysis for Torsional Stresses

As explained in forgoing section the torsion acting on structural elements causes shear stress over the cross-section. For the circular section of radius R subjected to pure torsion T, the maximum shear stress at the circumference and the angle twist θ are known to be related by

$$\frac{T}{J} = \frac{G\theta}{L} = \frac{\tau_{t,\max}}{R} \tag{4.15}$$

or

$$\tau_{t,\max} = \frac{T}{(J/R)}$$

where

J = Polar moment of inertia of cross-section = $\dfrac{\pi R^4}{4}$

G = Shear modulus = $\dfrac{E}{2(1+\mu)}$

E = Young's modulus of elasticity
μ = Poisson's ratio
L = length of the member

For a rectangular section subject to pure torsion, the maximum shear stress occurs at the middle of the wider (or longer) side D and is given by

$$\tau_{t,\max} = \frac{T}{(kb^2 D)} \tag{4.16}$$

where

kb^2D = Polar or torsional section modulus
b and D = The shorter and longer sides of the rectangular section, respectively
k = Torsion constant

The value of k depends upon the ratio (D/b) and is given in Table 4.3.

TABLE 4.3 Variation of Coefficient k with the Ratio D/b

D/b	1.0	1.5	2.0	2.5	3.0	4.0	5.0	6.0	8.0	10.0	∞
k	0.208	0.231	0.246	0.258	0.267	0.282	0.292	0.299	0.307	0.313	0.333

For the flange section shown in Fig. 4.7, the torsional shear stress may be determined by approximating the section into several rectangular components with the ratio D/b being large enough, so that the value of k may be taken as 1/3. The maximum shear occurs at the midpoint of the longer side D of the rectangle having greater value of b equals to b_m.

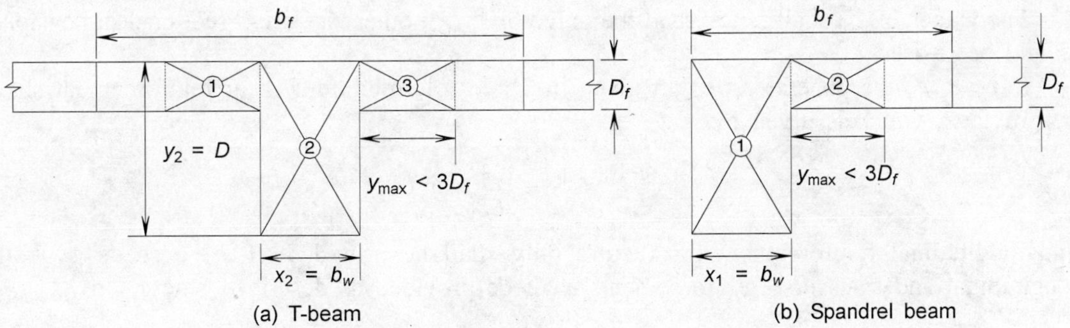

Fig. 4.7 Subdivision of flange sections for torsion analysis.

Thus,

$$\tau_{t,\max} = \frac{3Tb_m}{\Sigma b^3 D} \tag{4.17}$$

4.3.2 Design of the Beam Section for Torsion

The simplified approach recommended by IS:456 for the design of the rectangular beam section, subjected to torsion combined with flexure and shear, does not require determination of torsional reinforcement separately from that required for flexure and shear. Instead, the total longitudinal reinforcement is determined for a fictitious equivalent bending moment which is a function of flexural moment and torsion. Similarly, the transverse reinforcement is determined for a fictitious equivalent shear which is obtained from actual shear and torsion. The approach further simplifies the design by neglecting the contribution of flanges in the flanged sections and considering the rectangular web portion alone. The flange contribution is negligible and hence the preceding simplification may be justified.

Equivalent shear: The equivalent shear V_{ue} shall be calculated as follows:

$$V_{ue} = V_u + 1.6\,\frac{T_u}{b} \tag{4.18}$$

where

V_u = Factored shear force
T_u = Factored torsional moment
b = Width of the beam (= b_w for the flanged section)

The equivalent nominal shear stress shall be calculated as:

$$\tau_{ve} = \frac{V_{ue}}{bd} \quad \text{for beams of uniform depth}$$

$$= \frac{[V_{ue} \pm (M_u/d)\tan\beta]}{bd} \quad \text{for beams of varying depths} \tag{4.19}$$

where d and β are the effective depth and the angle between the top and bottom edges of the beam, respectively. The –ve sign shall be used when the moment M_u increases numerically in the same direction as the effective depth.

The values of τ_{ve} shall not exceed the value of $\tau_{c,\max}$, otherwise cross-sectional dimensions should be revised.

If $\tau_{ve} < \tau_c$, the concrete strength given in Table 4.1, minimum shear reinforcement shall be provided, which is given by

$$\frac{A_{sv}}{bS_v} = \frac{0.4}{f_y} \tag{4.20}$$

and longitudinal reinforcement for flexure only shall be provided. If τ_{ve} exceeds τ_c, both longitudinal and transverse reinforcements shall be provided as described below.

Longitudinal reinforcement: Longitudinal reinforcement shall be provided to resist equivalent bending moment M_{e1} given by

$$M_{e1} = M_u + M_t \tag{4.21}$$

where

M_u = Factored bending moment at the cross-section
M_t = Contribution of torsional moment T_u in the bending moment

$$= T_u\,\frac{(1 + D/b)}{1.7} \tag{4.22}$$

where D and b are overall depth and breadth of the cross-section, respectively.

If the numerical value of M_t as defined above, exceeds the numerical value of moment M_u, longitudinal reinforcement shall be provided on the flexural compression face to resist an equivalent moment $M_{e2} = M_t - M_u$ taken as acting in the sense opposite to the moment M_u to take care of reversal of moment.

Transverse reinforcement: The cross-sectional area A_{sv} of two-legged closed hoops or stirrups enclosing in the corner longitudinal bars is given by:

$$A_{sv} = \frac{T_u S_v}{b_1 d_1 (0.87 f_y)} + \frac{V_u S_v}{2.5 d_1 (0.87 f_y)} \tag{4.23}$$

But the transverse steel shall not be less than

$$A_{sv} = \frac{(\tau_{ve} - \tau_c) b S_v}{(0.87 f_y)} \tag{4.24}$$

where

S_v = Spacing of stirrup reinforcement

b_1 = Centre-to-centre distance between corner bars in the direction of width

d_1 = Centre-to-centre distance between corner bars in the direction of depth

b = Breadth of the member

f_y = Characteristic strength of transverse steel

Detailing Requirements for Torsion:

1. The *longitudinal steel* shall be placed as close as practicable to the corners of the cross-section. The spacing of longitudinal bars ($\not< 10$ mm ϕ) distributed around the periphery of closed stirrups shall not exceed 300 mm. In all cases, there shall be at least one longitudinal bar in each corner of the stirrup. In case the cross-sectional dimension of the member exceeds 450 mm, additional longitudinal bars shall be provided along the two faces. The total area of such reinforcement shall not be less than 0.1 per cent of the web area and shall be distributed equally on the two faces at a spacing not exceeding 300 mm or web width, whichever is less.
2. The transverse steel for torsion shall be in the form of rectangular closed stirrups placed perpendicular to the axis of the member. The spacing of stirrups shall not exceed the least of x_1, $(x_1 + y_1)/4$, and 300 mm, where x_1, y_1 are the short and long dimensions of the stirrups, respectively, as shown in the Fig. 4.5(a).
3. The torsion steel shall be provided at least for a distance $(d + b_t)$ beyond the point where it is theoretically required, where b_t is the width of the part of the cross-section having closed stirrups resisting torsion.
4. Stirrups and other bars shall be anchored adequately to develop the design yield strength of bars.

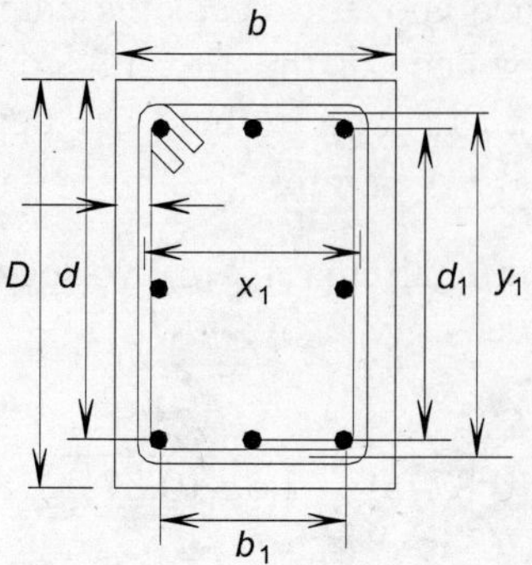

(a) Cross-sectional dimensions

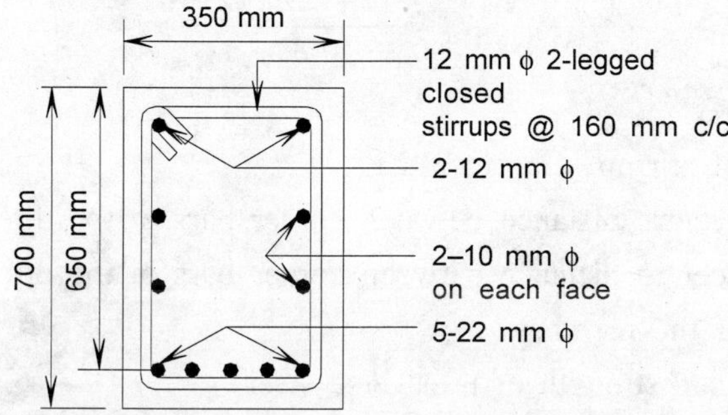

(b) Reinforcement details of beam of Example 4.4

Fig. 4.8 Dimensions and reinforcement details of a rectangular section subjected to flexure, shear and torsion.

Example 4.4 Design a rectangular reinforced concrete beam section to carry a factored bending moment of 200 kNm, factored shear force of 120 kN, and a factored torsional moment of 75 kNm. Concrete mix of grade M20 and HYSD steel of grade Fe415 are to be used in construction.

Solution For the given materials:

$$f_{ck} = 20 \text{ MPa and } f_y = 415 \text{ MPa}$$

The forces acting on the section are:

$$M_u = 200 \text{ kNm}, \; V_u = 120 \text{ kN and } T_u = 75 \text{ kNm}$$

Consider the case $\tau_{ve} > \tau_c$; wherein both longitudinal and transverse reinforcements shall be provided for resisting equivalent moment and equivalent shear force, respectively.

Longitudinal reinforcement: Equivalent moment is given by

$$M_e = M_u + M_t$$

where

$$M_t = \frac{T_u(1 + D/b)}{1.7}$$

Consider the ratio D/b = 2.0, then

$$M_t = \frac{75(1+2)}{1.7} = 132.35 \text{ kNm}$$

Since M_t is less than M_u, no steel on the compression face is required for equivalent moment of $M_t - M_u$.

$$M_e = M_u + M_t$$

$$= 200 + 132.35 = 332.35 \text{ kNm}$$

The design for flexure may result in either singly reinforced balanced, under-reinforced, or doubly reinforced section. For a balanced section,

$$x_{u,\max} = \frac{0.0035d}{\left(0.0055 + \dfrac{0.87 f_y}{E_s}\right)}$$

$$= \frac{0.0035d}{\left[0.0055 + \dfrac{(0.87 \times 415)}{(2 \times 10^5)}\right]} = 0.479d$$

From the internal force equilibrium condition,

$$0.87 f_y A_{st} = 0.362 f_{ck} b x_{u,\max}$$

or

$$\frac{A_{st}}{bd} = p_{t,\lim} = \frac{0.362 f_{ck} \times 0.4791}{0.87 f_y}$$

$$= \frac{0.362 \times 20 \times 0.4791}{0.87 \times 415} = 0.00961, \text{ i.e., } 0.961 \text{ per cent}$$

To ensure the beam to be under-reinforced, limiting p_t to (3/4)(0.00961) = 0.0072(say 0.0075).

If x_u be the depth of the actual neutral axis, then from the internal force equilibrium condition, we have

$$0.36 f_{ck} b x_u = 0.87 f_y A_{st}$$

or

$$x_u = \frac{0.87 f_y A_{st}}{0.362 f_{ck} b} = \frac{0.87 f_y}{0.362 f_{ck}} \left[\frac{A_{st}}{bd}\right] d$$

$$= \frac{0.87 \times 415 \times 0.0075}{0.362 \times 20} d = 0.3740d$$

and

$$M_u = 0.362 f_{ck} b x_u (d - 0.416 x_u)$$

$$= 0.362 \times 20b \times 0.3740d(d - 0.416 \times 0.3740d) = 2.286bd^2$$

Therefore,

$$2.286bd^2 = 332.35 \times 10^6$$

Consider $b = 0.5d$. Therefore,

$$d^3 = \frac{332.35 \times 10^6}{2.286 \times 0.5}, \text{ giving } d = 662.50 \text{ mm}$$

Adopt $b = 350$ mm, $d = 650$ mm and $D = 700$ mm, with the ratio $D/b = 700/350 = 2.0$. The value is the same as assumed and, hence, the section is satisfactory.

$$\text{Area of steel required } A_{st} = 0.0075 \times 350 \times 650 = 1706.25 \text{ mm}^2$$

Provide 5-22 mm ϕ bars as tension reinforcement (A_{st} = 1900 mm^2) and 2–12 mm ϕ bars at top in compression zone as hanger bars.

Side face reinforcement: As the depth of the section is more than 450 mm, side face reinforcement of 0.10 per cent of the web section is to be provided.

$$\text{Area of additional reinforcement} = \frac{0.10}{100} \times 350 \times 700 = 245 \text{ mm}$$

Provide 2 × 10 mm ϕ bars on each face.

Transverse reinforcement: The shear strength of concrete τ_c depends on p_t.

$$p_t = \frac{A_{st}}{bd}$$

$$= \frac{1900}{350 \times 650} = 0.00835, \text{ i.e., } 0.835 \text{ per cent}$$

From Tables 4.1 and 4.2, τ_c = 0.583 MPa and $\tau_{c,\max}$ = 2.8 MPa (for M20 concrete).

$$\text{Equivalent shear force } V_e = V_u + \frac{1.6T_u}{b}$$

$$= 120 + \frac{(1.6 \times 75)}{0.35} = 462.86 \text{ kN}$$

Equivalent nominal shear stress is given by

$$\tau_{ve} = \frac{V_e}{bd} = \frac{(462.86 \times 10^3)}{(350 \times 650)} = 2.035 \text{ MPa}$$

Since $\tau_c < \tau_{ve} < \tau_{c,\max}$ the section is acceptable with shear reinforcement.

Consider transverse reinforcement consisting of 12 mm ϕ 2-legged vertical stirrups (A_{sv} = 226 mm^2), the spacing of stirrups is given by

$$S_v = \frac{0.87 f_y A_{sv}}{\left(\dfrac{T_u}{b_1 d_1}\right) + \left(\dfrac{V_u}{2.5 d_1}\right)}$$

where

b_1 = c/c distance between corner bars in the direction of the width of the beam
= 350 – 25 – 25 – 22 = 278 mm

d_1 = c/c distance between corner bars in the direction of the depth of the beam
= 650 – 25 – 6 = 619 mm

Thus,

$$S_v = \frac{0.87 \times 415 \times 226}{\left[\dfrac{75 \times 10^6}{(278 \times 619)}\right] + \left[\dfrac{120 \times 10^3}{(2.5 \times 619)}\right]} = 158.9 \text{ mm}$$

Minimum reinforcement to be provided is given by

$$A_{sv} > \frac{(\tau_{ve} - \tau_c) S_v b}{0.87 f_y}$$

Thus,

$$S_v < \frac{0.87 f_y A_{sv}}{(\tau_{ve} - \tau_c) b} = \frac{0.87 \times 415 \times 226}{(2.035 - 0.583) \times 350} = 160.56 \text{ mm}$$

The spacing of stirrups S_v is restricted to the least of the following:

1. x_1 = shorter dimension of stirrups = 312 mm,
2. $(x_1 + y_1)/4$ = (Shorter + Longer dimensions of stirrup)/ 4 = (312 + 648)/ 4 = 240 mm
3. $0.75d$ = 487.5 mm
4. 300 mm

Hence provide 12 mm ϕ 2-legged vertical stirrups @ 160.56 mm rounded to 160 mm c/c. The reinforcement details are shown in Fig. 4.8(b).

Example 4.5 A reinforced concrete beam section of size 250 × 500 mm deep (effective) is subjected to a factored moment of 100 kNm and factored shear force of 140 kN. The grade of concrete used is M20. Design shear reinforcement using Fe500 grade steel bars. The effective cover to tension reinforcement is 50 mm.

If the beam is also subjected to factored torsional moment of 20 kNm, redesign the reinforcement for bending and shear.

Solution For the given cross-section and materials:

$$b = 250 \text{ mm, } d = 500 \text{ mm and } D = 550 \text{ mm}$$

$$f_{ck} = 20 \text{ MPa, } f_y = 500 \text{ MPa and } \tau_{c,\max} = 2.8 \text{ MPa}$$

CASE I $M_u = 100$ kNm, $V_u = 140$ kN, and $T_u = 0$

Consider the section to be a balanced section:

$$x_{u,\max} = \frac{0.0035d}{\left(0.0055 + \dfrac{0.87 f_y}{E_s}\right)}$$

$$= \frac{0.0035d}{0.0055 + \dfrac{(0.87 \times 500)}{2 \times 10^5}} = 0.456d$$

From the equation of equilibrium of internal forces:

$$0.362 f_{ck} b x_{u,\max} = 0.87 f_y A_{st}$$

$$\frac{A_{st}}{bd} = \frac{0.362 f_{ck} x_{u,\max}}{0.87 f_y d}$$

$$= \frac{0.362 \times 20 \times 0.456}{0.87 \times 500} = 0.00759$$

Thus, $p_{t,\lim} = 0.759$ per cent. Consider 0.75 per cent reinforcement steel consisting of 3 bars of 20 mm φ ($A_{st} = 942.48$ mm^2) at an effective cover of 50 mm, and 2 bars of 12 mm φ at the top as hanger bars. Also provide one 8 mm φ bar on each face.

$$\text{Nominal shear stress } \tau_v = \frac{V_u}{bd} = \frac{140 \times 10^3}{250 \times 500} = 1.12 \text{ MPa}$$

Shear strength of concrete τ_c for 0.75 per cent tension reinforcement and M20 grade concrete is 0.56 MPa. Since $\tau_c < \tau_v < \tau_{c,\max}$, section is acceptable with shear reinforcement. Shear reinforcement is required for net shear force:

$$V_{us} = V_u - bd$$

$$= 140 - (0.56 \times 250 \times 500) \times 10^{-3} = 70 \text{ kN}$$

Provide 8 mm φ 2-legged vertical stirrups ($A_{sv} = 100.6$ mm^2) at a spacing given by

$$S_v = \frac{0.87 f_y A_{sv} d}{V_{us}}$$

$$= \frac{0.87 \times 415 \times 100.6 \times 500}{70 \times 10^3} = 259.44 \text{ mm}$$

(f_y = 415 MPa has been used for Fe500 grade HYSD steel as recommended by IS:456). Check from minimum shear reinforcement criteria:

$$\frac{A_{sv}}{bS_v} \geq \frac{0.4}{f_y}$$

Assuming equality, we have

$$S_v = \frac{A_{sv} f_y}{0.4b}$$

$$= \frac{100.6 \times 415}{0.4 \times 250} = 417.49 \text{ mm}$$

Hence provide 8 mm φ 2-legged Fe500 grade steel bar stirrups at 250 mm c/c. The reinforcement details are shown in Fig. 4.9.

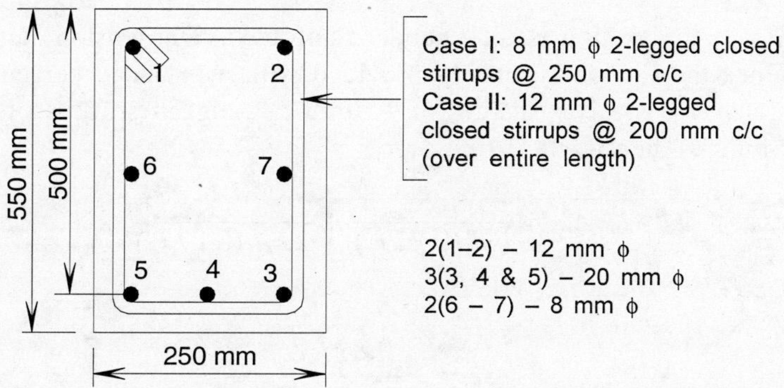

Fig. 4.9 *Reinforcement details for the beam section of Example 4.5.*

CASE II With introduction of the torsional moment T_u = 20 kNm in the beam, the equivalent shear is given by

$$V_e = V_u + \frac{1.6T_u}{b}$$

$$= 140 + \frac{1.6 \times 20}{0.25} = 268 \text{ kN}$$

Equivalent nominal shear stress is:

$$\tau_{ve} = \frac{V_e}{bd}$$

$$= \frac{268 \times 10^3}{250 \times 500} = 2.144 \text{ MPa}$$

Since $\tau_c < \tau_v < \tau_{c,\max}$, the section is acceptable with shear reinforcement. However, the longitudinal reinforcement required for equivalent moment is to be provided.

Equivalent moment taking into account the torsional effect:

$$M_{e1} = M_u + M_t$$

$$= 100 + \left[\frac{20 \times \left(1 + \frac{550}{250}\right)}{1.7}\right]$$

$$= 100 + 37.65 = 137.65 \text{ kNm}$$

Since $M_t < M_u$, longitudinal reinforcement is not required at the compression face. For checking the adequacy of the tension reinforcement, compute $M_{u,\lim}$ of the section. For Fe500 grade steel,

$$M_{u,\lim} = 0.1338\ f_{ck}bd^2$$

$$= 0.1338 \times 20 \times 250 \times 500^2 = 167.25 \text{ kNm}$$

Since $M_{e1} < M_{u,\lim}$, the section is to be under-reinforced. The tension steel does not require any revision, since the section is already provided with a balanced percentage of steel.

Consider transverse steel consisting of 12 mm ϕ 2-legged stirrups of Fe500 grade steel (A_{sv} = 226.20 mm^2). Then

$$A_{sv} = \frac{T_u S_v}{b_1 d_1 (0.87 f_y)} + \frac{V_u S_v}{2.5 d_1 (0.87 f_y)}$$

or

$$S_v = \frac{0.87 f_y A_{st}}{\left[\frac{T_u}{b_1 d_1} + \frac{V_u}{2.5 d_1}\right]}$$

where

b_1 = c/c distance between corner bars in the direction of b

= 250 – 25 – 25 – ϕ (= 20) = 180 mm

d_1 = c/c distance between corner bars in the direction of d

= 500 – 25 – 6 = 469 mm

Therefore,

$$S_v = \frac{0.87 \times 415 \times 226.2}{\frac{20 \times 10^6}{180 \times 469} + \frac{140 \times 10^3}{2.5 \times 469}} = 229.21 \text{ mm}$$

Minimum reinforcement to be provided is given by

$$A_{sv} \geq \frac{(\tau_{ve} - \tau_c) S_v b}{(0.87 f_y)}$$

or

$$S_v \ngtr \frac{(0.87 f_y) A_{sv}}{(\tau_{ve} - \tau_c) b}$$

Thus,

$$S_v \ngtr \frac{0.87 \times 415 \times 226.2}{(2.144 - 0.56) \times 250} = 206.24 \text{ mm}$$

The spacing of stirrups, S_v is restricted to the least of the following:

(i) x_1 = Shorter dimension of stirrup = 212 mm

(ii) $\dfrac{x_1 + y_1}{4} = \dfrac{\text{Shorter + Longer Dimensions of Stirrup}}{4}$

$$= \frac{212 + 497}{4} = 354.5 \text{ mm}$$

(iii) $0.75d = 0.75 \times 500 = 375$ mm

(iv) 300 mm

Hence provide 12 mm ϕ 2-legged stirrups of Fe500 grade steel at 200 mm centre to centre. The reinforcement details are shown in Fig. 4.9.

4.4 LIMIT STATE OF COLLAPSE—BOND

For reinforcement concrete to behave effectively, it is essential that stresses be transferred from one material to the other. In other words, there should be no slippage or relative movement between concrete and the reinforcement. The grip of concrete on reinforcement due to *adhesion* or *bearing* is termed as the **bond**. It resists any force tending to cause slippage of the bar relative to the surrounding concrete. The resulting shear stress at the interface of the bar and concrete is known as **bond stress** and is defined as force per unit nominal surface area of reinforcing bar. The bond between concrete and steel should be perfect at service loads since the validity of reinforced concrete theory is based on the fulfillment of this condition. The successful performance of a reinforced concrete structural element is based on adequate composite action between steel and concrete. This composite action is termed as **interface shear stress** which transfers forces from one material to the other.

The bond stress developed at the steel-concrete interface consists of three components:

1. due to pure adhesion between the steel bar and surrounding concrete.
2. due to frictional resistance mobilized by shrinkage of concrete which grips the reinforcement steel.
3. due to mechanical resistance provided by bearing of concrete on the lugs of deformed or twisted bars. Frictional and mechanical resistances are mobilized for transfer of force from concrete to steel when the pure adhesion fails and the bar starts slipping.

The failure in bond in concrete with plain bars occurs when adhesion and fractional resistance are overcome and the bar is pulled leaving a round hole in concrete. To prevent this, end anchorage is provided mostly in the form of bends or hooks. If the end anchorage is adequate, such a beam will not collapse even if the bond is broken over the entire length. On the other hand, a deformed bar increases the bond capacity due to mechanical resistance in addition to adhesion and frictional resistance. When adequate embedment length and end anchorage, L_o is provided, bond failure due to pulling of bar does not occur and failure will be normally by splitting of concrete along the bar either in the vertical plane or horizontal plane because of the proximity of reinforcing bars to the periphery. When the splitting has spread all the way to the end of anchorage, complete bond failure occurs, i.e., sliding of steel bars relative to concrete leads to immediate collapse of the beam.

The bond stress generated in reinforced concrete members due to anchorage of bars in case of tension or compression member, and change in bar force along its length due to change in bending moment along the span of member are termed as **anchorage bond** and **flexural bond**, respectively. The anchorage bond develops in the anchorage zone at the ends of a beam or at the cut-off point of a bar within the span of the beam. It causes slippage of the bar at the ends, resulting in cracking of the surrounding concrete. The bond anchorage is the average bond stess over the length of the bar required to transmit the bar force to concrete in a pull-out test, i.e., tension being changed from T to zero. The anchorage bond depends upon the surface condition of the bar, strength of concrete and on the condition whether the bar is in tension or compression. The flexural bond stress, on the other hand, is the average bond stress over a short length of bar because of the occurrence of cracks at discrete intervals along the length of the member. The cracks cause large local variation in tensile stress in the steel; the stress varies from a local maximum at the cracked section to a local minimum at the middle of the uncracked region. Thus, flexural bond stress does not predict accurately the actual bond stress.

Limit states design concepts are based on the criterion that the steel reinforcement will develop its yield strength before premature failure occurs due to inadequate bond. Adequacy of the bond must be ensured under a variety of loading situations such as tension, compression and flexure.

According to IS:456 the calculated tension and compression in any bar at any cross-section shall be developed (by flexural bond) on each side of the cross-section by appropriate bar length or end anchorage or a combination of both. The extended length of the bar required to transmit the bar force to concrete is known as **development length** L_d. It is obtained by equating the bond resistance of concrete to the strength of the reinforcing bar, as illustrated in Fig. 4.10. The values of **design bond stress** for different grades of concrete and steel are given in Table 4.4.

4.4.1 Development Length

The development length concept replaces the dual requirements for anchorage bond and flexural bond. To ensure that the full tensile or compressive strength of a reinforcing bar is realized at the various critical sections of a structural element, an adequate development length is provided beyond these critical sections. The concept of development length is based on the

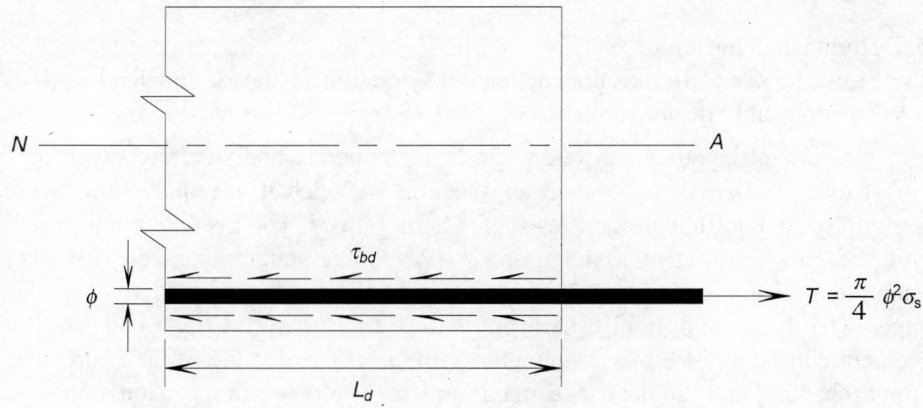

Fig. 4.10 Generation of anchorage and flexural bonds.

attainable average bond stress over the *length of embedment of reinforcement*. The consideration of average bond resistance is more meaningful, particularly because all bond tests consider an average bond stress over the length of embedment.

TABLE 4.4 Design Bond Strength and L_d/ϕ Values for Bars in Tension and Compression

f_y, MPa	Tension bars					Compression bars				
	Grade of concrete					Grade of concrete				
	M20	M25	M30	M35	M40	M20	M25	M30	M35	M40
Design bond stress, MPa	**1.2**	**1.4**	**1.5**	**1.7**	**1.9**	**1.5**	**1.75**	**1.875**	**2.125**	**2.375**
240*	43.5	37.3	34.8	30.7	27.5	34.8	29.8	27.8	24.6	22.0
250	45.3	38.8	36.3	32.0	28.6	36.3	31.1	29.0	25.6	22.9
Design bond stress, MPa	**1.92**	**2.24**	**2.4**	**2.72**	**3.04**	**2.4**	**2.8**	**3**	**3.4**	**3.8**
415	47.0	40.3	37.6	33.2	29.7	37.6	32.2	30.1	26.5	23.8
500	56.6	48.5	45.3	40.0	35.8	45.3	38.8	36.3	32.0	28.6

Notes:

1. The value of design bond stress for bars in compression is 25 per cent greater than for tension bars.
2. For deformed bars conforming to IS: 1786–1979 or IS: 1139–1966, the design bond stress is 60 per cent greater than that for plain bars.
3. For bars up to 20 mm diameter f_y = 250 MPa and for bars over 20 mm diameter f_y = 240 MPa.

The development length L_d is given by

$$L_d = \frac{\phi\sigma_s}{4\tau_{bd}} \tag{4.25}$$

where

ϕ = Nominal diameter of bar

σ_s = Stress in bar at the section under consideration at the design load (= $0.87f_y$)

τ_{bd} = Design bond stress

The development length for different grades of concrete and steel are given in Table 4.3. A bar must extend a length L_d beyond any section at which it is required to develop its full yield strength so that sufficient bond resistance is mobilized. The average bond stress along the length of the bar is assumed to be uniform. When the length available for anchorage is insufficient, the reinforcement can be bent or a hook may be formed to provide adequate anchorage. The bend and hook shall be made, as shown in Fig. 4.11(a). For tension reinforcement consisting of a bar of diameter ϕ, the anchorage value of the bend shall be taken as 4ϕ for each 45° bend, subject to a maximum of 16ϕ. For compression reinforcement the projected length of the hook, and straight length beyond bend shall be considered for development length.

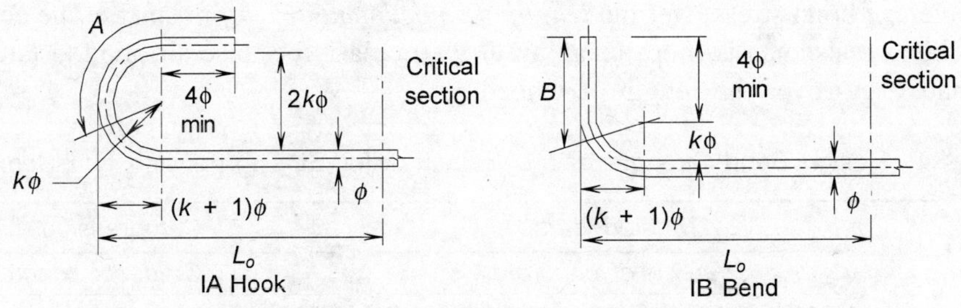

k > 2 for mild steel bars and k > 4 for HYSD steel bars

(a) IS:2502–1963 hooks and bends for reinforcement bars

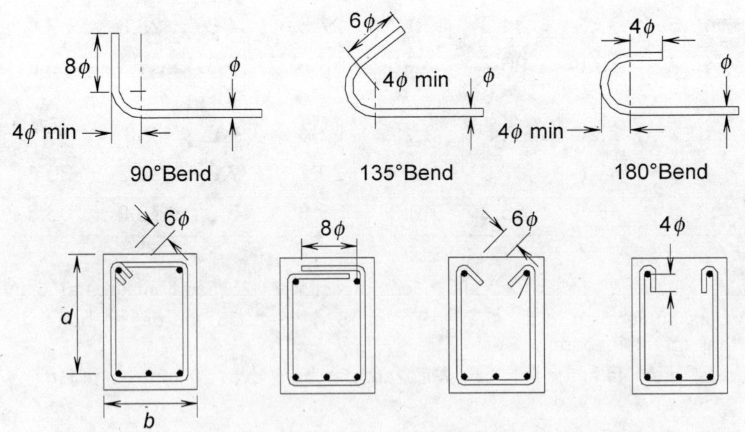

(b) IS:2502–1963 hooks and bends for stirrups

Fig. 4.11 Anchorage in shear reinforcement and standard hooks and bends.

For inclined bars in the tension zone, the development length is measured from the end of the sloping portion of the bar, whereas in the compression zone, it is measured from mid-depth of the beam.

For the development of the full strength of the stirrup, the ends of the stirrup should be bent through an angle of 90° or 135° or 180° around a bar of at least its own diameter and continued beyond the end of curve for a length of 8ϕ or 6ϕ or 4ϕ, respectively, as shown in Fig. 4.11(b).

IS:456 have recommended additional anchorage length equal to L_o to ensure anchorage bond stress to be within its maximum limit, such that

1. $L_d \le (M_1/V_u) + L_o$, when ends of bars lie in the tension zone
2. For the portions of the member where bending moment is nearly zero, viz. near the supports of simply supported beams and in case the ends of reinforcement are confined by compressive reaction, the value of (M_1/V_u) is to be increased by 30 per cent, i.e.,

$$L_d \le 1.3\left(\frac{M_1}{V_u}\right) + L_o \tag{4.26}$$

where

M_1 = Moment of resistance with respect to tension reinforcement

V_u = Factored shear force

The values of L_o at critical sections for development length are as follows and are illustrated in Fig. 4.12:

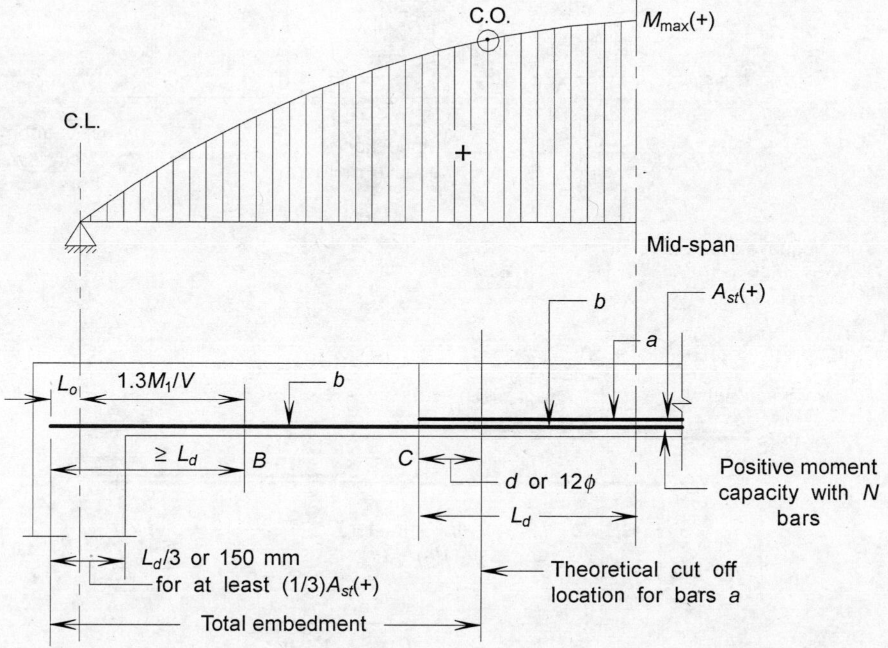

Fig. 4.12 *Curtailment of reinforcement in a simply supported beam.*

1. Simply supported end:

$$L_o = \text{Embedment length beyond the centre of the support} + \text{Equivalent embedment length of hook or bend}$$

2. Point of inflection and point of curtailment:

$$L_o = 12\phi \text{ or } d, \text{ whichever is greater.}$$

4.4.2 Continuation of Reinforcement

The longitudinal bars should continue into the support or beyond the points of inflection in case of continuous beams, as shown in Fig. 4.13. According to IS:456:

1. At least one-third of the positive moment reinforcement in simply supported members and one-fourth in continuous members should extend along the same face of the member into the support to a length equal to $L_d/3$ or 150 mm whichever is greater.
2. At least one-third of total reinforcement for negative moment at support should extend beyond the point of inflection for a distance not less than the effective depth of the member or 12ϕ or (1/16) of the clear span, whichever is greater.

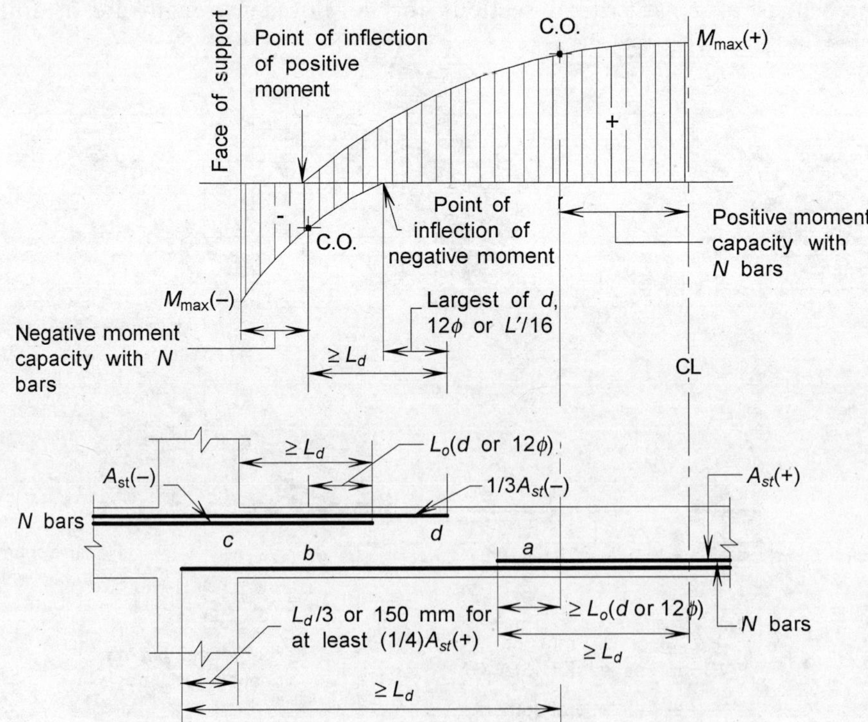

Fig. 4.13 Curtailment of negative moment reinforcement.

4.5 ANCHORING THE REINFORCEMENT

***In tension*:** Deformed bars may be used without end anchorages, provided the development length requirement is satisfied. Hooks should normally be provided for plain bars in tension. The anchorage value of the bend shall be taken to be 4ϕ (ϕ = diameter of the bar) for each 45° bend, subject to a maximum of 16ϕ. The anchorage value of the standard U-type hook shall be equal to 16ϕ.

***In compression*:** The anchorage length of a straight bar in compression shall be equal to the development length of bars in compression. The projected length of hooks, bends and straight lengths beyond the bends, if provided for bars in compression, shall be considered for development length.

***In shear*:** In case of inclined bars, the development length for bars in the tension zone is measured from the end of the sloping or inclined portion of the bar, whereas in the compression zone, it is measured from the mid-depth of the beams.

In case of stirrups in the beam and transverse ties in a column, complete development length and anchorage shall be deemed to have been provided when the bar is bent through an angle of at least 90° round another bar of at least its own diameter and is continued beyond the end of the curve for a length of at least eight diameters (8ϕ) or when the bar is bent through an angle of 135° and is continued beyond the end of the curve for a length of atleast 6ϕ or when the bar is bent through an angle of 180° and is continued beyond the end of the curve for a length of at least 4ϕ. The above provisions are illustrated in Fig. 4.11.

4.6 CURTAILMENT OF TENSION REINFORCEMENT

It is a common practice to cut off bars where they are no longer required to resist moment. In case of continuous beams, however, the tension reinforcement at the bottom may be bent up to provide some of the tension reinforcement required on the top face of the beam.

For curtailment, the reinforcement is extended beyond the point at which it is no longer required to resist flexure for a distance equal to the effective depth d of the member or 12 times the diameter of the bar (12ϕ), whichever is greater except at a simple support or cantilever, as illustrated in Fig. 4.12. In addition the following should also be satisfied:

1. The flexural steel, when terminated in the tension zone, a stirrup area in excess of that required for shear and torsion, is provided along each terminated bar over a distance from the cut-off point equal to three-fourth the effective depth of the member. The stirrup area shall not be less than $0.4bS_v/f_y$, where b is the width of the beam, S_v is the spacing of stirrups and f_y is the characteristic strength of reinforcement in MPa. The resulting spacing shall not exceed $d/(8\beta_b)$, where β_b is the ratio of the cut-off to total area of bars at the section.
2. In case of positive moment reinforcement, at least one-third of it in a simple member and one-fourth in a continuous member shall extend along the same face of the member by L_d beyond the theoretical cut-off point of curtailed bars into the

support to a length equal to $L_d/3$. In case of a frame member positive reinforcement extended into the support shall be anchored to develop design stress in tension at the face of the support.

3. At simple supports and at the points of inflection, positive moment (tension) steel shall be limited to a diameter such that development length $L_d (= \phi\sigma_{st}/4\tau_{bd})$ does not exceed $(M_1/V) + L_o$, where M_1 is the moment of resistance of the section assuming all reinforcement at the section to be stressed to σ_{st}, L_o is the sum of anchorage beyond the centre of support and the equivalent anchorage value of any hook or mechanical anchorage at the simple support; and at the point of inflection, L_o is limited to the effective depth of the member or 12ϕ, whichever is greater. The value of M_1/V may be increased by 30 per cent when the ends of the reinforcement are confined by a compressive reaction.
4. For negative moment reinforcement at the support the bars shall extend beyond the point of inflection to a distance not less than d or 12ϕ or one-sixteenth of the clear span, whichever is greater and also the bars must extend a full development length L_d beyond the face of the support. In addition, the curtailed bars shall extend by a distance d or 12ϕ, whichever is greater, beyond the theoretical cutoff point of the negative bending moment diagram. The remaining bars (at least one-third of the total reinforcement provided) must extend at least L_d beyond the theoretical cutoff point of curtailed bars, as illustrated in Fig. 4.13.

4.7 REINFORCEMENT SPLICING

As far as possible the splices provided shall be away from the section of maximum stress and be staggered. As per IS:456 recommendations, the splices in the flexural members should not be at the sections where bending moment is more than 50 per cent of the moment of resistance, and not more than half the bars shall be spliced at a section. When this is not avoidable, special precautions shall be taken such as increasing the length of the lap and/or using a spiral around the length of the splice. The different types of splices are as follows:

Lap splices: The lap length including anchorage value of hooks in flexural tension shall be L_d or 30ϕ, whichever is greater, and for direct tension $2L_d$ or 30ϕ, whichever is greater. The straight length of the lap shall not be less than 15ϕ or 200 mm. The lap length in compression shall be equal to L_d or 24ϕ, whichever is greater. When bars of two different diameters are to be spliced, the lap length shall be calculated on the basis of the diameter of the smaller bar.

The splices shall be considered staggered if the centre-to-centre distance of splices is more than the lap length. Lap splices are not used for bars larger than 36 mm ϕ. In such a case the bars may be welded.

Welded splices and mechanical connections: The design strength of the welded splices or mechanical connection shall be taken as equal to 80 per cent of the design strength of the bar for tension splices, and 100 per cent of the design strength for compression splices. However, 100 per cent of the design strength may be assumed in tension when the spliced area forms not more than 20 per cent of the total area of steel at the section and the splices are staggered at least 600 mm from each other.

The following examples will illustrate the procedure and the types of problems encountered in practice:

Example 4.6 A simply supported reinforced concrete beam of size 300 × 500 mm effective is reinforced with 4 bars of 16 mm ϕ HYSD steel of grade Fe415. Determine the anchorage length of the bars at the simply supported end if it is subjected to a factored shear force of 350 kN at the centre of 300 mm wide masonry support. The concrete mix of grade M20 is to be used.

Solution The cross-sectional dimensions and material properties are:

b = 300 mm, d = 500 mm, $A_{st}(4 \times 16 \text{ mm } \phi) = 804 \text{ mm}^2$

f_{ck} = 20 MPa and f_y = 415 MPa

$$\text{Development length } L_d = \frac{0.87 f_y \phi}{4\tau_{bd}}$$

where

τ_{bd} = Design anchorage bond stress

= 1.92 MPa (for M20 concrete and Fe415 grade steel)

Thus,

$$L_d = \frac{0.87 \times 415 \times 16}{4 \times 1.92} = 752.2 \text{ mm}$$

For anchorage bond stress not to exceed its design value, the following condition at the simply supported end of the beam should be satisfied:

$$L_d \le \left(\frac{1.3\, M_1}{V_u}\right) + L_o$$

where

M_1 = Ultimate moment of resistance of the section at the support

V_u = Factored shear force at the centre of support

L_o = Anchorage length

To compute the value of M_1, consider the section to be under-reinforced. The depth of the neutral axis of such a section is determined from

$$x_u = \frac{0.87 f_y A_{st}}{0.36 f_{ck} b}$$

$$= \frac{0.87 \times 415 \times 804}{0.36 \times 20 \times 300} = 134.39 \text{ mm}$$

The limiting depth of the neutral axis for a balanced section is given by

$$x_{u,\max} = \frac{0.0035d}{0.0055 + (0.87f_y)/E_s}$$

$$= \frac{0.0035 \times 500}{[0.0055 + (0.87 \times 415)/(2 \times 10^5)]} = 239.6 \text{ mm}$$

Since $x_u < x_{u,\max}$, it is an under-reinforced section,

$$\text{Lever arm } z = d - 0.416x_u$$

$$= 500 - 0.416 \times 134.39 = 444.09 \text{ mm}$$

Limiting moment of resistance is given by

$$M_1 = 0.87f_yA_{st}z$$

$$= 0.87 \times 415 \times 804 \times 444.09 \times 10^{-6} = 128.91 \text{ kNm}$$

Thus,

$$752.2 \le 1.3 \times \frac{128.91 \times 10^6}{350 \times 10^3} + L_o \le 478.80 + L_o$$

Therefore,

$$L_o \ge 273.4 \text{ mm}$$

The minimum length of the bar to be extended beyond the centre of support is:

$$= \frac{L_d}{3} - \frac{\text{Width of support}}{2}$$

$$= \frac{752.2}{3} - \frac{300}{2} = 100.73 \text{ mm} < L_o \ (= 273.4)$$

Hence, provide L_o = 273.4 mm (≈ 275 mm), as shown in Fig. 4.14.

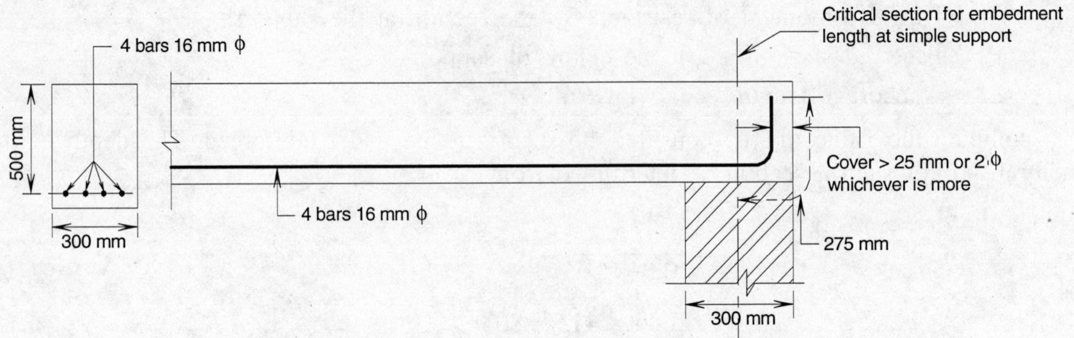

Fig. 4.14 Development length in the beam of Example 4.6.

Example 4.7 A reinforced concrete continuous beam with section of size 300 × 500 mm effective is reinforced with 4 bars of 16 mm ϕ. It is subjected to a factored shear force of 300 kN at the point of inflection. Check the adequacy of reinforcement at the point of inflection with regard to bond stress when concrete and steel used are of grades M20 and Fe415, respectively.

Solution The cross-sectional dimensions and material properties are:

$$b = 300 \text{ mm}, \; d = 500 \text{ mm}, \; A_{st}(4 \times 16 \text{ mm } \phi) = 804 \text{ mm}^2$$

$$f_{ck} = 20 \text{ MPa and } f_y = 415 \text{ MPa}$$

Design anchorage bond stress is:

$$\tau_{bd} = 1.92 \text{ MPa} \qquad \text{(for M20 concrete and Fe415 steel)}$$

Development length is given by

$$L_d = \frac{0.87 f_y \phi}{4\tau_{bd}}$$

$$= \frac{0.87 \times 415\phi}{4 \times 1.92} = 47.0\phi$$

From Example 4.6,

$$M_1 = 128.91 \text{ kNm}$$

Anchorage length is:

$$L_o = d \text{ or } 12\phi \text{ whichever is greater}$$

$$= 500 \text{ mm or } 12 \times 16 \; (= 192 \text{ mm}) = 500 \text{ mm}$$

For anchorage bond stress not to exceed the design value, the following condition at the point of inflection should be satisfied:

$$L_d \leq \left(\frac{M_1}{V_u}\right) + L_o$$

or

$$47\phi = \frac{128.91 \times 10^6}{300 \times 10^3} + 500$$

$$\phi \leq 19.78 \text{ mm}$$

Hence the bar diameter is satisfactory to ensure the bond stress at point inflection to be with in its permissible value.

Review Questions

4.1 The critical section for shear in a beam is located at a distance of effective depth d away from the face of support. Explain the circumstances when this is not permitted.

4.2 Why is the shear strength of concrete depend upon the percentage of steel?

4.3 What is the effect of axial force on the shear strength of concrete?

4.4 Why is it mandatory to provide minimum stirrup reinforcement (even if not required theoretically) in reinforced concrete beams?

4.5 Why is the maximum value of shear strength of concrete limited in a reinforced concrete beam with shear reinforcement?

4.6 Sketch the shape of stirrups to be used for *L*- and *T*- beams.

4.7 Why is it necessary to provide stirrups to take advantage of the shear resistance of bent-up bars?

4.8 What are the mechanisms by which bond resistance is mobilized in reinforced concrete?

4.9 Explain the difference between flexural bond and development bond.

4.10 Define development length and explain its significance.

4.11 Discuss the main factors that influence the bond strength.

4.12 Is the bond resistance different in tension and compression?

4.13 Describe the situations where check on development bond is necessary.

4.14 How is the requirement of development length of bars reduced in tension?

4.15 What is the criterion for deciding minimum turning radius in a bend of a reinforcing bar?

4.16 What is the purpose of splicing of reinforcement? What are the different types of splices used in flexural members?

4.17 Why is welded splice generally preferred to a lap splice?

4.18 How is the location of splice determined in the tension reinforcement in a flexural member?

4.19 Explain the terms: (a) bond and anchorage, (b) development length, lap length and anchorage length.

4.20 What are high bond bars? Why is it necessary to provide projections on bars made from Fe415 and Fe500 steels whereas such projections are not mandatory for Fe250 steel?

4.21 Explain the terms average bond stress and local bond stress. Derive expressions for same.

4.22 Explain why the local bond not checked for high bond bars.

4.23 What is meant by full development length? What are the approximate values for tension and compression steels in terms of the diameter of the bar?

4.24 If much more steel than required theoretically has been provided, can lesser development length for the bars be provided than the theoretical value?

4.25 Why is the bond stress in compression bars assumed to be more than that in tension bars?

4.26 Derive expression for the theoretical development length of: (a) tension bars and (b) compression bars?

4.27 What is meant by equivalent length of hooks and bends? Can hooks and bends be used to increase the theoretical development length of: (a) tension bars and (b) compression bars?

4.28 What are the specifications regarding development length of bundled bars? What is meant by bundling of bars? What is the maximum number of bars that can be bundled?

4.29 Up to what diameter of steel bars are allowed to be extended by splicing?

4.30 Explain the difference between primary and secondary torsion. Give two examples of each.

4.31 What are the stresses produced by torsion?

4.32 How is steel detailed for torsional moment in: (a) rectangular section and (b) T-section?

4.33 Explain IS:456 method of design of reinforced concrete members subjected to torsional moment.

4.34 What is equivalent shear as specified in IS:456 for members subjected to torsion and shear?

4.35 Explain the term torsional stiffness of a beam.

Tutorial Problems

T.4.1 (a) Explain the terms shear stress, diagonal tension, bond stress and development length with reference to reinforced concrete beams.

(b) Discuss the utility of bent-up bars in reinforced concrete beams in resisting the shear.

(c) Derive the expressions for computation of bond stress and shear stress in case of reinforced concrete beams of rectangular cross-section with tension reinforcement of diameter ϕ. Also obtain the relationship between shear stress and bond stress.

T.4.2 A reinforced concrete beam of cross-section of size 300 × 550 mm effective carries a uniformly distributed service load of 40 kN/m inclusive of self-weight over a clear span of 5 m. If the tensions reinforcement consists of 6 bars of 20 mm ϕ, determine the shear reinforcement. The materials used are M20 grade concrete and mild steel reinforcement.

T.4.3 A simply supported reinforced concrete beam of rectangular cross-section of size 300 × 700 mm overall carries a concentrated load of 125 kN at 1.0 m from one support, in addition to a uniformly distributed load of 20 kN/m over a clear span of 6 m. The section is reinforced with 6–20 mm ϕ bars of HYSD steel of grade Fe415. The concrete used is of grade M20. Design shear reinforcement when:
(a) only vertical stirrups are provided
(b) two of the six bars are bent up at 45° at the same cross-section
(c) four bars are bent up at 45° in two groups, each consisting of 2 bars. The two groups are separated by 500 mm.

T.4.4 A 125 mm thick simply supported one-way slab, reinforced with 12 mm ϕ bars @ 100 mm c/c at nominal cover of 15 mm, carries a uniformly distributed service load of 10 kN/m^2 inclusive of its self-weight. The materials used are M20 grade concrete and Fe415 grade steel. Examine shear stress, if the span of the slab is 4 m.

T.4.5 A reinforced concrete cantilever beam projects 1.2 m from the face of the support with a bearing on the support of 1.50 m. The tension reinforcement consists of 3 bars of 20 mm ϕ provided straight on the top face of the beam. Check the adequacy of the anchorage. If inadequate, suggest the maximum size of the bar that can be used as reinforcement. The materials used are M20 grade concrete and Fe415 grade steel. The side cover is 50 mm.

T.4.6 A reinforced concrete beam of rectangular cross-section of size 300 × 600 mm effective is provided with 6 bars of 20 mm ϕ on its tension face. Two of the bars on the tension face are bent up at the support for shear. Determine the development length required for the bars from the face of the supporting square column of size 300 × 300 mm in tension and compression zones. The materials used are M20 grade concrete and mild steel of grade Fe250.

T.4.7 A 120 mm thick reinforced concrete cantilever slab projects 1.10 m beyond the face of lintel of cross-section 225 × 200 mm. The slab is to carry a uniformly distributed service load of 7 kN/m^2. It is reinforced with 10 mm ϕ bars @ 250 mm c/c. Determine the development length required from the face of lintel. The materials used are M20 grade concrete and HYSD steel of grade Fe500.

T.4.8 A simply supported reinforced concrete beam of clear span of 7.5 m has a cross-section of size 200 × 500 mm effective depth. It is reinforced with 5 bars of 16 mm ϕ on the tension side. The beam carries a uniformly distributed service load of 20 kN/m including its own weight. Determine: (a) the length over which vertical stirrups are to be designed and their spacing and (b) spacing of stirrups when 3 bars are bent up at the same section.

T.4.9 A reinforced concrete rectangular beam section of size 300 × 1000 mm effective depth is reinforced with six mild steel bars of 22 mm ϕ. The beam carries a factored load of 67.5 kN/m over a clear span of 7 m. Design the shear reinforcement consisting of mild steel bar stirrups if the grade of concrete mix used is M20.

T.4.10 Design the transverse or shear reinforcement for a reinforced concrete L-shaped beam with a flange width of 600 mm, a flange thickness of 120 mm and a web width of 250 mm. The beam is reinforced with 8 bars of 25 mm ϕ at an effective depth of 400 mm. The beam supports a uniformly distributed factored load of 30 kN/m over a simply supported clear span of 9 m. The concrete used is of grade M20 and steel is of grade Fe250.

(**Hint:** Design it as a rectangular section of width equal to the width of the web).

(***Ans.*** Provide 6 mm ϕ 2-legged stirrups @ 120 mm c/c at a distance of 3 m from the face of support at each end and in the remaining portion provide stirrups at 300 mm c/c).

T.4.11 A simply supported reinforced concrete beam of cross-section 300 × 750 mm overall has to carry a uniformly distributed factored load of 137.50 kN/m over a clear span of 8 m. The beam is reinforced with 4 bars of 25 mm ϕ in tension at an effective cover of 50 mm throughout the beam. Design the shear reinforcement, if the grades of concrete mix and steel are M20 and Fe415, respectively.

(***Ans.*** Provide 12 mm ϕ 2-legged stirrups @ 135 mm c/c, gradually increasing to 300 mm c/c at 3.06 m from the support at each end. In the rest of the beam provide the stirrups @ 300 mm c/c).

T.4.12 A reinforced concrete beam section of size 300 × 400 mm effective is subjected to a factored torsion of 6 kNm in addition to a factored moment and a factored shear force of 40 kNm and 80 kN, respectively. Design the torsion reinforcement, if the grades of concrete and steel are M20 and Fe415, respectively.

(***Ans.*** Provide 4 bars of 12 mm ϕ as tension steel at an effective cover of 50 mm with 8 mm ϕ 2-legged stirrups at 160 mm centre-to-centre).

T.4.13 A reinforced concrete rectangular beam of cross-section of size 250 × 500 mm effective is subjected to a factored moment of 50 kN.m, factored shear force of 50 kN and a factored torsional moment of 20 kNm. Design the reinforcement to be provided in the beam, if the concrete mix of grade M20 and HYSD steel of grade Fe415 are to be used in the construction. The effective cover to tension reinforcement may be taken as 50 mm.

T.4.14 A simply supported reinforced concrete beam of rectangular section of size 250 × 500 mm effective is reinforced with 3 bars of 20 mm ϕ on the tension side. Determine the anchorage length of bars at the simply supported end of the beam when it is subjected to a factored shear force of 300 kN at the centre of a 345 mm wide support. The concrete of grade M20 and mild steel of grade Fe250 are used in construction.

T.4.15 A continuous reinforced concrete beam of width 200 mm and effective depth of 300 mm is reinforced with 3 bars of 12 mm ϕ on the tension side. Check the bond stress when the beam is subjected to a factored shear force of 80 kN at the point of inflection. The grade of concrete is M20 and that of steel is Fe415.

T.4.16 A simply supported reinforced concrete beam of rectangular cross-section of size 300 × 600 mm effective is reinforced with 6 bars of 20 mm ϕ. Determine the shear reinforcement required to resist a uniformly distributed service load of intensity: (a) 20 kN/m, (b) 35 kN/m, (c) 75 kN/m, over a clear span of 6 m. The materials used are M20 grade concrete and Fe415 grade HYSD steel. The stirrups are of mild steel.

T.4.17 The reinforced concrete cantilever beam of span 3.5 m has a section 300 mm wide and 550 mm deep (effective) at the face of the support. The depth is reduced uniformly to 250 mm (effective) at the free end. The beam is reinforced on the tension face with 6 bars of 20 mm ϕ at the support. Two of these tension bars are curtailed at a distance of 1.25 m from the support. Design the shear reinforcement for the beam when it carries a uniformly distributed service load of 60 kN/m. The materials used are M20 grade concrete and HYSD steel of grade Fe415.

T.4.18 A simply supported reinforced concrete rectangular beam of size 250 × 500 mm effective is reinforced with six bars of 20 mm ϕ at the bottom and two bars of 16 mm ϕ at the top of the beam. The beam carries uniformly distributed service load of 55 kN/m (inclusive of self-weight) over an effective span of 6 m. The beam rests on 300 mm wide supports. If the materials used are M20 grade concrete and HYSD steel of grade Fe415, determine the development lengths required for the bars on tension and compression sides, from the face of the support and also at mid-span.

T.4.19 A cantilever slab projecting 1.6 m beyond the face of the beam of rectangular cross-section of size 250 × 500 mm effective carries a uniformly distributed superimposed service load of 3,875 kN/m^2. The slab has a thickness of 125 mm with 10 mm ϕ mild steel bars @ 100 mm c/c at a nominal cover of 15 mm. Determine the development length required for steel from the face of the support. The concrete used is of grade M20. The unit weight of concrete is 25 kN/m^3.

T.4.20 A reinforced concrete beam of rectangular cross-section of size 300 × 600 mm effective depth is reinforced with five bars of 20 mm ϕ on the tension side. The beam carries a uniformly distributed service load of 12 kN/m over an effective span of 8 m. From the considerations of flexure, anchorage and bond, determine the distance on the span where (a) 2 bars can be curtailed and (b) 3 bars can be bent up. The materials used are M20 grade concrete and HYSD steel reinforcement of grade Fe415.

CHAPTER

5

Limit State of Collapse—Compression

5.1 INTRODUCTION

A structural element subjected predominantly to compressive force with or without bending moments is termed as **compression member**. When a compression member is vertical, it is called a **column** and, when inclined or horizontal, it is termed as **strut**. The vertical members of a multi-storey building that transmit the loads on floors and beams to the foundation, are the columns of the building. These are important elements in the sense that their failure may endanger the whole structure.

5.2 CLASSIFICATION OF COLUMNS

Depending upon the architectural requirements, the columns may have cross-sections of regular shapes such as rectangular, square, hexagonal, octagonal or circular. The cross-shaped (swastika), T-shaped, and L-shaped columns have been extensively used. Typical cross-sections are shown in Fig. 5.1(b).

5.2.1 Classification According to Transverse Reinforcement

A concrete column is reinforced with longitudinal bars (also called **main reinforcement**) held in position by separate closed loops called **ties** spaced at equal close intervals along the length, as shown in Fig. 5.2(a). Such a column is called a **tied column**. The longitudinal bars contribute, to the load-carrying capacity of the section and the transverse ties provide lateral

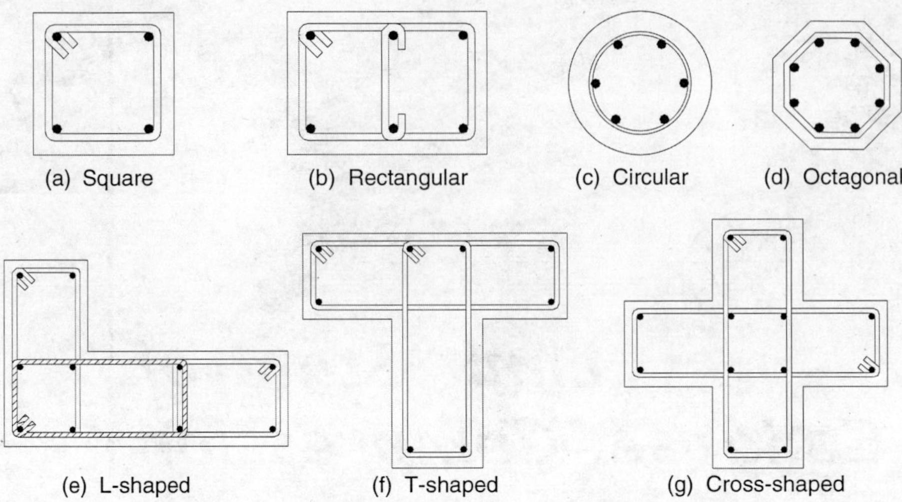

Fig. 5.1 Typical cross-sections of columns.

support to the longitudinal bars and also confines the concrete. The other function of the lateral ties is to prevent the buckling of main longitudinal bars. Sometimes, in a circular column, the longitudinal bars are wrapped by a closely spaced wire or bar helix called **spiral**. Such a column is called a **spiral column** as shown in Fig. 5.2(b). In some situations the columns may have either embedded rolled steel sections or in-filled cast iron or steel pipes with both longitudinal and transverse reinforcements, as illustrated in Fig. 5.2(c). The former is known as a **composite column** and the latter as an **in-filled column**. This chapter deals with the analysis and design of tied and spiral columns.

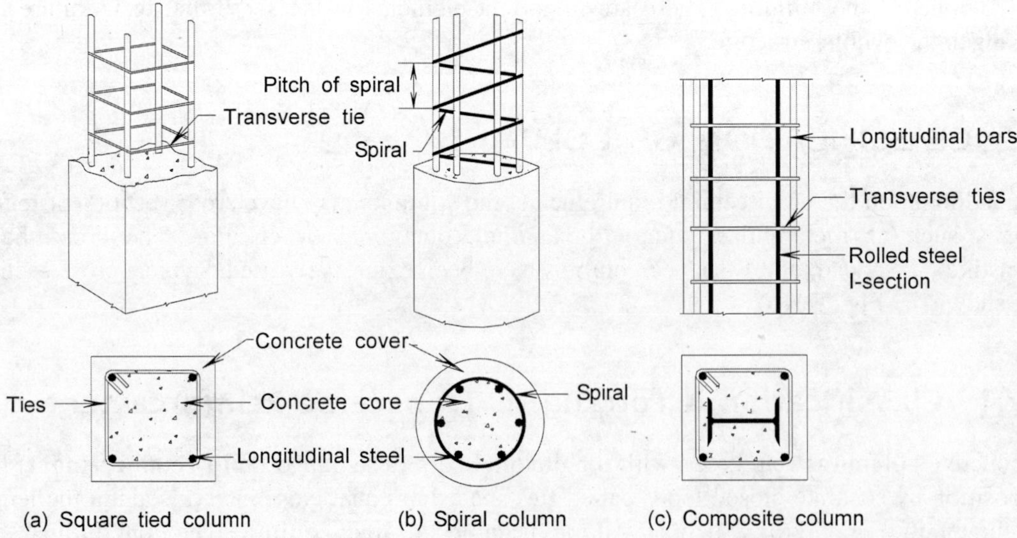

Fig. 5.2 Types of columns based on the form and types of reinforcement.

5.2.2 Classification According to Dimensions and Support Conditions

A column is classified as either a pedestal, short or slender column depending upon its length, lateral dimensions and support conditions. The support conditions affect the effective length of the column which is defined as the distance between the points of inflection of the deformed shape of the column under the axial loads.

A very short column having the ratio of effective length (height l_{eff}) to least lateral dimension (side d) as less than three is called **pedestal** or **stub column**. It is generally of plain concrete, i.e., un-reinforced. A column of intermediate height ($3 < l_{eff}/d \leq 12$) is called **short column**. The load-carrying capacity of a short column is governed by the strength of the constituent materials.

On the other hand the column which is relatively long with respect to its least lateral dimension d is termed as **long or slender column** ($l_{eff}/d > 12$). A long column is susceptible to buckling, i.e., its load-carrying capacity is influenced by the slenderness effect which produces additional moments resulting from transverse deformation. The slenderness ratio l_{eff}/d, in any case should be restricted to 60. As the columns are under primary compression, the failure is usually sudden with a warning given through the spalling of the cover.

5.2.3 Classification According to the Type of Loading

The column shown in Fig. 5.3(a), which carries a purely axial load, is termed as **concentrically loaded column**. Such an ideal column is rarely encountered in practice. The columns in industrial buildings are not only subjected to high compression but also to reasonably large bending and shear forces.

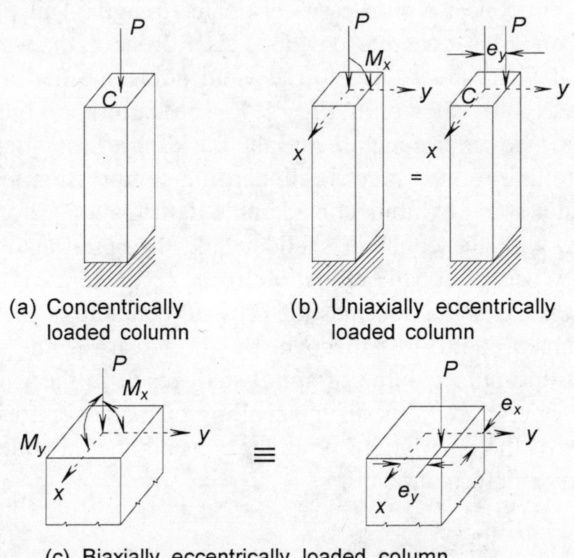

Fig. 5.3 Classification of columns based on the position of the load on the cross-section.

If a column carries an axial load and bending moment about either the x- or y-axis only, it is classified as a uniaxially eccentrically loaded column, as shown in Fig. 5.3(b). The peripheral columns located on the sides of a building are of this category.

A column subjected to an axial load along with moments about both the axes is termed as **biaxial eccentrically loaded column**, e.g. corner columns of a building carry axial loads along with moments about the x- and y-axes, as illustrated in Fig. 5.3(c).

5.2.4 Braced and Unbraced Columns

A braced column is one in which the side sway is insignificant, i.e., there is no significant relative displacement between the two ends of the column in the direction under consideration. Such columns do not resist any significant horizontal load like wind or earthquake loads. Columns of water towers with bracings at different heights and columns in tall buildings provided with shear walls are examples of such columns. On the other hand, columns resisting lateral loads in addition to vertical loads are considered unbraced. An unbraced or partially braced column restrained from rotation is subjected to sidesway or lateral drift i.e., there is significant lateral displacement between the ends.

5.3 BEHAVIOUR OF REINFORCED CONCRETE COLUMNS

When a column is subjected to an axial load within elastic limits, just like any other composite section, the stresses induced in steel and concrete are in proportion to their moduli of elasticity, E_s and E_c, respectively. However, the subsequent creep and shrinkage result in an increase in stress on steel and a decrease in stress on concrete. With an increase in load, the steel will attain its yield strength before the concrete reaches its ultimate strength. Thus, the column will carry a small additional load until the concrete develops its full strength, while the stress in steel remains the same (i.e., at yield stress). Up to the yield point, the tied and spirally reinforced columns behave identically, as shown in Fig. 5.4. The failure of the tied column occurs suddenly with the breaking down of concrete and the buckling of longitudinal bars between the ties in a pattern similar to that for a concrete cylinder in a compression test. On the other hand, a column reinforced with a spiral exhibits considerable deformation before complete failure on reaching the yield point, with the concrete shell outside the spiral spalling off. This reduces the load-carrying capacity because of the reduction of the concrete area, but the spiral prevents buckling of the longitudinal bars and confines the crushed concrete in the core. Thus, the spiral may offset the loss sustained due to loss of cover by an increase in the load-carrying capacity of the concrete core. An optimum volume of spiral shall result in the value of the failure load to be equal to the load carried at the time of the spalling of the cover concrete. Thus, the spiral adds little to the strength of the column but provides considerable ductility until the spiral steel yields and undergoes large deformations.

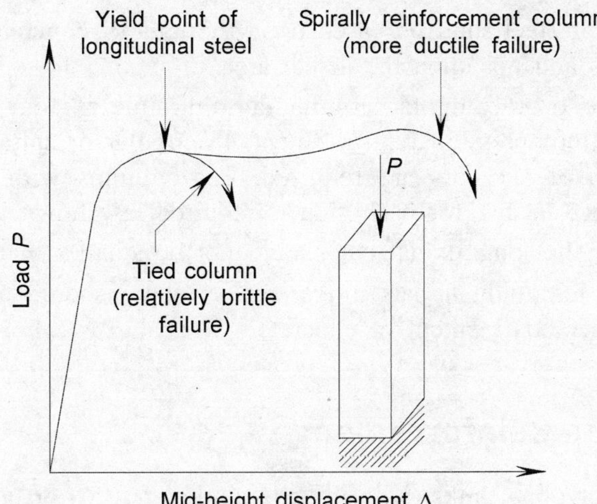

Fig. 5.4 Load-deformation behaviour of a column.

5.4 GENERAL DESIGN PRINCIPLES

The design of a column consists of proportioning the column section, along with the determination of the areas of longitudinal steel, and transverse steel in the form of ties, and their placement. The transverse steel is in the form of either circular rings or polygonal links (called lateral ties) with the internal angle not exceeding 135°. Two important considerations in the design of a reinforced concrete column are the economy and the limitations on the size of the column due to architectural and other considerations. The economy requires that a rich concrete mix be used and the quantity of compression steel is minimum. A limit on the size of a column may necessitate a provision of a higher amount of compression steel. Since the concrete is weak in tension, a minimum amount of steel is always required to resist any unvisualized flexure. IS:456 have recommended a minimum reinforcement of 0.8 per cent of the gross cross-sectional area of concrete. Longitudinal bars are usually located close to the periphery (for better flexural resistance), but may be placed in the interior of the column when eccentricities in loading are minimal.

5.4.1 Longitudinal or Main Reinforcement

IS:456 stipulate that the main reinforcement shall satisfy the following requirements:

1. The minimum reinforcement shall not be less than 0.8 per cent of the gross cross-sectional area and the maximum reinforcement should not exceed 6 per cent of the gross cross-sectional area. However, in practice, the maximum amount of steel is restricted to 4 per cent of the gross cross-sectional area provided. In any column that has a larger cross-sectional area than that required to support the load, the minimum

percentage of steel shall be based upon the area of concrete required to resist the direct stress and not upon the actual area.

2. In a square or rectangular column a minimum of four bars are provided as longitudinal compression reinforcement. In circular or spirally reinforced columns a minimum of six bars are required. For columns with five or more sides a minimum of one bar for each side is required, as shown in Fig. 5.1(d).
3. The size of the longitudinal bar should not be smaller than 12 mm diameter.
4. Spacing of longitudinal bars along the periphery should not exceed 300 mm to ensure proper confinement of concrete.

5.4.2 Transverse Reinforcement

The diameter of the polygonal links (with internal angles < 135°) or lateral ties should not be less than one-fourth of the diameter of the largest longitudinal bar, and in no case less than 6 mm. The diameter of the tie bar should not be more than 20 mm.

The pitch or spacing of lateral ties should not be more than the least of the following distances:

1. The least lateral dimension of the compression member
2. Sixteen times the smallest diameter of longitudinal reinforcement to be tied, and
3. 300 mm

5.4.3 Helical Reinforcement

The spiral shall be of a regular formation with the turns of helix spaced evenly and anchored properly. The pitch of the spiral in the spirally reinforced column shall neither be more than the smaller of the distances: (i) 75 mm and (ii) one-sixth of the core diameter of the column nor be less than the larger of the: (a) 25 mm and (b) three times the diameter of the steel bar forming the helix. The diameter of the helix bar shall be based on the criterion used for the ties.

5.4.4 Cover

The requirements of the cover are important from durability considerations. The IS:456 code requirements are:

1. For a longitudinal reinforcing bar in a column the nominal or clear cover should be more than the larger of the following:
   (i) 40 mm and
   (ii) Diameter of the longitudinal bar.

For a column of minimum dimension less than 200 mm, where the diameter of the reinforcing bar does not exceed 12 mm, a cover of 25 mm may be provided.

2. For the hoops and ties the nominal or clear cover shall not be less than the diameter of the bar used in making the hoops or ties, nor less than 20 mm for mild exposure.

5.5 EFFECTIVE LENGTH

The effective height or length l_{eff} of a column is generally defined as the length which corresponds to the length of a pin-jointed column that carry the same axial load as the given column, i.e., it is the length between the points of contraflexure, real or imaginary of a buckled column. The effective length depends upon bracings and the degree of restraint at the ends. On the other hand, the unsupported height or length of a column, L is generally taken as the clear height of column. For a braced column effective column height is less than the clear height and vary generally between $0.5L$ to L, whereas, for unbraced and partially braced columns the effective height is larger than the clear height and vary from L to a very large value.

The effective length l_{eff} of a column or compression member in a given plane is generally expressed as:

$$l_{eff} = \beta L$$

where β is the effective length coefficient and L the unsupported length. IS:456 has recommended the following values of effective length coefficient, β for an isolated column:

1. *Column braced against sidesway:*

(a) Both end effectively held in position and restrained against rotation, $\beta = 0.65$.

(b) Both ends effectively held in position and one of the ends restrained against rotation, $\beta = 0.80$.

(c) Both ends effectively held in position, but not restrained against rotation, $\beta = 1.0$.

2. *Column not braced against sidesway:*

(a) One end effectively held in position and restrained against rotation, while the other end is effectively restrained against rotation, $\beta = 1.20$.

(b) One end effectively held in position and restrained against rotation, while the other end is partially restrained, $\beta = 1.50$.

(c) One end effectively held in position and restrained against rotation, while the other end is free, $\beta = 2.00$.

BS:8110 has recommended the values of effective length coefficients for framed columns as given in Table 5.1. The coefficients depend on the degree of restraints at the ends of column expressed in terms of combinations of two basic end conditions: (a) effectively held or not held in position and (b) effectively restrained or not restrained against rotation. The end conditions may be different in two plan directions. For a stable compression member combination of these basic conditions of end restraints simulate different column boundary conditions:

1. ***Fixed end condition***: This condition is considered to exist at an end of the column which is connected monolithically to beams on either side which are atleast as deep as the overall dimension of column in the plane under consideration.

2. ***Partially-fixed end condition***: This condition includes the end of a column which is connected monolithically to beams or slabs on either side which are shallower than the overall dimension of column in the plane under consideration.

3. ***Hinged end condition***: This condition includes the end of a column which is connected to members that are not specifically designed to provide restraint to the rotation of column; however, they still provide nominal restraint.

4. ***Free end condition***: This includes free end of a cantilever column in an unbraced member, i.e., the end of column not restrained against both lateral movement and rotation.

TABLE 5.1 Effective Length Coefficients for Braced and Unbraced Columns

End condition at the top	*End condition at the bottom*					
	Braced columns			*Unbraced column*		
	1	*2*	*3*	*1*	*2*	*3*
1	0.75	0.80	0.90	1.20	1.30	1.60
2	0.80	0.85	0.95	1.30	1.50	1.80
3	0.90	0.95	1.00	1.60	1.80	—
4	—	—	—	2.20	—	—

5.5.1 Columns in Building Frames

Effective height of a column of a framed structure depends on relative stiffness of the column and the beams framing in to the column at its two ends. BS:8110 has recommended that the effective length coefficient β should be taken to be the lower value obtained from the following equations for:

1. *Braced column*

$$\beta = [0.70 + 0.05\ (\alpha_1 + \alpha_2)] < 1.0$$

$$\beta = [0.85 + 0.05\alpha_{min}\ ] < 1.0$$

2. *Unbraced column*

$$\beta = 1.00 + 0.15\ (\alpha_1 + \alpha_2)$$

$$\beta = 2.00 + 0.30\alpha_{min}$$

where

$$\alpha_1 = \frac{\sum k_c}{\sum k_b} = \frac{\sum\left(\frac{I_c}{h_c}\right)}{\sum\left(\frac{I_b}{l_b}\right)_{\text{lower}}}$$

$$\alpha_2 = \frac{\sum k_c}{\sum k_b} = \frac{\sum\left(\frac{I_c}{h_c}\right)}{\sum\left(\frac{I_b}{l_b}\right)_{\text{upper}}}$$

$$\alpha_{\min} = \text{Lesser of } \alpha_1 \text{ and } \alpha_2$$

Special conditions: For a fixed end condition, α = 0.0 and for hinged end or nominal moment condition, α =10.0 may be considered. For example,

1. For a foundation designed to resist moment, $k_f = k_c$, or width and depth > $4D$, α = 1.0.
2. When column and base are designed to resist only nominal moment, α =10.0.
3. For the column end which meets the beams designed as simply supported, α = 10.0.

Example 5.1 An interior reinforced concrete column of size 330 × 460 mm is connected with 330 × 660 mm deep beams from either side. The storey height is 4.25 m and spans of beams are 5.5 and 7.5 m on either side. Determine the effective height coefficient of the column, if (a) column is braced and its base is designed to resist moment, and (b) column is unbraced and its base is designed for nominal moment.

Solution The cross-sectional properties are:

Column: L = 4250 mm, $b \times D$ = 330 × 460 mm, and $I_c = 330 \times \dfrac{460^3}{12} = 2.68 \times 10^9$ mm^4

Beam-1: L = 5500 mm, $b \times D$ = 330 × 660 mm, and $I_{b1} = 330 \times \dfrac{660^3}{12} = 7.91 \times 10^9$ mm^4

Beam-2: L = 7500 mm, $b \times D$ = 330 × 660 mm, and $I_{b2} = 330 \times \dfrac{660^3}{12} = 7.91 \times 10^9$ mm^4

Stiffnesses are given by

$$k_c = \frac{I_c}{L_c} = \frac{2.68 \times 10^9}{4250} = 6.31 \times 10^5 \text{ mm}^3$$

$$k_{b1} = \frac{I_{b1}}{L_{b1}} = \frac{7.91 \times 10^9}{5500} = 14.38 \times 10^5 \text{ mm}^3$$

$$k_{b2} = \frac{I_{b2}}{L_{b2}} = \frac{7.91 \times 10^9}{7500} = 10.55 \times 10^5 \text{ mm}^3$$

CASE I ***Braced column:***
At (braced) bottom,

$$\alpha_1 = 1.0$$

and at top

$$\alpha_2 = \frac{\sum k_c}{\sum k_b}$$

$$= \frac{6.31 + 6.31}{14.38 + 10.55} = 0.506$$

Then

$$\beta = [0.70 + 0.05(\alpha_1 + \alpha_2)]$$
$$= 0.70 + 0.05 \times (1.0 + 0.506) = 0.7753 < 1.0$$

and

$$\beta = [0.85 + 0.05\alpha_{min}]$$
$$= 0.85 + 0.05 \times 0.506 = 0.8753 < 1.0$$

Therefore,

$$\beta = 0.78$$

CASE II ***Unbraced column:***
At hinged bottom,

$$\alpha_1 = 10.0$$

and at top,

$$\alpha_2 = 0.506$$

Then

$$\beta = 1.00 + 0.15(\alpha_1 + \alpha_2)$$
$$= 1.00 + 0.15 \times (10.0 + 0.506) = 2.576$$

and

$$\beta = 2.00 + 0.30\alpha_{min}$$
$$= 2.00 + 0.30 \times 0.506 = 2.152$$

Therefore,

$$\beta = 2.15$$

5.6 ASSUMPTIONS

In addition to the assumptions made for the limit states design of flexural members in Chapter 3, the following assumptions are made for the members in compression:

1. The maximum compressive strain in concrete in axial compression (i.e., a member under concentric load) is taken as 0,002.

2. The maximum strain at the highly compressed edge of the concrete section subjected to axial load and flexure, when there is no tension on the section, shall be taken as 0.0035 minus 0.75 times the strain on the least compressed edge of the section. Thus, if ε_{cb} is the strain in the least compressed edge of the section, the strain at the most compressed edge will be $0.0035 - 0.75\varepsilon_{cb}$.

The later assumption is based on the consideration that the peak stress occurs at a strain of about 0.002 and visible crack does not occur until the strain attains a value of about 0.0035. For the entire section in compression under axial load and moment, the transition of strain from 0.002 for the pure axial load condition to 0.0035 for pure bending condition governs the failure of the column section. The strain distribution line passes through the point of intersection O of strain distribution lines of the two extreme conditions; first of uniform strain of 0.002 for axial load case, and the second having a strain at the least compressed edge as zero and a strain at a highly compressed edge as 0.0035 with the neutral axis lying at the least compressed edge. The strain distribution lines for these two cases intersect each other at a depth of $3D/7(= 0.429D)$ from the highly compressed edge. This is illustrated in Fig. 5.10(b), hence the stipulated maximum strain occurs at the most compressed edge of the section, which is given by

$$\varepsilon_c = 0.002\left[1 + \frac{3D/7}{x_u - 3D/7}\right]$$

where $x_u \geq D$.

5.7 DESIGN OF AXIALLY LOADED SHORT COLUMN (with negligible eccentricity, i.e., $e = 0$ to e_{min})

A compression member shall be considered as a short column when slenderness ratios (l_{ey}/D) and (l_{ex}/b) are less than or equal to 12, where l_{ex} and l_{ey} are effective lengths in respect of major and minor axes, respectively; D is the depth in respect of the major axis and b is the width of the member.

The members carrying bending moments which are quite small as compared to the direct compressive load are termed as **axially loaded columns.** The limiting load-carrying capacity of a column with lateral ties subjected to a concentric load is attained when it develops a limiting strain of 0.002 resulting in uniform stress of $0.447f_{ck}$ in concrete. The stresses induced in steel are obtained from the stress-strain curve corresponding in strain $\varepsilon_{cu} = 0.002$ and are:

1. For mild steel, $f_s = 0.87f_y$
2. For high yield strength deformed bars, $f_s = 0.75f_y$ (5.1)

Therefore, the limiting load-carrying capacity of the section is calculated as follows:

With mild reinforcement:

$$P_u = 0.447f_{ck}A_c + 0.87f_yA_s = 0.447f_{ck}(A_g - A_s) + 0.87f_yA_s \tag{5.2}$$

With high strength deformed bars as reinforcement:

$$P_u = 0.447 f_{ck} A_c + 0.75 f_y A_s = 0.447 f_{ck}(A_g - A_s) + 0.75 f_y A_s \tag{5.3}$$

$$\frac{P_u}{f_{ck} A_g} = \frac{P_u}{f_{ck} bD} = 0.447(1 - 0.01p) + 0.75\left(\frac{p}{100}\right)\left(\frac{f_y}{f_{ck}}\right)$$

where

A_c = Area of concrete, excluding that of steel

$A_g = bD$, Gross area of concrete section

$A_s = (p/100)A_g$, Area of longitudinal steel

p = Percentage of steel

In practical problems, it is very rare that a column is subjected to a truly concentric load. Therefore, IS:456 stipulates that all compression members shall be designed for a minimum eccentricity of load in two principal directions. The minimum eccentricity may arise due to support or load conditions or tolerances. The code specifies a minimum eccentricity $e_{\min}$ for the design of column as:

$$e_{\min} = \frac{L}{500} + \frac{D}{30} \text{ subject to a minimum of 20 mm} \tag{5.4}$$

where

L = Unsupported length of the column

D = Lateral dimension of the column in the direction under consideration

When the calculated eccentricity due to applied loads is larger, the minimum eccentricity should be ignored. When the eccentricity does not exceed 0.05 times the lateral dimension, the axial load-carrying capacity is reduced by 11 per cent, and is given by

$$\begin{aligned} P_u &= 0.40 f_{ck} A_c + 0.67 f_y A_s = 0.40 f_{ck}(A_g - A_s) + 0.67 f_y A_s \\ &= 0.40 f_{ck}(1 - 0.01p)A_g + 0.67 f_y (0.01p) A_g \end{aligned} \tag{5.5}$$

The short column formula given by Eq. (5.5) takes into account the accidental eccentricity to the extent of $e = 0.05D$. However, if accidental eccentricity given by the expression, $e_{\min} = (L/500) + (D/30)$ or 20 mm (greater of the two) exceeds $0.05D$, the short column formula shall not be applied for design of such columns. The short column formula is used for determining the load capacity as well as for the design of column sections subjected to axial loads.

5.8 PROCEDURE FOR DESIGN

The following procedure may be adopted for the design of a reinforced concrete column:

1. Assume some suitable percentage p of longitudinal reinforcement A_s (normally between 1 and 2) in terms of gross cross-sectional area A_g of the column,

$$A_s = \frac{pA_g}{100} = 0.01 p A_g$$

2. Apply the load-carrying capacity formula:

$$P_u = 0.40f_{ck}A_g + (0.67f_y - 0.40f_{ck})A_s$$
$$= 0.40f_{ck}A_g + (0.67f_y - 0.40f_{ck})(0.01pA_g) \qquad (5.6)$$

and calculate the gross area A_g of concrete. Equation (5.6) can be expressed in different forms:

$$A_s = \frac{(P_u - 0.40\, f_{ck}A_g)}{(0.67\,f_y - 0.40\,f_{ck})} \qquad (5.7a)$$

$$A_g = \frac{P_u}{0.40\,f_{ck} + (0.67\,f_y - 0.40\,f_{ck})(0.01p)} \qquad (5.7b)$$

$$\left(\frac{p}{100}\right) = \left[\left(\frac{P_u}{A_g}\right) - 0.40\,f_{ck}\right]\left[\frac{1}{(0.67\,f_y - 0.40\,f_{ck})}\right] \qquad (5.7c)$$

3. Knowing A_g, proportion the dimensions of the column of the desired shape (square, rectangular, circular, etc.).
4. Check for the minimum eccentricity of load in two principal directions.
5. The area of longitudinal steel required can be obtained as:

$$A_s = \frac{pA_g}{100}$$

Choose the diameter and determine the number of bars to be adopted and distribute the bars suitably around the periphery with an appropriate cover.
6. Design the transverse steel.

The following examples illustrate the procedure and types of problems encountered in practice.

Example 5.2 A short reinforced concrete column of size 250 × 300 mm is reinforced with 4 bars of 20 mm φ. Determine the safe load that the column can carry. The grades of concrete mix and mild steel reinforcement used are M20 and Fe250, respectively.

Solution The cross-sectional properties are:

$$A_s(4 \times 20 \text{ mm } \phi) = 1257 \text{ mm}^2$$
$$A_c = 250 \times 300 - 1257 = 73743 \text{ mm}^2$$

The limiting or factored load carrying capacity of a column is given by

$$P_u = 0.40f_{ck}A_c + 0.67f_yA_s$$
$$= 0.40 \times 20 \times 73743 + 0.67 \times 250 \times 1257$$

$$= 589944 + 210547.5 = 800491.5 \text{ N}$$

$$= 800.49 \text{ kN}$$

$$\text{Safe service load} = \frac{800.49}{1.5} = 533.66 \text{ kN}$$

Example 5.3 A reinforced concrete column of unsupported length of 3.0 m is to be designed for a factored axial load of 2500 kN. Determine the cross-sectional dimensions of the column and the reinforcement required for the following two cases:

1. There is no restriction on the cross-sectional dimensions.
2. One side of the column is restricted to 300 mm.

The grades of concrete and steel to be used are M20 and Fe415, respectively.

Solution The cross-sectional dimensions will depend on the percentage of reinforcement. Consider the area of steel to be 1.0 per cent of the gross area of concrete, A_g; then,

$$A_s = 0.01A_g \text{ and } A_c = A_g - A_s = A_g - 0.01A_g = 0.99A_g$$

Substitute these values in ultimate load capacity formula, i.e.,

$$P_u = 0.40f_{ck}A_c + 0.67f_yA_s$$

$$2500 \times 10^3 = 0.4 \times 20 \times 0.99A_g + 0.67 \times 415 \times 0.01A_g$$

$$= 10.7005A_g$$

Therefore,

$$A_g = \frac{2500 \times 10^3}{10.7005} = 233634 \text{ mm}^2$$

CASE I Adopt 550 × 425 mm column. It is a short column with area of reinforcement,

$$A_s = 0.01 \times 550 \times 425 = 2338 \text{ mm}^2$$

For the selected column dimensions, it is necessary to place intermediate bars in addition to corner on the larger faces.
Provide 8 bars of 20 mm ϕ (A_s = 2513 mm^2).

Check for minimum eccentricity:

(a) In the direction of the longer dimension

$$e_{min} = \frac{L}{500} + \frac{D}{30} = \frac{3 \times 10^3}{500} + \frac{550}{30}$$

$$= 24.33 \text{ mm}$$

$$\frac{e_{min}}{D} = \frac{24.33}{550} = 0.0442 < 0.05, \text{ i.e., } e_{min} < 0.05D$$

(b) In the direction of the shorter dimension

$$e_{\min} = \frac{3 \times 10^3}{500} + \frac{425}{30} = 20.17 \text{ mm}$$

$$\frac{e_{\min}}{b} = \frac{20.17}{425} = 0.047 < 0.05, \text{ i.e., } e_{\min} < 0.05b$$

Since the minimum eccentricity ratio is less than 0.05 along both the dimensions, the section provided is satisfactory.

Ties Diameter of ties shall not be less than 20/4 = 5 mm. Use 8 mm φ mild steel ties. Spacing should not exceed the least of the following:

(a) The least lateral dimension = 425 mm

(b) 16 × 20 = 320 mm, and

(c) 300 mm

Hence provide 8 mm φ mild steel ties at 300 mm c/c. The reinforcement details are shown in Fig. 5.5(a).

CASE I When one dimension is restricted to 300 mm, the other side is:

$$= \frac{233634}{300} = 778.78 \text{ mm (say 780 mm)}$$

Adopt a column of size 780 × 300 mm, considering it to be a short column with the area of reinforcement given by

$$A_s = 0.01 \times 780 \times 300 = 2340 \text{ mm}^2$$

The dimensions of column necessitate the placement of two intermediate bars on each of the longer faces in addition to corner bars. Provide 8 bars of 20 mm φ (A_s = 2513 mm^2). As in Case-I provide 8 mm φ mild steel ties at 300 mm c/c. The reinforcement details are shown in Fig. 5.5(b).

Check for minimum eccentricity:

(a) Along the long dimension

$$e_{\min} = \frac{3 \times 10^3}{500} + \frac{780}{30} = 32 \text{ mm}$$

Thus,

$$\frac{e_{\min}}{D} = \frac{32}{780} = 0.041 < 0.05$$

(b) Along the short dimension

$$e_{\min} = \frac{3 \times 10^3}{500} + \frac{300}{30} = 16 \text{ mm}$$

$$\frac{e_{\min}}{b} = \frac{16}{300} = 0.053 > 0.05$$

Thus, the column is satisfactory in the direction of the longer dimension but not in the

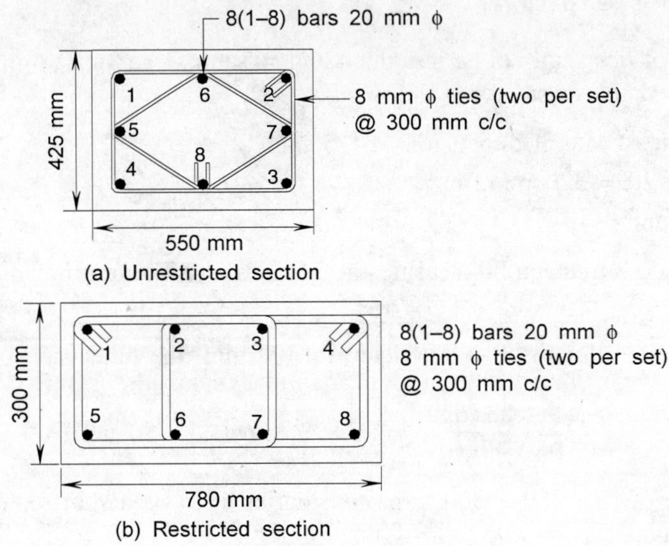

Fig. 5.5 *Detailing schemes for reinforcement in the columns of Example 5.3.*

direction of the shorter dimension. As such the column should be checked for eccentricity in the short direction.

5.9 COMPRESSION MEMBER WITH HELICAL REINFORCEMENT

The ultimate load for an axially loaded spirally reinforced column without partial safety factors may be expressed as:

$$P_u = P_{uc} + P_{us} + P_{ush} = 0.67\, f'_{ck} A_c + f_y A_{sc} + k_s f_{ysp} V_{sp} \tag{5.8}$$

where k_s is the experimental constant varying between 1.5 and 2.5, with an average value of 2.0. Thus, the contribution of spiral to column strength is $(2f_{ysp}V_{sp})$. Defining ρ_s as the ratio of volume of spiral reinforcement, V_{sp} to the volume of the concrete core out-to-out of spiral V_c, per unit length of column, i.e.,

$$\rho_s = \frac{V_{sp}}{V_c}$$

or

$$V_{sp} = \rho_s V_c = \rho_s (A_c \times 1.0\ ) = \rho_s A_c$$

The spiral is designed so as to increase the capacity of core by an amount approximately equal to 90 per cent of the capacity of shell, thus maintaining the column capacity when shell spalls off, i.e.,

$$(2 f_{ysp} V_{sp}) = (2 f_{ysp} \rho_s A_c) \not< 0.90 \times [0.67 f_{ck} (A_g - A_c)]$$

or

$$(f_{ysp} \rho_s A_c) \not< [0.30 f_{ck} (A_g - A_c)]$$

or

$$\rho_s \not< 0.30 \left(\frac{A_g}{A_c} - 1 \right) \left(\frac{f_{ck}}{f_{ysp}} \right)$$

An additional factor of safety of 1.2 is provided to assure that the spiral effect exceeds the shell capacity,

$$\rho_s \geq 0.36 \left(\frac{A_g}{A_c} - 1 \right) \left(\frac{f_{ck}}{f_{ysp}} \right) \tag{5.9}$$

If this condition is satisfied the strength of a compression member with helical reinforcement shall be taken as 1.05 times the strength of a similar member with lateral ties. Since the spiral completes one loop in a pitch length s_v,

$$\rho_s = \frac{a_{sp} \pi (D_c - \phi_{sp}) / s_v}{(\pi D_c^2 / 4)} \approx \frac{4 a_{sp}}{D_c s_v}$$

Thus,

$$\left(\frac{4 a_{sp}}{D_c s_v} \right) \geq 0.36 \left(\frac{D^2}{D_c^2} - 1 \right) \left(\frac{f_{ck}}{f_{ysp}} \right)$$

or

$$s_v \leq \left(\frac{11.11 a_{sp} D_c}{D^2 - D_c^2} \right) \left(\frac{f_{ysp}}{f_{ck}} \right) \tag{5.10}$$

where

A_g = Gross area of the cross-section

A_c = Area of concrete core measured to the outside diameter of the helix

a_{sp} = Area of helix wire

D = Diameter of column cross-section

D_c = Diameter of concrete core

D_{sp} = Diameter of helix coil

f_{ck} = Characteristic compressive strength of concrete

f_{ysp} = Characteristic strength of the helical reinforcement bar ≤ 415 MPa
s_v = Pitch of helix

Example 5.4 A reinforced concrete short column of 480 mm diameter is reinforced with 6 longitudinal bars of 20 mm φ of Fe415 grade HYSD steel and 8 mm φ helix with 75 mm pitch. Compute the maximum load-carrying capacity of the column if concrete used is of grade M25.

Solution For the given materials:

$$f_{ck} = 25 \text{ MPa}, f_y = 415 \text{ MPa and } A_s = 1885 \text{ mm}^2$$

Assume a clear cover of 40 mm to the helix.

Diameter of the core $D_c = 480 - 40 - 40 = 400$ mm

Diameter of the helix $D_{sp} = 480 - 40 - 40 - 8 = 392$ mm

Length of the spiral bar per pitch length,

$$h = [(\pi D_{sp})^2 + s_v^2]^{1/2}$$
$$= [(\pi \times 392)^2 + (75)^2]^{1/2} = 1234 \text{ mm}$$

Volume of the helix per mm height,

$$V_{sp} = \frac{\frac{\pi}{4} \times 8^2 \times 1234}{75} = 827 \text{ mm}^3$$

Volume of the core per mm height $V_c = \frac{\pi}{4} \times 400^2 \times 1 = 125664 \text{ mm}^3$

Ratio of the volume of the helix to the volume of the core $\rho_s = \frac{827}{125664} = 0.0066$

Therefore,

$$\text{Helix parameter} = \frac{0.36 f_{ck} (A_g / A_c - 1)}{f_{ysp}}$$

$$= 0.36 \times 25 \times \frac{(480^2/400^2 - 1)}{415} = 0.00954$$

As the ratio of the volume of the helical reinforcement to the volume of the core is less than the helix parameter, $[0.36 f_{ck}(A_g/A_c - 1)/f_{ysp}]$ no enhancement of strength over the strength of similar member with lateral ties is permitted.
Alternatively, from Eq. (5.10),

$$\left(\frac{11.11 a_{sp} D_c}{D^2 - D_c^2}\right)\left(\frac{f_{ysp}}{f_{ck}}\right) = \left(\frac{11.11 \times (\pi \times 8^2/4) \times 400}{480^2 - 400^2}\right)\left(\frac{415}{25}\right)$$

$$= 54.87 \text{ mm} \le s_v \ (= 75 \text{ mm})$$

Therefore, the code does not permit enhancement of strength over the strength of similar member with lateral ties, i.e.,

$$P_u = (0.40 f_{ck} A_c + 0.67 f_y A_s)$$

$$= \left[0.40 \times 25 \times \pi \times \frac{400^2}{4} + (0.67 \times 415 \times 1885)\right]$$

$$= 1780761 \text{ N} = 1780.76 \text{ kN}$$

Example 5.5 Design a reinforced concrete spiral column of 390 mm diameter subjected to an axial factored load of 1750 kN. The column is braced against sidesway and has unsupported length of 3.3 m. The concrete mix and steel to be used in construction are of grade M25 and Fe415, respectively.

Solution For the given column:

$$f_{ck} = 25 \text{ MPa}, f_y = 415 \text{ MPa and } P_u = 1750 \text{ kN}$$

$$l_{\text{eff}} = 1.0l = 3300 \text{ mm}, D = 390 \text{ mm}$$

Slenderness ratio

$$\frac{l_{eff}}{D} = \frac{3300}{390} = 8.462 < 12, \text{ the column is to be designed as a short column.}$$

Minimum eccentricity:

$$e_{\min} = \frac{l_{eff}}{500} + \frac{D}{30} = \frac{3.3 \times 10^3}{500} + \frac{390}{30}$$

$$= 19.6 \text{ mm } (< 20 \text{ mm})$$

$$\frac{e_{\min}}{D} = \frac{19.6}{390} = 0.0503 \approx 0.05 \text{ i.e., } e_{\min} \approx 0.05D$$

Design of longitudinal steel. Since the minimum eccentricity ratio is approximately equal to 0.05, the column can be designed as short axially loaded column, i.e.,

$$P_u = 1.05(0.40 f_{ck} A_c + 0.67\, f_y A_s)$$

$$\frac{P_u}{1.05} = 0.40 f_{ck} A_g (1.0 - 0.01p) + 0.67 f_y A_g (0.01p)$$

where

p = Percentage of longitudinal steel

$A_g = \pi \times 390^2/4$

Therefore,

$$\left(\frac{p}{100}\right) = \left[\left(\frac{P_u}{1.05 A_g}\right) - 0.40 f_{ck}\right][0.67 f_y - 0.40 f_{ck}]^{-1}$$

$$p = 100 \times \left[\left(\frac{1750 \times 10^3}{1.05 \times (\pi \times 390^2/4)}\right) - 0.40 \times 25\right][0.67 \times 415 - 0.40 \times 25]^{-1}$$

$$= 1.474 \text{ per cent} > A_{s,\min} \; (= 0.8 \text{ per cent})$$

Therefore,

$$A_s = \frac{\pi \times 390^2}{4} \times \frac{1.474}{100} = 1760.83 \text{ mm}^2$$

Provide 6 bars of 20 mm $\phi(A_s = 1885 \text{ mm}^2)$ at a clear cover of 40 mm over the helix.

Design of spiral steel. Consider a bar diameter of 6 mm and pitch s_v.

Diameter of the core $D_c = 390 - 40 - 40 = 310$ mm

Diameter of the helix $D_{sp} = 390 - 40 - 40 - 6 = 304$ mm

Ratio of the volume of the helix to the volume of the core per unit length of column is given as:

$$\frac{V_{sp}}{V_c} = \frac{(\pi \times 6^2/4) \times \pi \times 304/s_v}{\pi \times 310^2/4} = \frac{0.3578}{s_v}$$

For the enhancement of strength of a compression member with helical reinforcement over the strength of similar member with lateral ties, the ratio of the volume of the helical reinforcement to the volume of the core, i.e., (V_{sp}/V_c) shall be more than the helix parameter, $[0.36 f_{ck}(A_g/A_c - 1)/f_{ysp}]$. Thus,

$$\left(\frac{V_{sp}}{V_c}\right) \geq \left[0.36\left(\frac{f_{ck}}{f_{ysp}}\right)\left(\frac{A_g}{A_c} - 1\right)\right]$$

$$\left(\frac{0.3578}{s_v}\right) \geq \left[0.36\left(\frac{25}{415}\right)\left(\frac{\pi \times 390^2/4}{\pi \times 310^2/4} - 1\right)\right] = 0.01264$$

$$s_v \leq 28.31 \text{ mm}$$

Alternatively, from Eq. (5.10),

$$s_v \leq \left(\frac{11.11 a_{ysp} D_c}{D^2 - D_c^2}\right)\left(\frac{f_{sp}}{f_{ck}}\right) = \left(\frac{11.11 \times (\pi \times 6^2/4) \times 310}{390^2 - 310^2}\right)\left(\frac{415}{25}\right) = 28.87 \text{ mm}$$

However, IS:456 has imposed restrictions on the pitch as follows:

$$s_v > \begin{cases} 25 \text{ mm} \\ 3\phi_{sp} = 18 \text{ mm} \end{cases}$$

$$s_v < \begin{cases} 75 \text{ mm} \\ 310/6 = 51.67 \text{ mm} \end{cases}$$

Provide 6 mm ϕ spiral at 28 mm c/c pitch.

5.10 SHORT COLUMN SECTION SUBJECTED TO COMBINED AXIAL LOAD AND UNIAXIAL BENDING

In a majority of the cases, the column elements are subjected to combined axial load and bending moments. Columns in framed structures, arches, chimneys, silos and bunkers are some of the common examples where such a combination of loading occurs. A similar effect is obtained when the columns are eccentrically loaded.

The design of a section for a given axial load P_u and moment M_u is made by pre-assigning or estimating the section and testing its adequacy. The section may be considered as acted upon by a load P_u at an eccentricity e equal to M_u/P_u. The strain compatibility condition of plane section hypothesis is used to establish the strain profile in the section. The strain distribution is governed by the position of the neutral axis, as shown in Fig. 5.6. The neutral axis may have various positions depending upon the eccentricity of load. Two limiting cases, viz., zero eccentricity resulting in concentric axial load condition, and the infinite eccentricity resulting

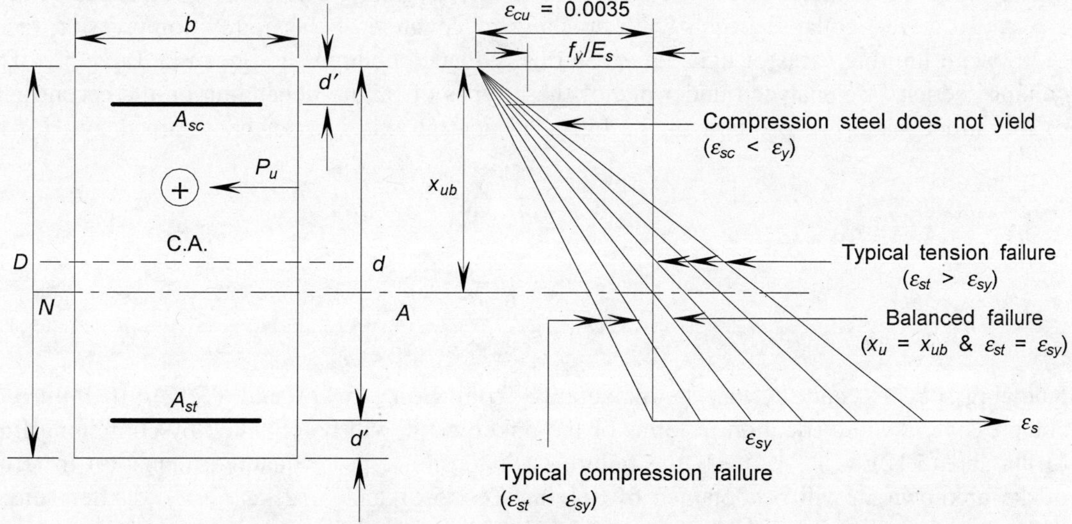

Fig. 5.6 *Strain diagrams of eccentrically loaded columns: types of failure.*

in pure bending condition, have been considered for covering all the cases. A true position of neutral axis will satisfy the requirement of the resultant internal force acting at the same eccentricity $e = M_u/P_u$ of the external load. The section is considered safe if the given load does not exceed the ultimate or limiting load capacity of the column.

The analysis of a concrete section of given dimensions, with known failure strain in concrete and stress block parameters, consists in evaluating the value of P_u for any given value of eccentricity e and vice-versa. The quantities P_u and e can be related by the equilibrium of internal forces and their moments about the centroid of the tension reinforcement as given by

$$P_u = C_c + C_{sc} + C_{st} \tag{5.11}$$

$$M_u = P_u\left(e + d - \frac{D}{2}\right) \tag{5.12}$$

These expressions are valid for all the sections that are symmetric about the axis from which the eccentricity is measured. A column subjected to axial load and uniaxial moment can fail in any one of the following three modes of failure:

***Balanced failure*:** This type of failure occurs when the tension steel reaches yield stress just as the concrete in compression attains failure strain. The corresponding values of load and eccentricity are represented by P_b and e_b, respectively.

***Compression failure*:** This mode of failure is initiated by crushing of concrete on the compression face of the column, i.e., $P_u > P_b$ or $e < e_b$. The steel on the tension face of the column does not yield. The compression reinforcement may or may not yield, depending upon the grade of steel and its proximity to the highly compressed edge.

***Tension failure*:** The failure is initiated by yielding of steel on the tension face of the column. This type of the failure occurs when the eccentricity is greater than that for balanced failure, $e > e_b$. However, collapse occurs with crushing of concrete at the highly compression face.

Special limiting cases of uniaxial eccentrically loaded columns are given in Table 5.2. The column sections are analyzed under one of these types of failure depending on the eccentricity of the direct load. The position of the balanced neutral axis x_{ub} can be obtained by

$$\frac{\varepsilon_{cu}}{x_{ub}} = \frac{\varepsilon_{sy}}{d - x_{ub}}$$

or

$$x_{ub} = \frac{\varepsilon_{cu} d}{\varepsilon_{cu} + \varepsilon_{sy}} \tag{5.13}$$

Knowing x_{ub}, P_b and e_b can be calculated from Eqn. (5.11) and (5.12). If failure is compressive, a cubic equation in terms of the unknown x_u will be obtained by combining Eqs. (5.11) and (5.12) as $\varepsilon_{st} < \varepsilon_{sy}$. If the failure is of tension type, a quadratic equation in terms of the unknown x_u will be obtained by solving Eqs. 5.11 and 5.12, as $f_{st} = f_{sy}$. After getting the value of x_u, the magnitudes of P_u and M_u can be obtained for the given eccentricity of the load.

TABLE 5.2 Distribution of Strains in Special Cases of Uniaxially Eccentrically Loaded Column

Case	*Loading eccentricity, M_u/P_u*	*Loading conditions*	*Strain profile (Stress and strain distributions along the column section)*	*Remarks*
1.	$e = 0$	Axial load, $M_u = 0$ $P_u \neq 0$	Entire section is under compression with uniform strain distribution, $\varepsilon_{cu} = 0.002$, $f_{cc} = 0.447 f_{ck}$ and $x_u = \infty$	Limit state of collapse in compression
2.	$0 < e < e_D$	$M_u \neq 0$ $P_u \neq 0$	Entire section is under compression with non-uniform linear strain distribution, Neutral axis lies outside the section, $x_u > D$ Strain in highly compressed edge, $0.002 < \varepsilon_{cu} \leq 0.0035$ $\varepsilon_{cu} = 0.0035 - 0.75\,\varepsilon_{c,\min}$ $\varepsilon_{c,\min}$ is strain in least compressed edge, $\varepsilon_{cu} = 0.002\left[1 + \frac{3D/7}{x_u - 3D/7}\right]$, $f_{cc} = 0.447 f_{ck}$	Compression failure
3.	$e = e_D$	$M_u \neq 0$ $P_u \neq 0$	Neutral axis coincides with the edge farthest from highly compressed edge. $x_u = D$, and $\varepsilon_{cu} = 0.0035$	Compression failure
4	$e_D < e = e_b$	$M_u \neq 0$ $P_u \neq 0$	Non-uniform linearly varying strain and neutral axis lies within section between compression and tensile steels, $\varepsilon_{cu} = 0.0035$ and $x_u = x_{u,\max}$ $\varepsilon_y = 0.87 f_y/E_s$ for Fe250 $\varepsilon_y = 0.87 f_y/E_s + 0.002$ for Fe415/Fe500	Balanced failure
5.	$e_b < e < \infty$	$M_u \neq 0$ $P_u \neq 0$	Non-uniform linearly varying strain and neutral axis lies within section between compression and tensile steels, Maximum strain at highly compressed edge, $0.002 < \varepsilon_{cu} \leq 0.0035$	Tension failure
6.	$e = \infty$	Pure flexure $P_u = 0$ $M_u \neq 0$	Non-uniform linearly varying strain and neutral axis lies within section between compression and tensile steels, $0.002 < \varepsilon_{cu} \leq 0.0035$ $x_u \leq x_{u,\max}$	Tension failure

Note: Point of intersection of strain profile cases $e = 0$ and $e = e_D$ occurs at distance $3D/7$ from highly compressed edge.

5.11 PROCEDURE FOR ANALYSIS OF THE SECTION

The following steps should be followed for the analysis:

1. Find P_b and e_b from the given dimensions of the section with known reinforcement details and the material properties.
2. Compare P_u(or e_u) with P_b(or e_b) to determine the type of failure.
3. Determine the position of the neutral axis x_u by solving the cubic or quadratic equation depending on whether the failure is of compression or tension type, respectively.
4. Compute the required P_u and M_u from the corresponding equations based on the type of failure.

Balanced failure: For the case of balanced failure, since both tension and compression steels yield, the neutral axis falls between tension and compression steels.

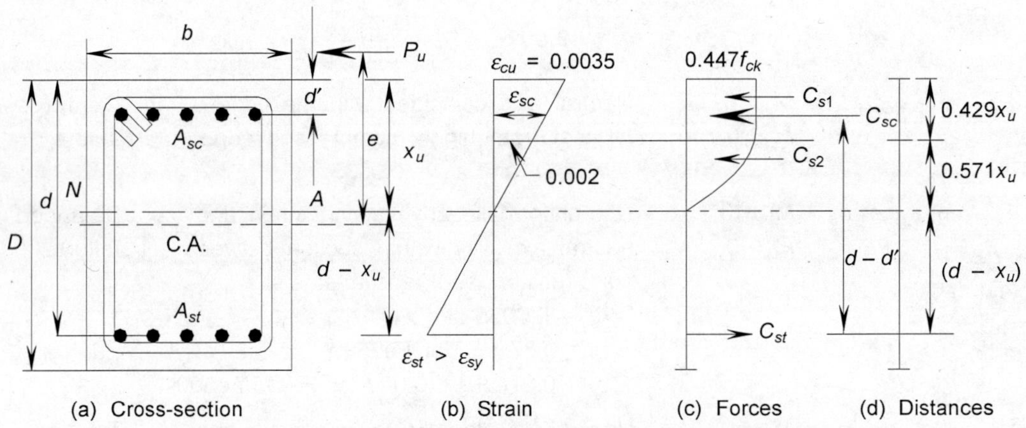

Fig. 5.7 *Analysis of a section with tension failure.*

Let x_u be the depth of the neutral axis from the compression edge, and e be the eccentricity of applied direct load from the centroidal axis. Due to similarity with the doubly reinforced concrete section subjected to flexure, consider the parabolic-rectangular compression stress block shown in Fig. 5.7, the depth of the rectangular portion being given by

$$x_1 = \left(x_u - \frac{0.0020}{0.0035}x_u\right) = \left(1 - \frac{4}{7}\right)x_u = 0.429x_u$$

Depth of the parabolic portion $x_2 = x_u - 0.429x_u = 0.571x_u$

Area of the stress block: $\bar{C}_c = C_{c1} + C_{c2}$

$$= 0.447f_{ck} \times 0.429x_u + \left[\frac{2}{3} \times 0.447\, f_{ck} \times 0.571x_u\right] \approx 0.362f_{ck}x_u$$

The distance of the centroid of the compression force in concrete from the top edge is equal to $0.416x_u$.

Compressive force in concrete $C_c = \bar{C}_c b = 0.362 f_{ck}\, bx_u$

Acting at a distance of $(d - 0.416x_u)$ from tension steel.

Force in compression steel A_{sc} (assuming that it yields):

$$C_{sc} = 0.87 f_{yd} A_{sc}$$

Acting at a distance of $(d - d')$ from the tension steel.

Force in tension steel $C_{st} = -0.87 f_{yd} A_{st}$

From the internal force equilibrium condition,

$$P_u = C_c + C_{sc} + C_{st} = 0.362 f_{ck} bx_u + 0.87 f_{yd}(A_{sc} - A_{st}) \tag{5.14}$$

For moment equilibrium take moment about the tension steel,

$$M_u = P_u\left(e + d - \frac{D}{2}\right)$$

$$= 0.362 f_{ck} bx_u(d - 0.416x_u) + 0.87 f_{yd} A_{sc}(d - d') \tag{5.15}$$

Equations (5.14) and (5.15) are the basic equations for the uniaxially loaded column which can further be simplified if A_{sc} and A_{st} are equal.

$$P_u = 0.362 f_{ck}\, bx_u \tag{5.16}$$

$$M_u = 0.362 f_{ck}\, bx_u(d - 0.416x_u) + 0.87 f_{yd} A_{sc}(d - d') \tag{5.17}$$

For the Eqs. (5.14) and (5.15) to be valid, it must be ensured that the compression and tension steels have yielded. For the compression steel to yield,

$$\varepsilon_{sc} = \left(\frac{x_u - d'}{x_u} \times 0.0035\right) \geq \frac{0.87 f_{yd}}{2 \times 10^5}$$

or

$$1 - \frac{d'}{x_u} \geq \frac{0.87 f_{yd}}{0.0035 \times 2 \times 10^5} (= 0.001243 f_{yd}) \tag{5.18}$$

Similarly, for the tension steel to yield,

$$\varepsilon_{st} = \left(\frac{d - x_u}{x_u} \times 0.0035\right) \geq \frac{0.87 f_{yd}}{2 \times 10^5}$$

$$\frac{d}{x_u} - 1 \geq 0.001243 f_{yd} \tag{5.19}$$

1. *For mild steel:* f_{yd} *= 250 MPa.*

$$\left(1-\frac{d'}{x_u}\right) \geq 0.3107 \qquad \text{or} \qquad \left(\frac{d'}{x_u}\right) \leq 0.6893$$

Hence,

$$x_u \geq 1.451d'$$

and

$$\frac{d}{x_u} - 1 \geq 0.3107$$

or

$$x_u \leq 0.763d$$

For $d' = 0.1D$, $d = D - d' = 0.9D$, the preceding equations reduce to

$$x_u \geq 0.1451D$$

$$x_u \geq 0.6867D$$

Hence for tension and compression steels to yield the following condition should be satisfied:

$$0.1451D \leq x_u \leq 0.6867D \tag{5.20}$$

2. *For cold worked HYSD steel of grades Fe415 and Fe500.*

The stress is proportional up to $f_{yd} = 0.80f_y$

With Fe415:

$$1 - \left(\frac{d'}{x_u}\right) \geq 0.001243 \times 0.8 \times 415 \ (= 0.4127)$$

$$x_u \geq 1.7026\, d'$$

and

$$\left(\frac{d}{x_u}\right) - 1 \leq 0.4127$$

$$x_u \leq 0.7079d$$

Thus, for the validity of the preceding analysis:

$$1.7026\, d' \leq x_u \leq 0.7079\, d$$

For $d' = 0.1D$, the domain of validity is:

$$0.1703D \leq x_u \leq 0.6371D$$

With Fe500:

The domain of validity can be shown as:

$$1.989d' \leq x_u \leq 0.6679d$$

For $d' = 0.1D$, the validity range reduces to:

$$0.1989D \leq x_u \leq 0.6011D \tag{5.21}$$

For the values of (d'/d) up to 0.2, f_{cc} is equal to $0.447f_{ck}$ and f_{sc} is equal to $0.87f_y$, i.e.,

$$f_{sc} = 0.87f_y = 0.87 \times 250 \times 217.5 \text{ MPa}.$$

For HYSD steel of grades Fe415 and Fe500, the stress in compression steel, f_{sc} can be obtained from Table 3.3 for the given values of d'/d.

Example 5.6 A reinforced concrete column section of size 225 × 500 mm is reinforced with 4 × 20 mm ϕ mild steel bars on each of the smaller faces at an effective cover of 50 mm. Determine the load-carrying capacity of the section if it is subjected to a direct load at an eccentricity of 400 mm. The concrete used is of grade M25.

Solution For the given section and materials:

$$b = 225 \text{ mm}, D = 500 \text{ mm}, d' = 50 \text{ mm}, d = D - d' = 450 \text{ mm}.$$

$$A_{sc} = A_{st} = (4 \times 20 \text{ mm } \phi\ ) = 1257 \text{ mm}^2$$

$$f_{ck} = 25 \text{ MPa and } f_{yd} = f_y = 250 \text{ MPa (for mild steel)}$$

This is a case of large eccentricity. Consider both compression and tension steels to yield at collapse, and thus the neutral axis lies between these two types of steel at a depth of x_u from the compression face. From Eqs. (5.16) and (5.17), we have

$$P_u = 0.362f_{ck}bx_u = 0.362 \times 25 \times 225x_u$$

$$= 2036.25x_u$$

$$M_u = P_u\left(e + d - \frac{D}{2}\right) = P_u\left(400 + 450 - \frac{500}{2}\right) = 600P_u$$

Hence,

$$600P_u = 0.362f_{ck}bx_u(d - 0.416x_u) + 0.87f_{yd}A_{sc}(d - d')$$

$$= 0.362 \times 25 \times 225x_u(450 - 0.416x_u) + 0.87 \times 250 \times 1257 \times (450 - 50)$$

$$P_u = 3.39375x_u(450 - 0.416x_u) + 182265$$

Equating the two values of P_u,

$$2036.25x_u = 3.39375x_u(450 - 0.416x_u) + 182265$$

or

$$1.4118x_u^2 + 509.06x_u - 182265 = 0$$

Solving, $x_u = 283.14$ mm

This value of x_u satisfies the condition

$$1.451\ d' \leq x_u \leq 0.763d$$

Hence the assumption that both steels yield is satisfactory:

$$P_u = 2036.25x_u$$

$$= 2036.25 \times 283.14 = 576543.82 \text{ N} = 576.54 \text{ kN}$$

Compression failure: The failure is initiated by crushing of concrete on the compression face of the column. Compression steel yields but tension steel does not. This may include the cases where the axial load acts at moderate to small eccentricity.

1. The neutral axis falls between tension and compression steels, i.e., $x_u < d$ as shown in Fig. 5.8. In the worst case the neutral axis is at the level of tension steel, i.e., $x_u = d$.

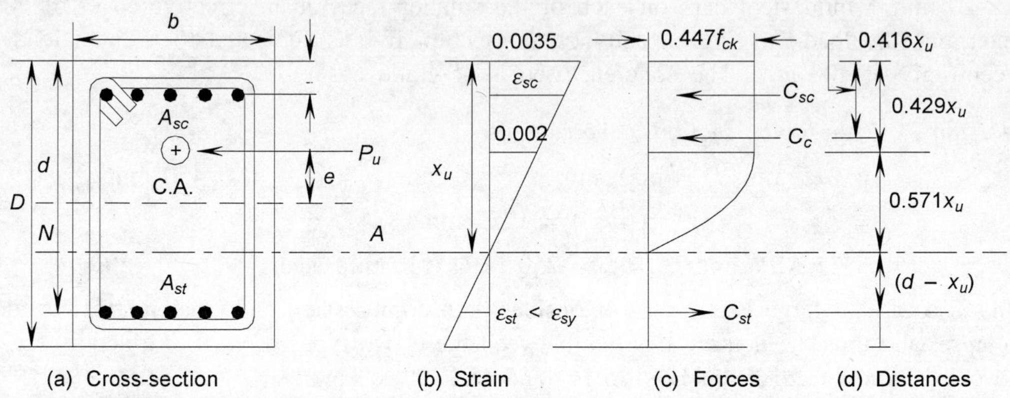

Fig. 5.8 *Compression failure with neutral axis falling between compression and tension steels.*

The internal force equilibrium equation for this case is

$$P_u = C_c + C_{sc} + C_{st}$$

$$= 0.362f_{ck}bx_u + 0.87f_{yd}A_{sc} - A_{st}\left(\frac{d - x_u}{x_u}\right) \times 0.0035 \times E_s \tag{5.22}$$

The moment equilibrium equation about tension steel is

$$M_u = P_u\left(e + d - \frac{D}{2}\right)$$

$$= 0.362f_{ck}bx_u(d - 0.416x_u) + 0.87f_{yd}A_{sc}(d - d') \tag{5.23}$$

It can be shown that the following conditions should be satisfied:

$$0.763d \le x_u \le d \qquad \text{for mild steel}$$

$$0.7079d \le x_u \le d \qquad \text{for Fe415 steel}$$

$$0.6679d \le x_u \le d \qquad \text{for Fe500 steel} \tag{5.24}$$

For the limiting case when the neutral axis is at the level of tension steel, i.e., $x_u = d$, Eqs. (5.22) and (5.23) reduce to:

$$P_u = 0.362f_{ck}bd + 0.87f_{yd}A_{sc} \tag{5.25}$$

and

$$P_u\left(e+d-\frac{D}{2}\right) = 0.362f_{ck}bd(d - 0.416d) + 0.87f_{yd}A_{sc}(d - d')$$

$$= 0.21148f_{ck}bd^2 + 0.87f_{yd}A_{sc}(d - d') \tag{5.26}$$

2. The neutral axis is at the edge of section, i.e., $x_u = D$, as shown in Fig. 5.9.

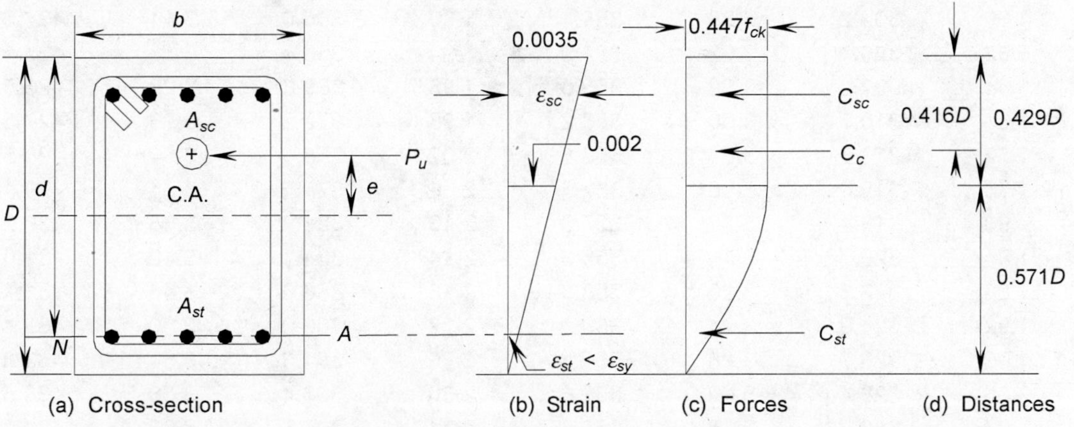

Fig. 5.9 *Compression failure with neutral axis at the edge of the section.*

This is a special case when the strain in the bottom compression fibre is zero. The internal force equilibrium equation for this case is:

$$P_u = C_c + C_{sc} + C_{st}$$

$$= 0.362f_{ck}bD + 0.87f_{yd}A_{sc} + A_{st}\left[\frac{D-d}{D}\right] \times 0.0035 \times E_s \tag{5.27}$$

and the moment equilibrium equation is:

$$P_u\left(e+d-\frac{D}{2}\right) = 0.362f_{ck}bD(d - 0.416D) + 0.87f_{yd}A_{sc}(d - d') \tag{5.28}$$

For HYSD steel of grades Fe415 and Fe500, the stress in steel f_{yd} can be obtained from Table 5.3 for the given values of strain.

TABLE 5.3 Stress and Strain Values in the Nonlinear Portion of Stress-strain Curve

Fe415				*Fe 500*			
Strain, $\varepsilon_y \times 10^{-3}$	*Stress f_s, MPa*	*Strain, $\varepsilon_y \times 10^{-3}$*	*Stress f_s, MPa*	*Strain, $\varepsilon_y \times 10^{-3}$*	*Stress f_s, MPa*	*Strain, $\varepsilon_y \times 10^{-3}$*	*Stress f_s, MPa*
1.44	**288.7**	2.72	350.8	**1.74**	**347.8**	3.02	421.0
1.48	293.0	**2.76**	**351.7**	1.78	352.5	3.06	422.2
1.52	297.0	2.80	352.6	1.82	356.9	3.10	423.4
1.56	300.7	2.84	353.4	1.86	361.1	**3.12**	**423.9**
1.60	304.2	2.88	354.1	1.90	365.0	3.14	424.5
1.63	**306.7**	2.92	354.9	1.94	368.7	3.18	425.5
1.64	307.5	2.96	355.6	**1.95**	**369.6**	3.22	426.5
1.68	310.5	3.00	356.2	1.98	372.2	3.26	427.5
1.72	313.3	3.04	356.8	2.02	375.4	3.30	428.5
1.76	315.9	3.08	357.3	2.06	378.5	3.34	429.4
1.80	318.4	3.12	357.8	2.10	381.4	3.38	430.2
1.84	320.7	3.16	358.3	2.14	384.1	3.42	431.0
1.88	322.8	3.20	358.7	2.18	386.6	3.46	431.7
1.92	**324.8**	3.24	359.0	2.22	389.0	3.50	432.4
1.96	326.7	3.28	359.3	**2.26**	**391.3**	3.54	433.0
2.00	328.4	3.32	359.6	2.30	393.4	3.58	433.5
2.04	330.1	3.36	359.8	2.34	395.5	3.62	434.0
2.08	331.7	3.40	360.0	2.38	397.4	3.66	434.5
2.12	333.2	3.44	360.2	2.42	399.3	3.70	434.8
2.16	334.7	3.48	360.3	2.46	401.1	3.74	435.1
2.20	336.1	3.52	360.4	2.50	402.8	3.78	435.4
2.24	337.5	3.56	360.5	2.54	404.4	3.82	435.6
2.28	338.8	3.60	360.6	2.58	406.0	3.86	435.7
2.32	340.0	3.64	360.6	2.62	407.6	3.90	435.8
2.36	341.3	3.68	360.7	2.66	409.1	3.94	435.8
2.40	342.5	3.72	360.7	2.70	410.5	3.98	435.7
2.41	**342.8**	3.76	360.7	2.74	412.0	4.02	435.7
2.44	343.6	**3.80**	**360.7**	**2.77**	**413.0**	4.06	435.5
2.48	344.8	3.84	360.7	2.78	413.3	4.10	435.4
2.52	345.8	—	—	2.82	414.7	4.14	435.2
2.56	346.9	—	—	2.86	416.0	**4.17**	**435.0**
2.60	347.9	—	—	2.90	417.3	4.18	434.9
2.64	348.9	—	—	2.94	418.6	4.22	434.7
2.68	349.9	—	—	2.98	419.8	4.24	434.6

Example 5.7 A reinforced concrete column section of size 400 × 400 mm is provided with 4 bars of 22 mm ϕ Fe415 steel on each of the two opposite faces at an effective cover of 50 mm. Find the limiting eccentricity for zero compression at the (bottom) edge and the corresponding load-carrying capacity. The concrete used is of grade M25.

Solution For the given cross-section and materials:

$$b = 400 \text{ mm}, D = 400 \text{ mm}, d' = 50 \text{ mm}, d = 350 \text{ mm}$$

$$A_{sc} = A_{st} = (4 \times 22 \text{ mm } \phi) = 1520 \text{ mm}^2$$

$$f_{ck} = 25 \text{ MPa}, f_y = 415 \text{ MPa and } f_{yd} = 0.8 \times 415 \text{ MPa}$$

The equilibrium equations are:

$$P_u = 0.362 f_{ck} bD + 0.87 f_{yd} A_{sc} + A_{st}\left[\frac{D-d}{D}\right] \times 0.0035 \times E_s$$

$$= 0.362 \times 25 \times 400 \times 400 + 0.87 \times (0.80 \times 415) \times 1520$$

$$+ 1520\left[\frac{50}{400}\right] \times 0.0035 \times 2 \times 10^5$$

$$= 1448000 + 439036.8 + 133000 = 2020036.8 \text{ N} = 2020.04 \text{ kN}$$

and

$$M_u = P_u\left(e + d - \frac{D}{2}\right) = (e + 150)P_u$$

$$= 0.362 f_{ck} bD(d - 0.416D) + 0.87 f_{yd} A_{sc}(d - d')$$

$$= 0.362 \times 25 \times 400 \times 400 \times (350 - 0.416 \times 400)$$

$$+ 0.87 \times (0.8 \times 415) \times 1520 \times (350 - 50)$$

$$= 265852800 + 131711040 = 397563840 \text{ Nmm}$$

Thus,

$$\left(\frac{Mu}{Pu}\right) = e + 150 = \left(\frac{397563840}{2020036.8}\right) = 196.81 \text{ mm}$$

Here,

$$e = \text{Limiting eccentricity} = 46.81 \text{ mm}$$

Thus,

$$\text{Permissible factored load} = 2020.04 \text{ kN}$$

3. The neutral axis falls outside the section, i.e., $x_u \geq D$, as shown in Fig. 5.10.

In this case, compression steel yields and bottom steel is in compression. The eccentricity is greater than 0.05 times the lateral dimension. The main difficulty lies in determining the area and centroid of incomplete parabolic compressive stress distribution. IS:456 has suggested Eq. (5.29) for parabolic part of the stress-strain curve of concrete, shown in Fig. (5.11).

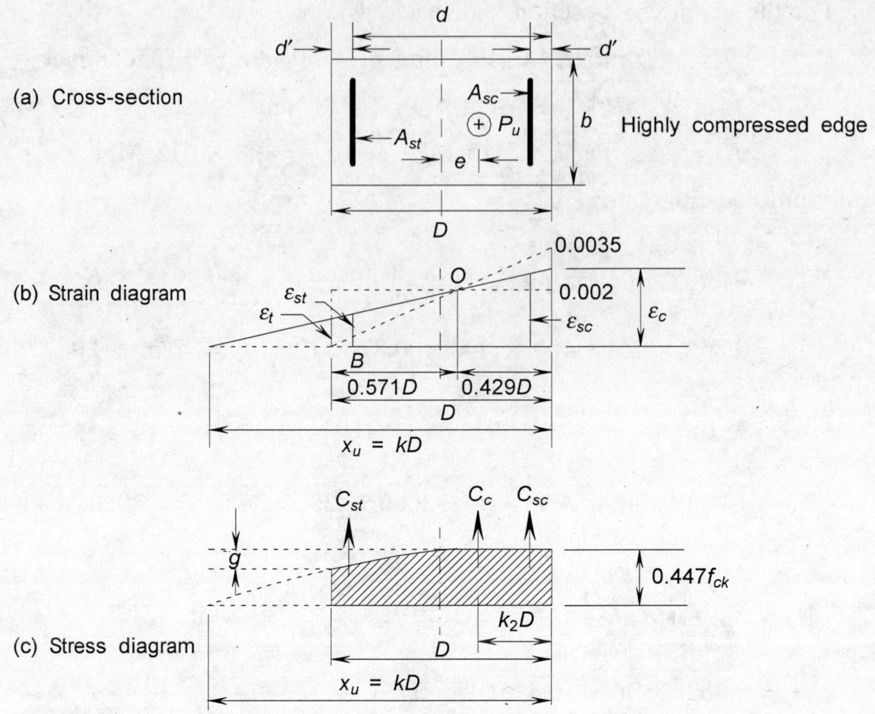

Fig. 5.10 *Compression failure with neutral axis outside the section.*

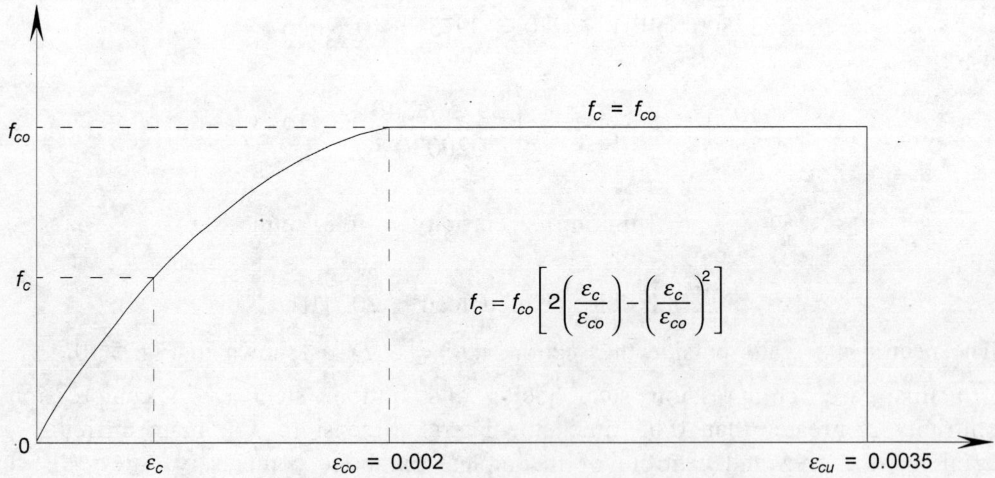

Fig. 5.11 *Stress-strain curve for concrete: stress block.*

Now,

$$f_c = f_{co}\left[2\frac{\varepsilon_c}{\varepsilon_{co}} - \left(\frac{\varepsilon_c}{\varepsilon_{co}}\right)^2\right]$$

$$= 0.447 f_{ck}\left[2\left(\frac{\varepsilon_c}{0.002}\right) - \left(\frac{\varepsilon_c}{0.002}\right)^2\right]$$

$$= \frac{0.447}{0.002} f_{ck}[2\varepsilon_c - 500\varepsilon_c^2] = 447 f_{ck}[\varepsilon_c - 250\varepsilon_c^2] \tag{5.29}$$

The failure strain for the axially loaded column has already been defined as 0.002 for the entire cross-section. However, in the limiting case when the neutral axis is at the least compressed edge of the column, i.e., $x_u = D$, the strain at the compression edge is 0.0035 and zero at the other end. The strain distributions for these two cases intersect each other at a depth $3D/7$ (= $0.429D$) from the highly compressed edge, as shown in Fig. 5.10. As per SP-16 the intersection point is assumed to act as a fulcrum for the strain distribution line when the neutral axis falls outside this section. Thus, whatever be the depth of the neutral axis below the least compressed edge, the rectangular portion of the stress block extends up to $0.429D$ with a uniform stress of $0.447f_{ck}$ and thereafter the stress diagram is parabolic. Let g be the difference between the stresses at the highly compressed and the least compressed edges. From the geometric properties of the parabola,

$$g = 0.447f_{ck}\left[\frac{4D/7}{x_u - 3D/7}\right]^2 = 0.447 f_{ck}\left[\frac{4}{7k-3}\right]^2 \tag{5.30}$$

where $x_u = kD$.
Area of the compressive stress block k_1 is expressed as:

$$= 0.447f_{ck}D - \left(\frac{1}{3}\right)g\left(\frac{4D}{7}\right) = 0.447 f_{ck}D\left[1 - \frac{4}{21}\left(\frac{4}{7k-3}\right)^2\right] \tag{5.31}$$

The centroid of the stress block can be determined by taking moment about the highly compressed edge. This moment is:

$$= 0.447f_{ck}D \times \frac{D}{2} - \left(\frac{4}{21}\right)gD \times \left[\left(\frac{3}{7}\right)D + \left(\frac{3}{4}\right)\left(\frac{4}{7}\right)D\right]$$

$$= \left(\frac{0.447 f_{ck}D^2}{2}\right) - \left(\frac{8gD^2}{49}\right) \tag{5.32}$$

The position of the centroid is obtained by dividing the preceding moment by the area of the stress block.

$$\bar{y} = \frac{0.2235 f_{ck} D^2 - \dfrac{8gD^2}{49}}{0.447 f_{ck} D - \dfrac{4gD}{21}} = \left(\frac{0.2235 f_{ck} - \dfrac{8g}{49}}{0.447 f_{ck} - \dfrac{4g}{21}} \right) D$$

For different values of k, the area of the stress block and the position of its centroid are given in Table 5.4.

The equilibrium conditions for the axial forces and moment of forces about the centre line of the section are given, in that order by

$$P_u = C_c + C_{sc} + C_{st} = k_1 b + A_{sc}(f_{sc} - f_{cc}) + A_{st}(f_{st} - f_{cc}) \tag{5.33}$$

$$M_u = C_c\left(\frac{D}{2} - \bar{y}\right) + C_{sc}\left(\frac{D}{2} - d'\right) - C_{st}\left(\frac{D}{2} - d'\right) \tag{5.34}$$

TABLE 5.4 Properties of Rectangular-Parabolic Stress Block for Different Positions of Neutral Axis when Neutral Lies Outside the Section

$k = (x_u/D)$	*Area of stress block* k_1	*Distance of c.g. from highly compressed edge,* $\bar{y}$
1.00	$0.362 f_{ck} D$	$0.416D$
1.05	$0.374 f_{ck} D$	$0.432D$
1.10	$0.384 f_{ck} D$	$0.443D$
1.20	$0.399 f_{ck} D$	$0.458D$
1.30	$0.409 f_{ck} D$	$0.468D$
1.40	$0.417 f_{ck} D$	$0.475D$
1.50	$0.422 f_{ck} D$	$0.480D$
2.00	$0.435 f_{ck} D$	$0.491D$
2.50	$0.440 f_{ck} D$	$0.495D$
3.00	$0.442 f_{ck} D$	$0.497D$
4.00	$0.444 f_{ck} D$	$0.499D$

5.12 COLUMN INTERACTION DIAGRAM

In the case of eccentrically loaded column, there are infinite combinations of M_u and P_u which can cause failure of a given section. For the given load (P_u or M_u) the other load (M_u or P_u) can be determined by trial and error. Thus the analysis is generally cumbersome. The difficulty is overcome by constructing *P-M* or *P-e* interaction diagrams of columns which represent the column behaviour under axial load, moment and eccentricity. It is convenient to construct interaction diagrams in a non-dimensional format by expressing the axial load P_u and moment M_u as $(P_u/f_{ck}bD)$ and $(M_u/f_{ck}bD^2)$, respectively. The non-dimensional quantities $(P_u/f_{ck}bD)$ and

$(M_u/f_{ck}bD^2)$ are plotted along *y*- and *x*-axes, respectively. This form has been adopted in the design curves given in BS handbook: SP-16. The interaction curves are convenient for routine design. They represent strength envelops for the reinforced concrete sections subjected to eccentric load. Points lying outside these curves represent failure of the column.

A column subjected to varying magnitudes of P_u and M_u has neutral axis at varying depths. Following cases arise:

CASE I Neutral axis lying outside the section i.e., $x_u > D$ with limiting case $x_u = D$.

CASE II Neutral axis lying at level of tension steel, i.e., $x_u = D - d'$.

CASE III Neutral axis lying between compression and tension steels, i.e., $d' < x_u < D - d'$. The values of P_u and M_u corresponding to a given position of neutral axis x_u can be computed from the equilibrium equation:

$$P_u = C_{uc} + C_{usc} + C_{ust}$$

$$M_u = C_{uc}\left(\frac{D}{2} - \bar{y}\right) + C_{usc}\left(\frac{D}{2} - d'\right) - C_{ust}\left(\frac{D}{2} - d'\right)$$

Here, M_u is the moment of force components about an axis about which eccentricity is measured.

The case when concrete and tension steel reach their ultimate and yield strains simultaneously, is called **balanced failure condition**. The balanced failure point on interaction curve represents a position of neutral axis beyond which *compression failure* occurs where the failure is initiated by concrete reaching its ultimate strain first. On the other hand, shallower depth of neutral axis $(x_u < x_{ub})$ represent *tension failure* condition where the steel on tension face reaches the yield point before failure.

The following procedure can be adopted to construct the interaction diagrams:

1. Estimate or predetermine the column section and the amount of reinforcing steel.
2. Consider different values of the depth of neutral axis x_u, eccentricities '*e*' and determine P_u and M_u.
3. Calculate the quantities $(P_u/f_{ck}bD)$ and $(M_u/f_{ck}bD^2)$. Plot $(P_u/f_{ck}bD)$ on the *y*-axis and $(M_u/f_{ck}bD^2)$ on the *x*-axis. Draw a curve joining all these points to obtain interaction diagrams.

Example 5.8 A column of size 300 × 500 mm overall is reinforced with 3 bars of 25 mm ϕ of Fe415 grade steel on each of the compression and tension sides at an effective cover of 52.5 mm. Construct the *P-M*, and *P-e* interaction curves. The concrete used is of grade M20.

Solution For the given cross-section and materials:

$$b = 300 \text{ mm}, D = 500 \text{ mm}, d' = 52.5 \text{ mm and } d = 447.5 \text{ mm}$$

$$A_{sc} = A_{st}(3 \times 25 \text{ mm } \phi) = 1473 \text{ mm}^2$$

$$f_{ck} = 20 \text{ MPa and } f_y = 415 \text{ MPa}$$

The solution is obtained for various eccentricities which are arbitrarily considered to construct the interaction curve.

CASE I Concentric load $e = 0$

$$P_u = 0.447 f_{ck} A_c + 0.75 f_y A_{sc}$$
$$= 0.447 \times 20 \times (300 \times 500 - 2 \times 1473) + 0.75 \times 415 \times 2 \times 1473$$
$$= 2231.605 \text{ kN}$$

$$\frac{P_u}{f_{ck} bD} = \frac{2231.61 \times 10^3}{20 \times 300 \times 500} = 0.744 \qquad \text{(Point 1)}$$

CASE II Limiting eccentricity of $0.05D$ (= 25 mm).

$$P_u = 0.40 f_{ck} A_c + 0.67 f_y A_{sc}$$
$$= 0.40 \times 20 \times (300 \times 500 - 2 \times 1473) + 0.67 \times 415 \times 2 \times 1473$$
$$= 1995567.3 \text{ N} = 1995.57 \text{ kN}$$

$$\frac{P_u}{f_{ck} bD} = \frac{1995567.3}{20 \times 300 \times 500} = 0.6652 \qquad \text{(point 2)}$$

and

$$M_u = P_u e = 1995567.3 \times (0.05 \times 500) = 49889182.5 \text{ Nmm}$$

$$\frac{M_u}{f_{ck} bD^2} = \frac{49889182.5}{20 \times 300 \times 500^2} = 0.0333$$

In this case the expression for P_u gives conservative value and the point lies inside the design curve and hence is ignored.

CASE III Small eccentricity with the neutral axis falling outside the section.
Consider the depth of the neural axis from the extreme compression fibre x_u to be $1.1D$. The strain at highly compressed edge is:

$$\varepsilon_c = 0.002\left[1 + \frac{\frac{3D}{7}}{x_u - \frac{3D}{7}}\right]$$

$$\varepsilon_c = 0.002\left[1 + \frac{\frac{3D}{7}}{1.1D - \frac{3D}{7}}\right] = 0.00328$$

Strain at the level of compression steel is given by

$$\varepsilon_{sc} = \frac{\varepsilon_{cu}}{x_u}(x_u - d') = \frac{\varepsilon_{cu}}{(1.1D)} \times (1.1D - d')$$

$$= \frac{0.00328}{550} \times (550 - 52.5) = 0.002967 > 0.00144$$

Therefore, for $\varepsilon_{sc} = 0.002967$ the stress in steel from the Table 5.3:

$$f_{sc} = 355.7 \text{ MPa}$$

Stress in concrete f_{cc} corresponding to strain $\varepsilon_c = \varepsilon_{sc}$ is determined from the design stress-strain relation of concrete:

As $\varepsilon_c > 0.002$ $\quad f_{cc} = 0.447 \times 20 = 8.94$ MPa

The tension steel is at 102.5 (= 50 + 52.5) mm from the neutral axis and corresponding strain is:

$$\varepsilon_{st} = \frac{0.00328}{550}(50 + 52.5) = 0.0006113 < 0.00144$$

Therefore,

$$f_{st} = \varepsilon_{st} E_s = 0.0006113 \times 2 \times 10^5 = 122.26 \text{ MPa}$$

Stress in concrete f_{ct} corresponding to strain $\varepsilon_{st} < 0.002$ is:

$$f_{ct} = 447 f_{ck}(\varepsilon_{st} - 250\varepsilon_{st}^2)$$

$$= 447 \times 20 \times (0.0006113 - 250 \times 0.0006113^2) = 4.63 \text{ MPa}$$

Axial load capacity is given by:

$$P_u = k_1 b + (f_{sc} - f_{cc})A_{sc} + (f_{st} - f_{ct})A_{st}$$

For $k = 1.1$, from Table 5.1,

$$k_1 = 0.384 f_{ck} D \text{ and } \overline{y} = 0.443D$$

Thus,

$$P_u = 0.384 \times 20 \times 300 \times 500 + (355.7 - 8.94) \times 1473$$

$$+ (122.26 - 4.63) \times 1473 = 1836046.5 \text{ N} = 1836.05 \text{ kN}$$

$$\frac{P_u}{f_{ck} bD} = \frac{1836046.5}{20 \times 300 \times 500} = 0.613 \qquad \text{(Point 3)}$$

The moment about the centre of the section is:

$$M_u = k_1 b(0.5D - \overline{y}) + (f_{sc} - f_{cc})A_{sc}(0.5D - d') - (f_{st} - f_{ct})A_{st} \times (0.5D - d')$$

Thus,

$$M_u = 0.384 \times 20 \times 300 \times 500 \times (250 - 0.443 \times 500)$$

$$+ (355.7 - 8.94) \times 1473 \times (250 - 52.5) - (122.26 - 4.63) \times 1473 \times (250 - 52.5)$$

$$= 99489927 \text{ Nmm} = 99.49 \text{ kNm}$$

$$\frac{M_u}{f_{ck}bD^2} = \frac{99489927}{20 \times 300 \times 500^2} = 0.0663$$

$$\text{Eccentricity of load } e = \frac{M_u}{P_u} = \frac{99489927}{1836046.5} = 54.19 \text{ mm}$$

CASE IV Limiting eccentricity when the neutral axis is at the level of bottom reinforcement. For this case $x_u = d$ and corresponding strain in the top reinforcement is:

$$\varepsilon_{sc} = \frac{0.0035}{d}(d - d') = 0.0035 \times \frac{(500 - 52.5 - 52.5)}{(500 - 52.5)} = 0.003089 > 0.00144$$

The values of stress in top steel f_{sc} and stress in concrete f_{cc} are determined from their respective stress-strain curves. For $\varepsilon_{sc} = 0.003089$ the stress in steel from the Table 5.3,

$$f_{sc} = 357.4 \text{ MP}$$

and for $\varepsilon_c > 0.002$,

$$f_{cc} = 0.447 \times 20 = 8.94 \text{ MPa}$$

Axial load capacity is:

$$P_u = 0.362 f_{ck} b(D - d') + (f_{sc} - f_{cc})A_{sc}$$

$$= 0.362 \times 20 \times 300 \times (500 - 52.5) + (357.4 - 8.94) \times 1473$$

$$= 1485251.6 \text{ N} = 1485.25 \text{ kN}$$

Therefore,

$$\frac{P_u}{f_{ck}bD} = \frac{1485251.6}{20 \times 300 \times 500} = 0.495 \qquad \text{(point 4)}$$

Moment of all forces about the centre of the section gives:

$$M_u = 0.362 f_{ck} b(D - d')(0.5D - 0.416x_u) + (f_{sc} - f_{cc})A_{sc}(0.5D - d')$$

$$= 0.362 \times 20 \times 300 \times (500 - 52.5) \times (250 - 0.416 \times 447.5)$$

$$+ (357.4 - 8.94) \times 1473 \times (250 - 52.5)$$

$$= 163423680 \text{ Nmm} = 163.42 \text{ kNm}$$

Therefore,

$$\frac{M_u}{f_{ck}bD^2} = \frac{163423680}{20 \times 300 \times 500^2} = 0.1089$$

Eccentricity of load, $e = M_u/P_u = 110.03$ mm.

CASE V Limiting eccentricity for the neutral axis within the section giving the balanced failure load.

The depth of the neutral axis for the balanced condition is given by

$$x_u = \frac{0.0035}{0.0055 + \dfrac{0.87 f_y}{E_s}}(D - d')$$

$$= \frac{0.0035}{0.0055 + \dfrac{0.87 \times 415}{2 \times 10^5}} \times (500 - 52.50) = 214.40 \text{ mm}$$

$$\varepsilon_{sc} = \frac{0.0035}{x_u}(x_u - d')$$

$$= \frac{0.0035 \times (214.40 - 52.5)}{214.40} = 0.002643$$

The corresponding stresses in steel and concrete are:

For ε_{sc} = 0.002643, the stress in steel from Table 5.3, f_{sc} =348.9 MP

For $\varepsilon_c > 0.002$, f_{cc} = 0.447 × 20 = 8.94 MPa

Axial load capacity is:

$$P_u = 0.362 f_{ck} b x_u + (f_{sc} - f_{cc})A_{sc} - 0.87 f_y A_{st}$$

$$= 0.362 \times 20 \times 300 \times 214.4 + (348.9 - 8.94) \times 1473 - 0.87 \times 415 \times 1473$$

$$= 434611.23 \text{ N} = 434.61 \text{ kN}$$

Therefore,

$$\frac{P_u}{f_{ck} bD} = \frac{434611.23}{20 \times 300 \times 500} = 0.145 \qquad \text{(point 5)}$$

$$\text{Moment capacity } M_u = 0.362 f_{ck} b x_u (0.5D - 0.416 x_u) + (f_{sc} - f_{cc})A_{sc}(0.5D - d') + 0.87 f_y A_{st}(0.5D - d')$$

$$= 0.362 \times 20 \times 300 \times 214.40 \times (0.5 \times 500 - 0.416 \times 214.4) + (348.9 - 8.94) \times 1473 \times (0.5 \times 500 - 52.5) + 0.87 \times 415 \times 1473 \times (0.5 \times 500 - 52.5)$$

$$= 278821380 \text{ Nmm} = 278.82 \text{ kNm}$$

Therefore,

$$\frac{M_u}{f_{ck} bD^2} = \frac{278821380}{20 \times 300 \times 500^2} = 0.1859$$

$$\text{Eccentricity of load } e = \frac{M_u}{P_u} = 641.54 \text{ mm}$$

CASE VI *Pure flexure.* In this case the tension steel would have definitely yielded. The depth of the neutral axis is determined from the equation of equilibrium of forces, i.e. $C_u = T_u$,

$$0.362f_{ck}bx_u + (f_{sc} - f_{cc})A_{sc} = 0.87f_yA_{st}$$

$$f_{sc} = \varepsilon_{sc}E_s = 0.0035 \times (x_u - 52.5) \times \frac{2 \times 10^5}{x_u}$$

$$= \frac{700(x_u - 52.5)}{x_u}$$

$$f_{cc} = 0.447f_{ck} = 0.446 \times 20 = 8.94 \text{ MPa}$$

Substituting values in the equilibrium equation,

$$0.362 \times 20 \times 300x_u + \left[\frac{700(x_u - 52.5)}{x_u} - 8.94\right] \times 1473 = 0.87 \times 415 \times 1473$$

or

$$2172x_u^2 + (691.06x_u - 36750) \times 1473 - 531827x_u = 0$$

$$x_u^2 + 223.8x_u - 24923 = 0$$

or

$$x_u = 81.61 \text{ mm}$$

Therefore,

$$\varepsilon_{sc} = 0.0035 \times \frac{x_u - 52.5}{x_u}$$

$$= 0.00124844 < 0.00144$$

and hence

$$f_{sc} = 0.00124844 \times 2 \times 10^5 = 249.69 \text{ MPa}$$

$$f_{cc} = 447 \times 20 \times (0.001248 - 250 \times 0.001248^2) = 7.68 \text{ MPa}$$

$$P_u = 0.362f_{ck}bx_u + (f_{sc} - f_{cc})A_{sc} - 0.87f_yA_{st}$$

$$= 0.362 \times 20 \times 300 \times 81.61 + (249.69 - 7.68) \times 1473 - 0.87 \times 415 \times 1473$$

$$= 533.74 - 531.83 = 1.91 \text{ kN}$$

As the value of P_u is very small, value of x_u can be adopted. For moment capacity, take moment about tension steel:

$$M_u = 0.362f_{ck}bx_u(d - 0.416x_u) + (f_{sc} - f_{cc})A_{sc}(d - d')$$

$$= 0.362 \times 20 \times 300 \times 81.61 \times (447.5 - 0.416 \times 81.61)$$

$$+ (249.69 - 7.68) \times 1473 \times (447.5 - 52.5)$$

$$M_u = 214114530 \text{ Nmm} = 214.114 \text{ kNm}$$

Hence,

$$\frac{M_u}{f_{ck}bD^2} = \frac{214114530}{20 \times 300 \times 500^2} = 0.1427 \qquad \text{(point 6)}$$

The points are summarized below:

Case	P_u	e/D	$P_u/f_{ck}bD$	$M_u/f_{ck}bD^2$	x_u/D
I	2231.6	0.000	0.744	0.000	∞
II	1995.6	0.050	0.665	0.033	∞
III	1836.1	0.108	0.613	0.066	**1.10**
IV	1485.3	0.220	0.495	0.109	**0.90**
V	851.6	0.655	0.284	0.186	**0.80**
VI	1.910	∞	0.000	0.143	**0.16**

The plot between non-dimensional quantities $P_u/(f_{ck}bD)$ and $M_u/(f_{ck}bD^2)$ is shown in Fig. 5.12(a) and that between load P and eccentricity e is shown in Fig. 5.12(b). The P-M interaction curve is more meaningful since it clearly identifies the compression and tension failure modes and also locates the balanced failure mode (P_b, M_b).

Interaction curves for rectangular column sections with reinforcement of grade Fe415 distributed equally on two sides parallel to the axis of bending and placed at cover-to-depth ratio d'/D of 0.04, 0.06, 0.08, 0.10, 0.12, 0.14, 0.16, 0.18 and 0.20, are shown in Figs. C.1 to C.9, respectively, in Appendix-C. In this case each outer row of bars has an area of 0.5 A_s and there are no inner rows of bars. The steel ratios presented by the parameter p/f_{ck} on the design curves vary from 0.00 to 0.24 at an interval of 0.01. Similar curves can be developed for other grades of steel and types of distribution of reinforcement. In the design Handbook SP:16, following three cases of symmetrically arranged reinforcement are covered:

1. Rectangular column sections with reinforcement distributed equally on two parallel faces as explained earlier.
2. Rectangular column sections with reinforcement distributed equally on four sides. The interaction curves of this category are developed considering arrangement of steel having two outer rows with area of $0.3A_s$ each, and four inner rows with area of $0.1A_s$ each.

   However, these curves can be used without significant error in the cases of not less than two inner rows with a minimum area of $0.3A_s$ of the each of outer rows.
3. The curves of circular column sections are applicable to circular sections with at least six bars of equal diameter uniformly spaced circumferentially. It should be noted that the stress block for rectangular section is not applicable to circular sections. Hence circular sections are usually divided into rectangular strips, and forces on these strips are summed up for determining the total forces and moments.

In the parameter p/f_{ck}, the term p represents percentage of total steel with respects to bD distributed symmetrically on two faces or on all faces in case of rectangular sections, and with respect to $\pi D^2/4$ for circular sections. SP:16 has covered three grades of steel Fe250, Fe415 and Fe500 with d'/D ratios of 0.05, 0.10, 0.15 and 0.20. Some of the typical curves are given in Figs. (5.13) to (5.20).

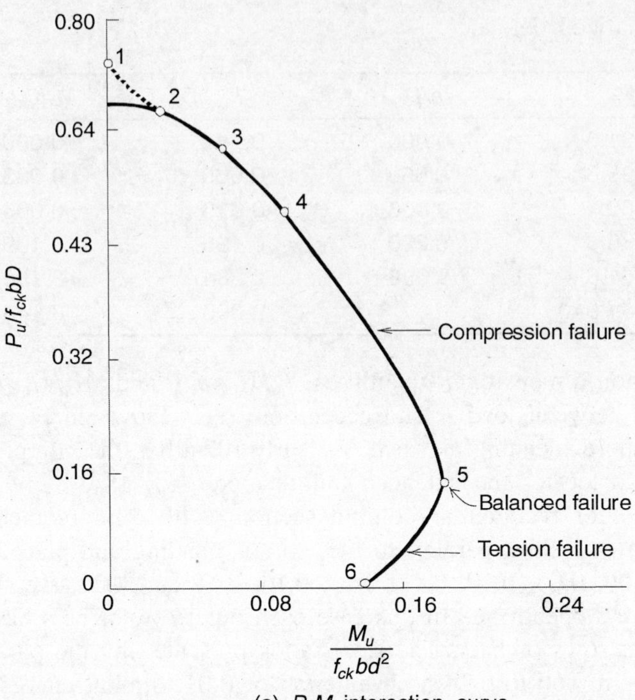

(a) P-M interaction curve

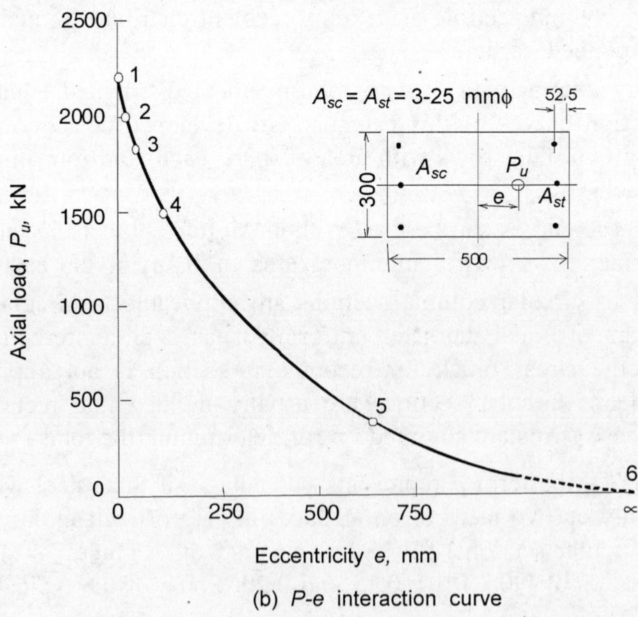

(b) P-e interaction curve

Fig. 5.12 Interaction curves for rectangular column with steel distributed on two sides of example 5.8.

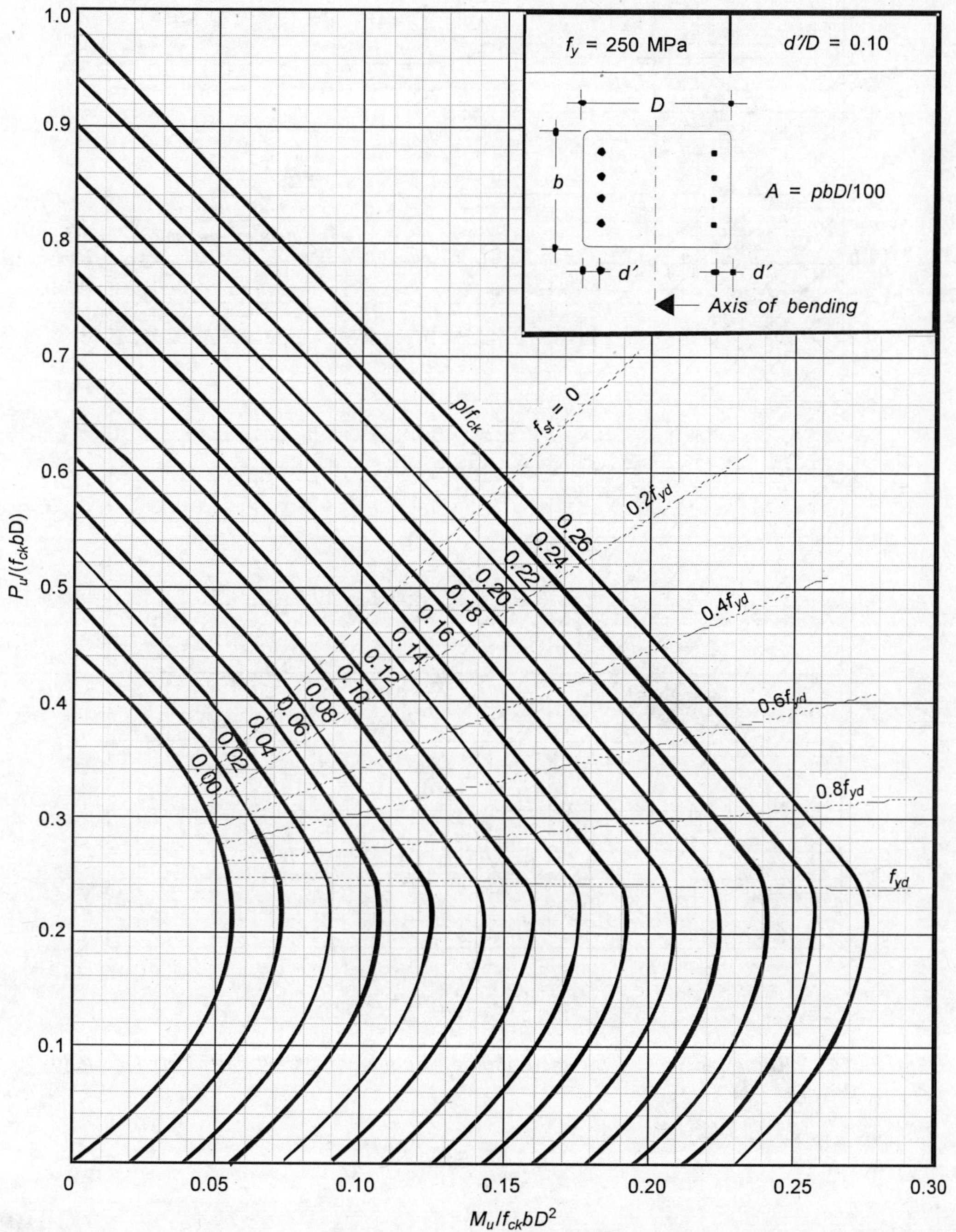

Fig. 5.13 Interaction curves for rectangular column with steel distributed equally on two sides.

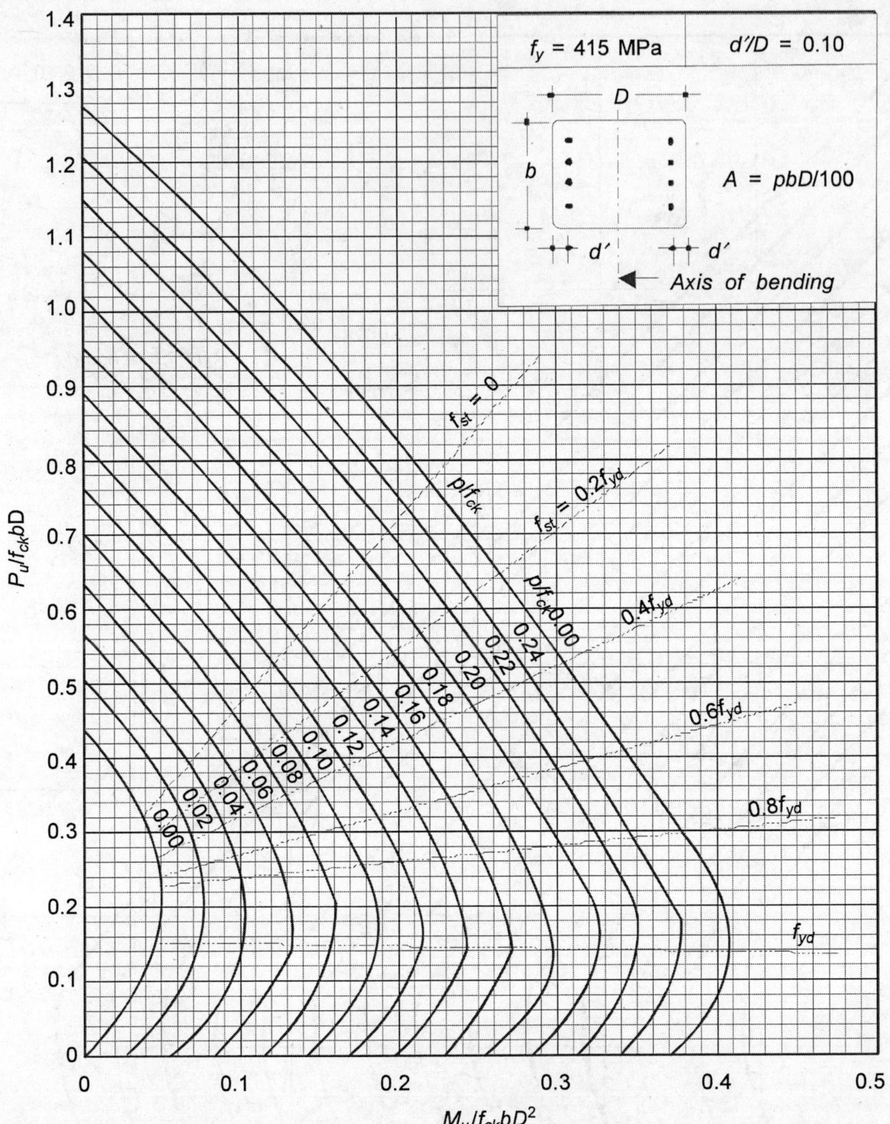

Fig. 5.14 *Interaction curves for rectangular column with steel distributed equally on two sides.*

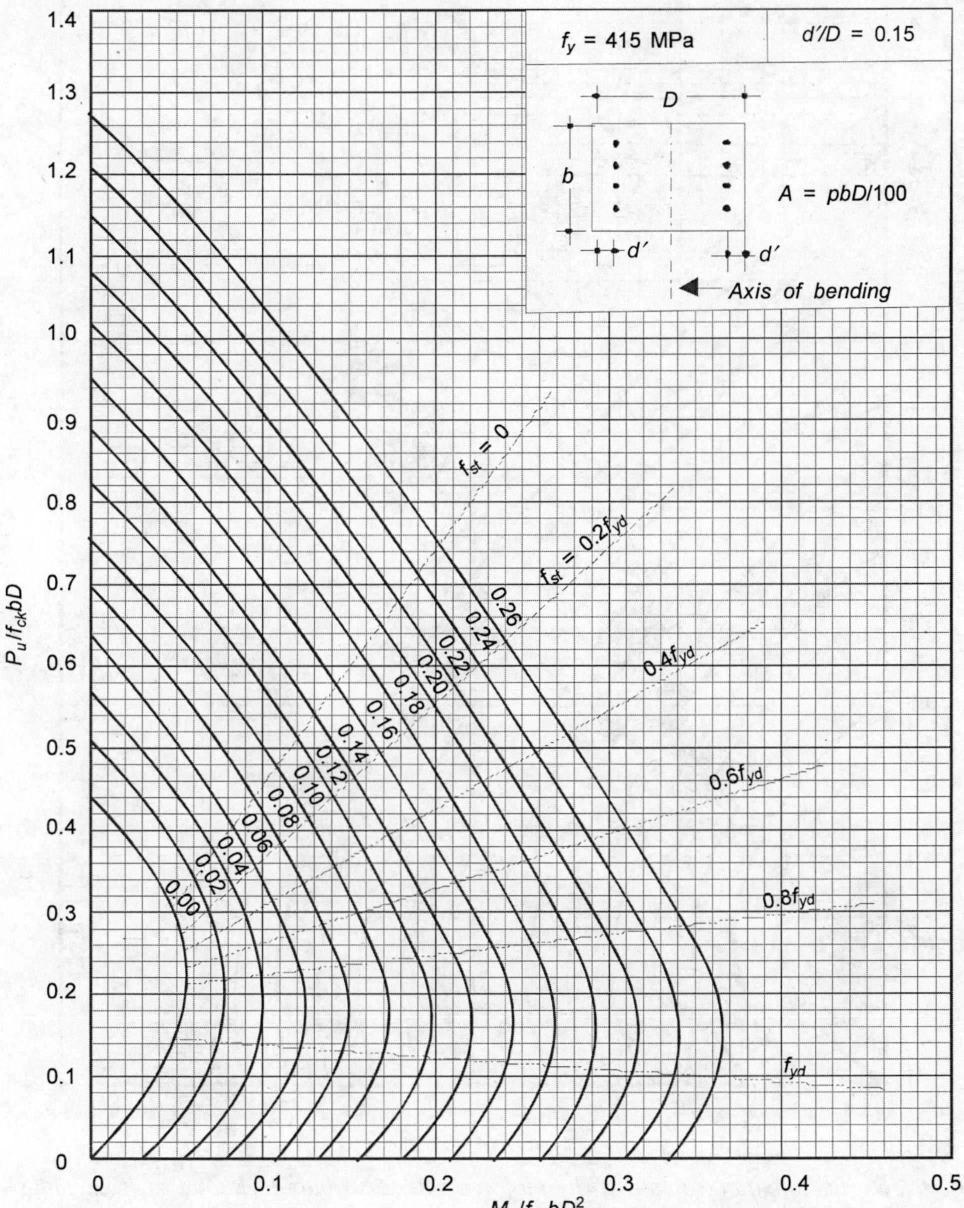

Fig. 5.15 *Interaction curves for rectangular column with steel distributed equally on two sides.*

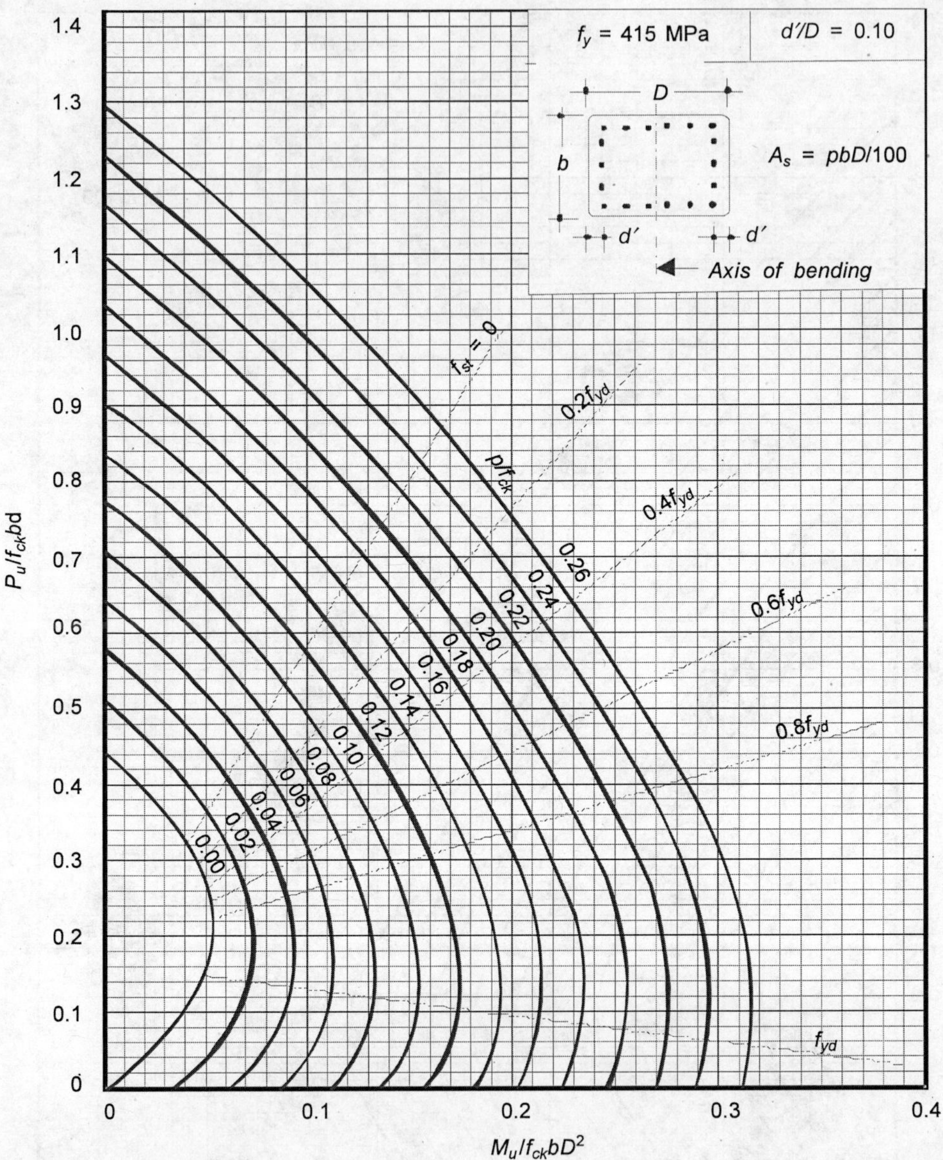

Fig. 5.16 *Rectangular column with reinforcement distributed equally on four sides.*

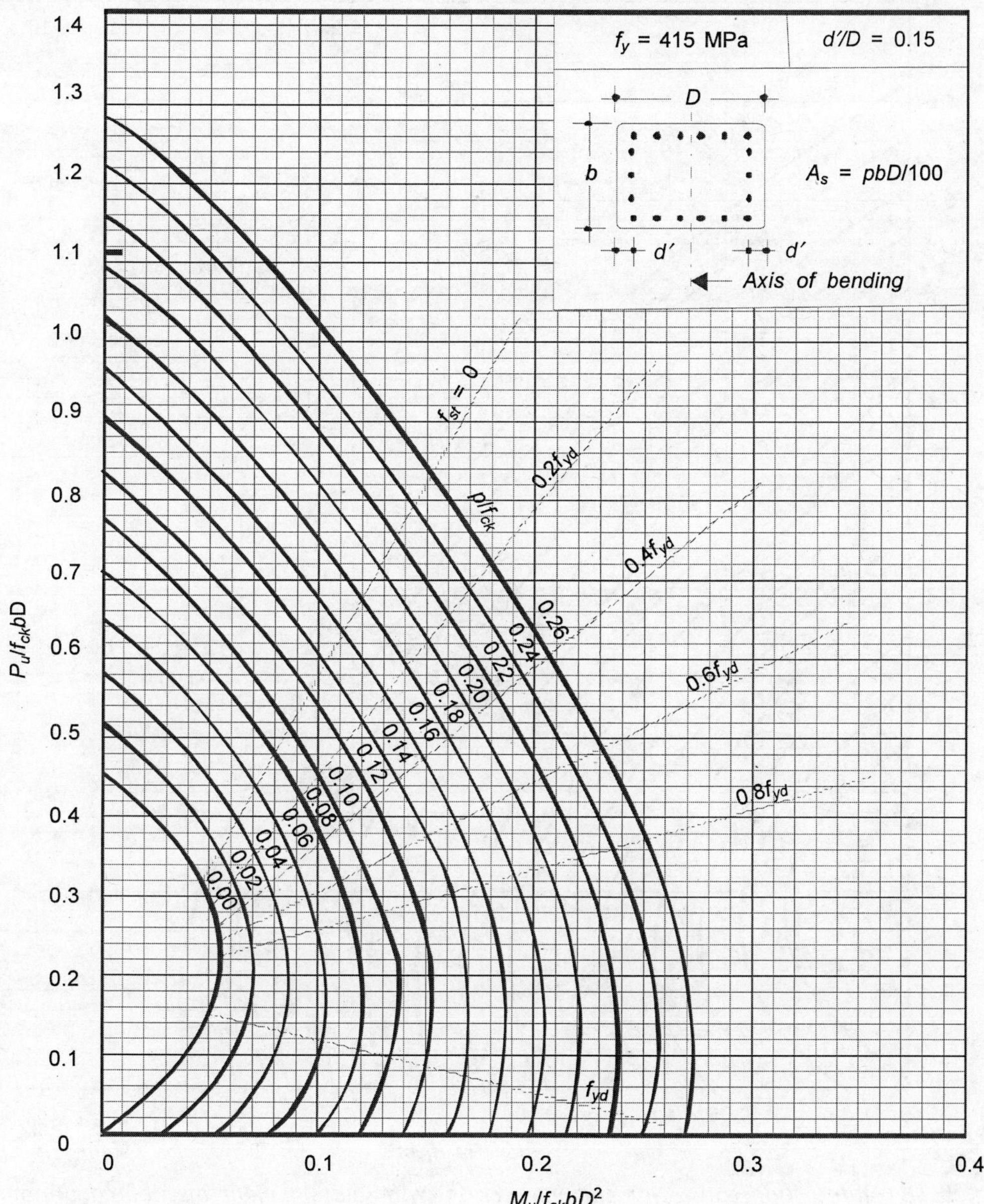

Fig. 5.17 *Rectangular column with reinforcement distributed equally on four sides.*

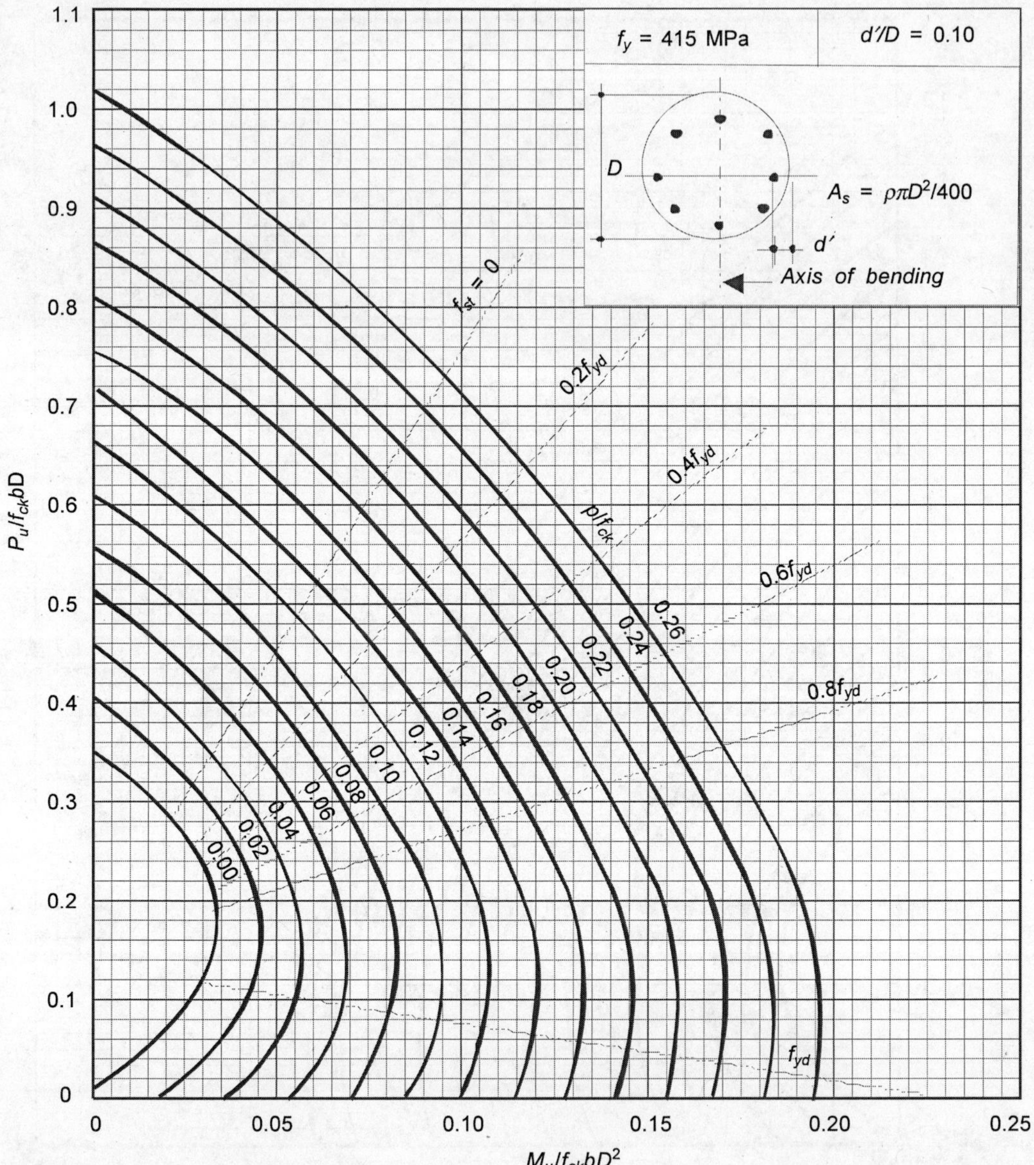

Fig. 5.18 *Interaction curves for circular section with reinforcement distributed uniformly.*

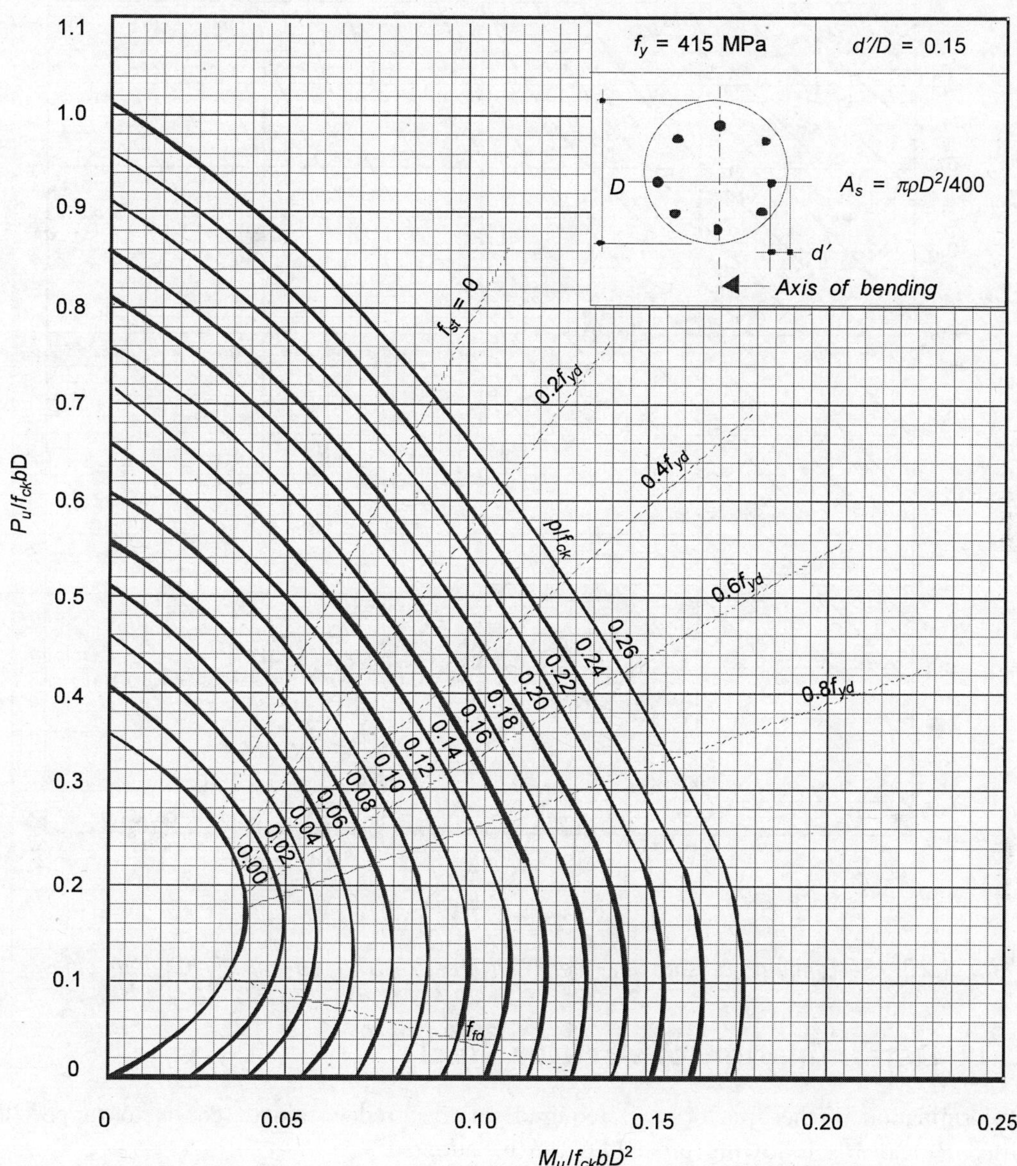

Fig. 5.19 Interaction curves for circular section with reinforcement distributed uniformly.

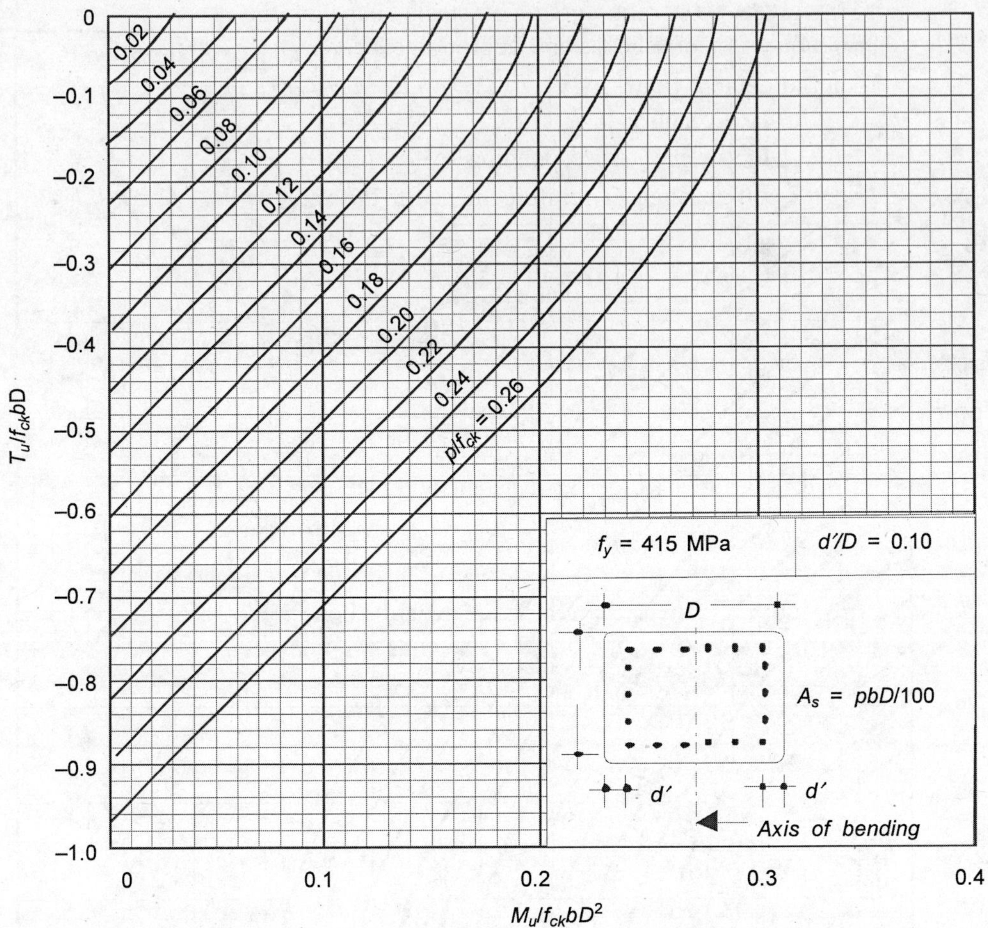

Fig. 5.20 Rectangular tension member with steel distributed equally on four sides.

5.12.1 Determination of Area of Steel

For determination of the area of steel required for the predetermined section to support the specified P_u and M_u, following procedure can be adopted:

1. Pre-determine the trial cross-sectional dimensions, distribution of reinforcement and its effective cover. Check whether the column is short or long. If short proceed as follows:
2. Check the eccentricity:

$$e = \frac{M}{P} > e_{\min} = \frac{L}{500} + \frac{D}{30} \text{ or } 20 \text{ mm}$$

Compute the design parameters d'/D, $P_u/(f_{ck}bD)$ and $M_u/(f_{ck}bD^2)$.

3. For the given d'/D ratio, grade of steel, shape of the section and the type of distribution of steel, choose the appropriate curve.
4. For the computed values of $P_u/(f_{ck}bD)$ and $M_u/(f_{ck}bD^2)$ mark the point s on the curve selected in step 3.
5. Find the value of p/f_{ck} corresponding to the point s plotted in step 4.
6. Calculate the total area of reinforcement for the value of (p/f_{ck}) obtained in step 5 as follows:

$$\frac{p}{f_{ck}} = \frac{100A_s}{f_{ck}bD}$$

or

$$A_s = \left(\frac{p}{f_{ck}}\right)\frac{f_{ck}bD}{100}$$

Distribute total area of steel consistent with the curve used.

Example 5.9 Determine the reinforcement of grade Fe415 steel required for a square column of size 400 × 400 mm subjected to a factored direct load and a factored moment of 2000 kN and 250 kNm, respectively, when the reinforcement is to be placed: (a) on two sides and (b) on four sides with nominal cover of 30 mm. The concrete mix to be used is of grade M20.

Solution The pre-determined cross-sectional dimensions are:

$$b = 400 \text{ mm}, \; D = 400 \text{ mm}$$

$$P_u = 2000 \text{ kN and } M_u = 250 \text{ kNm}$$

Consider 25 mm ϕ longitudinal bars with 30 mm nominal cover.

$$d' = 30 + 12.5 = 42.5 \text{ mm}$$

and

$$\frac{d'}{D} = \frac{42.5}{400} = 0.106$$

The design charts for Fe415 grade steel and $d'/D = 0.1$ will be used.
The dimensionless quantities are:

$$\frac{P_u}{f_{ck}bD} = \frac{2000 \times 10^3}{20 \times 400 \times 400} = 0.625$$

$$\frac{M_u}{f_{ck}bD^2} = \frac{250 \times 10^6}{20 \times 400 \times 4002} = 0.195$$

CASE I *Reinforcement placed on two sides:*
Refer to the design chart given in Fig. 5.14. For $f_y = 415$ MPa and $d'/D = 0.10$, $p/f_{ck} = 0.20$.

Thus, the area of reinforcement is given by

$$A_s = \frac{0.20 \times f_{ck}bD}{100} = \frac{0.20 \times 20 \times 400 \times 400}{100} = 6400 \text{ mm}^2$$

Provide 8 bars of 32 mm ϕ (A_s = 6934 mm^2) equally distributed on the two sides.

CASE II *Reinforcement placed on four faces*:

Refer to the design chart given in Fig. 5.16.

$$\frac{p}{f_{ck}} = 0.23$$

Area of reinforcement is:

$$A_s = \frac{0.23\, f_{ck}bD}{100} = \frac{0.23 \times 20 \times 400 \times 400}{100} = 7360 \text{ mm}^2$$

Provide 28 mm ϕ 12 bars (A_s = 7389 mm^2) equally distributed on four faces.

5.12.2 Adequacy of the Column Cross-section

For a column of given cross-sectional dimensions, distribution of reinforcement and its effective cover, to determine the load carrying capacity following procedure can be adopted:

1. Check whether the column is short or long. If short proceed as follows:
2. Compute the design parameters: d'/D, p/f_{ck}, $P_u/(f_{ck}bD)$ and $M_u/(f_{ck}bD^2)$.
3. For the given d'/D, f_y, p/f_{ck}, shape of the section and the type of distribution of steel, choose the appropriate curve.
4. For the computed values of $P_u/(f_{ck}bD)$ and $M_u/(f_{ck}bD^2)$ mark the point s on the curve selected in step (3). If the point is on or inside the interaction curve, the column is safe. If the point is outside the interaction curve, it is unsafe.

Example 5.10 (a) Determine the limiting load carrying capacity of a reinforced concrete column of size 250 × 450 mm overall, reinforced with 6 × 25 mm ϕ bars of Fe250 grade mild steel equally distributed on two smaller faces at an effective cover of 50 mm for a limiting eccentricity of 200 mm. The grade of concrete mix used is M20. (b) What is the maximum eccentricity at which it can support a factored load of 550 kN?

Solution For the given cross-section and materials:

$$b = 250 \text{ mm}, \; D = 450 \text{ mm}, \; d' = 50 \text{ mm},$$

$$A_{sc} = A_{st}(3 \times 25 \text{ mm } \phi) = 1473 \text{ mm}^2$$

$$f_{ck} = 20 \text{ MPa and } f_y = 250 \text{ MPa}$$

Percentage of steel is given as:

$$p = \frac{100A_s}{bD} = \frac{100 \times (2 \times 1473)}{250 \times 450} = 2.619$$

$$\frac{p}{f_{ck}} = \frac{2.619}{20} = 0.131$$

$$\frac{d'}{D} = \frac{50}{450} = 0.111 \approx 0.1$$

Design chart for $d'/D = 0.10$ given in Fig. 5.13 will be used.

(a) The value of P_u can be determined by establishing the line of given eccentricity $e_{\max} = 200$ mm on the appropriate P-M interaction design chart:

$$\frac{[M_u /(f_{ck}bD^2)]}{[P_u /(f_{ck}bD)]} = \frac{e}{D} = \frac{200}{450} = \frac{4}{9}$$

Draw a line passing through the origin at a slope 9/4 with the axis of $M_u/(f_{ck}bD^2)$. The point of intersection of this line with the curve $p/f_{ck} = 0.131$ will give the value of $P_u/(f_{ck}bD)$ as:

$$\frac{P_u}{f_{ck}bD} = 0.48$$

or

$$P_u = 0.48 \times 20 \times 250 \times 450 = 1080 \text{ kN}$$

(b) To obtain the maximum eccentricity draw a horizontal line at $P_u/(f_{ck}bD) = 0.244$. From the point of intersection of this line with the appropriate p/f_{ck} curve, drop a perpendicular to meet the x-axis and note $M_u /(f_{ck}bD^2) = 0.163$.

Therefore,

$$e = \frac{M_u}{P_u} = \frac{0.163 \times f_{ck}bD^2}{P_u}$$

$$= \frac{0.163 \times 20 \times 250 \times 450^2}{550 \times 10^3} = 300.07 \text{ mm}$$

5.13 SECTION SUBJECTED TO COMBINED AXIAL LOAD AND BIAXIAL BENDING

The design of a column section for a given axial load P_u and biaxial moments M_{ux} and M_{uy} about the x-and y-axes, respectively, is made by predetermining the trial section and testing its adequacy. The section may be considered to be acted upon by the axial load at eccentricities $e_x = M_{uy} /P_u$ and $e_y = M_{ux} /P_u$. The section is considered safe if the given axial load acting at eccentricities e_x and e_y does not exceed its ultimate load capacity for the same eccentricities. The trial cross-sectional dimensions of a column are predetermined during structural analysis for computing design forces. Thus, the design consists only of determining the reinforcement details. This can be achieved by designing the trial section for uniaxial bending moment which is 10 to 15 per cent higher than the resultant moment:

$$M_u = 1.15\sqrt{M_{ux}^2 + M_{uy}^2}$$

The moment is considered to act with respect to the axis which provides larger of M_{uy}/d and M_{uy}/b'.

5.13.1 Breslar Method

Breslar has suggested a simple method based on the reciprocal failure surface to estimate ultimate load for biaxial bending load system by using the expression:

$$\frac{1}{P_{uxy}} = \frac{1}{P_{ux}} + \frac{1}{P_{uy}} - \frac{1}{P_{uz}} \tag{5.35}$$

where,

P_{uxy} = Ultimate load for biaxial bending

P_{ux} = Ultimate load under uniaxial bending $e_x(e_y = 0)$

P_{uy} = Ultimate load under uniaxial bending $e_y(e_x = 0)$

P_{uz} = Ultimate axial load capacity ($e_x = e_y = 0$)

The method gives results within 10 per cent of the correct values for smaller values P_{uz}. The method is suitable for analysis.

5.13.2 Equivalent Uniaxial Bending Moment Method

BS:8110 has recommended a simplified procedure for design of symmetrically reinforced rectangular column sections subjected to biaxial bending. The method consists of transforming biaxial bending case to a uniaxial bending case with increased moment about the governing axis determined as follows.

Consider a column subjected to a biaxial load system P_u, M_{ux} and M_{uy}. This column can be designed for uniaxial bending system, (P_u, M'_{ux}) or (P_u, M'_{uy}) depending on the conditions:

1. For the case $(M_{ux}/d) \geq (M_{uy}/b')$, design load is P_u and

$$\left[M'_{ux} = M_{ux} + \alpha\left(\frac{d}{b'}\right)M_{uy} \right]$$

2. For the case $(M_{ux}/d) \leq (M_{uy}/b')$, uniaxial bending load system is P_u and

$$\left[M'_{uy} = M_{uy} + \alpha\left(\frac{b'}{d}\right)M_{ux} \right]$$

where

d = Effective depth w.r.t. major axis with total depth D

b' = Effective depth w.r.t. minor axis with total depth b

$$\alpha = \left[1 - \frac{7}{6}\left(\frac{P_u}{f_{ck} bD}\right)\right] \tag{5.36}$$

The trial area of steel can be predetermined from the resultant uniaxial moment acting about the controlling axis.

In case of circular section, due to dimensional symmetry defined by its diameter, design for longitudinal reinforcement can be made by single axis bending considering resultant of biaxial moments given by

$$M_u = \sqrt{M_{ux}^2 + M_{uy}^2}$$

5.13.3 Bresler's Load Contour Approach

IS:456 has recommended the formula proposed by Bresler for the design of columns subjected to biaxial bending. The method is based on interaction surface for the load system P_u, M_{ux} and M_{uy}.

$$\left(\frac{M_{ux}}{M_{ux1}}\right)^{\alpha} + \left(\frac{M_{uy}}{M_{uy1}}\right)^{\alpha} \leq 1.0 \tag{5.37}$$

where M_{ux1} and M_{uy1} are the uniaxial moment capacities combining the given axial load P_u with bending moments about x- and y-axes, respectively.

The value of α depends on the parameter P_u/P_{uz}. For P_u/P_{uz} values varying from 0.2 to 0.8, α varies linearly from 1.0 to 2.0:

for $0.2 < \dfrac{P_u}{P_{uz}} < 0.8$, $\qquad \alpha = \left(\dfrac{2}{3}\right) + \left(\dfrac{5}{3}\right)\left(\dfrac{P_u}{P_{uz}}\right)$ (5.38)

for $\dfrac{P_u}{P_{uz}} < 0.2$, $\qquad \alpha = 1.0$ square interface

for $\dfrac{P_u}{P_{uz}} > 0.8$, $\qquad \alpha = 2.0$ circle interface

where

P_u = Axial load

$P_{uz} = 0.447 f_{ck} A_c + 0.75 f_y A_{sc} (e = 0)$

$= 0.40 f_{ck} A_c + 0.67 f_y A_{sc}$ $\quad$ ($0 < e < 0.05$ times lateral dimension) (5.39)

In practical designs, α = 1.15 to 1.55 is satisfactory for rectangular sections and 1.5 to 2.0 for square sections.

The following procedure can be adopted for the design of biaxially loaded column:

1. Check that the eccentricities $e_x = (M_{ux}/P_u)$ and $e_y = (M_{uy}/P_u)$ are not less than the corresponding minimum design eccentricities.
2. Pre-determine or assign the trial cross-sectional dimensions and its effective cover. In practice, cross-sectional dimensions of column are, generally, tentatively fixed in advance for structural analysis or are governed by architectural considerations.
3. Compute M_{ux1}, and M_{uy1} corresponding to P_u such that they are significantly greater than M_{ux} and M_{uy}; otherwise revise the trial section.
4. For the given section, determine P_{uz} and hence α.
5. Check the adequacy of the section, redesign, if necessary.

Example 5.11 Design a suitable reinforced concrete section of a column of effective height 2.85 m to resist a factored axial load of 250 kN along with factored moments of 35 kNm about both the major and minor axes. The effective cover to reinforcement is 40 mm on all four sides. Concrete of grade M20 and HYSD steel of grade Fe415 are to be used in the construction.

Solution For the given materials:

$$f_{ck} = 20 \text{ MPa and } f_y = 415 \text{ MPa.}$$

$$P_u = 250 \text{ kN, } M_{ux} = M_{uy} = 35 \text{ kNm, and } l_{\text{eff}} = 2.85 \text{ m}$$

Since the moments about the major and minor axes are equal, consider a square section of size 250 mm. Therefore,

$$b = D = 250 \text{ mm, } \frac{d'}{D} = \frac{d'}{b} = \frac{40}{250} = 0.16 \approx 0.15$$

and

$$\frac{l_{\text{eff}}}{D} = \frac{2850}{250} \; 11.4 < 12$$

hence column may be designed as a short column.

Eccentricities.

$$\text{Applied eccentricities, } e_x = e_y = \frac{35 \times 10^6}{250 \times 10^3} = 140 \text{ mm}$$

$$\text{Minimum eccentricities, } e_{x,\min} = e_{y,\min} = \frac{2.85 \times 10^3}{500} + \frac{250}{30} = 14.03 \text{ mm} < 20 \text{ mm}$$

As the minimum eccentricities are less than the applied eccentricities, no modificatioin to, M_{ux} and M_{uy} values is required.

$$\frac{M_{ux}}{f_{ck}bD^2} = \frac{M_{uy}}{f_{ck}b^2D} = \frac{35 \times 10^6}{20 \times 250 \times 250^2} = 0.112$$

$$\frac{P_u}{f_{ck}bD} = \frac{250 \times 10^3}{20 \times 250 \times 250} = 0.20$$

Due to symmetry, consider the ratio of $M_{uy}/M_{uy1} = 0.50$ and hence

$$\frac{M_{uy1}}{f_{ck}bD^2} = \frac{0.112}{0.5} = 0.224.$$

Refer to design chart given in Fig. 5.17, for $P_u/(f_{ck}bD) = 0.2$ and $M_{uy1}/(f_{ck}bD^2) = 0.224$, giving

$$\frac{p}{f_{ck}} = 0.206$$

Thus,

$$p = \frac{100A_s}{dD} = 0.206 \times 20 = 4.12 \text{ per cent}$$

$$A_s = 4.12 \times 250 \times 250/100 = 2575 \text{ mm}^2$$

Therefore,

$$P_{uz} = 0.447 \times 20 \times (250 \times 250 - 2575) + 0.75 \times 415 \times 2575$$

$$= 1337198 \text{ N} = 1337.2 \text{ kN}$$

$$\frac{P_u}{P_{uz}} = \frac{250 \times 10^3}{1337.2 \times 10^3} = 0.1870$$

For P_u/P_{uz} (= 0.1870) < 0.20,

$$\alpha = 1.0$$

Therefore,

$$\left(\frac{M_{ux}}{M_{ux1}}\right)^{\alpha} + \left(\frac{M_{uy}}{M_{ux1}}\right)^{\alpha} = \left(\frac{M_{ux}}{M_{ux1}}\right)^{1.0} + (0.5)^{1.0} = 1.0$$

Thus, $M_{ux}/M_{ux1} = 0.50$ or $M_{ux} = 0.50M_{ux1}$, and same amount of steel is required to be symmetrically placed in the *y*-direction.
Provide 8 bars of 22 mm ϕ (A_s = 3041 mm^2).

Transverse reinforcement.
Use 8 mm ϕ ties at the spacing which is least of the following:

1. D = 250 mm
2. 16 × 22 = 352 mm,
3. 300 mm

Provide 8 mm ϕ ties at 250 mm c/c.

Example 5.12 A reinforced concrete column cross-section of size 230 × 345 mm is subjected to factored moments of 35 kNm about major axis bi-secting the depth and 10 kNm about minor axis bi-secting the width. The effective cover to reinforcement, consisting of four bars of 20 mm diameter, is 50 mm on all four sides. Determine the axial load the column can support safely if concrete of grade M25 and HYSD steel of grade Fe415 are used in the construction.

Solution For the given materials:

$$f_{ck} = 25 \text{ MPa and } f_y = 415 \text{ MPa}$$

$$M_{ux} = 35 \text{ kNm}, M_{uy} = 10 \text{ kNm, and } d' = 50 \text{ mm}$$

$$b = 230 \text{ mm}, D = 345 \text{ mm, and } A_{st} = 1257 \text{ mm}^2 \text{ i.e. } p = 1.584 \text{ per cent.}$$

Therefore,

$$\frac{p}{f_{ck}} = \frac{1.584}{25} = 0.0634$$

$$\frac{d'}{D} = \frac{50}{345} = 0.15$$

and

$$\frac{M_{ux}}{f_{ck}bD^2} = \frac{35 \times 10^6}{25 \times 230 \times 345^2} = 0.051$$

From *P-M* interaction curve given in Fig. 5.17,

$$\frac{P_{ux}}{f_{ck}bD} = 0.51$$

$$P_{ux} = 0.51 \times 25 \times 230 \times 345 \times 10^{-3} = 1011.7 \text{ kN}$$

$$\frac{p}{f_{ck}} = 0.0634, \quad \frac{d'}{D} = \frac{50}{230} = 0.217 \approx 0.20$$

and

$$\frac{M_{uy}}{f_{ck}b^2D} = \frac{10 \times 10^6}{25 \times 230^2 \times 345} = 0.022$$

Refer to the appropriate *P-M* interaction design chart of *SP*-16,

$$\frac{P_{ux}}{f_{ck}bD} = 0.59$$

$$P_{uy} = 0.59 \times 25 \times 230 \times 345 \times 10^{-3} = 1170.41 \text{ kN}$$

and

$$P_{uz} = [0.447 \times 25 \times (230 \times 345 - 1257) + 0.75 \times 415 \times 1257] \times 10^{-3}$$

$$= 1263.93 \text{ kN}$$

From Eq. (5.35),

$$\frac{1}{P_{uxy}} = \frac{1}{1011.70} + \frac{1}{1170.41} - \frac{1}{1263.93}$$

Therefore,

$$P_{uxy} = 950.88 \text{ kN}$$

5.14 SLENDER COLUMNS

A compression member is considered long or slender when the slenderness ration l_{ex}/D or l_{ey}/b about the major or minor axis, respectively, is more than 12, where l_{ex} is the effective length with respect to the major axis, l_{ey} is the effective length the respect to the minor axis, D is depth of the cross-section at right angles to the major axis, and b is the width of the member.

Slender columns are subjected to buckling which results in an increase in secondary moments due to lateral deflections. IS:456 has recommended the following additional secondary moments:

$$M_{xa} = \frac{P_u D}{2000}\left(\frac{l_{ex}}{D}\right)^2$$

and

$$M_{ya} = \frac{P_u b}{2000}\left(\frac{l_{ey}}{b}\right)^2 \tag{5.40}$$

where M_{xa}, M_{ya} are the additional moments about the major and minor axes, respectively and P_u is the axial load on the column. These additional moments are based on the balanced failure of the section, whereas it generally fails in compression (being primarily a compression member). In such cases, secondary moment is reduced due to decrease in curvature. Thus the values of M_{xa} and M_{ya} are reduced by a factor k, given by

$$k = \frac{P_{uz} - P_u}{P_{uz} - P_{ub}} \le 1 \tag{5.41}$$

where

P_u is the axial load to be supported on the compression member.

$P_{uz} = 0.447 f_{ck} A_c + 0.75 f_y A_{sc}$

P_{ub} is the axial load corresponding to the condition of the maximum compressive strain of 0.0035 in the extreme compression fibre of concrete and tension strain of 0.002 in the outermost layer of tensile steel.

5.14.1 Calculation of P_{ub}

SP:16 has proposed a method to calculate P_{ub}, when steel is distributed on all faces. The values are given by the follwing expressions:

For rectangular section

$$P_{ub} = \left[k_1 + k_2\left(\frac{p}{f_{ck}}\right)\right](f_{ck} bD) \tag{5.42}$$

For circular section

$$P_{ub} = \left[k_1 + k_2\left(\frac{p}{f_{ck}}\right)\right](f_{ck} D^2) \tag{5.43}$$

k_1 and k_2 can be obtained from Table 5.5, where p is the percentage of total steel in the column.

TABLE 5.5 Coefficients k_1 and k_2 for Computation of P_{ub} for Slender Compression Member

Constant	f_y, *MPa*	*Type of cross–section of column*	*d′/D* 0.05	0.10	0.15	0.20
k_1		*Rectangular*	0.219	0.207	0.196	0.184
		Circular	0.172	0.160	0.149	0.138
k_2	250	*Rectangular*				
		Equal steel on two sides	–0.045	–0.045	–0.045	–0.045
		Equal steel on four sides	0.215	0.146	0.061	–0.011
		Circular	0.193	0.148	0.077	–0.020
	415	*Rectangular*				
		Equal steel on two sides	0.096	0.082	0.046	–0.022
		Equal steel on four sides	0.424	0.328	0.203	0.028
		Circular	0.410	0.323	0.201	0.036
	500	*Rectangular*				
		Equal steel on two sides	0.213	0.173	0.104	–0.001
		Equal steel on four sides	0.545	0.425	0.256	0.040
		Circular	0.543	0.443	0.291	0.056

5.14.2 Design Moments in Slender Columns

As discussed earlier the columns in frames may be pin-ended, braced, and unbraced and may bend in single or double curvature. If the moments at the two ends, M_1 and M_2 are both are opposite in sign the columns bend in single curvature. This case is normally encountered in non-sway or braced frames. On the other hand If the moments on the two ends of the column are same in sign the columns bend in double curvature. This case generally occurs in sway or unbraced frames.

The design or total moment in slender columns, M_{ud} is governed by the combined effects of following three types of moments:

1. The initial moment M_{ui} caused by the end moments, M_{u1} and $M_{u2.}$ (The larger value is taken as positive and designated as $M_{u2.}$ If the bending is in double curvature, M_{u1} is taken as negative)
2. Moment due to accidental eccentricity usually designated by the term M_{umin}.
3. Additional moment M_{ua}.

The design or resultant moments are obtained by superposing the additional moments to the maximum initial or primary moments, obtained by a suitable combination of end moments of the member and accidental moments.

Design moment in braced columns: The initial moment M_{ui} for braced column due to M_{u2} and M_{u1} according to IS:456 is given by

$$M_{ui} = 0.60M_{u2} + 0.40M_{u1} \text{ (with appropriate signs)} \not< 0.40M_{u2} \tag{5.44}$$

where M_{u1} and M_{u2} are the smaller and larger end moments obtained from structural analysis, and M_{u1} is taken as negative if the column is bent in double curvature. Based on bending moment envelopes the maximum values possible at top, bottom and intermediate points are given in Table 5.6. The design moment should be largest of four values given in the table.

TABLE 5.6 Design Moments in Braced and Unbraced Frame Columns

Location	*Braced frame columns*		*Unbraced frame columns*	
	Bent in single curvature	*Bent in double curvature*	*Bent in single curvature*	*Bent in double curvature*
At M_{u2} end	M_{u2}	M_{u2}	$M_{u2} + M_{ua}$	$M_{u2} + M_{ua}$
Intermediate point	M_{ui} or $M_{umin} + M_{ua}$	M_{ui} or $M_{umin} + M_{ua}$		
At M_{u1} end	M_{u1}	$M_{u1} + 0.50M_{ua}$	M_{u1}	$M_{u1} + M_{ua}$

Example 5.13 Design a rectangular column of size 350 × 500 mm subjected to a factored axial load of 2000 kN and factored moments of 55 kNm at the top and 30 kNm at the bottom in the longer direction; 42 kNm at the top and 22 kNm at the bottom in the shorter direction. The unsupported length of the column (bent in double curvature) is 8.5 m with effective lengths of 7.25 and 6.05 m in the directions of the long and short dimensions of the section, respectively. The grades of concrete mix and reinforcing steel are M30 and Fe415, respectively. Consider the reinforcement to be equally distributed on all four sides.

Solution For the given materials and column cross-section:

$$f_{ck} = 30 \text{ MPa and } f_y = 415 \text{ MPa}$$

$$b \times D = 350 \times 500 \text{ mm and } L,\ l_{ex},\ l_{ey} = 8.5,\ 7.25,\ 6.05 \text{ m.}$$

Slenderness check:

$$\frac{l_{ex}}{D} = \frac{7.25 \times 10^3}{500} = 14.50 > 12$$

$$\frac{l_{ey}}{b} = \frac{6.05 \times 10^3}{350} = 17.29 > 12$$

The column is slender about both the axes. As the moments produce double curvature, M_1 is negative.

Primary moments due to applied loads.
As the column is braced and bent in double curvature,

$$M_{uxi} = [0.60 \times 55 - 0.40 \times 30] = 21.0 \text{ kNm} \not< 0.40M_2 (= 22.0 \text{ kNm})$$

$$M_{uyi} = [0.60 \times 42 - 0.40 \times 22] = 16.4 \text{ kNm} \not< 0.40M_2 (= 16.8 \text{ kNm})$$

Primary moments due to minimum eccentricities.
The primary eccentricities are greater than the minimum eccentricities, thus moments are:

$$M_{ux,\text{emin}} = P_u e_{x,\text{min}} = 2000 \times \left[\frac{8500}{500} + \frac{500}{30}\right] \times 10^{-3} = 67.33 \text{ kNm}$$

$$M_{uy,\text{emin}} = P_u e_{y,\text{min}} = 2000 \times \left[\frac{8500}{500} + \frac{350}{30}\right] \times 10^{-3} = 57.33 \text{ kNm}$$

Primary design moments.

$$M_{ux} = 67.33 \text{ kNm}$$

$$M_{uy} = 57.33 \text{ kNm}$$

Additional secondary moments.

$$M'_{ax} = \frac{P_u D}{2000}\left(\frac{l_{ex}}{D}\right)^2 = \frac{2000 \times 10^3 \times 500}{2000}(14.5)^2 \times 10^{-6} = 105.125 \text{ kNm}$$

$$M'_{ay} = \frac{P_u b}{2000}\left(\frac{l_{ey}}{b}\right)^2 = \frac{2000 \times 10^3 \times 350}{2000}(17.29)^2 \times 10^{-6} = 104.63 \text{ kNm}$$

The values of M_{xa} and M_{ya} are to be reduced by a factor k, given by

$$k_a = \frac{P_{uz} - P_u}{P_{uz} - P_{ub}} \leq 1.0$$

where

$$P_{ub} = \left[k_1 + k_2\left(\frac{p}{f_{ck}}\right)\right](f_{ck} bD)$$

As the value of p is not known at this stage, consider reduction factors $k_{ax} = k_{ay} = 0.50$.
Reduced additional moments.

$$M_{ax} = 0.50 \times 105.125 = 52.56 \text{ kNm}$$

$$M_{ay} = 0.50 \times 104.63 = 52.32 \text{ kNm}$$

Total factored moments.

$$M_{ux} = 67.33 + 52.56 = 119.89 \text{ kNm}$$

$$M_{uy} = 52.32 + 57.33 = 109.65 \text{ kNm}$$

Preliminary estimate of reinforcement. Using the method proposed by BS:8110

$$\frac{P_u}{f_{ck} bD} = \frac{2000 \times 10^3}{30 \times 350 \times 500} = 0.381$$

$$\frac{M_{ux}}{d} = \frac{119.89 \times 10^6}{447.5} = 268 \times 10^3$$

and

$$\frac{M_{uy}}{b'} = \frac{109.65 \times 10^6}{297.5} = 369 \times 10^3$$

As $(M_{ux}/d) \leq (M_{uy}/b')$, M_{uy} the moment about *y*-axis governs and equivalent uniaxial moment is:

$$\alpha = \left[1 - \frac{7}{6}\left(\frac{P_u}{f_{ck}bD}\right)\right] = \left[1 - \frac{7}{6}(0.381)\right] = 0.5555$$

and

$$\left[M'_{uy} = 109.65 + 0.5555 \times \left(\frac{297.5}{447.5}\right) \times 119.89 = 153.93 \text{ kNm}\right]$$

For d'/b = 0.15 and Fe415 from the *P-M* curve of Fig. 5.17,

$$\frac{P_u}{f_{ck}bD} = 0.381$$

and

$$\frac{M_{uy}}{f_{ck}b^2D} = \frac{153.93 \times 10^6}{30 \times 350^2 \times 500} = 0.084$$

As

$$p/f_{ck} = 0.06, \text{ therefore, } p = 0.06 \times 30 = 1.8 \text{ per cent}$$

Consider the section to be provided with 2.0 per cent steel distributed equally on four sides.

$$A_s = \frac{pbD}{100} = \frac{2.0 \times 350 \times 500}{100} = 3500 \text{ mm}^2$$

Adopt 25 mm bars with 40 mm nominal cover, and hence d' = 40 + (25/2) = 52.5 mm

$$\begin{aligned} P_{uz} &= 0.447 f_{ck} A_c + 0.75 f_y A_s \\ &= 0.447 \times 30 \times (0.98 \times 350 \times 500) + 0.75 \times 415 \times (0.02 \times 350 \times 500) \\ &= 3389.19 \times 10^3 \text{ N} = 3389.19 \text{ kN} \end{aligned}$$

P_{ub} for the effective cover ratios.

1. About *x*-axis, for $\frac{d'}{D} = \frac{52.5}{500} = 0.105 \approx 0.10$ and Fe415, from Table 5.3, k_1 = 0.207 and k_2 = 0.328. Then

$$P_{bx} = \left[0.207 + 0.328 \times \left(\frac{2.0}{30}\right)\right](30 \times 350 \times 500) \times 10^{-3} = 1201.55 \text{ kN}$$

and

$$k_{ux} = \frac{P_{uz} - P_u}{P_{uz} - P_{bx}} = \frac{3389.19 - 2000}{3389.19 - 1201.55} = 0.635 \leq 1.0$$

Reduced additional moment $M_{ax} = k_x M'_{ax} = 0.635 \times 105.125 = 66.75$ kN

2. About y-axis, for $\frac{d'}{b} = \frac{52.5}{350} = 0.15$ and Fe415 from the Table 5.3, $k_1 = 0.196$ and $k_2 = 0.203$.
Then

$$P_{by} = \left[0.196 + 0.203 \times \left(\frac{2.0}{30}\right)\right](30 \times 350 \times 500) \times 10^{-3} = 1100.05 \text{ kN}$$

and

$$k_{uy} = \frac{P_{uz} - P_u}{P_{uz} - P_{by}} = \frac{3389.19 - 2000}{3389.19 - 1100.05} = 0.607 \leq 1.0$$

$$M_{ay} = k_{uy} M'_{ay} = 0.607 \times 104.63 = 63.51 \text{ kN}$$

Design moments

The design moment $M_{tu,d}$ is larger of the following:

Combination of moments	M_{ux}	M_{uy}	*Location*
M_{u2}	55	42	at M_{u2} end
$M_{u1} + 0.50\, M_{ua}$	30 + 0.50 × 66.75 = 63.38	22+0.50 × 63.51 =53.76	at M_{u1} end
M_{ui} or $M_{ux,ein} + M_{ua}$	67.33 + 66.75 = **134.08**	57.33 + 63.51 = **120.84**	at intermediate points

Therefore,

$$M_{ux} = 134.08 \text{ kNm}$$

and

$$M_{uy} = 120.84 \text{ kNm}$$

Uniaxial moment capacities about the major and minor axes.

$$\frac{P_u}{f_{ck} bD} = \frac{2000 \times 10^3}{30 \times 350 \times 500} = 0.381$$

and

$$\frac{p}{f_{ck}} = \frac{2.0}{30} = 0.067$$

Refer to the column design interaction diagram of Figs. 7.16 and 7.17 corresponding to $f_y = 415$, for the $d'/D = 0.10$ and 0.15, respectively. For $P_u/(f_{ck}bD) = 0.381$ and $p/f_{ck} = 0.067$, the dimensionless load parameters are:

$$\text{For } \frac{d'}{D} = 0.10; \quad \frac{M_{ux}}{f_{ck}bD^2} = 0.093$$

Therefore, $M_{ux1} = 0.093(f_{ck}bD^2) = 0.093 \times (30 \times 350 \times 500^2) \times 10^{-6} = 244.125$ kNm

$$\text{For } \frac{d'}{D} = 0.15: \quad \frac{M_{ux}}{f_{ck}bD^2} = 0.083$$

and

$$M_{uy1} = 0.083\ (f_{ck}b^2D) = 0.083 \times (30 \times 350^2 \times 500) \times 10^{-6} = 152.51 \text{ kNm}$$

Check for the biaxial moment adequacy.

Pure axial load capacity of the column $P_{uz} = 3389.19$ kN

For the ratio,

$$\frac{P_u}{P_{uz}} = \frac{2000}{3389.19} = 0.590, \text{ the exponent,}$$

$$\alpha = \left(\frac{2}{3}\right) + \left(\frac{5}{3}\right)(0.590) = 1.65$$

Substitute the above values in the interaction formula:

$$\left(\frac{M_{ux}}{M_{ux1}}\right)^{\alpha} + \left(\frac{M_{uy}}{M_{uy1}}\right)^{\alpha} = \left(\frac{134.08}{244.125}\right)^{1.65} + \left(\frac{120.84}{152.51}\right)^{1.65} = 1.053 \geq 1.0$$

The section is just inadequate with 2.0 per cent steel, a little more steel will make it adequate. However, for second trial increase the amount of steel to 2.25 per cent, equally distributed on four sides and recheck the adequacy of the section.

Alternatively using the method proposed by BS:8110,

$$\frac{P_u}{f_{ck}bD} = \frac{2000 \times 10^3}{30 \times 350 \times 500} = 0.381$$

As $(M_{ux}/d) \leq (M_{uy}/b')$, M_{uy} the moment about y-axis governs and for equivalent uniaxial moment:

$$\alpha = \left[1 - \frac{7}{6}\left(\frac{P_u}{f_{ck}bD}\right)\right] = \left[1 - \frac{7}{6}(0.381)\right] = 0.5555$$

and

$$M'_{uy} = 120.84 + 0.5555 \times \left(\frac{297.5}{447.5}\right) \times 134.08 = 170.36 \text{ kNm}$$

From the *P-M* curve for $d'/b = 0.15$ and Fe415 given in Fig. 5.17:

For $\dfrac{P_u}{f_{ck}bD} = 0.381$

and

$$\frac{M_{uy}}{f_{ck}b^2D} = \frac{170.36 \times 10^6}{30 \times 350^2 \times 500} = 0.093$$

$$\frac{p}{f_{ck}} = 0.077,$$

Therefore,

$$p = 0.077 \times 30 = 2.3 \text{ per cent}$$

The trial reinforcement can also be computed from equivalent uniaxial moment obtained from resultant moment as follows:

$$M_u = 1.15\sqrt{M_{ux}^2 + M_{uy}^2} = 1.15\sqrt{(134.08)^2 + (120.84)^2} = 1.15 \times 180.50 \text{ kNm}$$

Since $(M_{ux}/d) \leq (M_{uy}/b')$ the moment acts about *y*-axis.

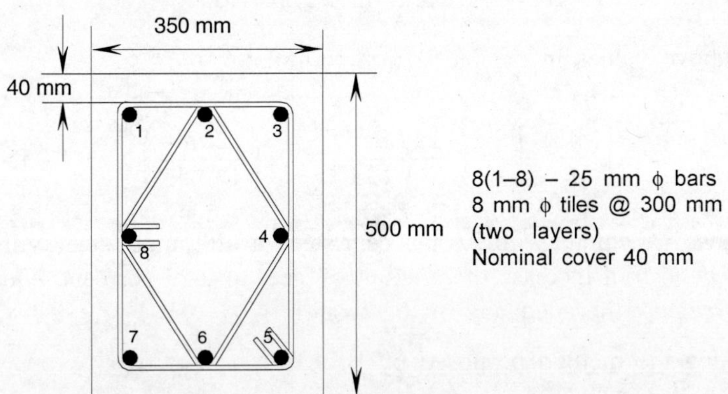

Fig. 5.21 *Detailing of reinforcement in the column of Example 5.13.*

Reinforcement.
Provide 8 bars of 25 mm ϕ bars (A_{sc} = 3927 mm^2) with 8 mm ϕ bar ties at 300 mm c/c. The reinforcement details are shown in Fig. 7.21.

Example 5.14 Design a biaxially loaded braced column section of size 400 × 550 mm subjected to a factored axial load of 2000 kN and factored moments about major and minor axes of 170 and 65 kNm at the top, and 215 and 125 kNm at the bottom. The unsupported length of the column (bent in single curvature) is 7.0 m with effective lengths about major

(x-) and minor (y-) axes of 6.0 and 5.75 m, respectively. The grades of concrete mix and reinforcing steel are M25 and Fe415, respectively. Consider the reinforcement to be equally distributed on all four sides.

Solution For the given materials and column cross-section:

f_{ck} = 25 MPa and f_y = 415 MPa

$b \times D$ = 400 × 550 mm and L, l_{ex}, l_{ey} = 7.0, 6.0, 5.75 m

P_u = 2000 kN, M_{ux1} = 215 kNm, M_{uy1} = 125 kNm, M_{ux2} = 170 kNm, and M_{uy2} = 65 kNm

Adopt 25 mm ϕ bars with 40 mm nominal cover, and hence d' = 40 + (25/2) = 52.5 mm

Therefore,

$$\frac{d'}{d} = \frac{52.5}{550} \approx 0.10$$

Slenderness check.

$$\frac{l_{ex}}{D} = \frac{6.0 \times 10^3}{550} = 10.91 \le 12$$

$$\frac{l_{ey}}{b} = \frac{5.75 \times 10^3}{400} = 14.375 > 12$$

Therefore, it is short column about x-axis and slender column about y-axis.

Primary moments due to applied loads.

As the column is braced and deformed in single curvature,

$$M_{uxi} = [0.60 \times 215 + 0.40 \times 170] = 197.0 \text{ kNm} \not< 0.40\ M_2\ (= 86 \text{ kNm})$$

$$M_{uyi} = [0.60 \times 125 + 0.40 \times 65] = 101.0 \text{ kNm} \not< 0.40\ M_2\ (= 50 \text{ kNm})$$

Primary moments due to minimum eccentricities.

$$e_{x,\min} = \left(\frac{L}{500} + \frac{D}{30}\right), \frac{D}{20} \text{ and } 20 \text{ mm whichever is greater}$$

$$= \left(\frac{7000}{500} + \frac{550}{30} = 32.33 \text{ mm}\right), 27.5 \text{ mm or } 20 \text{ mm}$$

and

$$e_{y,\min} = \left(\frac{7000}{500} + \frac{400}{30} = 27.33 \text{ mm}\right) \text{ or } 20 \text{ mm}$$

Adopt $e_{x,\min}$ = 32.33 mm and $e_{y,\min}$ = 27.33 mm

The primary eccentricities are greater than the minimum eccentricities, thus, moments are:

$$M_{ux,ein} = P_u e_{x,\min} = 2000 \times 32.333 \times 10^{-3} = 64.67 \text{ kNm}$$

$$M_{uy,ein} = P_u e_{y,\min} = 2000 \times 27.333 \times 10^{-3} = 54.67 \text{ kNm}$$

Primary design moments.

$$M_{uxi} = 197.0 \text{ kNm}$$
$$M_{uyi} = 101.0 \text{ kNm}$$

Additional secondary moments.

$$M'_{ax} = 0.0 \text{ kNm}$$

$$M'_{ay} = \frac{P_u b}{2000}\left(\frac{l_{ey}}{b}\right)^2 = \frac{2000 \times 10^3 \times 400}{2000} \times (14.375)^2 \times 10^{-6} = 82.66 \text{ kNm}$$

The values of M'_{xa} and M'_{ya} are to be reduced by a factor k_a, given by

$$k_a = \frac{P_{uz} - P_u}{P_{uz} - P_{ub}} \leq 1.0$$

where

$$P_{ub} = \left[k_1 + k_2\left(\frac{p}{f_{ck}}\right)\right](f_{ck}bD)$$

As the value of p is not known at this stage, consider reduction factor $k_{ay} = 0.50$.

Reduced additional moments.

$$M_{ax} = 0.0 \text{ kNm}$$
$$M_{ay} = 0.50 \times 82.66 = 41.33 \text{ kNm}$$

Total factored moments,

$$M_{ux} = 197.0 + 0.0 = 197.0 \text{ kNm} < M_{ux1} \ (= \mathbf{215\ kNm})$$
$$M_{uy} = 101.0 + 41.33 = \mathbf{142.33\ kNm} > M_{uy1} \ (= 125 \text{ kNm})$$

Preliminary estimate of reinforcement. Using the method proposed by BS:8110,

$$\frac{P_u}{f_{ck}bD} = \frac{2000 \times 10^3}{25 \times 400 \times 550} = 0.364$$

$$\frac{M_{ux}}{d} = \frac{215 \times 10^6}{497.5} = 432.2 \times 10^3$$

and

$$\frac{M_{uy}}{b'} = \frac{142.33 \times 10^6}{347.5} = 409.58 \times 10^3$$

As $(M_{ux}/d) > (M_{uy}/b')$, the moment M_{ux} about x-axis governs and for equivalent uniaxial moment:

$$\alpha = \left[1 - \frac{7}{6}\left(\frac{P_u}{f_{ck}bD}\right)\right] = \left[1 - \frac{7}{6}(0.364)\right] = 0.5753$$

and

$$M'_{ux} = 215.0 + 0.5753 \times \left(\frac{497.5}{347.5}\right) \times 142.33 = 332.23 \text{ kNm}$$

From the P-M curve for $d'/D = 0.10$ and Fe415 given in Fig. 5.16:

For $$\frac{P_u}{f_{ck}bD} = 0.364$$

and

$$\frac{M_{ux}}{f_{ck}bD^2} = \frac{332.23 \times 10^6}{25 \times 400 \times 5502} = 0.11$$

$$\frac{p}{f_{ck}} = 0.08$$

Therefore,

$$p = 0.08 \times 25 = 2.0 \text{ per cent}$$

Consider the section to be provided with 2.2 per cent steel distributed equally on four sides.

$$A_s = \frac{pbD}{100} = \frac{2.2 \times 400 \times 550}{100} = 4840 \text{ mm}^2$$

$$P_{uz} = 0.447\, f_{ck}A_c + 0.75 f_y A_s$$

$$= 0.447 \times 25 \times (0.978 \times 400 \times 550) + 0.75 \times 415 \times (0.022 \times 400 \times 550)$$

$$= 3927.0 \times 10^3 \text{ N} = 3910.86 \text{ kN.}$$

P_{uby} about y-axis:
For $d'/b \approx 0.15$ and Fe415, from Table 5.3, $k_1 = 0.196$ and $k_2 = 0.203$

$$P_{uby} = \left[0.196 + 0.203 \times \left(\frac{2.2}{25}\right)\right](25 \times 400 \times 550) \times 10^{-3} = 1176.25 \text{ kN}$$

$$k_y = \frac{P_{uz} - P_u}{P_{uz} - P_{uby}} = \frac{3910.86 - 2000}{3910.86 - 1176.25} = 0.699 \le 1.0$$

Reduced additional moment $M_{ay} = k_y M_{ay}$

$$0.699 \times 82.86 = 57.92 \text{ kN}$$

Modified design moments.

$$M_{ux} = 215 \text{ kNm}$$

$$M_{uy} = 101.0 + 57.92 = 158.92 \text{ kNm} > M_{uy1} \ (= 125 \text{ kNm})$$

Uniaxial moment capacities about the major and minor axes:

$$\frac{P_u}{f_{ck}bD} = 0.364$$

and

$$\frac{p}{f_{ck}} = \frac{2.2}{25} = 0.088.$$

Refer to the column interaction diagram of Figs. 7.16 and 7.17 corresponding to $f_y = 415$, for the $d'/D = 0.10$ and 0.15, respectively. For the dimensionless load parameters $P_u/(f_{ck}bD) = 0.364$ and $p/f_{ck} = 0.088$:

For $\dfrac{d'}{D} = 0.10$, $\quad \dfrac{M_{ux}}{f_{ck}bD^2} = 0.12$

Therefore,

$$M_{ux1} = 0.12(f_{ck}bD^2) = 0.12 \times (25 \times 400 \times 550^2) \times 10^{-6} = 363.0 \text{ kNm}$$

For $\dfrac{d'}{D} = 0.15$, $\quad \dfrac{M_{uy}}{f_{ck}b^2D} = 0.105$

Therefore,

$$M_{uy1} = 0.105\ (f_{ck}b^2D) = 0.105 \times (25 \times 400^2 \times 550) \times 10^{-6} = 231.0 \text{ kNm}$$

Check for the biaxial moment adequacy.
Pure axial load capacity of the column $P_{uz} = 3927.0$ kN
For the ratio,

$$\frac{P_u}{P_{uz}} = \frac{2000}{3910.86} = 0.5114$$

The exponent,

$$\alpha = \left(\frac{2}{3}\right) + \left(\frac{5}{3}\right)(0.5114) = 1.519$$

Substitute the above values in the interaction formula:

$$\left(\frac{M_{ux}}{M_{ux1}}\right)^{\alpha} + \left(\frac{M_{uy}}{M_{uy1}}\right)^{\alpha} = \left(\frac{215.0}{363.0}\right)^{1.519} + \left(\frac{158.92}{231.0}\right)^{1.519} = 1.018 \geq 1.0$$

The section is just inadequate with 2.2 per cent steel, a little more steel will make it adequate. However, for second trial increase the amount of steel to 2.3 per cent, equally distributed on four sides and recheck the adequacy of the section.

$$A_{sc} = 0.023 \times 400 \times 550 = 5060 \text{ mm}^2$$

Transverse reinforcement. Consider 8 mm φ bar ties at spacing which least of the following:

(i) 400 mm, (ii) 16 × 25 = 400 mm, and (iii) 300 mm

Provide 8–22 mm φ and 4-25 mm φ bars (A_{sc} = 5504 mm^2) with 8 mm φ bar ties at 300 mm c/c. The reinforcement details are shown in Fig. 5.22.

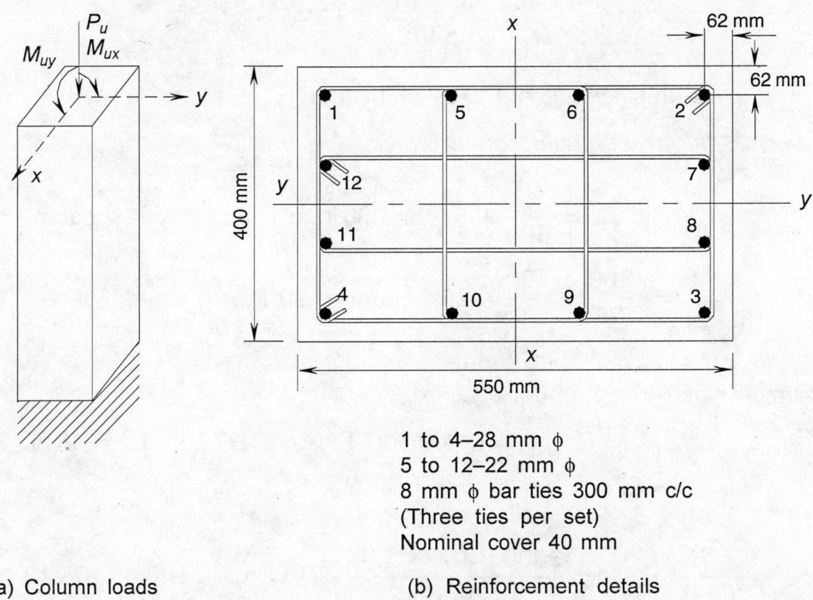

Fig. 5.22 Reinforcement details of column of Example 5.14.

Example 5.15 Design a biaxially loaded unbraced circular column section of 530 mm diameter with unsupported length of 3.30 m, subjected to a factored axial load of 2000 kN and factored moments about major and minor axes of 115 and 70 kNm at the top, and 180 and 90 kNm at the bottom. The effective lengths about major (x-) and minor (y-) axes are 7.5 and 6.6 m, respectively. The grades of concrete mix and reinforcing steel are M25 and Fe415, respectively.

Solution For the given materials and column cross-section:

f_{ck} = 25 MPa and f_y = 415 MPa

D = 530 mm and L, l_{ex}, l_{ey} = 3.3, 7.5, 6.6 m

P_u = 2000 kN, M_{ux1} =180 kNm, M_{uy1} = 90 kNm, M_{ux2} = 115 kNm, and M_{uy2} = 70 kNm

Adopt 25 mm bars with 40 mm nominal cover, and hence d' = 40 + (25/2) = 52.5 mm.

Therefore

$$\frac{d'}{D} = \frac{52.5}{530} \approx 0.10$$

Slenderness check.

$$\frac{l_{ex}}{D} = \frac{7.5 \times 10^3}{530} = 14.15 > 12$$

$$\frac{l_{ey}}{D} = \frac{6.6 \times 10^3}{530} = 12.45 > 12$$

Thus, it is a long column about both axes of bending.

Primary moments due to minimum eccentricities.

$$e_{x,\min} = e_{y,\min} = \frac{L}{500} + \frac{D}{30} \text{ and 20 mm whichever is greater}$$

$$= \frac{3300}{500} + \frac{530}{30} = 24.27 \text{ mm or 20 mm}$$

Adopt $e_{x,\min} = e_{y,\min} = e_{\min} = 24.27$ mm
Thus the moments due to minimum eccentricities are:

$$M_{ux,\min} = M_{uy,\min} = P_u e_{\min} = (2000 \times 10^3) \times 24.27 \times 10^{-6} = 48.54 \text{ kNm}$$

Additional secondary moments.

$$M'_{ax} = \frac{P_u D}{2000}\left(\frac{l_{ex}}{D}\right)^2 = \frac{2000 \times 10^3 \times 530}{2000} \times (14.15)^2 \times 10^{-6} = 106.12 \text{ kNm}$$

$$M'_{ay} = \frac{P_u D}{2000}\left(\frac{l_{ey}}{D}\right)^2 = \frac{2000 \times 10^3 \times 530}{2000} \times (12.45)^2 \times 10^{-6} = 82.15 \text{ kNm}$$

The values of M'_{xa} and M'_{ya} are to be reduced by a factor k_a, given by

$$k_a = \frac{P_{uz} - P_u}{P_{uz} - P_{ub}} \le 1.0$$

where

$$P_{ub} = \left[k_1 + k_2\left(\frac{p}{f_{ck}}\right)\right](f_{ck} D^2)$$

1. Column with lateral ties: Consider 2.0 per cent longitudinal steel in the form of 25 mm diameter bars uniformly distributed at nominal cover of 40 mm, with $d'/D \approx 0.10$.

$$A_g = \left(\frac{\pi D^2}{4}\right) = \left(\frac{\pi \times 530^2}{4}\right) = 220618.34 \text{ mm}^2$$

and

$$P_{uz} = 0.447\ f_{ck}A_c + 0.75 f_y A_s$$
$$= 0.447 \times 25 \times (0.98 \times 220618.34) + 0.75 \times 415 \times (0.02 \times 220618.34)$$
$$= 3789.45 \times 10^3 \text{ N} = 3789.45 \text{ kN}$$

For $d'/D \approx 0.10$ and Fe415, from Table 5.3, $k_1 = 0.160$ and $k_2 = 0.323$.
Then

$$P_{ub} = \left[0.160 + 0.323 \times \left(\frac{2.0}{25}\right)\right](25 \times 530^2) \times 10^{-3} = 1305.06 \text{ kN}$$

$$k_x = k_y = \frac{P_{uz} - P_u}{P_{uz} - P_{ub}} = \frac{3789.45 - 2000}{3789.45 - 1305.06} = 0.7202 \le 1.0$$

Reduced additional moment.

$$M_{ax} = k_x M'_{ax} = 0.7202 \times 106.12 = 76.43 \text{ kN}$$
$$M_{ay} = k_y M'_{ay} = 0.7202 \times 82.15 = 59.16 \text{ kN}$$

Primary moments due to applied loads.
As the column is unbraced,

$$M_{uxi} = [0.60 \times 180 + 0.40 \times 115] = 154.0 \text{ kNm} \nless 0.40 M_2\ (= 0.40 \times 180 = 72 \text{ kNm})$$
$$M_{uyi} = [0.60 \times 90 + 0.40 \times 70] = 82.0 \text{ kNm} \nless 0.40\ \text{M}_2\ (= 0.40 \times 90 = 36 \text{ kNm})$$

Total factored moments,

$$M_{ux} = 154.00 + 76.43 = \mathbf{230.43} \text{ kNm} > M_{ux1}\ (= 180 \text{ kNm})$$
$$M_{uy} = 82.00 + 59.16 = \mathbf{141.16} \text{ kNm} > M_{uy1}\ (= 90 \text{ kNm})$$

Preliminary estimate of reinforcement.
This can be achieved by designing the trial section for equivalent uniaxial bending moment.

$$M_u = \sqrt{M_{ux}^2 + M_{uy}^2} = \sqrt{(230.43)^2 + (141.16)^2} = 270.23 \text{ kNm}$$

Dimensionless load parameters are given as:

$$\frac{P_u}{f_{ck}D^2} = \frac{2000 \times 10^3}{25 \times 530^2} = 0.285$$

and

$$\frac{M_{ux}}{f_{ck}D^3} = \frac{270.23 \times 10^6}{25 \times 530^3} = 0.073$$

From the *P-M* curve corresponding to $d'/D = 0.10$ and Fe415 given in Fig. 5.18,

$$\frac{p}{f_{ck}} = 0.075, \text{ therefore, } p = 0.075 \times 25 = 1.875 \text{ per cent}$$

Then

$$A_g = \frac{0.01875 \times \pi \times 530^2}{4} = 4136.64 \text{ mm}^2$$

Transverse reinforcement. Consider 8 mm ϕ bar ties at spacing which is least of the following:

(i) 530 mm, (ii) $16 \times 25 = 400$ mm, and (iii) 300 mm

Provide 9×25 mm ϕ bars ($A_{sc} = 4417$ mm^2) with 8 mm ϕ bar ties at 300 mm c/c. The reinforcement details are shown in Fig. 5.23(a).

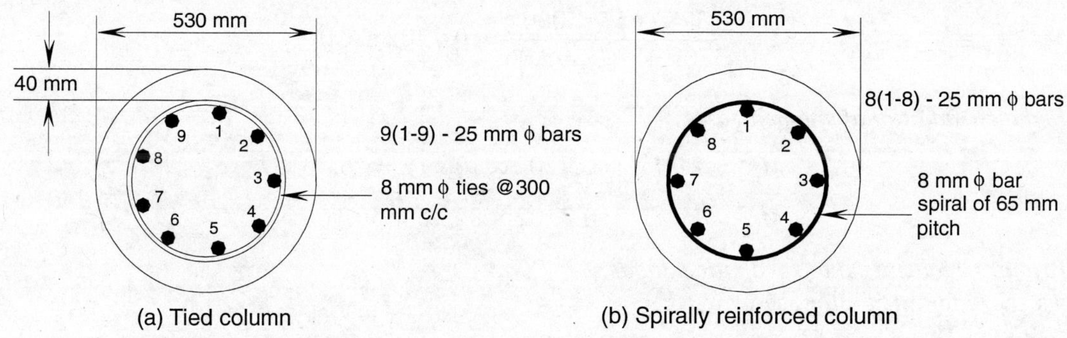

Fig. 5.23 Reinforcement details of circular column of Example 5.15.

2. *Column with spiral reinforcement:* Consider 1.75 per cent longitudinal steel in the form of 25 mm diameter bars uniformly distributed at nominal cover of 40 mm, with $d'/D \approx 0.10$. Then

$$A_g = \left(\frac{\pi D^2}{4}\right) = \left(\frac{\pi \times 530^2}{4}\right) = 220618.34 \text{ mm}^2$$

and

$$P_{uz} = 1.05 \times [0.447 f_{ck} A_c + 0.75 f_y A_s]$$

$$= 1.05 \times [0.447 \times 25 \times (0.9825 \times 220618.34) + 0.75 \times 415 \times (0.0175 \times 220618.34)]$$

$$= 3787.12 \times 10^3 \text{ N} = 3787.12 \text{ kN}$$

For $d'/D \approx 0.10$ and Fe415, from Table 5.3, $k_1 = 0.160$ and $k_2 = 0.323$.
Then

$$P_{ub} = 1.05 \times \left[0.160 + 0.323 \times \left(\frac{1.75}{25}\right)\right](25 \times 530^2) \times 10^{-3} = 1346.50 \text{ kN}$$

and

$$k_x = k_y = \frac{P_{uz} - P_u}{P_{uz} - P_{ub}} = \frac{3787.12 - 2000}{3787.12 - 1346.50} = 0.7322 \leq 1.0$$

Reduced additional moment.

$$M_{ax} = k_x M'_{ax} = 0.7322 \times 106.12 = 77.70 \text{ kN}$$

$$M_{ay} = k_y M'_{ay} = 0.7322 \times 82.15 = 60.15 \text{ kN}$$

Primary moments due to applied loads.
As in Case 1,

$$M_{uxi} = 154.0 \text{ kNm and } M_{uyi} = 82.0 \text{ kNm}$$

Total factored moments:

$$M_{ux} = 154.00 + 77.70 = \mathbf{231.70} \text{ kNm} > M_{ux1} \ (= 180 \text{ kNm})$$

$$M_{uy} = 82.00 + 60.15 = \mathbf{142.15} \text{ kNm} > M_{uy1} \ (= 90 \text{ kNm})$$

Preliminary estimate of reinforcement. This can be achieved by designing the trial section for equivalent uniaxial bending moment.

$$M_u = \sqrt{M_{ux}^2 + M_{uy}^2} = \sqrt{(231.70)^2 + (142.15)^2} = 271.23 \text{ kNm}$$

Dimensionless load parameters are given by

$$\frac{P_u}{1.05 f_{ck} D^2} = \frac{2000 \times 10^3}{1.05 \times 25 \times 530^2} = 0.271$$

$$\frac{M_{ux}}{1.05 f_{ck} D^3} = \frac{271.23 \times 10^6}{1.05 \times 25 \times 530^3} = 0.069$$

Refer the *P-M* curve of Fig. 5.18 corresponding to $d'/D = 0.10$ and Fe415:

$$p/f_{ck} = 0.07$$

Therefore,

$$p = 0.07 \times 25 = 1.75 \text{ per cent}$$

and

$$A_s = \frac{0.0175 \times \pi \times 530^2}{4} = 3860.82 \text{ mm}^2$$

Adopt 8 × 25 mm φ bars (A_s = 3929 mm^2).

Transverse reinforcement.
Consider 8 mm φ bar for helix,

$$D_c = 530 - 2 \times 40 + 2 \times 8 = 466 \text{ mm}$$

$$D_s = 530 - 2 \times 40 + 8 = 458 \text{ mm}$$

$$s_v = \left(\frac{11.11 a_{sp} D_c}{D^2 - D_c^2}\right)\left(\frac{f_{sp}}{f_{ck}}\right) = \left(\frac{11.11 \times (\pi \times 8^2/4) \times 458}{530^2 - 466^2}\right)\left(\frac{415}{25}\right) = 66.61 \text{ mm}$$

However, IS:456 has imposed restrictions on the pitch as follows:

$$s_v > \begin{cases} 25 \text{ mm} \\ 3\phi_{sp} = 24 \text{ mm} \end{cases}$$

$$s_v < \begin{cases} 75 \text{ mm} \\ 466/6 = 77.67 \text{ mm} \end{cases}$$

Provide 8 × 25 mm ϕ bars enclosed by 8 mm ϕ bar spiral of 65 mm pitch. The reinforcement details are shown in Fig. 5.23(b).

5.15 SHEAR FORCE IN COLUMNS SUBJECTED TO MOMENTS

The columns subjected to axial load and bending should be checked for shear. In the absence of axial load, shear along the entire height of column is given by

$$V = \frac{M_{\text{top}} + M_{\text{bottom}}}{\text{Storey height}}$$

However, due to presence of compression, the shear strength of concrete is larger than that in case of pure shear. BS:8110 has recommended enhanced shear which is smaller of the following:

(i) $\tau_c + 0.75\left(\frac{P}{A_c}\right)\left(\frac{Vd}{M}\right)$

(ii) $0.8\sqrt{f_{ck}}$ and

(iii) 5 MPa

Example 5.16 Check the adequacy in shear of a rectangular column of size 350 × 400 mm reinforced with 8 × 20 mm ϕ bars at an effective cover of 50 mm. The column is subjected to a factored axial load of 100 kN, and factored moments of 70 kNm at the top and 85 kNm at the bottom. The height of the column is 3.3 m. The grades of concrete mix and reinforcing steel are M30 and Fe415, respectively. Consider the reinforcement to be equally distributed on two small faces.

Solution For the given materials and column cross-section:

$$f_{ck} = 30 \text{ MPa and } f_y = 415 \text{ MPa}, A_s(8 \times 20 \text{ mm } \phi) = 2512 \text{ mm}^2$$

$$b \times D = 350 \times 400 \text{ mm}, d = 400 - 50 = 350 \text{ mm and } L = 3.3 \text{ m}$$

Calculation of nominal shear stress.

$$V_u = \frac{70 + 85}{3.3} = 46.97 \text{ kN}$$

$$\tau_v = \frac{V_u}{bd} = \frac{46.97 \times 10^3}{350 \times 350} = 3.83 \text{ MPa}$$

$$p_t = \frac{100A_s t}{bd} = \frac{100 \times (2512/2)}{350 \times 350} = 1.025 \text{ per cent}$$

From Table 4.1, shear strength of M30 grade concrete for $p_t = 1.025$ per cent is

$$\tau_c = 0.663 \text{ MPa}$$

Increased shear stress due to compression,

$$\tau'_c = \tau_c + 0.75\left(\frac{P}{A_c}\right)\left(\frac{Vd}{M}\right)$$

$$= 0.663 + 0.75\left(\frac{100 \times 10^3}{350 \times 400}\right) \times \left(\frac{46.97 \times 10^3 \times 350}{85 \times 10^6}\right) = 0.767 \text{ MPa}$$

Maximum permissible shear strength under compression:

$$\tau_{c,\max} = 0.8\sqrt{f_{ck}} = 0.8 \times \sqrt{30} = 4.38 < 5.0 \text{ MPa}$$

Since $\tau_v < \tau_c < \tau_{c,\max}$, the section is acceptable with nominal shear reinforcement.

Review Questions

5.1 Enumerate different criteria according to which columns are classified.

5.2 What is the minimum eccentricity specified for design of columns?

5.3 Derive the formula for calculating limiting load-carrying capacity of axially loaded tied column section. How much accidental eccentricity is assumed in this formula?

5.4 Why are high grade concrete mixes recommended for reinforced concrete columns? How does this practice lead to economy?

5.5 Outline the procedure for design of axially loaded reinforced concrete columns.

5.6 What is meant by slenderness ratio in compression member?

5.7 Define effective length of column.

5.8 What are the methods available in IS:456 to determine effective length of columns?

5.9 Outline the BS formulae to determine effective length of columns.

5.10 Distinguish between: (a) unsupported length and effective length of a compression member, and (b) braced and unbraced columns.

5.11 Why is it desirable that a column shall be able to resist a minimum eccentrically of loading?

5.12 Why does IS:456 limits the minimum and maximum reinforcements in columns?

5.13 Describe the functions of transverse reinforcement in a reinforced concrete column. What are IS:456 stipulations for placement of transverse reinforcement?

5.14 Compare the behaviour of tied column with spiral column subjected to axial load and tested to ultimate failure.

5.15 Under what circumstances helically reinforced columns preferred? What is helix parameter?

5.16 Explain the assumption regarding strain distribution across a section under eccentric loading at ultimate failure.

5.17 Explain briefly how the interaction diagram for columns under combined axial load and bending can be drawn.

5.18 What is meant by balanced failure of a column section subjected to *P* and *M*? How will you determine the combination of *P* and *M* that will cause balanced failure of a given section?

5.19 Describe salient points on a typical axial load-moment (*P-M*) interaction curve of a column.

5.20 Why is interaction diagram convex outwards from the origin?

5.21 Comment on the safety of uniaxially eccentrically loaded column when a point $[P_u/(f_{ck}bD),\ M_u/(f_{ck}bD^2)]$ on *P-M* interaction diagram lies: (a) marginally outside (b) marginally inside the design curve.

5.22 What is the difference between behaviour of a short and a long column?

5.23 Distinguish between the failures patterns of reinforced concrete short and slender columns.

5.24 What is meant by additional moments in slender columns?

5.25 Explain why a section under moderate axial load can carry more bending moment as compared to its limiting bending moment capacity in pure bending.

5.26 Give examples of columns in practice that are subjected to biaxial bending.

5.27 Explain the basic difference in shapes of the *P-M* interaction diagrams for Fe250 and Fe415 grade steels.

5.28 What is meant by moment of inertia of a concrete section? What methods are specified by IS code?

5.29 What is the difference between braced, i.e., the frame with restraint againstsway and unbraced, i.e., frame without restraint against sway? How does bracing affect effective length of columns?

5.30 Under what condition will the effective length be infinity and what is the meaning of this condition?

5.31 When will the effective length be greater than its actual length?

5.32 Can the effective lengths of the same column be different in the *x*- and *y*-planes?

5.33 What is the maximum slender ratio of reinforced columns allowed by IS:456? Give reason for specifying this limit.

5.34 What are the factors that affect the behaviour of slender columns?

5.35 How does bracing affect the behaviour of slender columns? Explain how bracing can be provided for columns in multi-storey buildings.

5.36 Show how the columns bend in single and double curvatures.

5.37 Derive the expression for the additional eccentricity in slender column as given in IS:456.

5.38 Explain how a column which is slender about both the axes designed?

Tutorial Problems

T.5.1 A reinforced concrete column of size 300 × 550 mm is reinforced with 4 bars of 20 mm ϕ of Fe415 grade steel placed at an effective cover of 50 mm. Determine the factored axial load that the column can carry, if the grade of concrete used is M25.

T.5.2 A reinforced concrete column is to support a service load of 950 kN inclusive of its own weight. The column is effectively held in position at both the ends and restrained against rotation at one of the ends. The unsupported length of the column is 5.5 m. The materials used are M25 grade concrete mix and HYSD steel of grade Fe415. Design the column section when its shape is: (a) square and (b) circular.

T.5.3 A reinforced concrete tied column is to support an axial load of 2250 kN. Calculate the dimensions of the column for an effective length of 4.0 m so as to obtain: (a) the smallest square column, (b) the most economical square column. Also determine the area of compression steel required for a column of size 350 × 500 mm. The materials used are M25 grade concrete and mild steel reinforcement.

T.5.4 Determine the limiting load-carrying capacity of a circular column section of 450 mm diameter reinforced with 8 bars of 20 mm ϕ adequately tied either with: (i) lateral ties, and (ii) spirals. The grades of concrete and HYSD steel used are M25 and Fe415, respectively.

T.5.5 Design braced reinforced concrete columns of rectangular, square and circular sections to support an axial load of 3500 kN. The unsupported length of the column is 3.5 m with effective lengths of 3.0 and 2.75 m in the direction of the long and short dimensions of the section, respectively. The grades of concrete and steel used are M25 and Fe415, respectively.

T.5.6 A reinforced concrete square column of side 300 mm is reinforced with 8 bars of 25 mm ϕ placed equally along the faces parallel to the axis of bending. Determine the maximum eccentricity at which the column can support a factored load of 1250 kN. The concrete mix and HYSD steel used are of grades M25 and Fe415, respectively.

T.5.7 A reinforced concrete tied square column of size 400 mm is reinforced with four bars of 25 mm ϕ placed at corners with an effective cover of 50 mm. It carries a direct load of 550 kN and a uniaxial moment of 44.0 kNm. Check the adequacy of the section. The materials used are M25 grade concrete and HYSD steel of grade Fe415.

T.5.8 A reinforced concrete column of circular cross-section of 300 mm diameter is reinforced with 8 × 22 mm ϕ mild steel longitudinal bars at a clear cover of 25 mm, and is provided with a spiral reinforcement of 6 mm diameter bar at a pitch of 50 mm. Concrete used is of grade M25. Determine the safe load the column can support if the effective length is (a) 3 m, and (b) 6 m.

T.5.9 A 300 mm wide column of a reinforced concrete framed structure is subjected to a factored load of 2500 kN and a factored moment of 200 kNm. Design a suitable section using concrete mix of grade M25 and steel reinforcement of grade Fe415 at an effective cover of 50 mm.

T.5.10 Design a rectangular column of size 350 × 550 mm subjected to a factored axial load of 1500 kN and factored moment of 400 kNm in the longer direction. The unsupported length of the column is 3.5 m with effective lengths of 3.0 and 2.75 m in the direction of the long and short dimensions of the section, respectively. The grades of concrete mix and reinforcing steel are M25 and Fe415, respectively.

T.5.11 A reinforced concrete column section of size 350 × 550 mm overall is reinforced with 4 × 25 mm ϕ bars of Fe415 grade steel on each of the short faces at an effective cover of 50 mm. Determine the maximum limiting moment the section can support in addition to a factored load of 1250 kN. The concrete mix to be used is of grade M25. Take $E_s = 2 \times 10^5$ MPa.

T.5.12 Design a reinforced concrete rectangular column to carry a factored load of 1500 kN at an eccentricity of 250 mm. The grades of concrete and steel to be used are M25 and Fe415, respectively.

T.5.13 A reinforced concrete column section of size 300 × 550 mm is reinforced with four bars of 28 mm ϕ placed at corners with an effective cover of 50 mm. The materials used are M25 grade concrete and HYSD steel of grade Fe415. Determine the maximum eccentricity at which a compressive load of 650 kN can be applied safely?

T.5.14 A reinforced concrete barrel arch is subjected to a factored moment of 1200 kNm/m width and a factored normal thrust of 7500 kN/m width. Design the arch section reinforced with mild steel bars at an effective cover of 50 mm. The concrete mix to be used is M30.

T.5.15 A reinforced concrete column section of size 300 × 450 m is subjected to a biaxial eccentrically applied load of 1500 kN. The eccentricities of the load from the centroid of the section in the directions of the larger and smaller dimensions are 75 mm and 50 mm, respectively. Determine the reinforcement required at a nominal cover of 40 mm along the faces. The grades of concrete mix and steel are M25 and Fe415, respectively.

T.5.16 A reinforced concrete tied column of size 350 × 350 mm is reinforced with four bars of 25 mm ϕ placed at the corners with an effective cover of 55 mm. The materials used are M25 grade concrete mix and HYSD steel of grade Fe415. (a) Check the adequacy of the section when it is subjected to a direct load of 500 kN placed at an eccentricity of 20 mm. (b) Determine the maximum eccentricity at which the load can be applied. (c) Calculate the eccentricity when the resulting minimum stress due to P = 500 kN becomes zero.

T.5.17 A reinforced concrete column of size 300 × 450 mm is reinforced with 8 bars of 25 mm ϕ equally distributed along the four sides at an effective cover of 50 mm. The grades of concrete mix and steel used are M25 and Fe250, respectively. Check the adequacy of the section for the following conditions:

(a) P_u = 400 kN acting at eccentricities of 200 mm and 150 mm from the centroid of the section, in the direction of smaller and larger dimensions, respectively, of the section.

(b) P_u = 550 kN acting at the eccentricities given in Part (a).

T.5.18 A reinforced concrete column section of size 350 × 500 mm is reinforced with 6 bars of 22 mm ϕ distributed equally on two short faces at an effective cover of 50 mm. The materials to be used are M25 grade concrete mix and HYSD steel of grade Fe415. Check the adequacy of the section if the column is to support a direct compressive load of 450 kN, and moments of 35 kNm, and 20 kNm about the axes parallel to the shorter and longer sides, respectively.

CHAPTER

6

Limit State of Serviceability

6.1 INTRODUCTION

A structure must fulfill three basic requirements, namely, structural, functional and aesthetic, during its life span under normal service conditions. Excessive deflection and cracking of the concrete adversely affect the appearance and efficiency of the structure and cause discomfort or psychologically alarm the users. Excessive deformation may lead to local damage to finishes and non-load-bearing members even though the limit state of local damage is not exceeded for the structure itself. Excessive cracking leads to corrosion and adversely affects the appearance. The serviceability criteria of concrete structures have become important factors in recent years.

6.2 DEFLECTION

Excessive deflection is likely to cause cracking and possible separation of plaster finishes, crushing of partition walls or cracking of glazing units. It may even make the structural system unserviceable. Deflection under the service load condition should be limited.

The serviceability requirement for the deflection should be such that neither the efficiency nor appearance of a structure should be affected by the deflection which will occur during its life span. The maximum deflections permitted by the code under service load conditions are given usually in terms of span or height. Experience has shown that the deflections are likely to be satisfactory if certain limiting effective spans to effective depth ratios are not exceeded. IS:456 has recommended the following limiting values for use in practical design of reinforced concrete structures under normal circumstances:

1. The total deflections taking place after the construction of partitions or application of finishes must not exceed span/350 or 20 mm, whichever is less, unless a provision is made for movement which may otherwise damage the partitions or finishes.
2. The final deflection (overall sag) of horizontal members due to all loads including the effects of temperature, creep and shrinkage, as measured below the level of casting (supports), should not exceed span/250. This will ensure both the satisfactory appearance and efficiency of the structure. The harmful effects of large deflection can be reduced by the provision of an initial camber. Such a provision must be restricted in general to span/250; otherwise a satisfactory minimum stiffness may not be achieved, and dynamic effects under variable loads could become objectionable.
3. The lateral deflections may be restricted to height/500, which will avoid difficulties with glazing panels, light openings, etc.

6.2.1 Deflection and its Control

The code has suggested two methods for checking the admissibility of deflection:

Effective span-to-effective depth ratios: In all normal cases, the vertical deflection of the flexural members, i.e., beams and slabs, may generally be assumed to be admissible provided that the effective span to effective depth ratios are not greater than the values given in Table 6.1.

TABLE 6.1 Basic Values of Effective Span-to-Effective Depth Ratio for Rectangular Beams and Slabs.

Type of beam or slab	*Basic effective span to effective depth ratio,* $(l/d)_{basic}$	
	Span ≤ 10 m	*Span* > 10 m
Cantilever beam	7	Deflection should be calculated
Simply supported beam	20	(20 × 10)/span
Continuous beam	26	(26 × 10)/span

The semi-empirical values given in Table 6.1 are based on a prismatic beam of rectangular section and slabs of uniform thickness with spans up to 10 m reinforced with 1.0 per cent tension reinforcement of grade Fe415 and they limit the deflection to span/250. These values shall be modified for: (a) the type and amount of tension steel, (b) the amount of compression steel and (c) the type of beam.

Depending on the area and stress in the tension reinforcement, the basic values given in Table 6.1 shall be modified by multiplying with the modification factor, m_{ft}, given in Table 6.3. The modification factor is given by:

$$m_{ft} = [0.225 + 0.00322 f_{st} + 0.625 \log_{10} (p_t)]^{-1} \leq 2.0 \tag{6.1}$$

It is observed that, for a given section, the modification factor decreases when a higher percentage of steel is used, i.e., lower ratio of effective span-to-effective depth is required. This is because

1. A higher percentage of tension reinforcement increases the depth of the neutral axis, thereby increasing the area of the compression zone which leads to a larger deflection due to creep.
2. The smaller area of concrete in the tension zone reduces the stiffness of the beam.

The deflection will increase with increase in stress, and therefore a lower ratio of effective span-to-effective depth ratio is required where steel of a higher grade is used.

Thus, the values of m_{ft} decrease with the use of higher percentages and higher service load stress levels of tension reinforcement. This is due to the fact that, under given service loads, higher values of p_t and f_{st} are indicative of smaller beam (or slab) cross-sections, resulting in lower flexural rigidity, and hence larger deflections. Alternatively, for given cross-sections, higher values of p_t and f_{st} are indicative of larger design loads and higher strains distributed across the cross-section, and hence larger curvatures and larger deflections. The use of mild steel reinforcement of grade Fe250, with relatively low allowable stress levels, is particularly effective in reducing deflections; the values of m_{ft} are invariably greater than unity, even at high p_t values. The stress in the tension steel f_{st} should be calculated considering the transformed cracked section properties. However, for convenience, the Code permits calculation of approximate value of f_{st} as follows:

$$f_{st} = (0.58 f_y)\left[\frac{\text{Area of steel required}}{\text{Area of steel provided}}\right]$$

For slabs spanning in two directions, the shorter of the two spans should be used for calculating the span to depth ratios. For two-way slabs of shorter spans up to 3.5 m, the span-to-overall depth ratios given below may generally be considered to satisfy the vertical deflection limits for loading class up to 3 kN/m^2.

TABLE 6.2 Span-to-Overall Depth Ratios to Satisfy the Vertical Deflection Limits

Type of slab	*Span-to-overall depth ratio Grade of steel*	
	Fe250	*Fe415*
1. Simply supported slabs	35	28
2. Continuous slabs	40	32

Note: For HYSD bars of grade Fe415, the mild steel values are multiplied by 0.8.

For a doubly reinforced beam, *effective span-to-effective depth ratio* is further modified by multiplying it with a factor depending on the area of compression reinforcement given in Table 6.3. It should be noted that an increase in compression steel reduces the deflection because reinforcement in the compression zone increases the stiffness of the beam. In addition the increase reduces differential shrinkage strains across the reinforced concrete section, thereby reducing long-term shrinkage deflections.

TABLE 6.3 Modification Factors m_{ft} and m_{fc} for Different Values of (p_t, f_{st}) and p_c [Ref. Figs. 4 and 5 of IS:456]

p_t or p_c = ($100A_{st}/bd$)	*For tension reinforcement, m_{ft}*										*For compression reinforcement, m_{fc}*
	f_{st}, MPa										
	120	*130*	*145*	*160*	*175*	*190*	*215*	*240*	*265*	*290*	
0.15									1.78	1.55	1.05
0.20								1.78	1.56	1.39	1.06
0.25							1.85	1.61	1.42	1.28	1.08
0.30						1.96	1.69	1.49	1.33	1.20	1.09
0.35					1.99	1.81	1.58	1.40	1.26	1.14	1.10
0.40					1.85	1.70	1.50	1.33	1.21	1.10	1.12
0.45				1.91	1.75	1.61	1.43	1.28	1.16	1.06	1.13
0.50			1.99	1.81	1.67	1.54	1.37	1.24	1.12	1.03	1.14
0.55			1.89	1.73	1.60	1.48	1.32	1.20	1.09	1.00	1.15
0.60		1.98	1.81	1.66	1.54	1.43	1.28	1.16	1.06	0.98	1.17
0.65		1.90	1.74	1.60	1.49	1.39	1.25	1.14	1.04	0.96	1.18
0.70	1.94	1.83	1.68	1.55	1.45	1.35	1.22	1.11	1.02	0.94	1.19
0.75	1.88	1.77	1.63	1.51	1.41	1.32	1.19	1.09	1.00	0.93	1.20
0.80	1.82	1.72	1.58	1.47	1.37	1.29	1.17	1.07	0.98	0.91	1.21
0.85	1.76	1.67	1.54	1.44	1.34	1.26	1.15	1.05	0.97	0.90	1.22
0.90	1.72	1.63	1.51	1.41	1.32	1.24	1.13	1.03	0.95	0.88	1.23
0.95	1.67	1.59	1.47	1.38	1.29	1.22	1.11	1.02	0.94	0.87	1.24
1.00	1.64	1.55	1.45	1.35	1.27	1.20	1.09	1.00	0.93	0.86	1.25
1.05	1.60	1.52	1.42	1.33	1.25	1.18	1.07	0.99	0.92	0.85	1.26
1.10	1.57	1.49	1.39	1.31	1.23	1.16	1.06	0.98	0.91	0.84	1.27
1.15	1.54	1.47	1.37	1.29	1.21	1.14	1.05	0.97	0.90	0.84	1.28
1.20	1.51	1.44	1.35	1.27	1.19	1.13	1.03	0.95	0.89	0.83	1.29
1.25	1.49	1.42	1.33	1.25	1.18	1.11	1.02	0.94	0.88	0.82	1.29
1.30	1.46	1.40	1.31	1.23	1.16	1.10	1.01	0.94	0.87	0.81	1.30
1.35	1.44	1.38	1.29	1.22	1.15	1.09	1.00	0.93	0.86	0.81	1.31
1.40	1.42	1.36	1.28	1.20	1.14	1.08	0.99	0.92	0.85	0.80	1.32
1.45	1.40	1.34	1.26	1.19	1.12	1.07	0.98	0.91	0.85	0.79	1.33
1.50	1.39	1.33	1.25	1.18	1.11	1.06	0.97	0.90	0.84	0.79	1.33
1.55	1.37	1.31	1.23	1.16	1.10	1.05	0.97	0.90	0.84	0.78	1.34
1.60	1.35	1.30	1.22	1.15	1.09	1.04	0.96	0.89	0.83	0.78	1.35
1.65	1.34	1.28	1.21	1.14	1.08	1.03	0.95	0.88	0.82	0.77	1.35
1.70	1.32	1.27	1.20	1.13	1.07	1.02	0.94	0.88	0.82	0.77	1.36
1.75	1.31	1.26	1.19	1.12	1.06	1.01	0.94	0.87	0.81	0.76	1.37
1.80	1.30	1.25	1.17	1.11	1.05	1.00	0.93	0.86	0.81	0.76	1.38
1.85	1.28	1.23	1.16	1.10	1.05	1.00	0.92	0.86	0.80	0.75	1.38

(*Contd.*)

TABLE 6.3 Modification Factors m_{ft} and m_{fc} for Different Values of (p_t, f_{st}) and p_c [Ref. Figs. 4 and 5 of IS:456] (*Contd.*)

p_t or p_c= ($100A_{st}/bd$)	For tension reinforcement, m_{ft}										For compression reinforcement, m_{fc}
	f_{st}, MPa										
	120	130	145	160	175	190	215	240	265	290	
1.90	1.27	1.22	1.15	1.09	1.04	0.99	0.92	0.85	0.80	0.75	1.39
1.95	1.26	1.21	1.15	1.09	1.03	0.98	0.91	0.85	0.79	0.75	1.39
2.00	1.25	1.20	1.14	1.08	1.02	0.98	0.90	0.84	0.79	0.74	1.40
2.05	1.24	1.19	1.13	1.07	1.02	0.97	0.90	0.84	0.79	0.74	1.41
2.10	1.23	1.18	1.12	1.06	1.01	0.96	0.89	0.83	0.78	0.74	1.41
2.15	1.22	1.17	1.11	1.05	1.00	0.96	0.89	0.83	0.78	0.73	1.42
2.20	1.21	1.17	1.10	1.05	1.00	0.95	0.88	0.83	0.77	0.73	1.42
2.25	1.20	1.16	1.10	1.04	0.99	0.95	0.88	0.82	0.77	0.73	1.43
2.30	1.19	1.15	1.09	1.03	0.99	0.94	0.87	0.82	0.77	0.72	1.43
2.35	1.19	1.14	1.08	1.03	0.98	0.94	0.87	0.81	0.76	0.72	1.44
2.40	1.18	1.13	1.08	1.02	0.97	0.93	0.87	0.81	0.76	0.72	1.44
2.45	1.17	1.13	1.07	1.02	0.97	0.93	0.86	0.81	0.76	0.71	1.45
2.50	1.16	1.12	1.06	1.01	0.96	0.92	0.86	0.80	0.75	0.71	1.45
2.55	1.16	1.11	1.06	1.01	0.96	0.92	0.85	0.80	0.75	0.71	1.46
2.60	1.15	1.11	1.05	1.00	0.95	0.91	0.85	0.80	0.75	0.71	1.46
2.65	1.14	1.10	1.05	1.00	0.95	0.91	0.85	0.79	0.74	0.70	1.47
2.70	1.14	1.10	1.04	0.99	0.95	0.90	0.84	0.79	0.74	0.70	1.47
2.75	1.13	1.09	1.03	0.99	0.94	0.90	0.84	0.79	0.74	0.70	1.48
2.80	1.12	1.08	1.03	0.98	0.94	0.90	0.84	0.78	0.74	0.70	1.48
2.85	1.12	1.08	1.02	0.98	0.93	0.89	0.83	0.78	0.73	0.69	1.49
2.90	1.11	1.07	1.02	0.97	0.93	0.89	0.83	0.78	0.73	0.69	1.49
2.95	1.10	1.07	1.01	0.97	0.92	0.88	0.83	0.77	0.73	0.69	1.50
3.00	1.10	1.06	1.01	0.96	0.92	0.88	0.82	0.77	0.73	0.69	1.50

The provision of compression steel can significantly contribute towards reducing deflections. For example, the modification factor m_{fc} takes values of 1.25 and 1.50 for values of p_c equal to 1 per cent and 3 per cent, respectively, for all grades of compression steel. Thus, the deflection may be controlled by increasing the compression reinforcement. The modification factor given in Table 6.3 is based on the following relation:

$$\text{Modification factor, } m_{fc} = \left(\frac{1.6p_c}{p_c + 0.275}\right) \le 1.5 \qquad \text{(SP:16)}$$

$$= 1 + \left(\frac{p_c}{3 + p_c}\right) \le 1.5 \qquad \text{(BS:8110)} \qquad (6.2)$$

where

$$p_c = \frac{100A_{sc}}{bd}$$

In a flange section, i.e., *T*- or *L*-beam, as the flange width increases, the neutral axis moves up and the concrete in the tension zone reduces with a consequent reduction in stiffness of the beam. Thus, a further modification factor given by Eqn. (6.3) must be used to reduce the permissible span to effective depth ratio. However, the basic span to effective depth ratios listed in Table 6.1 can also be used for a flanged section ignoring flange projections and considering the beam as a rectangular beam of size $b_w \times d$ for checking deflection. This treatment of flanged section as an equivalent rectangular section is always on the safer side.

$$m_{rf} = 0.80 \qquad \text{for } \frac{b_w}{b_f} \le 0.30$$

$$= 0.80 + \frac{2}{7}\left(\frac{b_w}{b_f} - 0.30\right) \qquad \text{for } \frac{b_w}{b_f} \ge 0.30 \tag{6.3}$$

Thus,

Allowable effective span-to-effective depth ratio

= Basic ratio × Modification factor for tension steel

× Modification factor for compression steel

× Reduction factor for the flange beam (6.4)

Example 6.1 The cross-section of a simply supported reinforced concrete beam of effective span of 6.5 m has been determined to be 250 × 400 mm effective. The tension and compression reinforcements have been estimated as 4 × 20 mm ϕ and 2 × 16 mm ϕ mild steel bars, respectively. Check the adequacy of the designed section with respect to limit state deflection if the concrete used is of grade M20.

Solution For the designed cross-section of the beam:

b = 250 mm, d = 400 mm, Effective span = 6500 mm

$A_{st}(4 \times 20 \text{ mm } \phi) = 1256 \text{ mm}^2$ and $A_{st}(2 \times 16 \text{ mm } \phi) = 402 \text{ mm}^2$

$$\text{Actual effective span to effective depth ratio} = \frac{6500}{400} = 16.25.$$

For a simply supported beam, basic (span/d) ratio is 20. The percentage of tension steel is:

$$p_t = \frac{1256 \times 100}{250 \times 400} = 1.256$$

Consider that the provided area of tension reinforcement is same as the required. Therefore, stress at the service load is $0.58f_y = 0.58 \times 250 = 145$ MPa, corresponding modification factor from Table 6.3 is 1.33.

$$\text{Percentage of compression steel, } p_c = \frac{402 \times 100}{250 \times 400} = 0.402$$

Corresponding modification factor = 1.12.

Allowable effective span to effective depth ratio = 20 × 1.33 × 1.12

= 29.79 > 16.25 (the actual ratio)

Thus, the designed beam section is adequate from limit state deflection considerations.

6.2.2 Calculation of Deflection

In the majority of cases, the deflection check is made by restricting the effective span to effective depth ratio to be within permissible limits. However, deflection calculations are necessary in the case where the designer wants to exceed the span to depth ratio or where the specific deflection control is required.

The total calculated deflection shall be taken as the sum of short-term and long-term deflections. The short-term deflection is instantaneous after the application of the load and is determined by using elastic theory, with short-term elastic modulus of elasticity of concrete E_c being taken as $5000\sqrt{f_{ck}}$ where f_{ck} is the characteristic strength of concrete. On the other hand, long-term deflection is due to shrinkage and creep effects under the sustained load. The effective moment of inertia I_{eff} should be based on uncracked concrete and an equivalent area of steel.

Short-Term Deflections: The short-term deflection may be calculated by the usual methods for elastic deflections using the short-term modulus of elasticity, $E_c = 5000\sqrt{f_{ck}}$, and an effective moment of inertia I_{eff} given by the equation:

$$I_{\text{eff}} = \frac{I_{cr}}{1.2 - \left(\dfrac{M_{cr}}{M}\right)\left(\dfrac{z}{d}\right)\left(1 - \dfrac{x}{d}\right)\left(\dfrac{b_w}{b}\right)} \tag{6.5}$$

subject to $I_{cr} \le I_{\text{eff}} \le I_g$, i.e. effective moment of inertia lies between that of gross section and cracked section.

Here

I_{cr} = Moment of inertia of the cracked section

I_g = Moment of inertia of the gross section about the centroid axis neglecting reinforcement

M_{cr} = Cracking moment of the section

M = Maximum moment under service loads

z = Lever arm

x = Depth of the neutral axis

d = Effective depth

b_w = Breadth of the web

b = Breadth of the compression face (= b_f for the flanged section and b_w for the rectangular beam)

The values of x and z are those obtained by elastic theory. The moment of inertia of the cracked rectangular section about the neutral axis, and that of the uncracked rectangular section (neglecting the reinforcement) about the centroid axis are given by

$$I_{cr} = \frac{bx^3}{3} + (1.5m - 1)A_{sc}(x - d')^2 + mA_{st}(d - x)^2$$

$$I_g = \frac{bD^3}{12} \tag{6.6}$$

The cracking moment of the section is given by

$$M_{cr} = \frac{I_g f_{cr}}{y_t} \tag{6.7}$$

where

f_{cr} = Modulus of rupture of concrete and is equal to $0.7\sqrt{f_{ck}}$ MPa

y_t = Distance of the extreme compression fibre from the centroid of the section

For continuous beams, deflections shall be calculated using the values of I_{cr}, I_g and M_{cr} modified by the following equation:

$$X_e = K\left(\frac{X_1 + X_2}{2}\right) + (1 - K)X_0 \tag{6.8}$$

where

X_e = Modified value of X

X = Value of I_{cr}, I_g and M_{cr}, as appropriate

X_1, X_2 = Values of X at the supports

X_0 = Value of X at mid-span

K = Coefficient given in Table 6.4

TABLE 6.4 Values of Coefficient K

$\frac{M_1 + M_2}{M_{f1} + M_{f2}}$	0.5 or less	0.6	0.7	0.8	0.9	1.0	1.1	1.2	1.3	1.4
K	0	0.03	0.08	0.16	0.30	0.50	0.73	0.91	0.97	1.00

Note: M_1 and M_2 represent support moments, and M_{f1} and M_{f2} the fixed end moments.

The short-term deflection δ_s is obtained by the usual method of elastic deflection as:

$$\delta_s = \frac{\alpha W l^3}{E_c I_{\text{eff}}} \tag{6.9}$$

where α is the coefficient depending on the support conditions of the beam and load distribution. For example, for a simply supported beam carrying a uniformly distributed load $\alpha = 5/384$; instead, if the ends of the beam are fixed, $\alpha = 1/384$.

Example 6.2 A reinforced concrete beam of cross-section 300 × 600 mm overall is reinforced on the tension side with 3 bars of 20 mm ϕ of Fe415 grade steel at an effective cover of 50 mm. Compute the short-term deflection of the beam at the mid-span under service loads consisting of a uniformly distributed load of 20 kN/m and a concentrated load of 25 kN at the centre, over a simply supported span of 5 m. The concrete mix to be used is of grade M20.

Solution For the given cross-section and materials:

$b = 300$ mm, $D = 600$ mm and $d = 600 - 50 = 550$ mm

$A_{st}(3 \times 20 \text{ mm } \phi) = 942 \text{ mm}^2$

$f_{ck} = 20$ MPa, $f_y = 415$ MPa and $E_s = 2 \times 10^5$ MPa

Bending moment at the service load is:

$$M = \frac{wl^2}{8} + \frac{Wl}{4} = \left(\frac{20 \times 5^2}{8}\right) + \left(\frac{25 \times 5}{4}\right)$$

$$= 93.75 \text{ kNm}$$

Factored moment = 1.5 × 93.75 = 140.625 kNm (say, 140 kNm)

Flexural tensile strength of concrete $f_{cr} = 0.7\sqrt{f_{ck}} = 0.7 \times \sqrt{20} = 3.13$ MPa

Short-term static modulus of elasticity of concrete is expressed as:

$$E_c = 5000\sqrt{f_{ck}} = 5000 \times \sqrt{20} = 2.236 \times 10^4 \text{ MPa}$$

$$\text{Modular ratio } m = \frac{E_s}{E_c} = \frac{2 \times 10^5}{2.236 \times 10^4} = 8.94$$

$$\text{Percentage of tension steel } p_t = \frac{100 A_{st}}{bd} = \frac{100 \times 942}{300 \times 550} = 0.571$$

$$\text{Factor } p = \frac{A_{st}}{bd} = \frac{p_t}{100} = \frac{0.571}{100} = 0.00571$$

Therefore,

$$pm = 0.00571 \times 8.94 = 0.0510$$

The depth of the neutral axis x can be obtained from elastic theory (Appendix A) as:

$$\frac{x}{d} = -pm + \sqrt{(pm)^2 + 2pm}$$

$$= -(0.0510) + \sqrt{(0.0510)^2 + 2 \times (0.0510)} = 0.272$$

$$x = 0.272 \times 550 = 149.6 \text{ mm}$$

$$\text{Lever arm } z = d - \left(\frac{x}{3}\right) = 550 - \left(\frac{149.6}{3}\right) = 500.13 \text{ mm (say, 500 mm)}$$

The moment of inertia of the cracked section is given by

$$I_{cr} = \frac{bx^3}{3} + mA_{st}(d - x)^2$$

$$= \left[\frac{300 \times (149.6)^3}{3}\right] + 8.94 \times 942 \times (550 - 149.6)^2$$

$$= 1.6849 \times 10^9 \text{ mm}^4$$

The moment of inertia of the uncracked section is obtained as:

$$I_g = \frac{bD^3}{12} = \frac{300 \times (600)^3}{12} = 5.4 \times 10^9 \text{ mm}^4$$

$$y_t = \frac{D}{2} = \frac{600}{2} = 300 \text{ mm}$$

Cracking moment of the section is given by

$$M_{cr} = \frac{I_g f_{cr}}{y_t}$$

$$= \frac{5.4 \times 10^9 \times 3.13}{300} = 5.634 \times 10^7 \text{ Nmm} = 56.34 \text{ kNm}$$

Effective moment of inertia from Eq. (6.3),

$$I_{\text{eff}} = \frac{I_{cr}}{1.2 - \left(\frac{M_{cr}}{M}\right)\left(\frac{z}{d}\right) \cdot \left(1 - \frac{x}{d}\right)\left(\frac{b_w}{b}\right)}$$

$$= \frac{1.6849 \times 10^9}{1.2 - \left(\frac{56.34}{93.75}\right) \times \left(\frac{500}{550}\right) \times \left(1 - \frac{149.6}{550}\right) \times 1}$$

$$= 2.1002 \times 10^9 \text{ mm}^4$$

The short-term deflection at the midpoint is:

$$\delta_s = \frac{5wl^4}{384E_c I_{\text{eff}}} + \frac{Wl^3}{48E_c I_{\text{eff}}}$$

$$= \frac{5 \times 20 \times (5 \times 1000)^4}{384 \times (2.236 \times 10^4) \times (2.1002 \times 10^9)} + \frac{25000 \times (5 \times 1000)^3}{48 \times (2.236 \times 10^4) \times (2.1002 \times 10^9)}$$

$$= 3.466 + 1.386 = 4.852 \text{ mm}$$

Long-term Deflections: The long-term deflections are due to-creep and shrinkage of concrete. Creep is time-dependent, inelastic deformation under sustained load. It increases with time but at a decreasing rate. Creep results in increase in strain at constant stress. The increase in compressive strain increases the depth of the neutral axis which, in turn, reduces the compressive stress. Within a span of two to three years the long-term deflections are largely completed. In the following section the method for computing deflection due to shrinkage and creep, as given in IS:456, is described.

Deflection due to shrinkage: The effect of shrinkage on a plain concrete member is to shorten its length, and shrinkage value can reach 0.0004 to 0.0007. In a reinforced concrete beam the bond between concrete and embedded reinforcement restrains the shrinkage and the shrinkage value reaches 0.0002 to 0.0003. Thus, in a singly reinforced beam, the difference in magnitude of restrained shrinkage on the reinforced face and unrestrained shrinkage at the unreinforced face causes curvature and hence deflection.

The deflection due to shrinkage, $\alpha_{cs,}$ may be computed from the equation:

$$\alpha_{cs} = k_3 \ \psi_{cs} \ l^2 \tag{6.10}$$

where

k_3 = Constant. The values for different structures are:

Cantilevers:	0.50
Simply supported members:	0.125
Members continuous at one end:	0.086
Fully continuous members:	0.063

l = Length of span

ψ_{cs} = Shrinkage curvature given by

$$= \frac{k_4 \varepsilon_{cs}}{D} \tag{6.11}$$

Here

ε_{cs} = Ultimate shrinkage strain of concrete, 0.0003

D = Overall depth of the section

k_4 = Constant given by

$$k_4 = \frac{0.72(p_t - p_c)}{\sqrt{p_t}} \le 1.0 \qquad \text{for } 0.25 < (p_t - p_c) \le 1.0$$

$$= \frac{0.65(p_t - p_c)}{\sqrt{p_t}} \le 1.0 \qquad \text{for } (p_t - p_c) \ge 1.0$$

where

$$p_t = \frac{100A_{st}}{bd}$$

$$p_c = \frac{100A_{sc}}{bd}$$

A_{st},A_{sc} = Area of tension and compression steels, respectively.
b, d = Width and effective depth of the beam, respectively.

Deflection due to creep: The creep of concrete in a reinforced concrete beam has an effect similar to that of shrinkage. While calculating long-term deflection, creep is taken into account by using the modified or reduced value of modulus of elasticity as given by

$$E_{cc} = \frac{\text{Stress in concrete}}{(\text{Elastic Strain + Creep Strain})} = \frac{\sigma_i}{(\varepsilon_i + \varepsilon_{cp})}$$

The creep strain can be expressed as:

$$\varepsilon_{cp} = \theta\varepsilon_i$$

Thus,

$$E_{cc} = \frac{\sigma_i}{\varepsilon_i + \theta\varepsilon_i} = \frac{\sigma_i}{\varepsilon_i(1+\theta)} = \frac{E_c}{1+\theta} \tag{6.12}$$

where θ is creep coefficient which is primarily a function of age at loading:

Age at loading	*Creep coefficient*
7 days	2.2
28 days	1.6
1 year	1.1

The long-term creep deflection due to permanent loads, δ_{cp} may be obtained as:

$$\delta_{cp} = \delta_{sc,p} - \delta_{s,p} \tag{6.13}$$

where

$\delta_{sc,p}$ = Initial short-term plus long-term creep deflection due to permanent loads obtained using E_{cc} in elastic analysis.

$\delta_{s,p}$ = Short-term deflection due to permanent loads using E_c

Thus, the total deflection of a beam is the summation of:

1. the short-term deflection under the total load
2. the deflection due to shrinkage
3. the deflection due to creep

Example 6.3 A reinforcement concrete cantilever beam of 4.0 m span has a rectangular cross-section of size 300 × 600 mm (overall). It is reinforced with 6 bars of 20 mm ϕ on the tension side and 2 bars of 22 mm ϕ on the compression side at an effective cover of 37.5 mm. Compute the total deflection of cantilever at the free end when it is subjected to a uniformly distributed service load of 25 kN/m. Sixty per cent of this load is permanent in nature. The grades of concrete mix and steel reinforcement used are M20 and Fe415, respectively.

Solution For the given cross-section and materials:

$$b = 300 \text{ mm}, D = 600 \text{ mm}, d' = 37.5 \text{ mm and } d = 562.5 \text{ mm}$$

$$A_{st}(6 \times 20 \text{ mm } \phi) = 1885 \text{ mm}^2, A_{sc}(2 \times 22 \text{ mm } \phi) = 760 \text{ mm}^2$$

$$f_{ck} = 20 \text{ MPa}, f_y = 415 \text{ MPa and } E_s = 2 \times 10^5 \text{ MPa}$$

Short-Term deflection.

$$\text{Flexural tensile strength } f_{cr} = 0.7\sqrt{f_{ck}}$$

$$= 0.7 \times \sqrt{20} = 3.13 \text{ MPa}$$

Short-term static modulus of elasticity is

$$E_c = 5000\sqrt{f_{ck}} = 5000\sqrt{20}$$

$$= 2.236 \times 10^4 \text{ MPa}$$

$$\text{Elastic modular ratio } m = \frac{E_s}{E_c} = \frac{2 \times 10^5}{2.236 \times 10^4} = 8.94$$

$$\text{Gross moment of inertia of section } I_g = \frac{bD^3}{12}$$

$$I_g = \frac{300 \times 600^3}{12} = 5.4 \times 10^9 \text{ mm}^4$$

$$y_t = \frac{D}{2} = \frac{600}{2} = 300 \text{ mm}$$

$$\text{Cracking moment } M_{cr} = \frac{f_{cr} \times I_g}{y_t} = \frac{3.13 \times 5.4 \times 10^9}{300} = 56.34 \text{ kNm}$$

$$\text{Bending moment at service loads } M = \frac{wl^2}{2}$$

$$M = \frac{25 \times 4^2}{2} = 200 \text{ kNm}$$

and

$$\text{Ratio } \frac{M_{cr}}{M} = \frac{56.34}{200} = 0.282$$

Let x be the depth of the neutral axis. Take moment of equivalent areas (Appendix A) about the neutral axis as:

$$\frac{bx^2}{2} + (m - 1)A_{sc}(x - 37.5) = mA_{st}(d - x)$$

$$\frac{300x^2}{2} + (8.94 - 1) \times 760 \times (x - 37.5) = 8.94 \times 1885 \times (562.5 - x)$$

$$x^2 + 40.23 \times (x - 37.5) = 112.35 \times (562.5 - x)$$

$$x^2 + 152.58x - 64705.5 = 0$$

which gives

$$x = 189.28 \text{ mm}$$

Approximate lever arm $z = d - \left(\frac{x}{3}\right) = 562.5 - \left(\frac{189.28}{3}\right) = 499.41$ mm

Moment of inertia of the cracked section is given by

$$I_{cr} = \frac{300 \times 189.28^3}{3} + (8.94 - 1) \times 760 \times (189.28 - 37.5)^2$$

$$+ 8.94 \times 1885 \times (562.5 - 189.28)^2$$

$$= 3.1645 \times 10^9 \text{ mm}^4$$

$$p_t = \frac{1885 \times 100}{300 \times 562.5} = 1.117$$

and

$$p_c = \frac{760 \times 100}{300 \times 562.5} = 0.450.$$

Alternatively, I_{cr} can be obtained from SP-16. Calculate

$$mp_t = 8.94 \times 1.117 = 9.99$$

$$\frac{(m-1)p_c}{mp_t} = \frac{7.94 \times 0.450}{9.99} = 0.358$$

Refer to Table 87 of SP-16:

$$I_{cr} = 0.722 \left(\frac{bd^3}{12}\right) = 0.722 \times \frac{300 \times (562.5)^3}{12} = 3.21 \times 10^9 \text{ mm}^4$$

The effective moment of inertia is given by

$$I_{eff} = \frac{I_{cr}}{1.2 - \left(\frac{M_{cr}}{M}\right) \cdot \left(\frac{z}{d}\right)\left(1 - \frac{x}{d}\right)\left(\frac{b_w}{b}\right)}$$

$$= \frac{3.1645 \times 10^9}{1.2 - 0.282 \times \left(\frac{499.41}{562.5}\right) \times \left(1 - \frac{189.28}{562.5}\right) \times 1} = 3.061 \times 10^9 \text{mm}^4$$

I_{eff} should satisfy the condition:

$$I_{cr} \le I_{\text{eff}} \le I_g$$

Therefore,

$$I_{\text{eff}} = I_{cr} = 3.1645 \times 10^9 \text{ mm}^4$$

Short-term deflection is obtained as:

$$\delta_s = \frac{Wl^3}{(8E_c I_{eff})}$$

$$= \frac{(25000 \times 4) \times (4000)^3}{8 \times (2.236 \times 10^4) \times (3.1645 \times 10^9)} = 11.31 \text{ mm}$$

Deflection due to shrinkage. As we know,

$$\delta_{cs} = k_3 \ \psi_{cs} \ l^2$$

where $k_3 = 0.50$ (for cantilever). The difference in percentage of tension and compression steels is:

$$p_t - p_c = 1.117 - 0.450 = 0.667 < 1.0$$

Therefore,

$$k_4 = \frac{0.72(p_t - p_c)}{\sqrt{p_t}}$$

$$= \frac{0.72 \times 0.667}{\sqrt{(1.117)}} = 0.454$$

In the absence of data, consider the value of the ultimate shrinkage strain as:

$$\varepsilon_{cs} = 0.0003$$

The shrinkage curvature is:

$$\psi_{cs} = \frac{k_4 \varepsilon_{cs}}{D} = \frac{0.454 \times 0.0003}{600} = 2.27 \times 10^{-7}$$

Thus, deflection due to shrinkage is expressed as:

$$\delta_{cs} = 0.50 \times 2.27 \times 10^{-7} \times (4000)^2 = 1.82 \text{ mm}$$

Deflection due to creep: In the absence of data, the age at loading is assumed to be 28 days and the value of the creep coefficient θ is taken as 1.6. Thus,

$$E_{cc} = \frac{E_c}{(1 + \theta)} = \frac{2.236 \times 10^4}{1 + 1.6} = 8.6 \times 10^3 \text{ MPa}$$

Total short-term plus long-term creep deflection due to permanent loads which is 60 per cent of total loads, is:

$$\delta_{sc,\,p} = \frac{W_p l^3}{8E_{cc} I_{\text{eff}}}$$

$$= \frac{(0.6 \times 25000 \times 4) \times (4000)^3}{8 \times 8.6 \times 10^3 \times 3.1645 \times 10^9} = 17.64 \text{ mm}$$

Short-term deflection $\delta_{s,p}$ due to permanent load obtained by using E_c is:

$$\delta_{s,p} = 0.6 \times \delta = 0.6 \times 11.31 = 6.80 \text{ mm}$$

Therefore,

$$\delta_{c,p} = 17.64 - 6.80 = 10.84 \text{ mm}$$

Total long-term deflection δ due to the initial load, shrinkage and creep is:

$$\delta = \delta_s + \delta_{cs} + \delta_{c,p} = 11.31 + 1.82 + 10.84 = 23.97 \text{ mm}$$

$$\text{Permissible deflection} = \frac{L}{250} = \frac{4000}{250} = 16 \text{ mm}$$

The calculated deflection is greater than the permissible value. Deflection occurring after the erection of partitions and application of finishes should not exceed L/350 or 20 mm, whichever is less:

$$\frac{L}{350} = \frac{4000}{350} = 11.43 \text{ mm}$$

Consider that half the total shrinkage strain occurs within the first 28 days; the calculated deflection occurring after 28 days will be 0.5 × 1.82 + 10.87 = 11.78 mm. This is slightly higher than the permissible value. The section should be revised or, alternatively, the following actions may be taken during construction:

1. Provide an upward camber of 10 mm at the free end of cantilever.
2. Delay the erection of partitions and application of finishes.

6.2.3 Control of Deflection at Site

The harmful effects of large deflections can be reduced by the following techniques applied on the site:

***Pre-cambering*:** It is the initial camber or upward curve of a flexural member made during construction such that on loading the member will straighten out and attain its correct shape. For example, long span simply supported beams, continuous beams, cantilever slabs and particularly cantilever beams, are usually cambered at the site. The precamber of horizontal members should not generally exceed span/250. Some designers provide precamber equal to the dead load deflection.

***Quality control*:** Lack of supervision during concreting can lead to larger deflections than expected; therefore, all possible precautions should be taken to produce low slump, well-compacted and effectively cured concrete.

***Adequate stripping time*:** The premature stripping of forms, i.e., removal of forms earlier than the specified time, may result in large initial creep deflection due to dead loads and superimposed construction loads. Thus, the forms shall not be removed until concrete has reached a strength corresponding to at least twice the stress to which the concrete may be subjected at the time of stripping.

***Controlling construction loads*:** Temporary construction loads on unpropped floors, e.g., storage of materials and equipment, can be several times more than the anticipated service loads. This may lead to larger short-term deflection and extensive premature cracking than expected. Precautions should be taken to control the temporary loads.

6.3 CRACK FORMATION AND ITS CONTROL

The limit states of local damage are concerned with the damage to the structure at service loads due to either excessive tension or compression in either the concrete or the steel. Tension in concrete can result in the limit state of crack formation and in various limit states of crack width. Cracking in concrete is initiated by either internal micro cracks or flexural micro cracks. Flexural micro cracks develop when the limiting tensile strain of concrete (0.002 to 0.0005) is reached. In case of high strength tensile steel, the strain in surrounding concrete will be as high as 0.001 even at the service state, thus leading to the formation of cracks.

Reinforced concrete members develop several types of cracks, depending upon the type of forces acting on it. In tension members cracks called **separation cracks** penetrate through the entire cross-section. In flexural members cracks, called **flexural cracks** develop in the tension zone and penetrate the cross-section up to the neutral axis only. Cracks in the form of a spiral are formed when a reinforced concrete member is subjected to a torsional moment. Bond cracks develop at the interface of reinforcement with concrete due to bond failure.

Excessively wide cracks affect the appearance and reduce the effectiveness of the cover, resulting in deterioration of reinforcement by corrosion. In the past due to the use of low yield strength steel (usually mild steel) the cracking of the member was limited and negligible. The increasing use of the limit states method of design and high yield strength deformed (HYSD) steel leads to wide cracks in the concrete structure, thus necessitating control of cracking. The maximum width of the crack is limited on the basis of the appearance of the structure, durability and corrosion resistance. The crack width can be controlled by proper detailing of reinforcement. A large number of smaller diameter bars well distributed in the tension zone reduce the width of the crack more effectively than the larger diameter bars for the same area.

As per the guidelines recommended by IS:456, the surface width of the crack should not, in general, exceed 0.3 mm for structures not subjected to an aggressive environment. However, if the structure is exposed to an aggressive environment the width of the crack near the main reinforcement should not exceed 0.004 times the nominal cover to the main reinforcement. The crack width in case of walls of bins as per IS:4995 shall not exceed 0.10 mm when water tightness is required and 0.20 mm in other cases.

ACI:318 has recommended two maximum (permissible values) crack widths: 0.4 mm for interior exposure and 0.33 mm for exterior exposure. The values for sea water or wetting and drying condition, and water retaining structures are 0.15 mm and 0.10 mm, respectively. On the other hand the limits recommended by the International Code (European Concrete Committee) are 0.30 mm for internal structural parts in a humid or aggressive atmosphere and external structural parts exposed to weather, and 0.10 mm for internal or external structural parts exposed to particular aggressive medium or where water tightness is needed. British Code CP:110 require, in general, that the surface crack widths at the service loads do not exceed 0.30 mm.

IS:456 has recommended the provisions for the control of flexural crack width by limiting the maximum distance between the tension reinforcement bars for beams and slabs. The rules shall be applied for the spacing control in flexural members in normal condition of exposure, unless the calculation of crack widths shows that a greater spacing is acceptable.

6.3.1 Beams

1. The spacing between bars or groups of bars, near the tension face of a beam shall not exceed the values given in Table 6.5, depending on the amount of redistribution carried out in analysis and the grade of reinforcement.
   It should be noted that the higher the grade of steel, the smaller is the clear spacing between the bars.
2. The clear distance from the corner of a beam to the surface of the nearest longitudinal bar should not exceed half the value obtained in (1).
3. When the depth of the web in a beam exceeds 750 mm, side face reinforcement shall be provided along the two faces. The total area of such reinforcement shall not be less than 0.1 per cent of the web area and shall be distributed equally on the two faces at a spacing not exceeding 300 mm or, the web width, whichever is less.

TABLE 6.5 Maximum Clear Spacing of Reinforcement Bars

Grade of steel f_y	*Spacing of bars, S mm, for redistribution to and from the section, per cent*				
	−30	*−15*	*0*	*+15*	*+30*
Fe250	215	260	300	300	300
Fe415	125	155	180	210	235
Fe500	105	130	150	175	195

6.3.2 Slabs

1. The nominal or clear spacing of main bars shall not be more than three times the effective depth of a solid slab or 450 mm, whichever is smaller.
2. The nominal spacing of secondary bars provided against shrinkage and temperature shall not be more than five times the effective depth of the slab or 450 mm, whichever is smaller.

In the preceding discussion the secondary reinforcement is provided perpendicular to main bars and is not resisting flexure. Its functions are to distribute the concentrated load uniformly, to keep the main reinforcement in position, and to resist stresses due to shrinkage and temperature.

Where the spacing limitations and minimum concrete cover are based on the bar diameter, a group of bars bundled in contact shall be treated as a single bar of diameter derived from the total equivalent area.

6.4 CALCULATION OF CRACK WIDTH

For the cases where strain in the tension steel does not exceed $0.8\,f_y/E_s$, the design surface crack width, W_{cr}, may be calculated from the following relation:

$$W_{cr} = \frac{(3a_{cr}\varepsilon_m)}{\left[1 + 2\left(\dfrac{a_{cr} - a'_{\min}}{D - x}\right)\right]} \tag{6.14}$$

where,

D = Overall depth

$a'_{\min}$ = Minimum nominal cover to longitudinal bar

x = Depth of neutral axis

a_{cr} = Distance from the point under consideration (for crack width determination) to the surface of the nearest longitudinal bar

ε_m = Average strain in steel at the level of point under consideration. It is given by

$$\varepsilon_m = e_1 - \frac{0.70 b_t D(a - x)}{A_{st} f_s (D - x)} \tag{6.15}$$

Here

a = Distance from the compression face to the point under consideration

A_{st} = Area of tension steel

b_t = Width of the section at centroid of tension steel

e_1 = Strain at the level of point under consideration

For the point on the tension face,

$$e_1 = \left(\frac{D - x}{d - x}\right)\left(\frac{f_s}{E_s}\right) \tag{6.16}$$

where f_s is the service stress in the tension steel and is expressed as:

$$f_s = m\left(\frac{M}{I_{cr}}\right)(d - x) \tag{6.17}$$

It should be noted that ε_m and hence the crack width increases with the increase in the service stress in the reinforcement, increase in the nominal cover and increase in the spacing of bars. Minimum value of nominal cover is limited by durability requirements; however, the bars should be as close to the surface as possible, i.e., a_{cr} should be minimum.

Example 6.4 Estimate the design surface crack width directly under the bar on the tension face at the location of maximum moment in the beam of Example 6.3.

Solution From Example 6.3,

$$D = 600 \text{ mm}, \; d = 562.5 \text{ mm}, \; d' = 37.5 \text{ mm}, \; b = 300 \text{ mm}$$

$$m = 8.94, \; x = 189.28 \text{ mm}, \; I_{cr} = 3.21 \times 10^9 \text{ mm}^4, \; E_s = 2 \times 10^5 \text{ MPa}$$

$$M = 200 \text{ kNm}, \; A_{st} = 1885 \text{ mm}^2$$

Clear or nominal cover $a'_{\min} = 37.5 - 10.0 = 27.5$ mm

Then

$$f_s = 8.94 \times \frac{200 \times 10^6}{3.21 \times 10^9} \times (562.5 - 189.28) = 207.89 \text{ MPa}$$

$$e_1 = \frac{600 - 189.28}{562.5 - 189.28} \times \frac{207.89}{2 \times 10^5} = 0.001144$$

$$\varepsilon_m = (0.001144) - \left[\frac{0.7 \times 300 \times 600 \times (600 - 189.28)}{1885 \times 207.89 \times (600 - 189.29)}\right] \times 10^{-3} = 0.0008225$$

$$a_{cr} = 27.5 \text{ mm}$$

and

$$W_{cr} = \frac{3 \times 27.5 \times 0.0008225}{1 + 2 \times \{(27.5 - 27.5)/(600 - 189.28)\}}$$

$$= 0.06786 \text{ mm}$$

6.5 SERVICEABILITY LIMIT STATE OF LATERAL STABILITY

The lateral instability of a beam due to the width of compression face being small as compared to the depth of the beam is prevented by restricting the slenderness ratio. A simply supported or continuous beam should be so proportioned that the clear distance between lateral restraints l, is such that

$$l \not> 60 \text{ b or } \frac{250b^2}{d} \text{ whichever is less} \tag{6.18}$$

where b and d are the breath of compression face mid-way between lateral restraints and effective depth of the beam, respectively. For a cantilever beam, the clear distance from the free end to the lateral restraint should not exceed $25b$ or $100\ b^2/d$ whichever is less.

6.6 OTHER SERVICEABILITY REQUIREMENTS

The other important serviceability requirements to ensure proper performance of a structural member during its service life are durability and fire resistance. However, serviceability requirements with respect to fatigue due to moving loads, thermal and sound insulation properties may be important depending upon the function of the structure.

Review Questions

6.1 Describe *serviceability failures* commonly encountered in practice. Suggest remedial repair measures also.

6.2 Why is the limitation on deflection in a structure called a serviceability condition? Name another serviceability condition commonly used in the limit states design.

6.3 Explain the following briefly: (a) Limit states of serviceability conditions, (b) shrinkage and creep, (c) short-term and long-term deflections, and (d) crack formation and its control.

6.4 Why is it necessary to limit deflections in the reinforced concrete flexural members?

6.5 State the allowable limits of deflection in slabs and beams: Does allowable depth depend on span length?

6.6 How can the deflection at site be controlled?

6.7 What are the major factors that affect deflection?

6.8 Distinguish between short-term and long-term deflections.

6.9 Write down the loading conditions for checking the serviceability.

6.10 Why is the short-term deflection due to live load alone estimated? What is its relevance?

6.11 How does shrinkage of concrete effect the deflections in reinforced concrete flexural members?

6.12 What is the affect of amount of compression reinforcement on the deflection due to: (a) shrinkage and (b) creep?

6.13 Describe the code procedure for calculating the deflection due to creep.

6.14 Explain with derivation how span-to-depth ratio can be used to control deflection in beams.

6.15 What is the basic span-to-depth ratio? Would the initial depth you assume in your design of a T-beam section be greater or less than that for a rectangular section from allowable deflection in beam?

6.16 How would you choose the depth for a reinforced beam?

6.17 In the two-way slabs which of the spans (the larger or shorter) is considered for checking the adequacy of the section for deflection?

6.18 What are the major factors which influence the crack-widths in flexural members?

6.19 Are the usual detailing requirements of the code adequate for ensuring crack-width control?

6.20 What is the magnitude of crack width allowed in concrete structures?

6.21 What are the recommendations of IS code for control of crack-width in RC members?

6.22 Discuss briefly the rules given in IS:456 regarding the following:
(a) Maximum and minimum steels in R.C.C. beams
(b) Minimum and maximum spacings of steel bars in R.C.C. slabs
(c) Stress level in steel at working load in structures designed by IS:456

6.23 When is side reinforcement provided in beams? What are the specifications regarding its placement?

6.24 Comment on the formula given in IS:456 for minimum area of reinforcement of beams.

Tutorial Problems

T.6.1 A simply supported doubly reinforced concrete beam of the rectangular section of size 250 × 400 mm (effective) is reinforced with 3 bars of 25 mm ϕ on the tension side and 2 bars of 20 mm ϕ on the compression side. Check the adequacy of the section for deflection. The grades of concrete mix and steel reinforcement are M20 and Fe415, respectively.

T.6.2 A simply supported reinforced concrete beam of T-shaped cross-section with a flange width of 1500 mm, flange thickness of 100 mm and web width of 300 mm is reinforced with 4 bars of 20 mm ϕ on the tension side at the effective depth of 500 mm. Check the adequacy of the section for the deflection of the beam to be used over an effective span of 7 m. M25 concrete mix and Fe415 steel are used in construction.

T.6.3 A simply supported T-beam section with a flange width of 1250 mm, web width of 250 mm and flange thickness of 100 mm is reinforced with 3 bars of 20 mm ϕ at an effective depth of 450 mm. The beam carries a uniformly distributed service load of 12 kN/m and a concentrated load of 25 kN at mid-point over an effective span of 6 m. The concrete mix used is of grade M20 and steel is of grade Fe415. Calculate the short-term and total deflections of the beam at the mid-point.

T.6.4 A reinforced concrete cantilever beam of rectangular section of size 300 × 600 mm (overall) is reinforced at the top with 3 bars of 20 mm ϕ at an effective cover of 50 mm. Determine the short-term deflection at the free end of the cantilever carrying a uniformly distributed service load of 5 kN/m over an effective span of 3.5 m.

What is the total deflection when an additional load of 15 kN is acting at the free end of the cantilever? The grades of concrete mix and steel used are M20 and Fe415, respectively.

T.6.5 A reinforced concrete slab of 3.5 m span and 125 mm thickness is reinforced with 10 mm ϕ bars @ 140 mm centre-to-centre as the main reinforcement. Check the adequacy of the section, if the slab is simply supported, and grades of concrete mix and steel reinforcement are M20 and Fe415, respectively. Calculate the deflections when the slab supports a uniformly distributed load of 12.5 kN/m^2.

T.6.6 A simply supported reinforced T-beam section consists of 1500 mm wide (effective) flange of thickness 100 mm. The width of web is 200 mm and the overall depth of the beam is 425. The beam is reinforced in compression with 2 × 16 mm ϕ and in tension with 4 × 25 mm ϕ bars placed in two layers of two each separated by 20 mm ϕ spacer bars. The nominal cover to compression steel is 35 mm and that to tension steel is 40 mm both from soffit and side faces. The grades of concrete and steel used are M20 and Fe415, respectively. Take $m = 13$ and $E_s = 2 \times 10^6$ MPa. Determine the width of cracks at a point on soffit below reinforcement and mid point, and at the corners of the beam when the maximum service moment is 10.5 kNm.

[**Hint:** Analyze it as a rectangular beam since the neutral axis lies in the flange.]

CHAPTER

7

Design of Key Building Elements

7.1 INTRODUCTION

This chapter is aimed at presenting designs of some of the basic building elements using the principles discussed in the preceding chapters. The structures considered include flexural and torsional members as well as axially loaded columns. Design of any structural component is always preceded by the analysis to compute the forces it has to withstand. Thus for an economical and safe design it is essential that the loads be computed correctly which requires good estimation or selection of member sizes. A designer should account for all the possible loads acting on the structure. Once the loads are estimated, the structure is analyzed to determine axial, flexural, shear and torsional forces in the structural element for which it is to be designed. In this chapter emphasis is laid on the design part.

7.2 GUIDELINES FOR SELECTION OF MEMBER SIZES

As explained earlier, the selection of cross-section dimensions of flexural members (thickness of slabs, in particular) from a structural view point is often dictated by serviceability criteria and the requirements related to the placement of reinforcement. The serviceability criteria requires the control of deflections and crack widths under service loads. For convenience in design, and as an alternative to the actual calculation of deflection, the Code recommends certain basic values of span/effective depth (l/d) ratios which are expected to satisfy the requirements of deflection control, i.e., $\Delta/l < l/250$.

For prismatic beams of rectangular cross-sections and slabs of uniform thickness with spans up to 10 m, the limiting or maximum l/d ratios specified by the IS:456 are:

$$\left(\frac{l}{d}\right)_{\max} = \left(\frac{l}{d}\right)_{\text{basic}} \times m_{ft} \times m_{fc}$$

or

$$d_{\min} = \frac{l}{(l/d)_{\text{basic}} \times m_{ft} \times m_{fc}} \tag{7.1}$$

where

$$\left(\frac{l}{d}\right)_{\max} = \begin{cases} 7 & \text{for cantilever spans} \\ 20 & \text{for simply supported spans} \\ 26 & \text{for continuous spans} \end{cases}$$

For larger spans refer to Table 6.1.

7.2.1 General Guidelines for Beam Sizes

In general, it is economical to opt for singly reinforced sections with moderate percentage of tension reinforcement, i.e., $p_t \approx 0.6$ to 0.8 times $p_{t,\text{lim}}$. The recommended ratio of overall depth, D to width, b in rectangular beam sections is generally in the range of 1.5 to 3, it may be higher for beams carrying very heavy loads. The width and depth of beams are also governed by the shear force on the section. If the sizes of beams as dictated by architectural or other considerations are too restrictive then the desired moment resisting capacity of the beam in flexure can be provided by making it doubly reinforced and/or by providing high strength concrete and steel. In the case of beam-slab systems in which slabs are cast monolithically with beams, the beams are advantageously modelled as flanged beams, as explained earlier. Given a choice between increasing either the width of a beam or its depth, it is always advantageous to resort to the latter.

In practice, the overall depths of beams are often fixed in relation to their spans. Span-to-overall-depth ratios of 10 to 16 are generally found to be suitable in the case of simply supported and continuous beams. However, in case of cantilevers, lower ratios are adopted, and the beams are generally tapered in depth along their lengths for economy.

7.2.2 General Guidelines for Slab Thicknesses

In practice, Fe415 grade steel is most commonly used, and for such steel a p_t value of about 0.4–0.5 per cent may be assumed for preliminary proportioning. This assumption gives m_{ft} value of about 1.25, the corresponding effective depth for preliminary design works out to about span/25 for simply supported slabs and about span/32 for continuous slabs. The calculated value of the thickness should be rounded off to the nearest multiple of 5 mm or 10 mm.

7.2.3 Deep Beams and Slender Beams

Where the depth of the beam becomes comparable to its span, the beam is referred to as **deep beam**. By definition, a deep beam is the one whose *L/D* ratio is less than 2.0 for a simply supported beam, and 2.5 for a continuous beam.

In the situations where slender beam having excessive length in comparison with its cross-sectional dimensions, particularly its width b, there is a possibility of instability of lateral buckling in the compression zone. IS:456 specifies certain slenderness limits to ensure lateral stability. In the case of simply supported and continuous beams, the clear distance between lateral restraints should not exceed $60b$ or $250b^2/d$, whichever is less. For a cantilever, the distance from the free end to the edge of the support should be limited to the least of the $25b$ and $100b^2/d$.

7.3 DESIGN OF SINGLY REINFORCED RECTANGULAR SECTIONS

The design of a reinforced concrete beam generally consists of determining the cross-sectional dimensions of the beam, viz. b and D and the area of tension steel A_{st} required to resist the design moment M_u. The material properties f_{ck} and f_y are generally prescribed/selected on the basis of exposure conditions, availability of materials and economy. For normal applications, Fe415 grade steel and either M20 or M25 grade concrete are generally used. For exposures conditions rated severe, very severe and extreme, the minimum concrete grades specified are M30, M35 and M40, respectively. As explained earlier, for under-reinforced sections, the influence of f_{ck} on the ultimate moment of resistance $M_{u,\text{lim}}$ is relatively small; hence the use of high strength concrete is not beneficial from the point of economy. However, higher grade concrete is desirable from the durability considerations.

The ultimate moment of resistance of rectangular section is given by

$$M_{u,\text{lim}} = Q_{u,\text{lim}} bd^2$$

where

$$Q_{u,\text{lim}} = \begin{Bmatrix} 0.1498 \\ 0.1388 \\ 0.1338 \end{Bmatrix} f_{ck} \quad \begin{matrix} \text{for Fe250} \\ \text{for Fe415} \\ \text{for Fe500} \end{matrix} \tag{7.2}$$

7.3.1 Fixing Dimensions of Rectangular Section

The problem is simplified if the values of b and D are predetermined either by architectural considerations or arrived at on some logical basis. In the case of slabs, b is taken as 1000 mm and d is governed by the limiting l/d ratios for the deflection control. The trial value of d, the effective depth, may be assumed as approximately $l/25$ for simply supported spans, $l/32$ for continuous spans and $l/8$ for cantilevers. The overall depth D may be taken as d plus effective cover. In the case of single layer of tension reinforcement, the effective cover will be sum of nominal/clear cover, the diameter of the stirrup and half the bar diameter. Assuming a stirrup diameter of 10 mm and a bar diameter of 20 mm, the effective cover will be in the range of 40–95 mm depending on the exposure condition. Effective cover may be taken as 40, 50, 65, 70 and 95 mm, respectively, for mild, moderate, severe, very severe and extreme exposure conditions.

In the case of flexural members, it is desirable to adopt under-reinforced sections with $p_t < p_{t,\text{lim}}$. For the beams, value of b may be suitably fixed as 200 mm, 250 mm, 300 mm and so on and the value of d corresponding to $M_u \leq M_{u,\text{lim}}$ is given by

$$d = \sqrt{\frac{M_u}{Q_u b}} \tag{7.3}$$

where M_u is the factored or design moment in Nmm and Q_u is given for the stipulated grade and percentage of steel. The minimum value of d corresponding to the limiting case $p_t = p_{t,\text{lim}}$ is obtained by substituting $M_u = M_{u,\text{lim}}$. It is desirable to provide a value of d which is larger than d_{min} in order to obtain an under-reinforced section. The overall depth of the beam may be taken as $D > d_{\text{min}} +$ effective cover and should be expressed in rounded figures for ease in formwork construction. Multiples of 25 or 50 mm are generally adopted in practice. However, as explained earlier, the resulting D/b ratio should neither be excessive nor too small; ideally, it should be in the range 1.5 to 2.0. If the resulting D/b ratio is not acceptable and needs to be modified, this can be accomplished by suitably modifying b, recalculating d and fixing D. Having fixed the rounded-off value of D, the correct value of the effective depth d can be obtained by assuming the reinforcing bars to be located in one layer as follows:

$$d = D - (\text{Clear cover}) - \phi_{\text{tie}} - \frac{\phi}{2} \tag{7.4}$$

where ϕ_{tie} and ϕ are the diameters of tie and flexural steels, respectively.

7.3.2 Determination of Area of Tension and Compression Steels

At this stage of the design process, b and d are known, and it is desired to determine the required A_{st} from Eq. (3.14b) so that the section has a moment resisting capacity $M_{u,R}$ equal to the factored moment M_u.

$$\left(\frac{p_t bd}{100}\right) \equiv (A_{st})_{\text{reqd}} = \left(\frac{f_{ck} bd}{2 f_y}\right)[1 - \sqrt{1 - 4.6 R_u}\,] \tag{7.5}$$

The above formula provides a convenient and direct estimate of the area of tension reinforcement in singly reinforced rectangular sections. For the case where cross-sectional details are known the moment resisting capacity $M_{u,R}$ can be determined by using the following relation:

$$\frac{M_{u,R}}{bd^2} = \left(\frac{f_{ck}}{4.6}\right)\left[1 - \left\{1 - \left(\frac{p_t f_y}{50 f_{ck}}\right)\right\}^2\right] \tag{7.6}$$

In case the dimensions of the beam section are predetermined, the design problem reduces to determination of reinforcements in tension or both in tension and compression zones. In the later case, i.e., for doubly reinforced section, total percentages of tension and compression steels required are given by

$$p_t = p_{t,\lim} + \frac{100\left[(M_u/bd^2) - (M_{u,\lim}/bd^2)\right]}{0.87 f_y (1 - d'/d)} \tag{7.7}$$

and

$$p_c = \frac{0.87 f_y (p_t - p_{t,\lim})}{f_{sc} - 0.447 f_{ck}} \tag{7.8}$$

The minimum percentage of tension steel should be 0.50 with Fe250 and 0.30 with Fe415 grade steels for small and medium size beams. For large size beams corresponding values may be taken as 0.35 and 0.25.

7.3.3 Determination of Shear Reinforcement

The required spacing S_v of the vertical stirrups for a selected diameter is given by

$$S_v = \frac{0.87 f_y A_{sv}}{V_{us}/d}$$

or

$$\frac{V_{us}}{d} = \frac{0.87 f_y A_{sv}}{S_v} \tag{7.9}$$

Equation (7.9) provides a convenient and direct design procedure for vertical stirrups for specified V_{us}/d.

7.3.4 Determination of Number of Tension and Compression Bars

The calculated area of steel A_{st} has to be expressed in terms of bars of specified nominal diameter ϕ and their number (or spacing). Familiarity with the standard bar areas ($a_{st} = \pi\phi^2/4$) renders this task easy. In some cases, it may be economical to select a combination of two different bar diameters (close to each other) in order to arrive at an area of steel as close as possible to the calculated A_{st}. As explained earlier, in the case of slab, the area of steel is expressed in terms of centre-to-centre spacing of bars, given by

$$s = \frac{1000a_{st}}{A_{st}} \tag{7.10}$$

The actual spacing provided should be rounded off to the nearest lower multiple of 5 or 10 mm.

The diameter of hanger bars should not be less than 10 mm and that of main bars 12 mm. The usual sizes of bars chosen for beams are: 10, 12, 16, 20, 22, 25 and 32 mm. For main steel bars, not more than two sizes should be used, that to close to each other. When using different size of bars in one layer, the largest diameter bars should be placed near the beam faces. The area of steel should be symmetrical about the centre line of the beam.

The width of the beams should not be more than the dimensions of the columns into which they frame. In T-beams the depth of slab is usually taken as 20 per cent of the overall depth of beam.

7.3.5 Use of Design Aids

In the limit states design method, the relationship between $R_u \equiv M_u/bd^2$ and p_t is generally expressed in the form of charts or tables for various combinations of f_y and f_{ck}. The charts and tables are available in design handbooks such as SP:16. The tabular format as given in the Appendix D is generally more convenient to deal with than the charts. Table D.1 has been developed for M20, M25 and M30 grades of concrete, covering the three grades of steel Fe250, Fe415 and Fe500. For a given value of R_u, and specified values of f_y and f_{ck}, the desired value of p_t can be read off. For intermediate values linear interpolation is used.

The percentage of tension and compression steels, p_t and $p_{c,}$ required for doubly reinforced section, as obtained from Eqs.(7.7) and (7.8), are given in Table D.4.

For convenience, Tables E.1 and E.3 may be referred to for a quick selection of bar diameter and its number, and spacing of bars, respectively. The values of mass per metre length of the bars which may be useful in the cost estimation are obtained from Table E.5.

7.3.6 Design Check for Strength and Deflection Control

To ensure a ductile failure at the ultimate limit state, i.e., p_t is less than $p_{t,\text{lim}}$, the actual A_{st} and d provided should be computed to check the validity of this proposition. It is good practice to calculate the actual moment resisting capacity, $M_{u,,R}$ of the designed section, and thereby ensure that $M_{u,R} \geq M_u$.

For flexural members the adequacy of the depth is checked for deflection control. In the case of singly reinforced beam sections, the limiting (l/d) ratio is generally satisfied. However, in the case of slabs, the criteria for deflection control are generally critical. It is therefore necessary to adopt a suitable value of d at the initial stage of the design itself. The section should be suitably redesigned if it is found to be inadequate.

7.4 DESIGN OF BASIC BUILDING COMPONENTS SUBJECTED TO FLEXURE

7.4.1 Estimation of Loads

The dead load on a flexural member may be due to self-weight, floor finish, partitions, false ceiling and other specified special loads. The live loads shall be different for different structures depending upon their functional use. These loads are given in IS:875– Loading Standards. For computation of dead load, unit weights of plain and reinforced concretes may be taken as 24 kN/m^3 and 25 kN/m^3, respectively.

In the limit state design of flexural members, the sections are either balanced or under-reinforced. Over-reinforced sections are not permitted because they may result in a brittle compression failure. The values of design parameters $M_{u,\text{lim}}$, $x_{u,\text{max}}$, and $p_{t,\text{lim}}$ are listed in Table 7.1 for ready reference.

TABLE 7.1 Design Parameters for Various Materials

Type or grade of reinforcement	$x_{u,max}$	$M_{u,lim}$	$p_{t,lim}$ *M20*	*M25*	*M30*	*M35*
Mild steel grade-I (Fe250)	0.5313d	$0.1498 f_{ck} bd^2$	1.769	2.211	2.653	3.095
Fe415	0.4791d	$0.1388 f_{ck} bd^2$	0.961	1.201	1.441	1.681
Fe500	0.4560d	$0.1338 f_{ck} bd^2$	0.759	0.949	1.138	1.328

7.4.2 Design Procedure

The following design steps may be adopted:

1. Calculate the design constants from the properties of the materials to be used in construction.
2. Estimate the loads to be carried by the member. The serviceability requirement of deflection in terms of the effective span-to-effective depth ratio may be used to obtain the trial section of the member which is used to calculate the self-weight.
3. From the loads at the service state obtained in Step 2, determine the factored loads, and hence the factored shear force and factored or design moment.
4. Using the design moment, calculate the cross-section of the member required for a balanced design. If the size of the member is specified, check whether it is singly reinforced or doubly reinforced.
5. Compare the computed section with the trial section and modify the calculations if required and repeat the Steps 2 to 4.
6. Compute the area of reinforcement required for the design moment.
7. If some of the bars are curtailed, check for curtailment using curtailment rules.

8. Check the shear stresses and development length of bars.
9. Check the serviceability requirements with respect to deflection and cracking.
10. Show the reinforcement details through the elevations and sections of the member.

7.4.3 Design of a Rectangular Beam

Following design example which has been worked out in detail will illustrate the basics of procedure involved in the design of a rectangular beam.

Example 7.1 Design a simply supported beam of a rectangular section with a clear span of 6.5 m and subjected to uniformly distributed live and superimposed loads of 20 kN/m and 15 kN/m, respectively, at the service state. The support width (i.e., bearing) is 500 mm. The materials to be used are: concrete mix of grade M20 and mild steel reinforcement of grade Fe250.

Solution For M20 concrete and mild steel reinforcement:

$$f_{ck} = 20 \text{ MPa}, \; f_y = 250 \text{ MPa}$$

The dimensions of the beam are proportioned to be near the balanced design. The depth of the balanced neutral axis is:

$$x_{u,\max} = \frac{0.0035d}{0.0055 + 0.87(f_y/E_s)}$$

$$= \frac{0.0035d}{0.0055 + 0.87 \times (250/200000)} = 0.5313 \text{ d}$$

and hence

$$M_{u,\lim} = 0.362 f_{ck} x_{u,\max} b(d - 0.416 x_{u,\max}) = 0.1498 f_{ck} bd^2$$

Alternatively, these values can be obtained from Table 7.1.

Cross-sectional dimensions of the beam. In this case,

Effective span = Clear span l_c + Width of support = 6.5 + 0.5 = 7.0 m

Minimum depth based on the limit state of serviceability is given as:

$$d = \frac{\text{Effective span}}{20 \times \text{Modification factor}}$$

Consider a modification factor of 1.0. Then

$$d = \frac{7000}{20 \times 1.0} = 350 \text{ mm}$$

Minimum width from stability considerations is the higher of the following:

$$b = \frac{l_c}{60} \quad \text{or} \quad \left(\frac{l_c \times d}{250}\right)^{1/2}$$

$$= \frac{6500}{60} \quad \text{or} \quad \left(\frac{6500 \times 350}{250}\right)^{1/2}$$

$$= 108.33 \text{ mm or } 95.39 \text{ mm (say 110 mm)}$$

Trial dimensions. For the trial section the self-weight may be ignored.

$$\text{Factored load } w_u = 1.5(w_l + w_s)$$

$$= 1.5 \times 35 = 52.5 \text{ kN/m}$$

$$\text{Factored bending moment } M_u = \frac{52.5 \times (7.0)^2}{8} = 321.563 \text{ kNm}$$

The width b, is generally, adopted in the range $0.35d$ to $0.50d$. Consider $b = 0.425d$. Equate M_u with $M_{u,\text{lim}}$:

$$0.1498 \times 20 \times (0.425d) \times d^2 = 321.563 \times 10^6$$

Therefore,

$$d = 632.09 \text{ mm, and hence } b = 268.64 \text{ mm}$$

Adopt b = 280 mm and increase the depth of the beam by about 10 per cent to account for the self-weight. Therefore,

$$d = 632.09 + 63.21 = 695.30 \text{ mm} \quad \text{(say 700 mm)}$$

$$\text{Overall depth } D = d + \text{Effective cover} = 700 + 50 = 750 \text{ mm}$$

Consider 280 × 750 mm overall section with d = 700 mm as a trial section.

$$\text{Self-weight of beam } w_g = 0.28 \times 0.75 \times 25 = 5.25 \text{ kN/m}$$

$$\text{Total factored load } w_u = 1.5\ (35 + 5.25) = 60.375 \text{ kN/m}$$

$$\text{Factored bending moment } M_u = \frac{60.375 \times 7.0^2}{8} = 369.80 \text{ kNm}$$

Hence,

$$0.1498 \times 20 \times 280 \times d^2 = 369.80 \times 10^6$$

and

$$d = 663.95 \text{ or } 670 \text{ mm}$$

Therefore,

$$D = 670 + 50 = 720 \text{ mm}$$

These dimensions are little smaller than that of the trial section. Hence there is no necessity for repeating the calculations.

Design for flexure:

Consider the section to be under-reinforced. Dimensionless parameter for flexural reinforcement is given by

$$R_u = \frac{4.6 \times 369.80 \times 10^6}{20 \times 280 \times 670^2} = 0.6767$$

Therefore,

$$A_{st} = \frac{f_{ck}bd}{2f_y} \times [1 - \sqrt{1 - R_u}]$$

$$= \frac{20 \times 280 \times 670}{2 \times 250} \times [1 - \sqrt{1 - 0.6767}] = 3237.3 \text{ mm}^2$$

$$\text{Minimum tension reinforcement} = \frac{0.85bd}{f_y} = \frac{0.85 \times 280 \times 670}{250} = 638 \text{ mm}^2$$

Provide 7 bars of 25 mm ϕ (A_{st} = 3436 mm^2) along the length of the beam.

Curtailment of reinforcement. Consider that the reinforcement is curtailed at a distance $0.125l$ from the support. Distance of the point of curtailment from the mid-span is:

$$= l/2 - 0.125l = 3.5 - 0.125 \times 7.0 = 2.625 \text{ m}$$

This should be greater than the development length for mobilizing full stress in reinforcement at mid-span.

Development length.

$$L_d = \frac{0.87 f_y \phi}{4\tau_{bd}} = \frac{0.87 \times 250 \times \phi}{4 \times 1.2} = 45.31\phi$$

$$= 45.31 \times 25 = 1132.8 \text{ mm} = 1.133 \text{ m} < 2.625 \text{ m}$$

Hence curtailment can be made at a distance of 2.625 m from the mid-span. Distance of the theoretical point of curtailment from the mid-span is given as:

$$= 2.625 - (d \text{ or } 12\phi\text{, whichever is more})$$

$$= 2.625 - (0.67 \text{ or } 12 \times 0.025)$$

$$= 2.625 - 0.67 = 1.955 \text{ m}$$

Distance from the support = 3.5 – 1.955 = 1.545 m

Moment capacity required at the theoretical point of curtailment is obtained as:

$$M_u = 60.375 \times \left[(3.5 \times 1.545) - \left(\frac{1.545^2}{2}\right)\right] = 254.42 \text{ kNm}$$

For area of flexural reinforcement required at this point, dimensionless parameter is:

$$R_u = \frac{4.6 \times 254.42 \times 10^6}{20 \times 280 \times 670^2} = 0.46555$$

Therefore,

$$A_{st} = \frac{20 \times 280 \times 670}{2 \times 250} \times [1 - \sqrt{1 - 0.46555}\ ]$$

$$= 2018.12 \text{ mm}^2$$

Hence curtail 2 × 25 mm ϕ bars and continue 5 × 25 mm ϕ bars, giving a total area of 2454 mm^2.

Checks for development length at the support.

1. Length of continuing bars beyond the cutoff point shall be at least equal to the development length. Development length is given by

$$L_d = \frac{0.87 f_y \phi}{4\tau_{bd}} = \frac{0.87 \times 250 \times 25}{4 \times 1.2} = 1133 \text{ mm}$$

   Length of the bar inside the support is:

$$= L_d - \text{Distance of the cutoff point from the face of the support}$$

$$= 1133 - (3250 - 2625) = 508 \text{ mm}$$

   Length of the bar embedded into the support is 575 mm, as shown in Fig 7.1(c).
2. Minimum length of bar embedment into the support is obtained as:

$$\frac{L_d}{3} = \frac{1133}{3} = 377.67 \text{ mm} \qquad \text{(say 380 mm)}$$

   A minimum length of 575 mm is embedded in the support as in (1).

$$L_d \le 1.3\left(\frac{M_1}{V}\right) + L_o$$

   where

$$L_d = \frac{0.87 f_y \phi}{4\tau_{bd}}$$

   or

$$\phi \le \frac{4\tau_{bd}\left[\left(\frac{1.3M_1}{V}\right) + L_o\right]}{0.87 f_y}$$

   where M_1 is the moment of resistance provided by continuing bars, and is given by

$$M_1 = 0.87 f_y A_{st}\left(d - \frac{0.416 \times 0.87 f_y A_{st}}{0.362 f_{ck} b}\right)$$

$$= 0.87 \times 250 \times 2454 \times \left(670 - \frac{0.416 \times 0.87 \times 250 \times 2454}{0.362 \times 20 \times 280}\right) \doteq 299.15 \text{ kNm}$$

V = Shear force at support = 60.375 × 3.25 = 196.22 kN

and

L_o = Anchorage length beyond the centre of support ≤ (d or 12ϕ, whichever is greater)

Consider L_o = 0. Therefore,

$$\phi \leq \frac{4 \times 1.2 \times [(1.3 \times 299.15 \times 10^6)/(196.22 \times 10^3)]}{0.87 \times 250} \leq 43.74 \text{ mm}$$

Since the diameter of the bar is less than 43.74 mm, the beam is safe in anchorage bond.

Design for shear

Longitudinal steel.

$$p_t = \frac{2454 \times 100}{280 \times 670} = 1.308 \text{ per cent}$$

Therefore,

$$\tau_{uc} = 0.684 \text{ MPa (from Table 4.1)}$$

Maximum shear strength of concrete with shear reinforcement.

$$\tau_{uc,\max} = 2.8 \text{ MPa} \qquad \text{(for M20 concrete)}$$

Shear force at the critical section at a distance d from the face of support.

$$V_u = 60.375 \times (3.25 - 0.67) = 155.77 \text{ kN}$$

$$\text{Nominal shear stress } \tau_v = \frac{155.77 \times 10^3}{280 \times 670} = 0.83 \text{ MPa}$$

Since $\tau_{uc} < \tau_v < \tau_{uc,\max}$, the section is acceptable with shear reinforcement. Consider 6 mm ϕ 2-legged stirrups (A_{sv} = 56 mm^2).

Shear resistance to be provided by the stirrups is given by

$$V_{us} = V_u - \tau_{uc}bd$$

$$= 155.77 - 0.684 \times 280 \times 670 \times 10^{-3} = 27.45 \text{ kN}$$

Required spacing of stirrups is:

$$S_v = \frac{0.87 f_y A_{sv} d}{V_{us}}$$

$$= \frac{0.87 \times 250 \times 56 \times 670}{27.45 \times 10^3} = 297.27 \text{ mm}$$

The spacing shall be the least of the following:

(a) $0.75d = 0.75 \times 670 = 502.5$ mm

(b) 450 mm

(c) $\dfrac{A_{sv} f_y}{0.4b} = \dfrac{56 \times 250}{0.4 \times 280} = 125$ mm

(d) 297.27 mm (required to carry shear force)

Provide 6 mm ϕ 2-legged vertical stirrups @ 125 mm c/c with 2-8 mm ϕ anchor bars. In the cut off regions, shear capacity is maintained by providing additional vertical stirrups over a distance $0.75d$ from the cutoff point. Area of additional vertical stirrups is given as:

$$A'_{sv} = \frac{0.40 b S_v}{f_y}$$

where $S_v \leq d/(8\beta_b)$ and β_b is the ratio of area of cutoff bars to area of the total bars (= 2/7 = 0.286).
Therefore,

$$S_v \leq \frac{670}{8 \times 0.286} \quad (= 300 \text{ mm})$$

Consider the spacing of additional stirrups at 125 mm c/c. Thus,

$$A'_{sv} = \frac{0.4 \times 280 \times 125}{250} = 56.0 \text{ mm}^2$$

Provide 6 mm ϕ 2-legged additional stirrups ($A_{sv} = 56$ mm^2) @ 125 mm c/c. The reinforcement details are shown in Fig. 7.1(c).

Check for deflection:

(a) Short-term deflection. Quantities required for short term deflection are:
Short-term modulus of elasticity of concrete

$$E_c = 5000\sqrt{f_{ck}} = 5000\sqrt{20} = 2.236 \times 10^4 \text{ MPa}$$

$$\text{Modular ratio } m = \frac{E_s}{E_c} = \frac{2 \times 10^5}{2.236 \times 10^4} = 8.94$$

$$\text{Modulus of rupture } f_{cr} = 0.7\sqrt{f_{ck}} = 0.7 \times \sqrt{20} = 3.13 \text{ MPa}$$

$$\text{Gross moment of inertia } I_g = \frac{280 \times 720^3}{12} = 87.09 \times 10^8 \text{ mm}^4$$

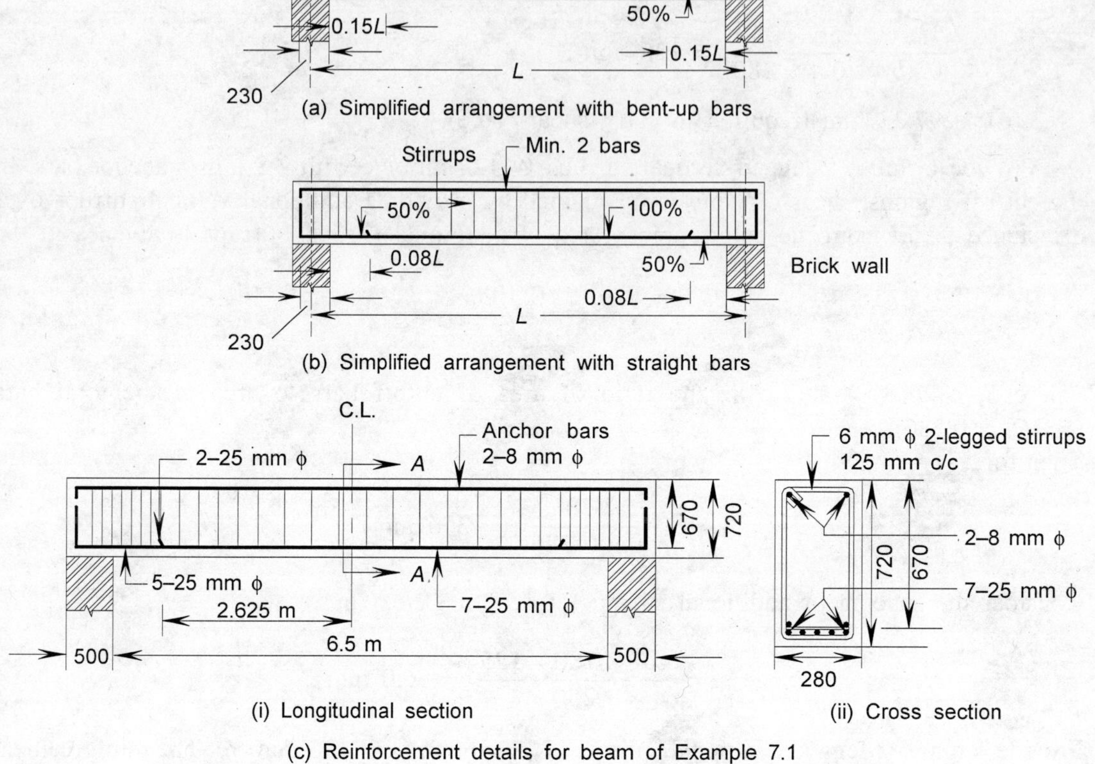

Fig. 7.1 Reinforcement details for simply supported beams.

Lever arm: Let the neutral axis lie at a distance x from the compression fibre; then take moment of the effective area about the neutral axis as:

$$\frac{bx^2}{2} = mA_{st}(d - x)$$

or

$$\frac{280x^2}{2} = 8.94 \times 3436 \times (670 - x)$$

or

$$x^2 + 219.41x - 147006.81 = 0$$

Thus

$$x = 289.1 \text{ mm}$$

and

$$\text{lever arm } z = d - \left(\frac{x}{3}\right) = 670 - \left(\frac{289.1}{3}\right) = 573.63 \text{ mm}$$

Moment of inertia of the cracked section

$$I_{cr} = \frac{[280 \times 289.1^3]}{3} + [8.94 \times 3436 \times (670 - 289.1)^2]$$

$$= 671.19 \times 10^7 \text{ mm}^4$$

Cracking moment $M_r = \dfrac{f_{cr} I_g}{y_t} = \dfrac{3.13 \times 87.09 \times 10^8}{(720/2)} = 75.72 \times 10^6$ Nmm

Maximum moment under service load $M = \dfrac{wl^2}{8} = \dfrac{40.25 \times 7^2}{8} = 246.53$ kNm

and

Effective moment of inertia $I_{\text{eff}} = I_{cr}\left[1.2 - \left\{\dfrac{M_r}{M} \cdot \dfrac{z}{d}\left(1 - \dfrac{x}{d}\right)\dfrac{b_w}{b}\right\}\right]^{-1}$

$$= (671.19 \times 10^7) \times \left[1.2 - \left\{\left(\frac{75.72 \times 10^6}{246.53 \times 10^6}\right) \times \left(\frac{573.63}{670}\right)\left(1 - \frac{289.1}{670}\right) \times 1\right\}\right]^{-1}$$

$$= \frac{671.19 \times 10^7}{1.0505} = 638.92 \times 10^7 \text{ mm}^4$$

Since $I_{cr} > I_{\text{eff}}$, adopt $I_{\text{eff}} = I_{cr} = 638.92 \times 10^7$ mm^4

Thus,

Short-term deflection $\delta_s = \left(\dfrac{5}{384}\right)\left(\dfrac{Wl^3}{E_c I_{\text{eff}}}\right)$

$$\delta_s = \left(\frac{5}{384}\right) \times \left[\frac{(40.25 \times 7 \times 10^3) \times (7000)^3}{2.236 \times 10^4 \times 638.92 \times 10^7}\right] = 8.81 \text{ mm}$$

(b) Long-term deflection. Percentage of tension and compression reinforcements are:

$$p_t = \frac{3436 \times 100}{280 \times 670} = 1.832, \text{ and } p_c = 0.000$$

(i) Deflection due to shrinkage:

$$\delta_{cs} = k_3 \psi_{cs} l^2$$

where

k_3 = 0.125 for a simply supported beam

ψ_{cs} = Shrinkage curvature = $k_4 \varepsilon_{cs}/D$ where k_4 = (0.65 × 1.832)/(1.832)$^{1/2}$ = 0.88, since $p_t - p_c > 1.0$

ε_{cs} = Ultimate shrinkage strain = 0.0003

Hence,

$$\psi_{cs} = \frac{0.88 \times 0.0003}{720} = 3.67 \times 10^{-7}$$

Therefore,

$$\delta_{cs} = 0.125 \times (3.67 \times 10^{-7}) \times (7000)^2 = 2.246 \text{ mm}$$

(ii) Deflection due to creep:

$$E_{cc} = \frac{E_c}{(1+\theta)} = \frac{2.236 \times 10^4}{1+1.6} = 8.60 \times 10^3 \text{ MPa}$$

Assuming the permanent load to be 60 per cent of the service load, we have

$$W_p = 0.6 \times (40.25 \times 7) \text{ kN}$$

$$\delta_{sc,p} = \frac{5W_p l^3}{384E_{cc} \times I_{\text{eff}}}$$

$$= \frac{5 \times (0.6 \times 40.25 \times 7 \times 10^3) \times (7000)^3}{384 \times 8.60 \times 10^3 \times 638.92 \times 10^7} = 13.74 \text{ mm}$$

$$\delta_{sp} = \frac{5 \times (0.6 \times 40.25 \times 7 \times 10^3) \times (7000)^3}{384 \times 2.236 \times 10^4 \times 638.92 \times 10^7}$$

$$= 5.28 \text{ mm}$$

and

$$\delta_{cp} = 13.74 - 5.28 = 8.46 \text{ mm}$$

Thus,

$$\text{total deflection } \delta = \delta_s + \delta_{cs} + \delta_{cp}$$

$$= 8.81 + 2.246 + 8.46 = 19.516 \text{ mm}$$

and

$$\text{permissible deflection } \delta = \frac{7000}{250} = 28 \text{ mm} > 19.516 \text{ mm}$$

The deflection occurring after the erection of partitions and finishes should not exceed (span/350) or 20 mm, whichever is less:

$$\frac{\text{Effective span}}{350} = \frac{7000}{350} = 20 \text{ mm}$$

Consider that half the total shrinkage strain occurs within 28 days; deflection after 28 days will be

$$(0.5 \times 2.246) + 8.46 = 9.583 < 20 \text{ mm}$$

Hence the design is satisfactory.

For the procedure for the design of a cantilever beam a reader should refer to Example 7.4.

7.4.4 Design of Slabs

The slab is a two-dimensional plane element with its depth D being much smaller than the other two dimensions, namely, the width and the span. It may be supported on masonry walls or on beams or directly on columns. It carries gravity loads acting normal to its surface and transfers the same to the supports primarily by flexure. Slabs supported on two parallel sides carry loads by flexure in the direction perpendicular to the supports; such slabs are known as **one-way slabs** and virtually act as wide, shallow rectangular beams. The slab supported on all four sides and with a ratio of longer span to shorter span being greater than 2 is also considered as a one-way slab.

Design of one-way slab: Analysis and design of a one-way slab are normally carried out by considering a 1.0 m wide strip acting as a shallow rectangular beam spanning in the direction of bending. Since the loads on slabs are frequently specified in terms of load per square metre, the unit load on a 1.0 m wide strip becomes the load per linear metre. The reinforcing steel is usually spaced uniformly over its width as illustrated in Fig. 3.8, and the effective area of reinforcement A_{st} may correspond to a fractional number of bars in the 1.0 m wide strip. In this chapter, the one-way slab is introduced simply to highlight its similarity with a rectangular beam.

For a simply supported one-way slab, the maximum bending moment at the mid-span is $(wl^2/8)$, while in the slab fixed at supports the sagging (i.e., positive) bending moment at the mid-span is $(wl^2/24)$ and the hogging (i.e., negative) moment at the supports is $wl^2/12$.

Like a simply supported one-way slab, a continuous one-way slab is idealized by a series of continuous long strips, and the moments and shear forces may be obtained by using the coefficients recommended by IS:456. In this approach, redistribution of moments is not permitted. The moment analysis may also be carried out by moment distribution or by any other conventional method. In the latter approach redistribution of moments is permitted whereby moments at the supports are reduced by 15 per cent and those at the centre are increased by 15 per cent.

For checking the development length L_o may be assumed as 8ϕ for HYSD bars for which end anchorage is not usually provided, and 12ϕ for mild steel with hooks.

Design procedure. The steps involved in the design of a one-way slab are:

1. For the stipulated materials determine the design parameters for a balanced section.
2. Estimate the effective span of the slab as being the least of the following:
   (a) Clear span + effective depth of the slab or width of support, whichever is less
   (b) Centre-to-centre of supports
3. Estimate the minimum thickness of the slab based on serviceability limit state criteria of the deflections, i.e., from the limiting value of *effective span to effective depth ratio*.
4. Consider a 1.0 m wide strip of slab and determine the total factored load due to live (or superimposed) load and dead weight. Calculate the factored design moment M_u and factored shear force, V_u.

5. Calculate the slab thickness d for the factored design moment by equating M_u to $M_{u,\lim}$. Select the diameter of the reinforcing bar and determine the overall depth D as:

$$D = d + (0.5 \times \text{Diameter of the bar}) + \text{Minimum clear or nominal cover}$$

If the value of depth so calculated is equal to or less than the value estimated in Step 3, the section is adequate. If not assume a greater depth and find revised M_u, and repeat Steps 4 and 5.

6. Determine the area of main reinforcement required for the design moment. Provide suitable spacing for the selected bar diameter Φ as:

$$s = \frac{(1000)(\pi\phi^2/4)}{A_{st}} < 3d \text{ or } 300 \text{ mm, whichever is less}$$

Based on the actual A_{st} and d adopted, it should be ensured that p_t is less than $p_{t,\lim}$ for ductile failure at the limit state.

7. Check the adequacy of depth from the deflection criteria by calculating the modification factor for the amount of steel provided in Step 6.
8. Check the adequacy of development length at the face of the wall or beam:

$$L_d = \frac{0.87 f_y \phi}{4\tau_{bd}}$$

9. Apply the check for shear at the critical section at a distance d from the free support.
10. Design secondary reinforcement. The spacing of selected bars should not be more than $5d$.
11. Draw reinforcement details.

Detailing simply supported one-way slabs. The main reinforcement is placed in the bottom layer at mid-span. At least 50 per cent of main reinforcement shall extend into the support to a length equal to $L_d/3$ along the same face. The development length L_d is based on full design stress in reinforced concrete. The remaining 50 per cent shall be extended to within $0.15l_x$ of the support and then cranked up and extended in the upper part of the slab to within $0.10l_x$ of the support. The secondary or distribution steel complying with IS:456 stipulations shall be placed above the main reinforcement. It shall be extended up to the end if the slab is supported on two parallel edges only. If the slab is supported on four sides then 50 per cent of the secondary steel shall extend along the same face of the slab into the support and balance 50 per cent may be cranked up and extended to within $0.10l_x$ of the support as illustrated in Figs. 7.2(b) and (c), where l_x is the effective span in the direction of bending .

Example 7.2 Design the floor slab for an interior room, with clear dimensions of 4.0×8.5 m, of a building located in a coastal town. The slab is assumed to be simply supported on 230 mm thick masonry walls. The specified floor loading consists of a live load of 4.0 kN/m^2 and dead load due to finish, partition, etc. of 1.5 kN/m^2. Fe415 grade HYSD reinforcement is to be used in construction.

Solution Although the building is located in a coastal town, but the floor slab being interior is not directly exposed to coastal environment and thereby is protected against weather. The exposure conditions may be taken as moderate requiring concrete of grade M25 and nominal cover of 30 mm. Design constants for M25 concrete and Fe415 steel are:

$$f_{ck} = 25 \text{ MPa}, f_y = 415 \text{ MPa}, M_{u,\text{lim}} = 0.1388\, f_{ck}bd^2 \text{ and } p_{t,\text{lim}} = 1.201 \text{ per cent}$$

Since the aspect ratio, i.e., the ratio of larger span to shorter span is more than 2, the slab will be designed as one-way slab. The minimum thickness will be based on serviceability criteria of deflections. For an under-reinforced section $p_t < p_{t,\text{ lim}}$, consider modification factor m_t of 1.2 (for $p_t = 0.55$ per cent). Thus,

$$\text{Effective depth } d = \frac{4000 + d}{20 \times 1.2}$$

or

$$d = \frac{4000}{20 \times 1.2 - 1} \approx 175 \text{ mm}$$

Provide 10 mm ϕ bars at nominal cover of 30 mm, the overall depth is:

$$D = 175 + 30 + 5.0 = 210 \text{ mm}$$

Thus, consider 210 mm thick slab with $d = 175$ mm.
Effective span is the lesser of the following:

1. 4000 + 230 = 4230 mm (c/c distance)
2. 4000 + 175 = 4175 mm

Therefore,

$$l_x = 4.175 \text{ m}$$

Analysis for moments and shear forces.

Dead load of slab w_g = 0.210 × 1.0 × 25 = 5.25 kN/m
Load due to finish, etc. w_f = 1.5 × 1.0 = 1.50 kN/m
Live load w_l = 4.00 kN/m
Total service load w = 10.75 kN/m
Factored load $w_u = 1.5w$ = 16.125 kN/m

Factored moment at mid span, i.e., maximum moment is expressed as:

$$M_u = \frac{w_u(l_x)^2}{8} = \frac{16.125 \times (4.175)^2}{8} = 35.134 \text{ kNm/m}$$

Thickness of slab from flexural considerations for a balanced design is:

$$0.1388 \times 25 \times 1000 \times d^2 = 35.134 \times 10^6, \text{ i.e., } d = 100.6 \text{ mm}$$

Overall depth 135.6 mm. However, from the serviceability limit state criterion of deflection, the depth required is 210 mm.

Reinforcement. The dimensionless flexural reinforcement parameter is:

$$R_u = \frac{4.6 \times 35.134 \times 10^6}{25 \times 1000 \times 175^2} = 0.2111$$

and

$$A_{st} = \left(\frac{25 \times 1000 \times 175}{2 \times 415}\right)[1 - \sqrt{1 - 0.2111}] = 589.28 \text{ mm}^2\text{/m}$$

Spacing of 10 mm ϕ (a_{st} = 78.5 mm^2) bars, $S = \dfrac{1000a_{st}}{A_{st}}$ = 133.21 mm

Maximum spacing limits is:

$$3d = 3 \times 175 = 525 \text{ mm or } 300 \text{ mm (whichever is smaller)}$$

Provide 10 mm ϕ bars @ 130 mm c/c (A_{st} = 603.85 mm^2) for main reinforcement.

$$p_t = \frac{100a_{st}}{Sd} = 0.345 \text{ per cent} < p_{t,\lim} \text{ (=1.201 per cent)}$$

Thus, the section is under-reinforced as presumed earlier.

Distribution steel. Steel to be provided at right angles to the main reinforcement is:

$$A_{st} = 0.0012 \times 1000 \times 210 = 252 \text{ mm}^2\text{/m}$$

Spacing of 8 mm ϕ bars (a_{st} = 50 mm^2),

$$S = \frac{1000 \times 50}{252} = 198.4 \text{ mm}$$

Maximum spacing limits is:

$$5d = 5 \times 175 = 875 \text{ mm or } 450 \text{ mm (whichever is smaller)}$$

Thus, provide 8 mm ϕ bars @ 195 mm c/c for the distribution steel.

Deflection check. For p_t = 0.345 per cent and f_s = 0.58 × 415 × (589.28/603.85) = 234.88 MPa, the Magnification factor from Table 6.3, m_{ft} = 1.44. Thus the ratio

$$\left(\frac{l}{d}\right)_{\max} = 20 \times 1.44 = 28.8$$

$$\left(\frac{l}{d}\right)_{\text{provided}} = \frac{4175}{175} = 23.86 < 28.8$$

Therefore, the slab is safe in deflection.

Check for shear. Maximum factored shear V_u = 0.5 × 16.125 × 4 = 32.25 kN

$$\text{Nominal shear, } \tau_v = \frac{32.25 \times 10^3}{1000 \times 175} = 0.184 \text{ MPa}$$

For a solid slab with M25 grade concrete having $p_t = 0.345$ per cent and (from Table 4.1) $\tau_c = 0.411$ MPa, the permissible shear strength $k_s\tau_c = 1.21 \times 0.411 = 0.497$ MPa > τ_v. Hence no shear reinforcement is required.

Check for development length.

$$\text{Development length } L_d = \frac{0.87 f_y \phi}{4\tau_{bd}} = \frac{0.87 \times 415\phi}{4 \times 2.24} = 40.3\phi$$

1. Minimum embedment length in to the support is $L_d/3$.

   For 10 mm ϕ bars,

$$\frac{L_d}{3} = \frac{40.3\phi}{3} = \frac{40.3 \times 10}{3} = 134.32 \text{ mm}$$

   Length of bar embedment into the support = Width of support – Clear side cover of 30 mm

$$= 230 - 30 = 200 \text{ mm} > \frac{L_d}{3}$$

2. At simple support

$$L_d \leq 1.3\left(\frac{M_{u1}}{V_u}\right) + L_o$$

   As alternate bars are bent up at support, moment of resistance of section may be taken as half of span-moment. Therefore,

$$M_{u1} = 0.5 \times 35.134 = 17.567 \text{ kNm}$$

$$V_u = 32.25 \text{ kN}$$

$$L_o = 8\phi \text{ (for 90° bend from the centre of support).}$$

   Thus,

$$1.3 \times \left[\frac{17.567 \times 10^6}{32.25 \times 10^3}\right] + 8\phi \geq 40.3\phi, \text{ i.e., } \phi \leq 21.9 \text{ mm}$$

Therefore, $\phi_{\text{provided}} = 10$ mm is satisfactory.

This slab can also be designed as a two-way slab using moment coefficients from Table 7.2.

The following example will illustrate the procedure for the design of a one-way continuous slab.

Example 7.3 A hall in a building is to be provided a floor consisting of a continuous slab cast monolithically with simply supported 300 mm wide beams spaced 3.6 m c/c as shown in Fig. 7.2(a). The clear span of the beam is 8.75 m. The floor is to support live, and partition loads of 3.0 kN/m^2 and 1.40 kN/m^2, respectively at the service state. The load due to finishes may be considered as 0.60 kN/m^2. Design the slab, if M20 grade concrete mix and HYSD steel of grade Fe415 are to be used.

Solution Design constants for the given materials are:

$$f_{ck} = 20 \text{ MPa},\ f_y = 415 \text{ MPa}$$

$$x_{u,\max} = 0.4791d,\ M_{u,\lim} = 0.1388 f_{ck} bd^2 \text{ and } p_{t,\lim} = 0.961$$

$$L_d = \frac{0.87 f_y \phi}{4\tau_{bd}} = \frac{0.87 \times 415 \times \phi}{4 \times 1.92} = 47.0\phi$$

Consider 1.0 m width of slab and base the minimum thickness of the slab on serviceability limit state criteria of deflection. A uniform thickness of slab shall be provided which is governed by the end span having one end simply supported and the other end continuous.
Let the effective span of slab is c/c distance between beams, i.e., 3.6 m and

$$d = \frac{\text{Effective span}}{0.5 \times (20 + 26) \times \text{Modification factor}}$$

Modification factor $m_f = m_{ft} \times m_{fc}$

Since there is no compression steel at mid-span, $m_{fc} = 1.0$ and as the slab is to be designed as an under-reinforced section, i.e., $p_t < p_{t,\lim}$, consider $p_t \approx 0.55$ per cent and $f_s = 0.58 \times 415 = 240.7$ MPa. The modification factor m_{ft} is 1.2 giving $m_f = 1.2$. Thus,

$$d = \frac{3600}{23 \times 1.2} = 130.43 \text{ mm.}$$

With 10 mm ϕ bars the overall depth is:

$$D = 130.43 + \text{Clear cover} + \frac{\phi}{2}$$

$$= 130.43 + 15 + \frac{10}{2} = 150.43 \text{ mm}$$

Consider 150 mm thick slab with $d = 130$ mm.
Design for moment.

Dead load of slab $w_s = 0.15 \times 1 \times 25$	= 3.75 kN/m
Load due to partition w_p	= 1.40 kN/m
Load due to finishes w_f	= 0.60 kN/m
Total dead load $w_d = w_s + w_p + w_f$	= 5.75 kN/m
Live load w_l	= 3.00 kN/m
Factored dead load $w_{u,d} = \gamma w_d = 1.5 \times 5.75$	= 8.625 kN/m
Factored live load $w_{u,l} = \gamma w_l = 1.5 \times 3.00$	= 4.500 kN/m

The maximum factored bending moments occur in the end span, their values as per IS: 456 are:

$$M_{u,\text{mid}} = \frac{w_{u,d}l^2}{12} + \frac{w_{u,l}l^2}{10}$$

$$= \frac{8.625 \times 3.6^2}{12} + \frac{4.5 \times 3.6^2}{10}$$

$$= 9.32 + 5.83 = 15.12 \text{ kNm}$$

and

$$M_{u,\text{support}} = \frac{w_{u,d}l^2}{10} + \frac{w_{u,l}l^2}{9}$$

$$= \frac{8.625 \times 3.6^2}{10} + \frac{4.5 \times 3.6^2}{9}$$

$$= 11.18 + 6.48 = 17.66 \text{ kNm}$$

Thus, the design moment M_u for calculation of thickness is 17.66 kNm. Consider that the thickness of the slab is proportioned to the moment near its balanced value:

$$0.1388 f_{ck} b d^2 = M_u$$

$$0.1388 \times 20 \times 1000 \times d^2 = 17.66 \times 10^6$$

or

$$d = 79.76 \text{ mm}$$

Consider 10 mm ϕ bars at 15 mm clear cover. Then, the overall depth D = 79.76 + 15 + (10/2) = 99.76 mm. However, from the serviceability limit state criterion of deflection the depth adopted is 150 mm.

Reinforcement.

$$A_{st} = \left(\frac{f_{ck}bd}{2f_y}\right)\left[1 - \sqrt{1 - \left(\frac{4.6 \times M_u}{f_{ck}bd^2}\right)}\right]$$

$$= \left(\frac{20 \times 1000 \times 130}{2 \times 415}\right)\left[1 - \sqrt{1 - \left(\frac{4.6 \times 17.66 \times 10^6}{20 \times 1000 \times 130^2}\right)}\right]$$

$$= 402.27 \text{ mm}^2$$

Adopt 10 mm ϕ bars at 195 mm c/c (A_{st} = 402.56 mm^2).

$$p_t = \frac{100 \times 402.56}{1000 \times 130} = 0.31 \text{ per cent}$$

Modification factor from Table 6.3 for p_t = 0.31 per cent is 1.47. The depth of the section required from the serviceability limit state criterion of deflection is:

$$d = \frac{\text{Effective span}}{[0.5 \times (26 + 20) \times \text{Modification factor}]}$$

$$= \frac{3600}{23 \times 1.47} = 106.48 \text{ mm} < 130 \text{ mm (provided).}$$

Therefore, the slab is safe in deflection.

Distribution steel.

$$A_{st} = \frac{0.12(bD)}{100} = \frac{0.12 \times 1000 \times 150}{100} = 180 \text{ mm}^2$$

Provide 8 mm ϕ bars at 270 mm c/c (< $5d$ or 450 mm); A_{st} = 186.15 mm^2.

Curtailment of positive steel. Alternative bars shall be curtailed at distances not more than $0.15l$ and $0.25l$ from discontinuous and continuous edges, respectively. The development length criteria are:

(a) At simply supported end

(i) Length beyond cut off point = 0.15 × 3600 – 0.50 × 230 + 150
= 575 mm > L_d (= 470 mm)

(ii) Length of curtailed bar from the mid-span
= $0.50l - 0.15l$ = 0.351 = 0.35 × 3600
= 1260 mm > L_d

(b) At interior support

(i) 0.25 × 3600 – 0.50 × 300 + 150 = 900 mm > L_d (= 470 mm)

(ii) 0.50 × 3600 – 0.25 × 3600 = 900 mm > L_d

Curtailment of negative steel. The length of the bars extending from the discontinuous and continuous edges shall not be less than L_d.

Length of bars from centre of end support = $0.10l$ or L_d, i.e., 470 mm

Length of bar from centre of interior support = $0.15l$ or L_d, i.e., 540 mm

Check for shear. Maximum factored shear force

$$V_u = 0.6 \times (w_{u,d} + w_{u,l}) \times l = 0.6 \times (8.625 + 4.5) \times 3.6 = 28.35 \text{ kN}$$

$$\text{Nominal shear stress } \tau_v = \frac{V_u}{bd} = \frac{28.35 \times 10^3}{1000 \times 130} = 0.218 \text{ MPa}$$

For M20 grade concrete with p_t = 0.31 per cent from Table 4.1, τ_c = 0.393 MPa. For solid slab permissible shear strength, τ_c =1.3 × 0.393 = 0.511 MPa. Since $\tau_c > \tau_v$, no shear reinforcement is required.

Check for development length.

$$L_d = \frac{0.87\, f_y \phi}{4\tau_{bd}} = \frac{0.87 \times 415 \times \phi}{4 \times 1.92} = 47.0\phi$$

The minimum length of the bar to be embedded in the support = $L_d/3$

For 10 mm ϕ bars, $\frac{L_d}{3} = \frac{47 \times 10}{3} \approx 160$ mm

For 8 mm ϕ bars, $\frac{L_d}{3} = \frac{47 \times 8}{3} \approx 125$ mm

Consider that the reinforcing bars are embedded up to the total width of the support with a clear side cover of 25 mm. Thus, the length of the bar embedded into the support is:

= Width of the support – Clear side cover of 25 mm

= 230 – 25 = 205 mm, more than the minimum length required

(a) At the simple supports

$$L_d \not> 1.3\left(\frac{M_{u,1}}{V_u}\right) + L_o$$

(b) At points of inflection

$$L_d \not> \left(\frac{M_{u,1}}{V_u}\right) + L_o$$

At the simply supported end, $M_{u,1}$, the moment of resistance of the section at the support may be taken as half the span moment as alternate bars have been bent up, i.e.,

$$M_{u,1} = 0.50 \times (15.12) = 7.56 \text{ kNm}$$

$$V_u = (0.40 \times 8.625 \times 3.6) + (0.45 \times 4.5 \times 3.6) = 19.71 \text{ kN}$$

L_o is the anchorage length beyond the centre of support which should not exceed d or 12ϕ, whichever is more.

Consider,

$$L_o = 8\phi \text{ (for } 90^\circ \text{ bend)}$$

Thus,

$$47\phi \not> 1.3 \times \frac{7.56 \times 10^6}{19.71 \times 10^3} + 8\phi \text{ or } \phi \not> 12.79$$

Since, $\phi_{provided}$ (= 10 mm) ≤ 12.79 mm the diameter of bars is satisfactory.

The preceding condition of development length will also be satisfied at the points of inflection since the value of $M_{u,1}$ remains the same as half of the span moment and V_u has a smaller value.

The reinforcement details are shown in Fig. 7.2. Figure 7.2(b) illustrates an arrangement with straight bars, and Fig. 7.2(c) an arrangement with bent-up bars. However, the later arrangement is preferable.

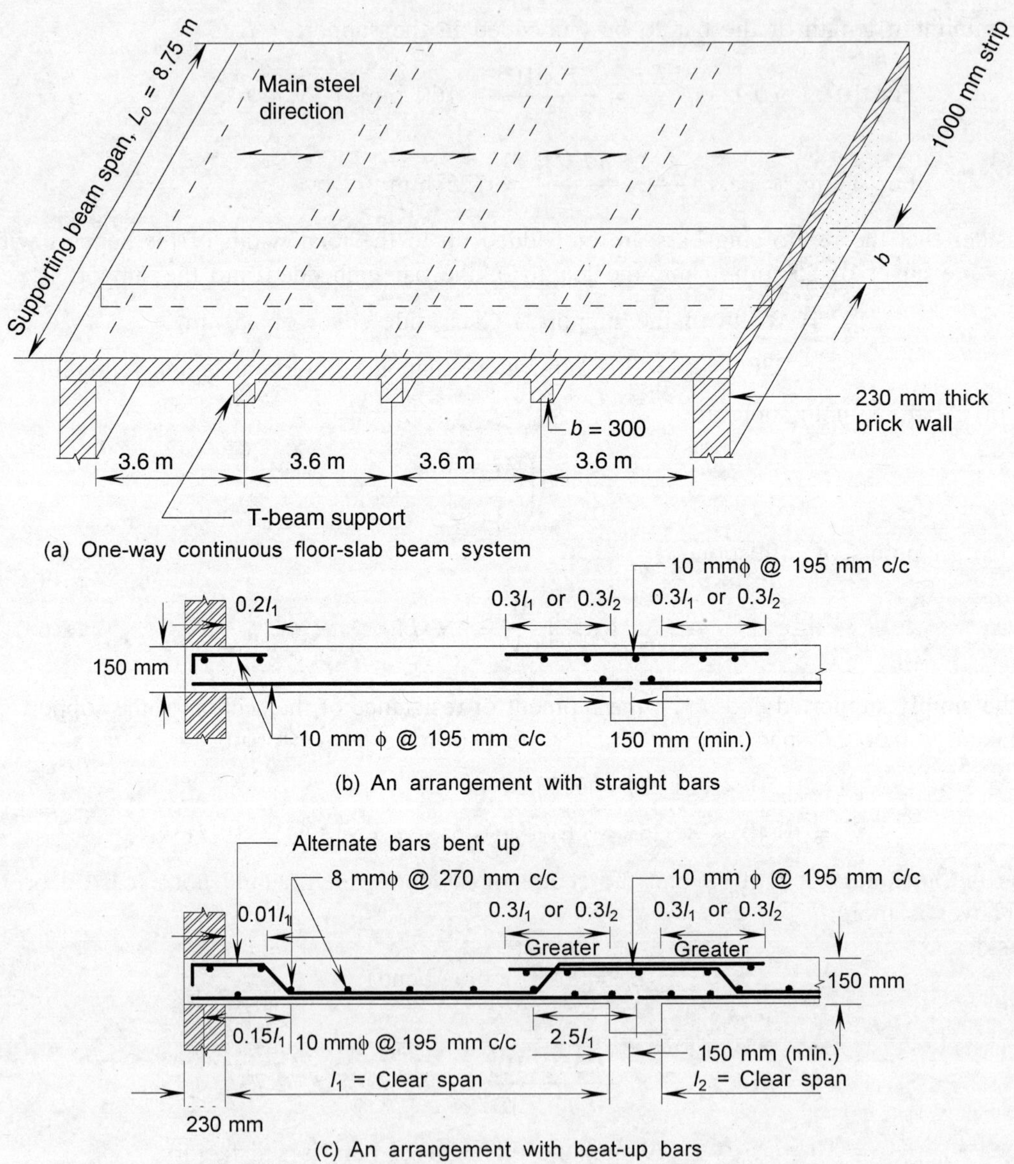

Fig. 7.2 *Simplified detailing of one-way continuous slab of Example 7.3.*

Design of a two-way slab: The values of Codal positive moment coefficient near mid-span obtained for different values of r (= l_y/l_x) in the range 1.0 to 2.0 for the given set of boundary conditions are listed in Table 7.2. For the value of l_y/l_x exceeding 2.0, the Code recommends that the slab should be treated as *one-way*. The Codal negative moment coefficient at a continuous support is approximately 1.33 times the positive span moment. However, at a discontinuous support, the negative moment developed is zero. The possibility of partial restraint must be considered at the time of detailing, the provision of secondary reinforcement in the long span direction is expected to take care of the nominal bending moments that may arise in this direction.

TABLE 7.2 Bending Moment Coefficients for Two-way Rectangular Panels with Provision for Torsion at Corners

Case No.	*Type of panel and moments considered*	*Short span coefficients, α_x (values of l_y/l_x)*								*Long span coefficients, α_y, for all values of l_y/l_x*
		1.0	*1.1*	*1.2*	*1.3*	*1.4*	*1.5*	*1.75*	*2.0*	
1	**Interior Panels:**									
	Negative moment at continuous edges	0.032	0.037	0.043	0.047	0.051	0.053	0.060	0.065	0.032
	Positive moment at mid-span	0.024	0.028	0.032	0.036	0.039	0.041	0.045	0.049	0.024
2	**One Short Edge Discontinuous:**									
	Negative moment at continuous edges	0.037	0.043	0.048	0.051	0.055	0.057	0.064	0.068	0.037
	Positive moment at mid-span	0.028	0.032	0.036	0.039	0.041	0.044	0.048	0.052	0.028
3	**One Long Edge Discontinuous:**									
	Negative moment at continuous edge	0.037	0.044	0.052	0.057	0.063	0.067	0.077	0.085	0.037
	Positive moment at mid-span	0.028	0.033	0.039	0.044	0.047	0.051	0.059	0.065	0.028
4	**Two Adjacent Edges Discontinuous:**									
	Negative moment at continuous edge	0.047	0.053	0.060	0.065	0.071	0.075	0.084	0.091	0.047
	Positive moment at mid-span	0.035	0.040	0.045	0.049	0.053	0.056	0.063	0.069	0.035
5	**Two Short Edges Discontinuous:**									
	Negative moment at continuous edges	0.045	0.049	0.052	0.056	0.059	0.060	0.065	0.069	—
	Positive moment at mid-span	0.035	0.037	0.040	0.043	0.044	0.045	0.049	0.052	0.035

(Contd.)

TABLE 7.2 Bending Moment Coefficients for Two-way Rectangular Panels with Provision for Torsion at Corners (*Contd.*)

Case No.	Type of panel and moments considered	Short span coefficients, α_x (values of l_y/l_x)								Long span coefficients, α_y, for all values of l_y/l_x
		1.0	1.1	1.2	1.3	1.4	1.5	1.75	2.0	
6	**Two Long Edges Discontinuous:**									
	Negative moment at continuous edges	—	—	—	—	—	—	—	—	0.045
	Positive moment at mid-span	0.035	0.043	0.051	0.057	0.063	0.068	0.080	0.088	0.035
7	**Three Edges Discontinuous (One Long Edge Continuous):**									
	Negative moment at continuous edge	0.057	0.064	0.071	0.076	0.080	0.084	0.091	0.097	—
	Positive moment at mid-span	0.043	0.048	0.053	0.057	0.060	0.064	0.069	0.073	0.043
8	**Three Edges Discontinuous (One Short Edge Continuous):**									
	Negative moment at continuous edge	—	—	—	—	—	—	—	—	0.057
	Positive moment at mid-span	0.043	0.051	0.059	0.065	0.071	0.076	0.087	0.096	0.043
9	**Four Edges Discontinuous:**									
	Positive moment at mid-span	0.056	0.064	0.072	0.079	0.085	0.089	0.100	0.107	0.056

The maximum moments obtained by using moment coefficients shall apply only to the middle strip and no redistribution shall be made.

In case of two-way slabs, the reinforcement in shorter direction is placed in the bottom layer. At least 50 per cent of the mid-span reinforcement in each direction shall extend into the supports up to a length of $L_d/3$. The remaining 50 per cent of the reinforcement shall be extended to within $0.15l_x$ or $0.15l_y$ of supports and then cranked up and extended to within $0.10l_x$ and $0.10l_y$ of the supports, where l_x and l_y are the effective spans in shorter and longer directions, respectively.

For the design of more general cases of two-way slabs the reader should refer to Vol. 2 of this book. Following example will illustrate the basic principles used to design a two-way slab.

Example 7.4 The drawing-cum-dining room of a residential building measures 4 × 6.5 m. The floor slab is assumed to be simply supported on 230 mm walls on all the four edges. The live load is 2.0 kN/m^2 and load due to finishes may be taken as 1.00 kN/m^2. Design the floor slab using M20 grade concrete and mild steel reinforcement.

Solution Design constants for M20 concrete and mild steel reinforcement are:

$$f_{ck} = 20 \text{ MPa}, f_y = 250 \text{ MPa}, M_{u,\text{lim}} = 0.1498 f_{ck} bd^2 \text{ and } p_{t,\text{lim}} = 1.769 \text{ per cent}$$

Since the aspect ratio, i.e., ratio of longer span to shorter span is less than 2, the slab shall be designed as a two-way slab. Consider 1.0 m width of slab in both directions. Minimum thickness of slab will be based on serviceability limit state criterion for deflection. For an under-reinforced section, $p_t < p_{t,\text{ lim}}$, consider modification factor m_t of 1.68 (for $p_t = 0.7$ per cent). Thus,

$$\text{Effective depth } d = \frac{\text{Effective shorter span}}{20 \times \text{Modification factor}}$$

i.e.,

$$d = \frac{4000 + d}{20 \times 1.68}$$

or

$$d = \frac{4000}{33.6 - 1} = 122.70 \text{ mm}$$

Using 10 mm ϕ bars,

$$\text{Overall depth } D = 122.70 + 15 + \left(\frac{10}{2}\right) = 142.70 \text{ mm}$$

Thus, consider 150 mm thick slab with d = 130 mm. The effective span of the slab in each direction is given by

Clear span + d (or width of support, whichever is smaller)

Thus, the effective span is the lesser of the following:

(i) 4000 + 130 = 4130 mm

(ii) 6500 + 130 =6630 mm

Therefore, l_x = 4.13 m

Analysis for moments and shear forces.

Dead load of slab $w_g = 0.15 \times 1.0 \times 25$	= 3.75 kN/m
Floor finish w_f	= 1.00 kN/m
Live load w_l	= 2.00 kN/m
Total service load w	= 6.75 kN/m
Factored load $w_u = 1.5w$	= 10.125 kN/m

Factored moment at mid-span (maximum).

$$\text{Ratio } \frac{l_y}{l_x} = \frac{6.63}{4.13} = 1.605$$

From Table 7.2, by interpolation,

$$\sigma_x = 0.1076 \text{ and } \alpha_y = 0.0424$$

Therefore,

$$M_{ux} = \alpha_x w_u (l_x)^2$$
$$= 0.1076 \times 10.125 \times (4.13)^2 = 18.583 \text{ kNm/m}$$

and

$$M_{uy} = \alpha_y w_u (l_x)^2$$
$$= 0.0424 \times 10.125 \times (4.13)^2 = 7.323 \text{ kNm/m}$$

Thickness of slab from flexural considerations for a balanced design is obtained as:

$$0.1498 \times 20 \times 1000 \times d^2 = 18.583 \times 10^6$$

i.e.,

$$d = 78.76 \text{ mm}$$

Overall depth $D = 78.76 + 15 + \left(\frac{10}{2}\right) = 98.76$ mm. However, from the serviceability limit state criterion of deflection, the depth adopted is 150 mm.

Reinforcement. The dimensionless flexural reinforcement parameters are:

$$R_{ux} = \frac{4.6M_{ux}}{f_{ck}bd_x^2}$$

$$= \frac{4.6 \times 18.583 \times 10^6}{20 \times 1000 \times 130^2} = 0.2529$$

and

$$R_{uy} = \frac{4.6M_{uy}}{f_{ck}bd_y^2}$$

$$= \frac{4.6 \times 7.323 \times 10^6}{20 \times 1000 \times 120^2} = 0.1170$$

$$\text{Reinforcement } A_{st} = \left(\frac{f_{ck}bd}{2f_y}\right)[1 - \sqrt{1 - R_u}]$$

$$A_{st,x} = \left(\frac{20 \times 1000 \times 130}{2 \times 250}\right)[1 - \sqrt{1 - 0.2529}] = 705.38 \text{ mm}^2$$

$$A_{st,y} = \left(\frac{20 \times 1000 \times 120}{2 \times 250}\right)[1 - \sqrt{1 - 0.1170}] = 289.53 \text{ mm}^2$$

Minimum (distribution) steel $A_{st,\min} = \dfrac{0.15 \times 1000 \times 150}{100} = 225 \text{ mm}^2$

Along short span provide 10 mm ϕ bars at 100 mm c/c (A_{st} = 785.4 mm^2 > 705.38 mm^2). Along long span provide 10 mm ϕ bars at 260 mm c/c (A_{st} = 302.07 mm^2). The bars cannot be bent or curtailed because if the alternate bars are curtailed, the remaining bars will be less than the minimum. However, the additional bars are provided at the top at the supports to take care of negative moments.

Check for serviceability requirements.

(i) Deflection This check is critical along the short span.

Basic span-to-depth ratio = 20

For

$$p_t = \frac{100A_{st}}{bd} = \frac{100 \times 785.4}{1000 \times 130} = 0.604 \text{ per cent}$$

the modification factor is 1.81.
Therefore,

$$d = \frac{4000 + 130}{20 \times 1.81} = 114.09 \text{ mm} < 130 \text{ mm}$$

Hence it is safe.

(ii) Cracking

(a) Maximum spacing permitted for short span steel is smaller of the: $3d(= 3 \times 130 = 390 \text{ mm})$ or 450, i.e., 390 mm.

Spacing provided = 100 mm c/c < 390 mm

(b) Maximum spacing permitted for long span steel is smaller of the: $3d(= 3 \times 120 = 360 \text{ mm})$ or 450, i.e., 360 mm

Spacing provided = 260 mm c/c < 360 mm

Check for shear. Shear is critical along supports parallel to long span at a distance of the effective depth from the face of support:

$$\text{Maximum factored shear } V_u = 10.125 \times \left(\frac{4.0}{2} - 0.13\right) = 18.93 \text{ kN}$$

$$\text{Nominal shear } \tau_v = \frac{18.93 \times 10^3}{1000 \times 130} = 0.146 \text{ MPa}$$

For solid slab of M20 grade concrete having p_t = 0.30 per cent with τ_c = 0.388 MPa, the permissible shear stress $k\tau_c = 1.3 \times 0.388 = 0.5044 \text{ MPa} > \tau_v$. Hence no shear reinforcement is required.

Check for development length.

$$\text{Development length } L_d = \frac{0.87 f_y \phi}{4\tau_{bd}}$$

$$= \frac{0.87 \times 250\phi}{4 \times 1.2} = 45.3\phi$$

At simple support, the following condition of development length should be satisfied:

$$L_d \leq 1.3\left(\frac{M_{u,1}}{V_u}\right) + L_o$$

Along the long edges,

$$V_u = 10.125 \times \frac{4.13}{2} = 20.91 \text{ kN}$$

$$M_{u,1} = \frac{18.763}{2} = 9.3815 \text{ kNm}$$

Assume, $L_o = 12\phi$.

$$45.3\phi \leq 1.3 \times \frac{9.3815 \times 10^6}{20.91 \times 10^3} + 12\phi$$

i.e.,

$$\phi \leq 17.52 \text{ mm}$$

Hence 10 mm ϕ bars are satisfactory.
Along the short edges,

$$V_u = 20.91 \text{ kN and } M_{u,1} = 7.39 \text{ kNm}$$

Assume $L_o = 12\phi$.

$$45.3\phi \leq 1.3 \times \frac{7.39 \times 10^6}{20.91 \times 10^3} + 12\phi$$

i.e., $\phi \leq 13.8$ mm. Thus, the bar size adopted is satisfactory. This indicates that the bond is usually critical along the short edge.
Minimum length of embedment into the support $L_e = L_d/3$
For 10 mm ϕ bars,

$$L_e = \frac{45.3\phi}{3} = \frac{45.3 \times 10}{3} = 151.00 \text{ mm}$$

Length of bar embedment available into the support is:

= Width of support – Clear side cover of 15 mm

$$= 230 - 15 = 215 \text{ mm} > \frac{L_d}{3} \ (=151.00 \text{ mm})$$

The reinforcement details are shown in Fig. 7.3.

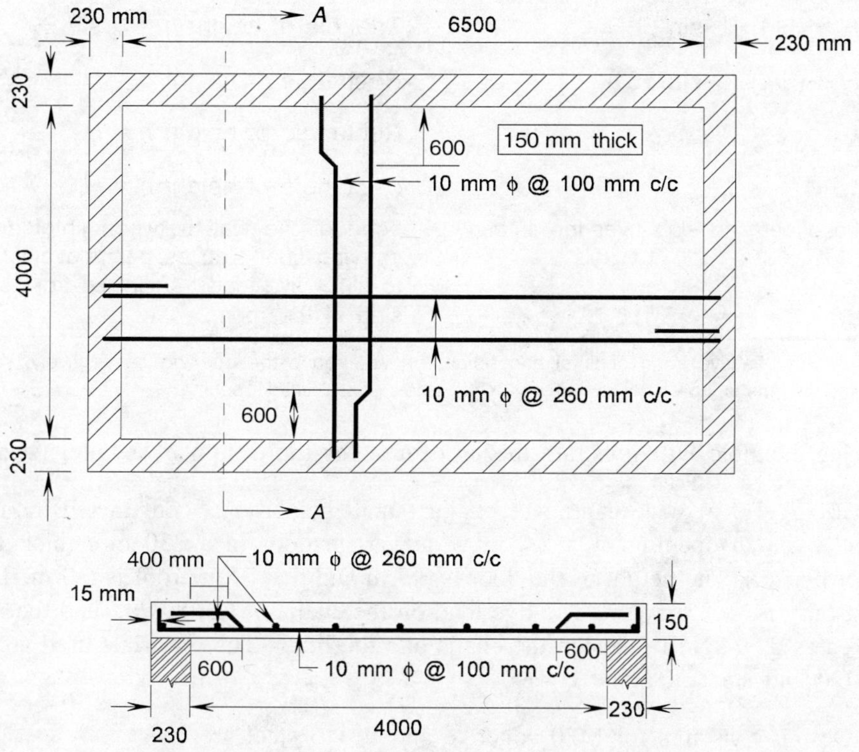

Fig. 7.3 Detailing of the two-way slab of Example 7.4.

7.4.5 Design of a Lintel

The lintels are the beams provided over the openings of doors, windows, etc. mainly to support the load of the wall over it. A good bond in brick masonry enables the load over the opening to be transmitted to the sides of the opening due to arch action. It is usually assumed that the masonry load transferred to the lintel is in the form of a triangle. The base angle of the triangle varies from 45° for good masonry to 60° for poor masonry. For the arch action it is necessary that the height of the wall above the opening is more than 1¼ times the height of the triangle described above, and the length of the wall on each side of opening is more than half the effective span of the lintel. If the height is less, full rectangular load over the opening is considered. Depending upon the location of lintel in the wall various load distributions are given in Table 7.3.

TABLE 7.3 Loading Patterns on Lintels

Case	*Location parameter*	*Loading distribution*
(i)	$l_{w1}, l_{w2} > ½\, l_e$	
(a)	$H > (5/4)\ (l_e \sin 60°)$	Triangle of height $h = l_e \sin 60°$
(b)	$H < (5/4)\ (l_e \sin 60°)$	Rectangle of height $h = H$
(ii)	l_{w1} or $l_{w2} < l_e$	Rectangle of height $h = l_e$
(iii)	$l_{w1}, l_{w2} < l_e$	Rectangle of height $h = H$
(iv)	Load carrying slab over the lintel	Load of the wall between lintel and load carrying floor slab as per relevant Case (i) to (iii) + live load transferred from the floor slab of the room.

Note: l_e and H are effective span of lintel and height of wall above the opening, respectively. Whereas, l_{w1} and l_{w2} are the lengths of the wall on left and right sides of the lintel opening.

The following example will illustrate the procedure for the design of a lintel beam.

Example 7.5 A 1.0 m wide cantilever chajja (small balcony) is constructed monolithically with a lintel over an opening of a 2.0 m wide garage door in a 230 mm thick brick wall (inclusive of plaster). The height of the door is 3.5 m and that of the roof is 6.0 m. The weight of brick masonry is 19.25 kN/m^3. The live load on the chajja is 1.50 kN/m^2 and the finish load may be taken as 0.50 kN/m^2. Design the chajja and the lintel. The materials used are: concrete of grade M20 and steel of grade Fe415.

Solution Design constants for M20 concrete and Fe415 steel are:

$$f_{ck} = 20 \text{ MPa},\ f_y = 415 \text{ MPa},$$

$$M_{u,\text{lim}} = 0.1388\ f_{ck}bd^2 \text{ and } p_t = 0.961 \text{ per cent}$$

Design of chajja. The chajja is designed as a cantilever beam. Consider a 1.0 m wide strip. The depth of the chajja based on the serviceability requirement of deflection is given by

$$d = \frac{\text{Span}}{7 \times \text{Modification factor}}$$

Consider modification factor $m_f = 1.78$ (for $p_t = 0.2$ and $f_s = 0.58f_y = 240.7$ MPa). Therefore

$$d = \frac{1000}{7 \times 1.78} = 80.26 \text{ mm}$$

$$\text{Total depth } D = d + \text{Nominal or clear cover} + \frac{\phi}{2}$$

$$= 80.26 + 15 + \left(\frac{10}{2}\right) = 100.26 \text{ mm}$$

Adopt a 100 mm thick slab at the face of the wall which is reduced to 75 mm at the free end.

Loads.

Self-weight $W_g = 0.5 \times (0.10 + 0.075) \times 1.0 \times 25 = 2.186$ kN

Weight of finish $W_f = 0.5 \times 1.0 \times 1.0 = 0.500$ kN

Live load $W_\ell = 1.5 \times 1.0 \times 1.0 = 1.500$ kN

Total load $W = 4.186$ kN

Factored load $W_u = 1.5 \times 4.186 = 6.279$ kN

$$\text{Factored moment } M_u = \frac{W_u l}{2} = 6.279 \times 0.5 = 3.1395 \text{ kN}$$

Cross-section. To compute the effective depth equate $M_{u,\text{lim}}$ to M_u:

$$0.1388 \times 20 \times 1000 \times d^2 = 3.1395 \times 10^6$$

Therefore,

$$d = 33.63 \text{ mm}$$

Consider 10 mm ϕ bars at 15 mm nominal cover, and hence

$$D = 33.63 + 15 + \left(\frac{10}{2}\right) = 53.63 \text{ mm}$$

However, from the serviceability consideration adopt $D = 100$ with

$$d = 100 - 15 - \left(\frac{10}{2}\right) = 80 \text{ mm}$$

Reduce the depth to 75 mm at the free end.

Reinforcement. Dimensionless parameter for flexural steel is:

$$R_u = \frac{4.6 \times 3.1395 \times 10^6}{20 \times 1000 \times 80^2} = 0.112826$$

$$\text{Therefore, } A_{st} = \left[\frac{20 \times 1000 \times 80}{2 \times 415}\right]\left[1 - \sqrt{1 - 0.112826}\right] = 112.00 \text{ mm}^2$$

Provide 6 mm ϕ bars at 230 mm c/c ($< 3d = 240$ mm and < 450 mm) with $A_{st} = 122.91 \text{ mm}^2$. These bars are embedded into the lintel up to a length equal to L_d given by

$$L_d = \frac{0.87 f_y \phi}{4\tau_{bd}} = \frac{0.87 \times 415 \times 6}{4 \times 1.92} = 282.07 \text{ mm}$$

The reinforcement bars are embedded into the lintel for a distance of 300 mm as shown in Fig. 7.4.

Secondary reinforcement. Distribution reinforcement is:

$$A_{st} = \frac{0.12 \times 1000 \times [0.5 \times (100 + 75)]}{100} = 105 \text{ mm}^2$$

Provide 6 mm ϕ bars at 260 mm c/c ($< 5d = 400$ mm and < 450 mm) in the longitudinal direction ($A_{st} = 109$ mm^2).

Check for shear. Factored shear force $V_u = W_u = 6.279$ kN

$$\text{Nominal shear stress } \tau_v = \frac{6.279 \times 10^3}{1000 \times 80} = 0.078 \text{ MPa}$$

Shear strength of concrete $\tau_{uc} = 0.326$ MPa for $p_t \leq 0.20$

For a solid slab, permissible strength of concrete is $k\tau_{uc}$, where $k = 1.3$ for $D < 150$ mm. Since $\tau_v < k\tau_{uc}$ (= 1.3 × 0.326 = 0.424 MPa), no shear reinforcement is required.

Design of lintel.

Clear span of the lintel beam = 2.0 m

Minimum effective depth from the serviceability requirement of deflection is:

$$d = \frac{\text{Effective span}}{20 \times \text{Modification factor}}$$

Consider a modification factor of 1.11 for $p_t = 0.70$ per cent, and $f_s = 0.58 f_y = 240.7$ MPa. Therefore,

$$d = \frac{2000 + d}{20 \times 1.11}$$

or

$$d = \frac{2000}{22.2 - 1} = 94.34 \text{ mm}$$

$$\text{Overall depth, } D = d + \text{Nominal cover} + (\phi/2)$$

$$= 94.34 + 15 + \frac{10}{2} = 114.34 \text{ mm}$$

or

$$D = 2 \text{ brick layers} + 1 \text{ mortar layer} = 2 \times 70 + 13 = 153 \text{ mm}$$

Provide a 155 mm deep lintel with one layer of 10 mm ϕ bars; thus

$$d = 155 - 15 - 5 = 135 \text{ mm}$$

Loads and moments. Width of lintel beam b = Thickness of the wall = 230 mm

$$\text{Effective span of lintel, } l_e = 2.0 + 0.135 = 2.135 \text{ m}$$

Consider perfect arch action and that the triangular portion of the masonry load is transferred to the lintel, as illustrated in Fig. 7.4. The base angle of the triangle may be considered as 60° for the average quality masonry work. For arch action, it is necessary that the clear height of the roof above the top of the lintel $H > 1.25h$, where $H = (6000 - 3500) - 155 = 2345$ mm, and

$$h = \frac{l_e \tan 60°}{2} = (\sqrt{3}/2)l_e$$

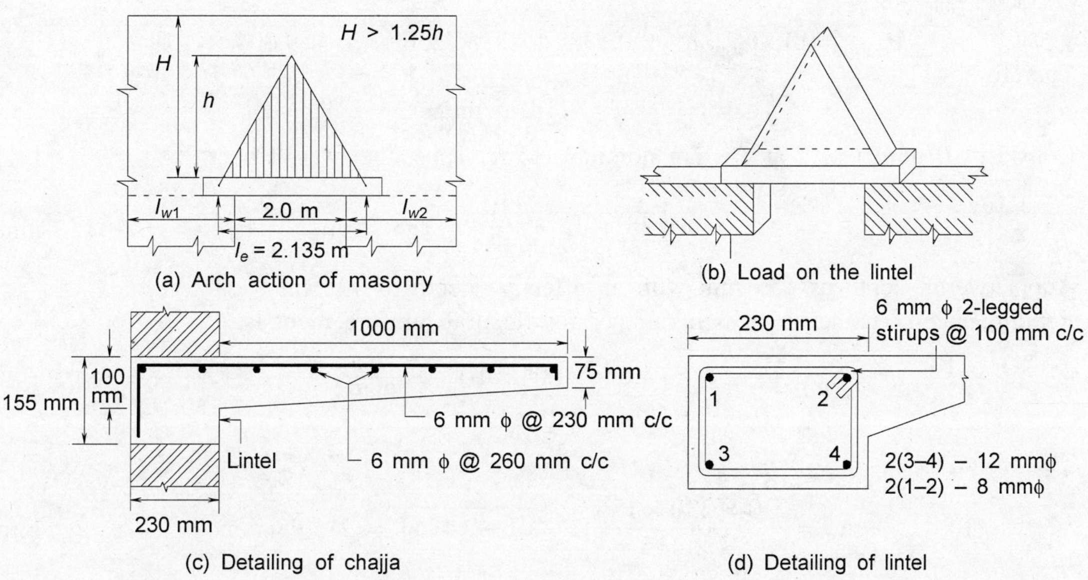

***Fig. 7.4** Detailing of the lintel supportjng a chajja of Example 7.5.*

Thus

$$1.25 \times l_e(\sqrt{3}/2) = 1.25 \times 2135 \times (\sqrt{3}/2) = 2311.21 \text{ mm} < H$$

Hence, arch action is possible. The loads acting on the lintel are:

Self-weight of lintel $w_g = 0.155 \times 0.230 \times 25 = 0.891$ kN/m

$$\text{Weight of masonry } W_m = \frac{1}{2} \times 2.135 \times \left(\frac{2.135 \tan 60°}{2}\right) \times 0.230 \times 19.25$$

$$= 8.739 \text{ kN}$$

Dead and live loads from chajja $w_c = 4.186$ kN/m

$$\text{Factored shear force } V_u = \frac{1.5[(w_g + w_c)l + W_m]}{2}$$

$$= 1.5 \times \frac{[(0.891 + 4.186) \times 2.135 + 8.739]}{2} = 14.684 \text{ kN}$$

Factored bending moment.

$$M_u = \frac{1.5(w_g + w_c)l^2}{8} + 1.5 \times \frac{W_m l}{6}$$

$$= \frac{1.5 \times (0.891 + 4.186) \times (2.135)^2}{8} + \frac{1.5 \times 8.739 \times 2.135}{6} = 9.004 \text{ kNm}$$

Cross-section. The effective depth of a singly reinforced section can be obtained by equating $M_{u,\lim}$ to M_u, i.e.,

$$M_{u,\lim} = 0.1388 f_{ck} b d^2 = 0.1388 \times 20 \times 230 \times d^2 = 9.004 \times 10^6$$

Therefore,

$$d = 118.75 \text{ mm}$$

Consider 10 mm ϕ bars at 15 mm nominal cover. Thus, the overall depth is:

$$D = 118.75 + 15 + \frac{10}{2} = 138.75 \text{ mm}$$

Adopt overall depth as 155 mm with an effective depth of 135 mm.

Reinforcement. Dimensionless parameter for flexural reinforcement is:

$$R_u = \frac{4.6 \times 9.004 \times 10^6}{20 \times 230 \times 135^2} = 0.494$$

Therefore,

$$A_{st} = \frac{20 \times 230 \times 135}{2 \times 415}[1 - \sqrt{1 - 0.494}] = 215.98 \text{ mm}^2$$

Provide 2 bars of 12 mm ϕ at the bottom (A_{st} = 226.19 mm^2) and 2 hanger bars of 8 mm ϕ as nominal reinforcement at the top, and hence

$$p_t = \frac{226.19 \times 100}{230 \times 135} = 0.728 \text{ per cent}$$

Check for shear. Factored shear force V_u = 14.684 kN

$$\text{Nominal shear stress } \tau_v = \frac{14.684 \times 10^3}{135 \times 230} = 0.473 \text{ MPa}$$

Shear strength of concrete from Table 4.1, τ_{uc} = 0.554 MPa (for p_t = 0.728).

Since $\tau_v < \tau_{uc}$, provide nominal shear reinforcement of 6 mm ϕ 2-legged vertical stirrups (A_{sv} = 56.5 mm^2) at a spacing of:

$$S_v = \frac{A_{sv} f_y}{0.4b} = \frac{56.5 \times 415}{0.4 \times 230}$$

$$= 254.86 \text{ mm} < 450 \text{ mm}$$

$$\not> 0.75d (= 101.25 \text{ mm})$$

Hence provide 6 mm ϕ 2-legged stirrups @ 100 mm c/c.

Check for development length.

$$\text{Development length } L_d = \frac{0.87 f_y \phi}{4\tau_{bd}} = \frac{0.87 \times 415\phi}{4 \times 1.92} = 47.0\phi$$

Since the ends of the lintel are considered simply supported,

$$L_d \leq \frac{1.3 M_{u,1}}{V_u} + L_o$$

where $M_{u,1} \approx 9.004$ kNm, $V_u = 14.684$ kN and consider L_o to be 8ϕ.
Thus,

$$47.0\phi \leq \left[\frac{1.3 \times 9.004 \times 10^6}{14.684 \times 10^3}\right] + 8\phi$$

or

$$\phi \leq 20.44 \text{ mm}$$

Hence the beam is safe in the anchorage bond. Minimum length of the bar embedment into the support is:

$$\frac{L_d}{3} = \frac{47.0 \times 12}{3} = 188 \text{ mm}$$

A length of 190 mm should be embedded into the support.

Check for serviceability requirement of deflection. Modification factor for $p_t = 0.728$ per cent

and $f_s = 0.58 \times 415 \times \dfrac{215.98}{226.19} = 229.83$ MPa is 1.15.

Then

$$d = \frac{2135}{20 \times 1.15} = 92.83 < 135 \text{ mm (adopted).}$$

Hence the section is adequate. The reinforcement details are shown in Fig. 7.4.

7.4.6 Design of a Flange Section

The following example will illustrate the procedure used for the design of a flanged section.

Example 7.6 Design a typical simply supported beam of a floor system where the beams are spaced 3.5 m apart and support a 150 mm thick continuous slab cast monolithically with them. The clear span of the beam is 9.5 m. The floor is to support live and partition loads of

3.0 kN/m^2 and 1.25 kN/m^2, respectively, at the service state. The load due to finish may be considered to be 0.75 kN/m^2. The width of the wall supporting the beams is 460 mm. The materials to be used are M20 concrete and HYSD steel of grade Fe415 for mild exposure conditions.

Solution Design constants for M20 concrete and Fe415 grade steel are:

$$f_{ck} = 20 \text{ MPa},\ f_y = 415 \text{ MPa},\ x_{u,\max} = 0.4791d,$$

$$M_{u,\lim} = 0.1388\, f_{ck}bd^2 \text{ and } p_{t,\lim} = 0.961 \text{ per cent}$$

Cross-section of the beam. Consider the trial width of the web $b_w = 300$ mm.

$$\text{Effective span } l_o = \text{Clear span} + \text{Width of support}$$

$$= 9.5 + 0.46 = 9.96 \text{ m}$$

The flange width b_f is governed by the least of the following:

(i) Centre-to-centre distance of beams = 3500 mm

(ii) $\left(\dfrac{l_o}{6}\right) + b_w + 6D_f = \dfrac{9.96 \times 10^3}{6} + 300 + 6 \times 150 = 2860$ mm

(iii) $b_w + 12D_f = 300 + 12 \times 150 = 2100$ mm

Therefore,

$$b_f = 2100 \text{ mm}$$

Minimum depth of beam, based on the requirements of the limit state of the serviceability of deflection is:

$$d = \frac{\text{Effective span}}{20 \times \text{Modification factor}}$$

For the trial depth, consider modification factor m_{ft} of 1.11 from Table 6.3 for a rectangular beam section with $p_t = 0.7$ per cent and $f_s = 0.58\, f_y$ (= 240.7 MPa).This factor is modified for the T-beam action:

$$\text{For } \frac{b_w}{b_f} = \frac{300}{2100} = 0.143 < 0.3,\ m_{rf} = 0.8$$

Therefore,

$$d = \frac{9960}{20 \times 1.11 \times 0.8} = 560.81 \text{ mm}$$

In the case of a rectangular beam section, the depth is generally governed by the limit state of collapse hence a higher value of depth than that required for the serviceability limit state may be considered. However, for the case of T-beam section consider overall depth D of 600 mm with effective depth d of 550 mm.

Check for the minimum width of the beam. The minimum width of the beam for stability requirement is given by

$$b_w = \left(\frac{\text{Clear span}}{60}\right) \text{ or } \sqrt{\left(\text{Clear span} \times \frac{d}{250}\right)} \text{ whichever is more}$$

$$= 158.33 \text{ mm or } 151 \text{ mm (whichever is more)} = 158.33 \text{ mm}$$

Hence the adopted value of b (= 300 mm > 158.33 mm) is satisfactory.

Loads:

Weight of the slab $w_s = 0.15 \times 2.1 \times 1.0 \times 25$ = 7.875 kN/m

Weight of the web
(projected portion of beam) $w_g = 0.30 \times (0.60 - 0.15) \times 1.0 \times 25$ = 3.375 kN/m

Load from partitions $w_p = 1.25 \times 1.0 \times 2.1$ = 2.625 kN/m

Weight of finish $w_f = 0.75 \times 1.0 \times 2.1$ = 1.575 kN/m

Live load $w_l = 3.0 \times 1.0 \times 2.1$ = 6.300 kN/m

Total load w = 21.75 kN/m

Factored load $w_u = 1.5 \times 21.75 = 32.625$ kN/m

$$\text{Factored moment } M_u = \frac{32.625 \times 9.96^2}{8} = 404.557 \text{ kNm}$$

Reinforcement. For a trial, consider the lever arm z to be

$$z = d - \frac{D_f}{2} = 550 - \frac{150}{2} = 475 \text{ mm}$$

Approximate reinforcement steel.

$$A_{st} = \frac{M_u}{0.87 f_y z} = \frac{404.557 \times 10^6}{0.87 \times 415 \times 475} = 2359 \text{ mm}^2$$

Provide 5 bars of 25 mm ϕ (A_{st} = 2454 mm^2). Consider the neutral axis to be in the flange;

$$0.362 f_{ck} b_f x_u = 0.87 f_y A_{st}$$

or

$$x_u = \frac{0.87 \times 415 \times 2454}{0.362 \times 20 \times 2100}$$

$$= 58.3 \text{ mm} < 150 \text{ mm}$$

Depth of the balanced neutral axis is:

$$x_{u,\max} = 0.4791 d = 0.4791 \times 550 = 263.51 \text{ mm}$$

Since $x_u < x_{u,\max}$, the section is under-reinforced, therefore,

$$M_u = 0.87 f_y A_{st}(d - 0.416 x_u)$$

$$= 0.87 \times 415 \times 2454 \times (550 - 0.416 \times 58.3) = 465.82 \text{ kNm} > 404.557 \text{ kNm}$$

Hence the design for flexure is satisfactory.

Alternatively

Dimensionless parameter for flexural reinforcement:

$$R_u = \frac{4.6 M_u}{f_{ck} b d^2} = \frac{4.6 \times 404.557 \times 10^6}{20 \times 2100 \times 550^2} = 0.146475$$

Therefore,

$$A_{st} = \left(\frac{20 \times 2100 \times 550}{2 \times 415}\right) \times [1 - \sqrt{1 - 0.146475}] = 2118.96 \text{ mm}^2$$

Provide 6 bars of 22 mm ϕ (A_{st} = 2281 mm^2). Therefore,

$$x_u = \frac{0.87 \times 415 \times A_{st}}{0.362 \times 20 \times 2100} = 0.02375 A_{st}$$

$$= 0.02375 \times 2281 = 54.17 \text{ mm}$$

and

$$M_u = 0.87 \times 415 \times 2281 \times (550 - 0.416 \times 54.17) \times 10^{-6} = 434.40 \text{ kNm}$$

This is an improvement over the previous design.

Curtailment of bars. Consider that fifty per cent of the bars are to be curtailed. Theoretical cut-off point from the support:

$$= 0.15 \times 9500 = 1425 \text{ mm}$$

Extension beyond the point of curtailment is:

$$= 12\phi \text{ or effective depth d, whichever is more}$$

$$= 12 \times 22 \text{ (= 264 mm) or 550 mm, whichever is more} = 550 \text{ mm}$$

Curtail 3–22 mm ϕ bars at a distance 1425 – 550 = 875 mm, say 850 mm from the face of the support.

Check for development length.

$$\text{Development length } L_d = \frac{0.87 f_y \phi}{4\tau_{bd}}$$

$$= \frac{0.87 \times 415\phi}{4 \times 1.92} = 47.0\phi$$

At the simply supported end, the following conditions shall be satisfied:

(a) Minimum length of the bar to be embedded into the support is obtained as:

$$\frac{L_d}{3} = \frac{47.0 \times 22}{3} = 344.67 \text{ mm} \qquad \text{(say 345 mm)}$$

Hence, 345 mm length of bars is to be embedded into the supports.

At the support, $A_{st}(3 \times 22 \text{ mm } \phi) = 1140 \text{ mm}^2$

and moment of resistance provided by continuing bars $M_{u,1}$

$$= 0.87 f_y A_{st}(d - 0.416 x_u)$$

where

$$x_u = \frac{0.87 f_y A_{st}}{0.362 f_{ck} b}$$

$$= \frac{0.87 \times 415 \times 1140}{0.362 \times 20 \times 300} = 189.50 \text{ mm}$$

Therefore,

$$M_{u,1} = 0.87 \times 415 \times 1140 \times (550 - 0.416 \times 189.50) \times 10^{-6} = 193.93 \text{ kNm}$$

Now,

$$\text{Shear force at support } V_u = \frac{32.625 \times 9.96}{2} = 162.47 \text{ kN}$$

As the ends of reinforcing bars are confined with compressive reaction, for safety in anchorage bond:

$$\frac{1.3 M_{u,1}}{V_u} + L_o \geq L_d$$

Consider $L_o = 8\phi$, therefore,

$$1.3 \times \frac{193.93 \times 10^6}{162.47 \times 10^3} + 8\phi \geq 47.0\phi$$

or

$$1551.73 + 8\phi \geq 47.0\phi$$

i.e.,

$$\phi \leq 39.79$$

Thus, the diameter of tension reinforcement bars provided (ϕ = 22 mm) is satisfactory.

Design for shear. At the support as the ends of reinforcement are confined with compressive reaction, shear at a distance d from the face of the support will be used. Distance of the critical section from the centre of the support is:

$$= 0.55 + 0.23 = 0.78 \text{ m}$$

Critical factored shear force $V_u = 32.625 \times (4.98 - 0.78) = 137.03$ kN

$$\text{Nominal shear stress } \tau_v = \frac{137.03 \times 10^3}{300 \times 550} = 0.83 \text{ MPa}$$

and

$$p_t = 100\frac{A_{st}}{bd} = \frac{100 \times 1140}{300 \times 550} = 0.691 \text{ per cent}$$

Therefore,

$$\tau_{uc} = 0.543 \text{ MPa (for } p_t = 0.691 \text{ per cent from Table 4.1)}$$

Maximum shear strength of concrete with shear reinforcement $\tau_{uc,\max}$ is 2.8 MPa. Since $\tau_{uc} < \tau_v < \tau_{uc,\max}$, the section is acceptable with shear reinforcement. Consider 8 mm ϕ 2-legged vertical stirrups (A_{sv} = 100.53 mm^2). The spacing S_v shall be the least of the following:

(i) Spacing for resisting the excess shear

$$S_v = \frac{0.87 f_y A_{sv}}{(\tau_v - \tau_{uc}) \times b}$$

$$= \frac{0.87 \times 415 \times 100.53}{(0.83 - 0.543) \times 300} = 421.56 \text{ mm}$$

(ii) Spacing based on the minimum shear reinforcement

$$S_v = \frac{A_{sv} f_y}{0.4b} = \frac{100.53 \times 415}{0.4 \times 300} = 347.67 \text{ mm}$$

(iii) Spacing based on the effective depth of the beam

$$S_v = 0.75d = 0.75 \times 550 = 412.5 \text{ mm}$$

Hence provide 8 mm ϕ 2–legged stirrups at 275 mm c/c throughout the beam. To maintain shear capacity in the cut-off region, the area of additional vertical stirrups required is:

$$A'_{sv} = \frac{0.4bS_v}{f_y}$$

where $S_v \leq d/(8\beta)$ in which

$$\beta = \frac{\text{Area of cutoff bars}}{\text{Area of total bars}} = 0.50$$

or

$$S_v \leq \frac{550}{8 \times 0.5} = 137.5 \text{ mm}$$

Consider that the spacing of additional stirrups is 275 mm so that the resultant spacing will be 137.5 mm. Therefore,

$$A'_{sv} = \frac{0.4 \times 300 \times 275}{415} = 79.52 \text{ mm}^2 < 100.53 \text{ mm}^2$$

Provide 8 mm φ 2-legged additional stirrups at 275 mm c/c.

Check for deflection. The basic effective span to effective depth ratio for simply supported beam is 20. The percentage of tension steel is:

$$p_t = \frac{100A_{st}}{b_f d} = \frac{100 \times 2281}{2100 \times 550} = 0.197 \text{ per cent}$$

The modification factor m_{ft} for $p_t = 0.197$ per cent and presumed $f_s = 0.58\, f_y$ (= 240.7 MPa) from Table 6.3 is 1.78.

For T-beam section with the ratio $b_w/b_f = 300/2100 = 0.143$, the modification factor m_{rf} from Eq. (6.3), is 0.8.

Therefore, the effective depth

$$d = \frac{9960}{20 \times 1.78 \times 0.8} = 349.72 \text{ mm} < \text{actual depth } (= 550 \text{ mm})$$

Hence the section is satisfactory. Reinforcement details are given in Fig. 7.5.

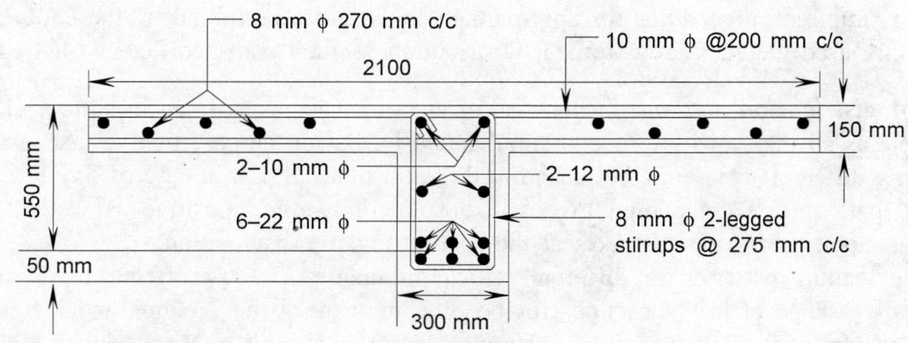

(a) Cross-section of beam

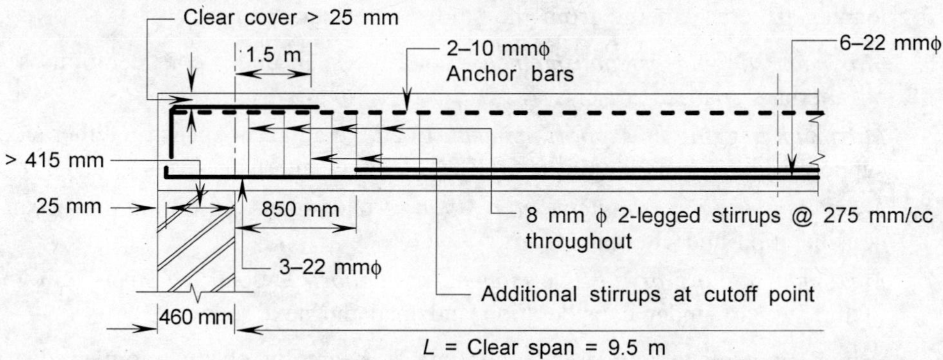

(b) Longitudinal section of beam

Fig. 7.5 Reinforcement details of a simply supported beam of Example 7.6.

7.4.7 Design of Continuous Beams

Continuous beams which form a part of RC building frames, are cast monolithically with floor slabs and columns. The continuous systems result in reduced moments and distribute the loads more evenly. Though, a framed building is a 3-D or a space structure, it is idealized as a system of interconnected 2-D vertical frames along two mutually perpendicular directions. These frames are analyzed independently. Presently, abundant software is available the for analysis and design of 3-D building frames. However, the manual approximate methods are useful for preliminary analysis and verification.

The building frames are designed for dead loads, live (imposed) loads and wind or earthquake loads. The dead loads are gravity loads which are permanent in nature and are fixed in position. On the other hand, live loads are random in nature, their distribution varies with time. Hence each member is designed for the worst combination of dead and live loads. For simplicity the effects of gravity loads and wind or earthquake loads are considered separately and their results are superimposed.

The analysis and design of building frames carrying gravity and lateral loads due to wind or earthquake are beyond the scope of this chapter. The analysis of a continuous beam framing into and supported by girders (assumed rigid supports) is discussed in this chapter.

The simplified procedures for approximate analysis are permitted by the code. The more commonly used methods are: Moment Distribution Method and Coefficient Method.

Moment distribution method: In the case of dead load which is fixed in nature, the analysis is simple as all the spans are loaded simultaneously, but in the case of live load which occur randomly different spans may be loaded at different times. Thus, in case of live load different loading patterns are considered to obtain absolute maximum positive as well as negative bending moments for worst effects at the section under consideration.

The loading patterns for different critical moments or worst effects are based on the quantitative shape of influence lines for bending moment at the sections under consideration drawn by *Müller-Breslau Principles*, wherein a fictitious hinge is first introduced at the section under consideration and a rotation is applied there in the direction of desired bending moment. Following loading patterns emerge from the study of influence lines:

1. *Maximum sagging or span moment.* Load the span under consideration as well as all alternate spans.
2. *Maximum hogging or support moment.* Load the adjacent spans on either side of the support under consideration as well as all other alternate spans.
3. *Maximum hogging moment in a span.* Load spans adjoining the span under consideration and alternate spans.
4. *Maximum sagging or positive moment at support.* Unload the spans on either side of the support under consideration and load the next spans.

The bending moments obtained in a continuous beam by elastic moment distribution method are significantly modified in inelastic phase wherein moments are transferred to less stressed sections. This is called *redistribution of moments.* Generally a moment redistribution upto 12 per cent is permitted in case of Fe415 steel and 6 per cent for Fe250 steel.

Method using coefficients specified by the code: The method is applicable to continuous beams having three or more spans and subjected to gravity loads. The method does not permit any redistribution in moment values. The moment and shear coefficients are listed in Table 7.6. Their use has been demonstrated earlier in case of continuous slabs.

The methods with total load w (= $w_d + w_l$) distributed uniformly on all spans generally underestimate moments to different degrees. The maximum under estimation occurs in the mid-span moment in the interior span. Whereas, the method of moment coefficients overestimates the mid-span positive moments in the interior span by significant amount and marginally in the end spans. Thus, the coefficient method results in a relative conservative design. On the other hand, elastic method using pattern loading results in moment values which lie between the values obtained by other two load cases. The maximum variation in the support moments which control the cross-sectional dimensions is of the order of 10 per cent and depends upon the method used.

Example 7.7 Design a 3-span continuous beam of a typical interior idealized plane frame of a building. The frames are spaced 5.5 m apart and in the typical floor 140 mm thick continuous slab is cast monolithically with beams. The thickness of floor finish is 40 mm. The beam has three equal spans of length 6.1 m. The floor is to support imposed load of 5.0 kN / m^2 at the service state. The unit weight of the finishing material is 20 kN/m^3. The materials to be used are: M20 grade concrete mix and HYSD steel of grade Fe415 for moderate exposure conditions.

Solution Design constants for M20 concrete and Fe415 grade steel are:

$$f_{ck} = 20 \text{ MPa}, \; f_y = 415 \text{ MPa}, \; x_{u,\max} = 0.4791d$$

$$M_{u,\lim} = 0.1388 f_{ck} b d^2 \text{ and } p_{t,\lim} = 0.961 \text{ per cent}$$

Preliminary design. Consider flanged beam of uniform depth throughout. The maximum moment in a beam can be determined for dead and live loads transferred from the slab, and dead load of the web of beam. The loads transferred from slab on to the beam have triangular or trapezoidal distribution depending upon the spacing of the beams. However, in this particular example uniform load from a 5.5 m wide strip of slab is considered to be transferred to the beam. The +ve moment in the end span is maximum where flange is in compression resulting in a T-section. On the other hand, the negative moment at the support where flange is in tension results in rectangular section, giving higher depth for the beam. Thus, the support moment shall be considered for the computation of the depth of the beam. For preliminary design dead weight of web-beam may be considered to be 5 to 10 per cent of the total load transferred from the slab.

Factored dead load from the slab and floor finish per metre run of beam is:

$$w_{us} = 1.5 \times (0.14 \times 25 + 0.040 \times 20) \times 5.5 = 35.48 \text{ kN/m}$$

Factored live load from the slab is given by

$$w_{ul} = 1.5 \times (5.0 \times 5.5) = 41.25 \text{ kN/m}$$

Factored dead load of the web beam:

$$w_{uw} = 0.05 \times (w_{us} + w_{ul}) = 0.05 \times (35.48 + 41.25) = 3.84 \text{ kN/m}$$

Total factored dead load per metre run of beam:

$$w_{ud} = w_{us} + w_{uw} = 39.32 \text{ kN/m}$$

Total factored load per metre run of beam:

$$w_u = w_{ud} + w_{ul} = 80.57 \text{ kN/m}$$

The continuous beam sections at supports are generally doubly reinforced. Thus, the depth of beam may be determined by considering it to be a doubly reinforced beam with about 60 to 65 per cent of design moment being resisted as a singly reinforced balanced section.

Analysis. For comparison, the moment have been obtained by the following methods:

(a) Elastic analysis with total load w_u on all spans Due to symmetry in the geometry and loading, one-cycle of moment distribution procedure will be enough to obtain the results. The results may also be obtained by using Table C.1 in Appendix.

Span moments: End span $M_1 = +\ 0.080\ w_u l^2 = 239.94$ kNm

Interior span $M_2 = +\ 0.025 w_u l^2 = 74.95$ kNm

Support moment: Interior support $M_3 = -0.100 w_u l^2 = -299.80$ kNm

(b) Elastic analysis with pattern loading In this method maximum moments are obtained by superimposing the contributions of live and dead loads acting separately. The moments due to uniform dead load w_{ud} on all spans is obtained as in Case (a) by substituting w_{ud} for w_u, the total moments are:

Span moments: End span $M_1 = 0.101 w_{ul} l^2 + 0.080 w_{ud} l^2 = 272.07$ kNm

Interior span $M_2 = 0.075 w_{ul} l^2 + 0.025 w_{ud} l^2 = 151.70$ kNm

Support moment: $M_3 = -0.117 w_{ul} l^2 - 0.100 w_{ud} l^2 = -325.89$ kNm

(c) Code specified coefficients

Span moments: $$M_1 = \left(\frac{w_{ud}}{12} + \frac{w_{ul}}{10}\right) l^2 = 275.42 \text{ kNm}$$

$$M_2 = \left(\frac{w_{ud}}{16} + \frac{w_{ul}}{12}\right) l^2 = 219.35 \text{ kNm}$$

Support moment: $$M_3 = \left(\frac{w_{ud}}{10} + \frac{w_{ul}}{9}\right) l^2 = -316.86 \text{ kNm}$$

The pattern loading method is more realistic and is considered to be accurate. It can be noted that the simplified consideration of w_u on all spans results in an underestimation of the moments and the code coefficients method overestimates the moments. The maximum error occurs in positive moment in the interior span. The error in other moments is small. For the design of continuous beam of this problem the moments obtained by using code specified coefficient will be used.

Consider the maximum negative moment of 316.86 kNm at the support, where section behaves as rectangular section, as the design moment. Consider b = 300 mm for singly reinforced section designed for a moment of 0.60 × 316.86 kNm. Therefore,

$$0.60 \times (316.86 \times 10^6) = 0.1388 f_{ck} b d^2$$

$$= 0.1388 \times 20 \times 300 \times d^2$$

or

$$d = 477.62 \text{ mm}$$

Consider 25 mm ϕ bars for reinforcement at nominal cover of 30 mm. Therefore,

$$D = 477.62 + 30 + \left(\frac{25}{2}\right) = 520.12 \text{ mm}$$

Consider the overall depth of 530 mm with the effective depth of 487.50 mm.
For lateral stability, the value of b shall not be less than $(l_c/60)$ or $(l_c d/250)^{1/2}$ whichever is larger, where l_c = (6100 – 300) = 5800 mm, i.e.,

$$b \not< \frac{5800}{60} \text{ or } \left(\frac{5800 \times 487.50}{250}\right)^{1/2} \text{ whichever is more}$$

$$= 96.67 \text{ mm or } 106.35 \text{ mm} < b \ (=300)$$

Hence the section of size 300 × 530 overall is satisfactory.
Design moments. Factored dead weight of web

$$w_{uw} = 1.5[0.300(0.530 - 0.140) \times 25] = 4.39 \text{ kN/m}$$

Which is close to the value of 3.84 kN/m considered in the preliminary design. Thus, there is no necessity of re-calculation of moments.

$$\text{Width of flange } b_f = \frac{l_o}{6} + b_w + 6D_f \not> 5500 \text{ mm}$$

$$\text{For end span } l_o = 0.85L = 0.85 \times 6100 = 5185 \text{ mm}$$

Therefore,

$$b_f = \frac{5185}{6} + 300 + 6 \times 140 = 2004 \text{ mm}$$

$$\text{For interior span } l_o = 0.70L = 0.70 \times 6100 = 4270 \text{ mm}$$

Thus

$$b_f = \left(\frac{4270}{6}\right) + 300 + 6 \times 140 = 1852 \text{ mm}$$

Design for support moment. The beam behaves as a rectangular section beam,

$$M_{u,\lim} = 0.1388 f_{ck} bd^2 = 0.1388 \times 20 \times 300 \times 487.50^2$$

$$= 197.92 \times 10^6 \text{ Nmm} = 197.92 \text{ kNm} < M_u$$

As $M_u > M_{u,\lim}$, the section has to be doubly reinforced with $p_t > p_{t,\lim}$ where

$$A_{st} = A_{st1} + A_{st2} = p_{t,\lim}\left(\frac{bd}{100}\right) + \frac{M_u - M_{u,\lim}}{0.87 f_y (d - d')}$$

$$= 0.961 \times \left(\frac{300 \times 487.5}{100}\right) + \frac{(316.86 - 197.92) \times 10^6}{0.87 \times 415 \times (487.5 - 42.5)}$$

$$= 1405.46 + 740.29 = 2145.75 \text{ mm}^2$$

From Table 3.3 for d'/d (= 42.5/487.5) = 0.087 and Fe415, f_{sc} = 353.75 MPa. Therefore,

$$A_{sc} = \frac{0.87 f_y A_{st2}}{(f_{sc} - 0.447 f_{ck})}$$

$$= \frac{0.87 \times 415 \times 740.29}{(353.75 - 0.447 \times 20)} = 775.16 \text{ mm}^2$$

Alternatively, design tables given in Appendix-D can be used to determine p_t and p_c for the calculated value of M_u/bd^2, accordingly A_{st} and A_{sc} can be suitably provided.

$$\frac{M_u}{bd^2} = \frac{316.86 \times 10^6}{300 \times 487.5^2} = 4.444 \text{ MPa}$$

From Table D.2 for M20 and Fe415, d'/d = 0.087. By linear interpolation,

$$p_t = 1.468,\ A_{st} = \frac{1.468 \times 300 \times 487.5}{100} = 2146.95 \text{ mm}^2$$

$$p_c = 0.532,\ A_{sc} = \frac{0.532 \times 300 \times 487.5}{100} = 778.05 \text{ mm}^2$$

Design for span moments. The trial area of tension steel can be estimated as follows:

$$A_{st} = \frac{M_u}{0.87 f_y z}$$

where lever arm z may be approximated by larger of $0.9d$ (= 439 mm) and $(d - D_f/2)$ (= 417.5 mm). Consider z = 439 mm, therefore,

(i) For end span,

$$A_{st} = \frac{275.42 \times 10^6}{0.87 \times 415 \times 439} = 1737.66 \text{ mm}^2$$

Provide 4 – 25 mm ϕ bars (A_{st} = 1963 mm^2) and 8 mm ϕ 2-legged stirrups.

To find the actual position of neutral axis, consider $x_u = D_f$ and compare the given design moment with moment of resistance.

$$\begin{aligned} M_{u,\lim} &= 0.362 f_{ck} b_f x_u (d - 0.416 x_u) \\ &= 0.362 \times 20 \times 2004 \times 140 \times (487.5 - 0.416 \times 140) \\ &= 435.97 \times 10^6 \text{ Nmm} = 435.97 \text{ kNm} > M_u (= 275.42 \text{ kNm}) \end{aligned}$$

Thus, the neutral axis lies within the flange ($x_u < D_f$). Therefore, T-section can be designed as a singly reinforced rectangular section with $b = b_f$. Then

$$R_u = \frac{4.6 M_u}{f_{ck} b d^2}$$

$$= \frac{4.6 \times 275.42 \times 10^6}{20 \times 2004 \times 487.5^2} = 0.133$$

and

$$A_{st} = \left(\frac{f_{ck} b d}{2 f_y}\right)\left[1 - \sqrt{1 - R_u}\right]$$

$$= \left(\frac{20 \times 2004 \times 487.5}{2 \times 415}\right)[1 - \sqrt{1 - 0.133}] = 1621.4 \text{ mm}^2$$

Provide 2–25 mm ϕ + 2–20 mm ϕ (A_{st} = 1741 mm^2).

(ii) For interior span,

$$M_u = 219.35 \text{ kNm}$$

$$A_{st} = \frac{219.35 \times 10^6}{0.87 \times 415 \times 439} = 1383.9 \text{ mm}^2$$

Provide 3–25 mm ϕ bars (A_{st} = 1472 mm^2).

Now,

$$\begin{aligned} M_{u,\lim} &= 0.362 f_{ck}\, b_f x_u (d - 0.416 x_u) \\ &= 0.362 \times 20 \times 1852 \times 140 \times (487.5 - 0.416 \times 140) \\ &= 402.9 \times 10^6 \text{ Nmm} = 402.9 \text{ kNm} > M_u \ (=219.35 \text{ kNm}) \end{aligned}$$

Reinforcement parameter $R_u = \dfrac{4.6 M_u}{f_{ck} b d^2}$

$$= \frac{4.6 \times 219.35 \times 10^6}{20 \times 1852 \times 487.5^2} = 0.1146$$

$$A_{st} = \left(\frac{f_{ck}bd}{2f_y}\right)[1 - \sqrt{1 - R_u}]$$

$$= \left(\frac{20 \times 1852 \times 487.5}{2 \times 415}\right)[1 - \sqrt{1 - 0.1146}] = 1284.78 \text{ mm}^2$$

Provide 2–25 mm ϕ + 1–20 mm ϕ (A_{st} = 1295 mm^2).

Check for serviceability requirement of deflection. Ignoring the contribution of flanges d for the end span which is critical need to be calculated from the serviceability requirements as:

$$d = \frac{6100}{0.5 \times (26 + 20)m_f}$$

where,

$$p_t = \frac{100 \times 1741}{300 \times 487.5} = 1.19 \text{ per cent}$$

$$f_t = 0.58 f_y \times \frac{A_{st,\text{required}}}{A_{st,\text{provided}}}$$

$$= \frac{0.58 \times 415 \times 1621.4}{1741} = 224.16 \text{ MPa}$$

Therefore, from Table 6.3, m_f = 1.01.
and

$$d_{\min} = \frac{6100}{0.5 \times (26 + 20) \times 1.01} = 241.11 \text{ mm} < 487.5 \text{ mm}$$

Therefore, the section is adequate.

Curtailment of reinforcement.

$$\text{Development length } L_d = \frac{0.87 \times 415 \times \phi}{4 \times 1.92} = 47\phi$$

The continuing tension reinforcement shall be embedded into the support of a minimum length given by

$$\frac{L_d}{3} = \frac{47\phi}{3} = \frac{47 \times 25}{3} \approx 392 \text{ mm}$$

Therefore, the tension steel is embedded into the support for a length 400 mm.

Check for development length.

(i) At the end support, length of continuing bar beyond cut-off point ≥ L_d.

$$0.15l - \frac{\text{Column depth}}{2} + 400 \geq 47\phi$$

or

$$0.15 \times 6100 - 200 + 400 \geq 47\phi$$

or

$$\phi \leq 23.72 \text{ mm}$$

Hence the bar diameter of 25 mm is not satisfactory, increase the embedment length.

(ii) At the interior support, length of curtailed bar on either side of support ≥ L_d.

$$0.30l\ (= 0.30 \times 6100) \geq 47\phi$$

i.e.,

$$\phi \leq 38.94 \text{ mm}$$

Hence the bar diameter is satisfactory.

At the support, area of continuing bars A_{st} (2 × 25 mm ϕ) = 982 mm^2.

In the case of span moment in the beam having under-reinforced section and neutral axis lying in the flange, moment of resistance provided by continuing bars is given as

$$M_{u,1} = 0.87 f_y A_{st}(d - 0.416 x_u)$$

where

$$x_u = \frac{0.87 f_y A_{st}}{0.362 f_{ck} b_f}$$

$$= \frac{0.87 \times 415 \times 982}{0.362 \times 20 \times 2004} = 24.44 \text{ mm}$$

Thus, $M_{u,1} = 0.87 \times 415 \times 982 \times (487.5 - 0.416 \times 24.44) \times 10^{-6} = 169.24$ kNm

and $V_u = (0.40 w_d + 0.45 w_l) l_c = (0.40 \times 39.32 + 0.45 \times 41.25) \times (6.1 - 0.4) = 195.46$ kN

For safety in anchorage bond,

$$\frac{M_{u,1}}{V_u} + L_o \geq L_d$$

i.e.,

$$\frac{169.24 \times 10^6}{195.46 \times 10^3} + L_o \geq 47.0 \times 25\ (= 1175 \text{ mm})$$

or

$$L_o \geq 309.15 \text{ mm (say 310 mm)}$$

Therefore, carry the bars to the outer face of the column and bend up for a distance of 110 mm.

At inflection points the condition:

$$\frac{M_{u,1}}{V_u} + L_o \geq L_d$$

will be satisfied because $M_{u,1}$ is equal to or greater than that at the end support and V_u has much smaller value and L_o has larger value than those at the end support.

Design for shear force.

End support

Factored shear force $V_u = (0.40 \times 39.32 + 0.45 \times 41.25) \times (6.1 - 0.4) = 195.46$ kN

The spans are divided into three segments each to provide uniform spacing of stirrups in each segment. Consider 8 mm ϕ 2-legged vertical stirrups ($A_{sv} = 100$ mm^2).

$$\text{Nominal shear stress } \tau_v = \frac{V_u}{bd} = \frac{195.46 \times 10^3}{300 \times 487.5} = 1.336 \text{ MPa}$$

and

$$p_t = \frac{982 \times 100}{300 \times 487.5} = 0.671 \text{ per cent}$$

For M20 grade concrete with $p_t = 0.671$ per cent, $\tau_{uc} = 0.537$ MPa.
Since $\tau_{uc} < \tau_v < \tau_{uc,\max}$, the section is acceptable with shear reinforcement. The spacing S_v is given by

$$S_v = \frac{0.87 f_y A_{sv}}{(\tau_v - \tau_{uc}) \times b}$$

$$= \frac{(0.87 \times 415 \times 100)}{(1.336 - 0.537) \times 300} = 150.63 \text{ mm}$$

$$\not> \frac{A_{sv} f_y}{0.4b} = \frac{100 \times 415}{0.4 \times 300} = 345.83 \text{ mm}$$

$$\not> 0.75d (= 0.75 \times 487.5 = 365.63 \text{ mm}) \text{ or } 450 \text{ mm}$$

Hence provide 8 mm ϕ 2-legged stirrups at 150 mm c/c in a distance of 2.0 m near support. To maintain shear capacity in cut-off region, the area of additional vertical stirrups required is:

$$A'_{sv} = \frac{0.4bS_v}{f_y}$$

where $S_v \geq \dfrac{d}{8\beta}$ in which

$$\beta = \frac{\text{Area of cut-off bars}}{\text{Area of total bars}}$$

$$= \frac{2 \times 20 \text{ mm } \phi}{2 \times 25 \text{ mm } \phi + 2 \times 20 \text{ mm } \phi} = \frac{628}{1609} = 0.39$$

or

$$S_v \ngtr \frac{487.5}{8 \times 0.39} = 156.25 \text{ mm}$$

Consider that the spacing of additional stirrups is 300 mm so that the resultant spacing will be 150 mm. Therefore,

$$A'_{sv} = 0.4 \times 300 \times \frac{300}{415} = 86.75 \text{ mm}^2 < 100 \text{ mm}^2$$

Provide 8 mm φ 2-legged additional stirrups at 300 mm c/c in the length of 370 mm.

Interior support

Outside the interior support $V_u = (0.60 \times 39.32 + 0.60 \times 41.25) \times (6.1 - 0.4) = 275.55$ kN

Inside the interior support $V_u = (0.55 \times 39.32 + 0.60 \times 41.25) \times (6.1 - 0.4) = 264.34$ kN

Then

$$\tau_v = \frac{V_u}{bd} = \frac{275.55 \times 10^3}{300 \times 487.5} = 1.884 \text{ MPa}$$

and

$$p_t = \frac{2277 \times 100}{300 \times 487.5} = 1.56 \text{ per cent}$$

For M20 grade concrete with $p_t = 1.56$ per cent, from Table 4.1, $\tau_c = 0.726$ MPa. Now,

$$S_v = \frac{0.87 \times 415 \times 100}{(1.884 - 0.726) \times 300} = 103.93 \text{ mm}$$

Provide 8 mm φ 2-legged stirrups at 100 mm c/c in the length of 2.0 m on both the sides of support. For additional steel,

$$\beta = \frac{\text{Area of cut-off bars}}{\text{Area of total bars}}$$

$$= \frac{2 \times 25 \text{ mm } \phi + 1 \times 20 \text{ mm } \phi}{4 \times 25 \text{ mm } \phi + 1 \times 20 \text{ mm } \phi} = \frac{1295}{2277} = 0.57$$

Therefore,

$$S_v \ngtr \frac{487.5}{8 \times 0.57} = 106.91 \text{ mm}$$

Consider that the spacing of additional stirrups is 200 mm so that the resultant spacing will be 100 mm. Therefore,

$$A'_{sv} = \frac{0.4 \times 300 \times 200}{415} = 57.83 \text{ mm}^2 < 100 \text{ mm}^2$$

Provide 8 mm φ 2-legged additional stirrups at 200 mm c/c in the length of 0.75d (= 370 mm). In the middle 2.1 m length provide 8 mm φ 2-legged stirrups at 345 mm c/c. A typical detailing scheme of the reinforcement recommended for continuous beams is given in the Fig. 7.6. The readers can workout the detailing arrangement of reinforcement for this example accordingly.

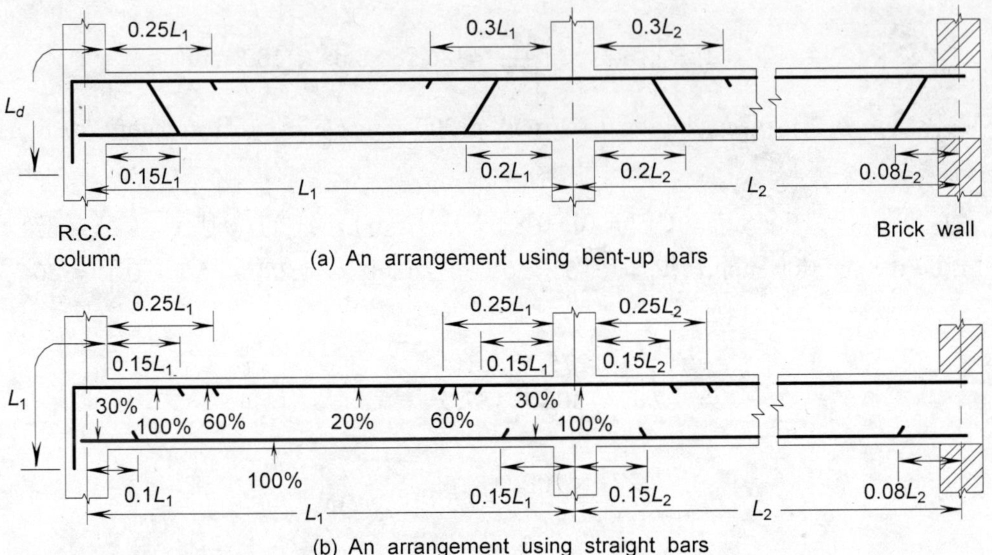

Fig. 7.6 *Simplified detailing scheme of reinforcement for a continuous beam.*

7.5 DESIGN OF STAIRCASES

The staircase provides a means of movement from one level to another in a structure. It consists of a number of steps arranged in a series of flights with landings at suitable intervals, to provide comfort, ease and safety. The number of steps in a flight may vary from 3 to 12 depending upon the type of structure and purpose of stair. The rise of a step and the tread should be so arranged as to give maximum comfort to the users. Generally, for maximum comfort the rise (R) and tread (T) should be such that: $(T + 2R) \approx 600$ and $TR = 40{,}000$ to $42{,}000$ where R and T are in mm. The width of stairs depends on the functional requirements, i.e., on the type of building.

7.5.1 Gravity Loads

The stair slabs are normally designed to resist gravity loads which include dead and live loads.

Live load: The live loads are generally uniformly distributed loads on the horizontal projection of the flight. The live load on stairs depends on the type of building where the staircase is located. As per IS:875 (Part II) the minimum value of uniformly distributed characteristic live load on the plan area of stairs for residential and public buildings should be taken as:

(a) Places not liable to overcrowding (residential, office and hospitals) 3 kN/m^2

(b) Places liable to overcrowding (public buildings) 5 kN/m^2

subject to minimum of 1.3 kN concentrated load at the unsupported end of each structurally independent cantilever step. A horizontal load of 0.75 kN/m is considered as acting on the balustrade or parapet.

Dead load: The components of dead load to be considered are the self-weights of stair slab in case of tread slab/tread-riser slab/waist slab; step in case of waist slab type stairs and finish.

Self-weight of the waist slab. The dead weight per unit area of the waist slab is first calculated at right angles to the slope and then converted to the corresponding load per unit horizontal area by multiplying it by the ratio $[(\sqrt{R^2 + T^2})/T]$, where R and T represent the rise and tread of the step, respectively.

Self weight of steps. The dead weight due to steps is not uniform. However, for design, the weight due to each step is converted into uniform load on the tread width T.

$$\text{Weight of one step } W_s = \frac{1}{2} \times R \times T \times 25 = 12.5RT \text{ kN/m width}$$

where R and T are in metre. When weight of step considered uniformly distributed on horizontal length T, the weight due to step may be expressed as:

$$w_s = \frac{W_s}{T} = 12.5\text{R kN/m}^2$$

Depth of the section. The depth of the section shall be taken as minimum thickness perpendicular to the soffit of the staircase.

The following examples will illustrate the procedure for the design of simple staircases. For more complex staircases the readers should refer to Volume 2 of this book.

Example 7.8 A straight staircase flight with the rise, tread and width of stairs of 160 mm, 280 mm and 1.25 m, respectively, consists of steps projecting from the face of a reinforced concrete wall. Design the staircase with: (i) independent steps (tread slab) projecting from the face of wall, (ii) waist slab supported between two stringer beams or walls along the edges and (iii) tread-riser arrangement cantilevering from the face of wall The stairs are liable to be overcrowded. The weight of finishes is 0.60 kN/m^2. The materials to be used are: M20 grade concrete and HYSD steel bars of grade Fe415.

Solution Design constants for M20 concrete and Fe415 steel are:

$$f_{ck} = 20 \text{ MPa}, f_y = 415 \text{ MPa}$$

$$M_{u,\lim} = 0.1388 f_{ck} bd^2 \text{ and } p_{t,\lim} = 0.961 \text{ per cent}$$

For the given staircase,

$$R = 160 \text{ mm}, T = 280 \text{ mm}, W = 1.25 \text{ m and effective span } l = 1.25 \text{ m}$$

CASE I Independent steps

Generally a marginal overlap of 10 mm is provided between adjacent tread slabs. Therefore, the actual or effective tread is 290 mm.

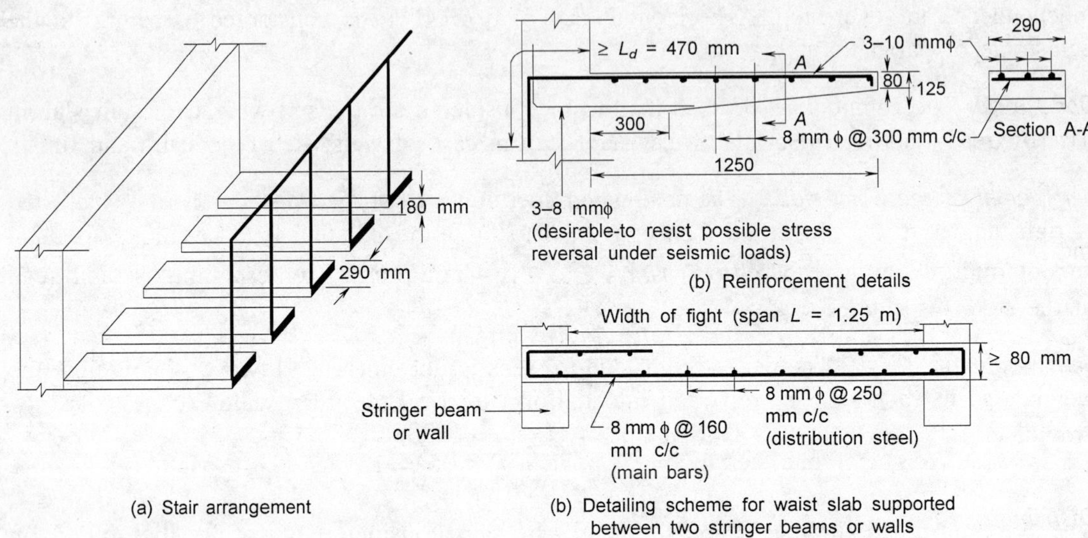

Fig.7.7 Detailing scheme of stairs of Example 7.8.

Consider a slab thickness of 125 mm ($\approx l/10$) at the fixed support and tapered to a minimum thickness of 80 mm, beyond a distance of 300 mm from the support, as shown in Fig. 7.7(b).

Dead loads.

Self weight of tread slab $w_g = (0.125 \times 0.29) \times 25 = 0.906$ kN/m

Weight of finish $w_f = 0.6 \times 0.29 = 0.174$ kN/m

Total dead load $w_d = 1.080$ kN/m

Factored dead load $w_{u,d} = 1.5 w_d = 1.5 \times 1.08 = 1.62$ kN/m

Live loads.

Factored u.d. live load $w_{u,l} = 1.5 \times (5.0 \times 0.28) = 2.10$ kN/m

Concentrated load at the free end $W_{u,l} = 1.5 \times 1.3 = 1.95$ kN

Design moment.

$$\text{Factored fixed end moment } M_{u,d} = 1.62 \times \frac{1.25^2}{2} = 1.266 \text{ kNm}$$

$$\text{Live load moment } M_{u,l} = 2.10 \times \frac{1.25^2}{2} = 1.641 \text{ kNm}$$

or

$$1.95 \times 1.25 = 2.438 \text{ kNm} = 2.438 \text{ kNm}$$

Design moment $M_u = 1.266 + 2.438 = 3.704$ kNm

Reinforcement.

Consider 10 mm φ bars at a nominal or clear cover of 15 mm. Then

$$\text{Effective depth } d = 125 - 15 - \frac{10}{2} = 105 \text{ mm}$$

$$\text{Reinforcement parameter } R_u = \frac{4.6M_u}{f_{ck}bd^2}$$

$$= \frac{4.6 \times 3.704 \times 10^6}{20 \times 290 \times 105^2} = 0.2668$$

and

$$\text{Reinforcement } A_{st} = \left(\frac{f_{ck}bd}{2f_y}\right)[1 - \sqrt{1 - R_u}\,]$$

$$= \left(\frac{20 \times 290 \times 105}{2 \times 415}\right)\left[1 - \sqrt{1 - 0.2668}\,\right] = 105.50 \text{ mm}^2$$

Provide 3–10 φ bars (A_{st} = 235.5 mm² > 105.31 mm²).

Anchorage length L_d = 47ϕ = 470 mm

Distribution steel.

$$A_{st} = 0.0012 \times (1000 \times 125)$$

= 150 mm²/m (assuming uniform thickness of 125 mm).

Provide 8 φ bars @ 300 c/c.

Check for shear.

Factored shear force at support V_u = (1.62 + 2.10) × 1.25 = 4.65 kN

∴

$$\tau_v = \frac{V_u}{bd}$$

$$= \frac{4.65 \times 10^3}{290 \times 105} = 0.153 \text{ MPa} < \tau_c (= 0.326 \times 1.3 = 0.424 \text{ MPa})$$

Hence, thickness is adequate. The reinforcement details of the tread slab are shown in Fig. 7.7(b). Two alternate arrangements for the independent tread slab type cantilever stairs are shown in Figs. 7.8(a) and (b).

CASE II *Waist slab type staircase, supported between two stringer beams along the sides*

For R = 160 mm, T = 280 mm

$$\text{Length of waist slab} = \sqrt{R^2 + T^2} = 322.5 \text{ mm}$$

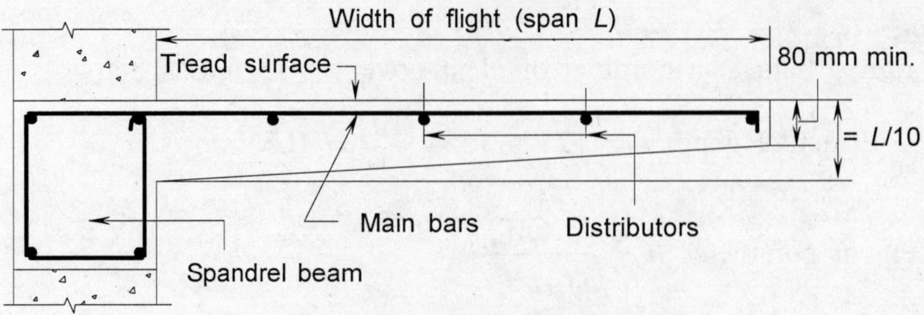

(a) Slab cantilevered from a spandrel beam

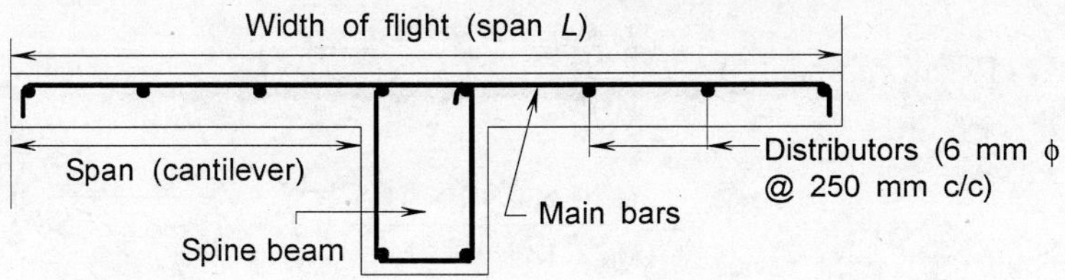

(b) Slab double cantilevered from a central spine beam.

Fig. 7.8 Detailing for alternate arrangements for independent tread slab of Example 7.8.

Consider a waist slab thickness of 80 mm (minimum) with 15 mm nominal cover and 10 ϕ bars. The moment of resistance is provided entirely by the waist slab. The effective depth is:

$$d = 80 - 15 - \frac{10}{2} = 60 \text{ mm}$$

Gravity loads over a tread width.

Self-weight of slab $w_g = (0.080 \times 0.0.323) \times 25 \quad = 0.646$ kN/m

Self-weight of step $w_s = \left(\frac{1}{2}\right) \times 0.16 \times 0.28 \times 25 = 0.560$ kN/m

Weight of finish $w_f = 0.6 \times 0.28 \quad = 0.168$ kN/m

Live loads $w_l = 5.0 \times 0.28 \quad = 1.400$ kN/m

Total service gravity load $w \quad = 2.774$ kN/m

Factored load causing flexure in the transverse direction

$$w_u = (1.5 \times w)\cos\theta = (1.5 \times 2.774) \times \left(\frac{280}{323}\right) = 3.61 \text{ kN/m}$$

$$\text{Load per metre width} = \frac{3.61}{0.323} = 11.18 \text{ kN/m}$$

Reinforcement.
Maximum sagging moment at mid-span is given as:

$$M_u = \frac{11.18 \times 1.25^2}{8} = 2.184 \text{ kNm/m}$$

$$\text{Reinforcement parameter } R_u = \frac{4.6M_u}{f_{ck}bd^2}$$

$$= \frac{4.6 \times 2.184 \times 10^6}{20 \times 10^3 \times 60^2} = 0.1394$$

$$\therefore \qquad A_{st} = \left(\frac{20 \times 1000 \times 60}{2 \times 415}\right)\left[1 - \sqrt{1 - 0.1394}\right] = 104.00 \text{ mm}^2\text{/m}$$

Maximum spacing = $3d$ = 3 × 60 = 180 mm. Provide 8 mm ϕ bars @ 160 mm (< 323/2 = 162 mm) c/c, as shown in Fig. 7.7(c).

Distribution or longitudinal steel.

$$A_{st,\min} = 0.0012 \times 1000 \times 80 = 96 \text{ mm}^2\text{/m}$$

Provide 8 mm ϕ @ 250 c/c, as shown in Fig.7.7(c).

CASE III *Tread-riser arrangement*
Consider a slab of nominal thickness of 100 mm reinforced with 10 mm ϕ bars and 8 mm ϕ ties at a nominal cover of 20 mm.

The effective depth of a typical tread-riser unit is given by

$$d = (160 + 100) - 20 - \left(\frac{10}{2}\right) - 8 = 227 \text{ mm}$$

Load on a unit.

(i) Dead load

Self-weight w_g = (0.28 + 0.16) × 0.10 × 25 = 1.10 kN/m

Weight of finish w_f = 0.6 × 0.28 = 0.168 kN/m

Total service dead load w_d = 1.268 KN/m

Factored dead load $w_{u,d}$ = 1.5 w_d = 1.5 × 1.268 = 1.902 kN/m

(ii) Live load

Factored uniformly distributed load $w_{u,l}$ = 1.5 × 5.0 × 0.28 = 2.100 kN/m

or Factored concentrated load at free end $w_{u,c}$ = 1.5 × 1.3 =1.95 kN whichever gives greater moment.

As in Case I concentrated live load at the free end results in a larger design moment.

Design moment.

Maximum moment at the support is calculated as:

$$M_{u,1} = 1.902 \times \frac{1.25^2}{2} + 1.95 \times 1.25 = 3.923 \text{ kNm}$$

For simplicity the moment is considered to be resisted by rectangular section with b = 100 mm and d = 227 mm.

The flexural reinforcement parameter:

$$R_u = \frac{4.6M_u}{f_{ck}bd^2}$$

$$= \frac{4.6 \times 3.923 \times 10^6}{20 \times 100 \times 227^2} = 0.1751$$

$$\therefore \qquad \text{Reinforcement } A_{st} = \left(\frac{20 \times 100 \times 227}{2 \times 415}\right)\left[1 - \sqrt{1 - 0.1751}\right]$$

$$A_{st} = 96.88 \text{ mm}^2$$

Provide 2-10 ϕ bars on top (A_{st} = 157 mm²) with anchorage length of 470 mm.

Distribution steel.

$$A_{st,\min} = 0.0012 \times 1000 \times 100 = 120 \text{ mm}^2\text{/m}$$

Provide 8 mm ϕ bars @ 200 c/c as distribution steel in the form of closed loops, with a 8 mm ϕ bar placed transversely at each bend. The reinforcement details are given in Fig. 7.9.

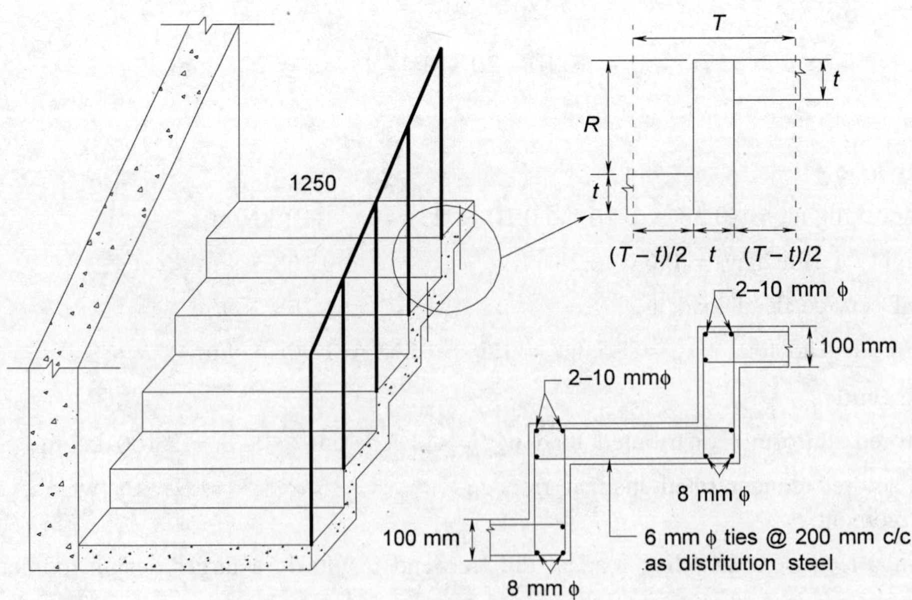

Fig.7.9 Detailing scheme of tread-riser cantilever stair of Example 7.8.

The following example will illustrate the design of stair slabs spanning longitudinally.

Example 7.9 Design a staircase to be provided in a residential building in two straight opposite flights of 1.0 m width connected by a landing for a floor height of 3.3 m. The landing which is 1.0 m wide spans in the same direction as the stair slab. The rise and tread shall be 150 mm and 270 mm, respectively. The weight of the finishes is 1.0 kN/m^2. The materials to be used are: concrete of grade M20 and steel of grade Fe415.

Solution For the given materials, the design constants are:

$$f_{ck} = 20 \text{ MPa}, f_y = 415 \text{ MPa},$$

$$M_{u,\text{lim}} = 0.1388 f_{ck} bd^2 \text{ and } p_{t,\text{lim}} = 0.961 \text{ per cent}$$

Geometric proportions are:

Floor height = 3.3 m, Width of flights = 1.0 m,

Rise = 150 mm and Tread = 270 mm

$$\text{Number of risers} = \frac{\text{Storey height}}{\text{Rise}} = \frac{3.3 \times 1000}{150} = 22$$

$$\text{Horizontal span of the flight} = \text{Number of steps} \times \text{Tread} + \text{Landing width}$$
$$= 10 \times 270 + 1000 = 3700 \text{ mm}$$

The geometry of the staircase is shown in Fig. 7.10(a). Since the span and loading are the same on both the flights, the same design will be adopted for both the flights.

Loads. The effective depth of the waist slab d from the serviceability requirement of deflection is given by

$$d = \frac{\text{Span}}{20 \times \text{Modification factor}}$$

For an under-reinforced section (say $p_t = 0.7$ per cent), consider $f_s = 0.58\, f_y (= 0.58 \times 415 = 240.7 \text{ MPa})$ giving trial modification factor of 1.11. Thus,

$$d = \frac{3700}{20 \times 1.11} = 166.67 \text{ mm}$$

Adopt 10 mm ϕ bars for reinforcement with 15 mm clear or nominal cover. Therefore, the overall depth:

$$D = 166.67 + 15 + \frac{10}{2} = 186.67 \text{ mm}$$

However, consider 175 mm as the overall depth and 155 mm as the effective depth for computation of gravity load.

$$\text{Inclined length of the waist slab for one step} = \sqrt{0.27^2 + 0.15^2} = 0.3089 \text{ m}$$

$$\text{Area of surface finish for one step} = (0.270 + 0.150) \times 1.0 = 0.420 \text{ m}^2$$

Weight of waist slab in plan $w_g = \dfrac{0.3089 \times 0.175 \times 25}{0.270}$ $= 5.01$ kN/m

Weight of floor finish $w_f = \dfrac{0.420 \times 1.0}{0.270}$ $= 1.556$ kN/m

Weight due to steps $w_s = \dfrac{(1/2) \times 0.15 \times 0.27 \times 1.0 \times 25}{0.270}$ $= 1.875$ kN/m

Live load w_l $= 3.0$ kN/m

Total load $w = w_g + w_f + w_s + w_l$ $= 11.441$ kN/m

Factored load $w_u = 1.5w \approx 1.5 \times 11.441$ $= 17.1615$ kN/m

Factored bending moment is obtained as:

$$M_u = \frac{w_u l^2}{8} = \frac{17.1615 \times 3.7^2}{8} = 29.37 \text{ kNm}$$

Depth of the slab can be computed by equating $M_{u,\lim}$ to M_u:

$$0.1388 f_{ck} b d^2 = M_u$$

or

$$0.1388 \times 20 \times 1000 \times d^2 = 29.37 \times 10^6$$

Thus,

$$d = 102.86 \text{ mm}$$

With 10 mm ϕ bars as tension reinforcement at a nominal cover of 15 mm,

$$D = 102.86 + 15 + \left(\frac{10}{2}\right) = 122.86 \text{ mm}$$

Consider revised overall depth $D = 165$ mm with $d = 145$ mm.

Reinforcement. Dimensionless flexure reinforcement parameter:

$$R_u = \frac{4.6 M_u}{f_{ck} b d^2}$$

$$= \frac{4.6 \times 29.37 \times 10^6}{20 \times 1000 \times 145^2} = 0.32129$$

Therefore,

$$A_{st} = \left(\frac{20 \times 1000 \times 145}{2 \times 415}\right) \times \left[1 - \sqrt{1 - 0.32129}\right] = 615.50 \text{ mm}^2$$

Provide 8 bars of 10 mm ϕ as tension reinforcement ($A_{st} = 628.32$ mm^2).

Distribution steel.

$$A_{st,\min} = \frac{0.12 \times 1000 \times 165}{100} = 198 \text{ mm}^2$$

Provide 8 mm ϕ bars at 250 mm c/c (A_{st} = 201.06 mm^2).

Check for shear.

$$\text{Factored shear } V_u = \frac{17.1615 \times 3.7}{2} = 31.749 \text{ kN}$$

$$\text{Nominal shear stress } \tau_v = \frac{31.749 \times 10^3}{1000 \times 145} = 0.219 \text{ MPa}$$

For

$$p_t = \frac{628.32 \times 100}{1000 \times 145} = 0.4333$$

$$\tau_{uc} = 0.451 \text{ MPa (from Table 4.1)}$$

$$\text{Permissible shear stress} = k_s \tau_{uc}$$

$$= 1.3 \times 0.451 = 0.586 \text{ MPa} > \tau_v$$

Hence the slab is safe in shear.

Check for development length.

$$\text{Development length } L_d = \frac{0.87 f_y \phi}{4\tau_{bd}}$$

$$= \left[\frac{0.87 \times 415}{4 \times 1.92}\right]\phi = 47\phi$$

At the simply supported end, the following condition shall be satisfied:

$$L_d \le \frac{1.3 M_{u,1}}{V_u} + L_o$$

$$M_{u,1} = 0.87 f_y A_{st}(d - 0.416 x_u)$$

where

$$x_u = \frac{0.87 \times 415 \times A_{st}}{0.362 \times 20 \times 1000}$$

$$= 0.04987 A_{st} = 0.04987 \times 628.32 = 31.33 \text{ mm}$$

Thus,

$$M_{u,1} = 0.87 \times 415 \times 628.32 \times (145 - 0.416 \times 31.33) \times 10^{-6}$$

$$= 29.92 \text{ kNm}$$

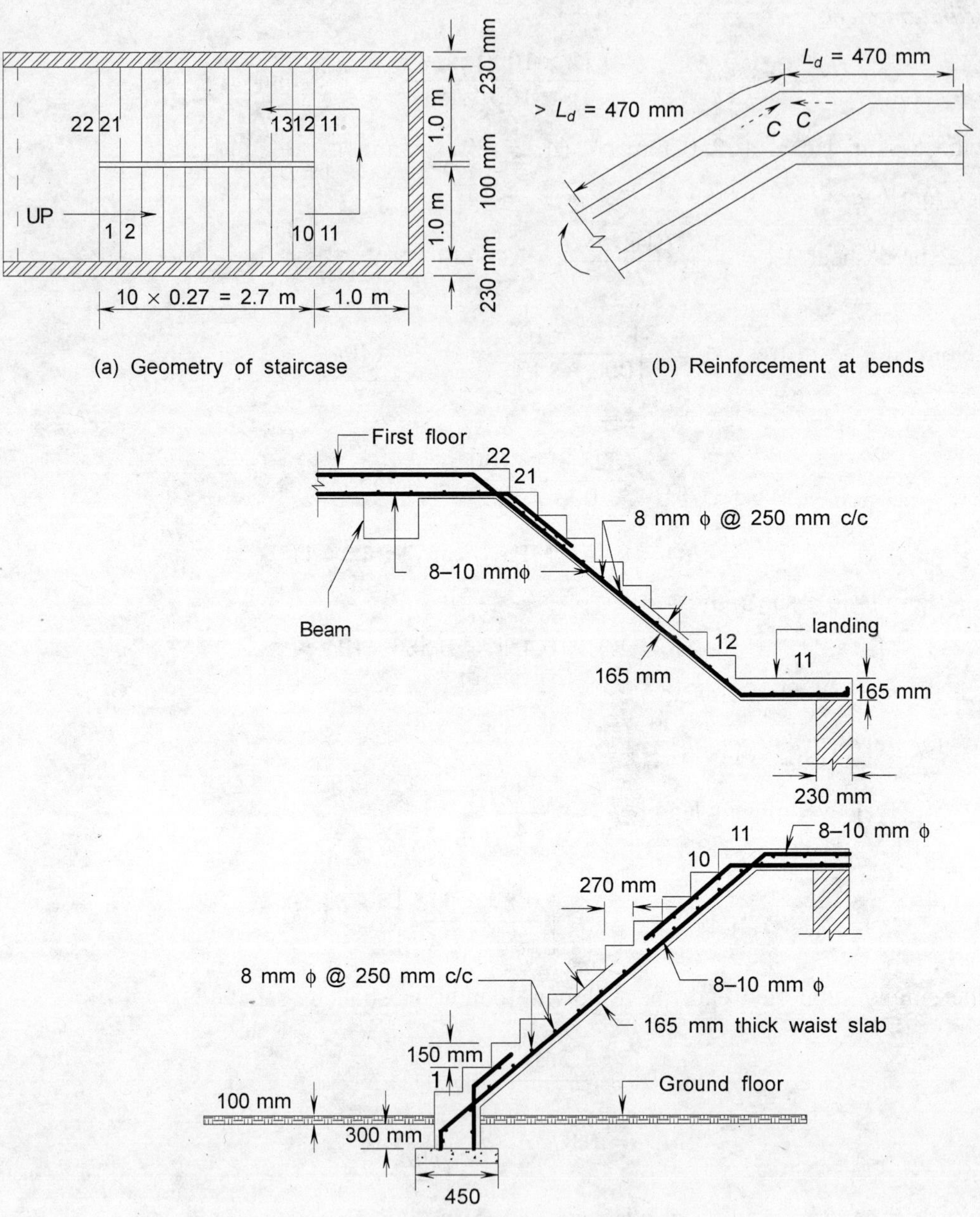

Fig. 7.10 Detailing of staircase slab of Example 7.9.

and

$$V_u = 31.749 \text{ kN}$$

Assume $L_o = 8\ \phi$ (for Fe415 steel).
Then

$$L_d \leq \frac{1.3M_{u,1}}{V_u} + \text{L}_\text{o}$$

or

$$47.0\phi \leq \frac{1.3 \times 29.92 \times 10^6}{31.749 \times 10^3} + 8\phi$$

or

$$\phi \leq 31.41 \text{ mm}$$

Hence the diameter of the bars provided is satisfactory.

Check for deflection. For serviceability requirements of deflection,

$$f_s = 0.58 f_y \left(\frac{A_{st,\text{required}}}{A_{st,\text{provided}}} \right)$$

$$= 0.58 \times 415 \times \frac{615.52}{628.32} = 236 \text{ MPa}$$

Modification factor for $p_t = 0.4333$ and $f_s = 236$ MPa, from Table 6.3, is 1.31. Thus,

$$d = \frac{\text{Span}}{20 \times \text{Modification factor}}$$

$$= \frac{3700}{20 \times 1.31} = 141.22 \text{ mm} < 145 \text{ mm}$$

Hence section is satisfactory. The reinforcement details are given in Fig. 7.10(c).

7.6 DESIGN OF COMPRESSION MEMBERS

7.6.1 Axially Loaded Short Column

As discussed in Chapter 5, all columns in practice are subjected to some bending moment due to both initial crookedness and unsymmetrical loading. Hence an axially loaded column is not a practical case. The code has recommended that the columns, though axially loaded, shall be designed for a minimum eccentricity of $e_{\min} = (L/500) + (D/30)$ subject to a minimum of 20 mm. In this expression, L is the unsupported length of the column, and D is the lateral dimension of the column in the direction under consideration. As a matter of fact, one could justify adding to all columns an additional eccentricity to allow for unforeseen effects that may increase the eccentricity of the loading.

When e_{min}, as calculated earlier, is less than 0.05 times the lateral dimension, the column may be designed according to the following equation:

$$P_u = 0.40f_{ck}A_c + 0.67f_yA_s = 0.40f_{ck}(A_g - A_s) + 0.67f_yA_s$$

$$= 0.40f_{ck}(1 - 0.01p)A_g + 0.67f_y(0.01p)A_g \tag{7.11}$$

where P_u is axially applied factored load on the column and p is percentage of longitudinal reinforcement.

7.6.2 Column Subjected to Combined Axial Load and Uniaxial Bending

The usual practice is to estimate the column section and the reinforcement, and modify the same by trial and adjustment method till an adequate economical section is obtained. In order to overcome lengthy calculations column design interaction diagrams may be used. The following procedure is recommended:

1. Consider the area of longitudinal steel, say, between 1 and 2 per cent of the gross cross-sectional area of concrete. With P_u, f_{ck} and f_y, being known, the trial section can be estimated from Eq. (7.11).
2. If there are no restrictions on the dimensions of the column, consider the dimensions of the column such that it is a short column, that is, $l_{ex}/D \leq 12$ and $l_{ey}/b \leq 12$.
3. Based on durability and fire resistance considerations, fix the cover d' to the longitudinal steel.
4. Calculate the quantities $(P_u/f_{ck}bD)$,$(M_u/f_{ck}bD^2)$ and (d'/D), and refer to the appropriate column design interaction diagram to determine the percentage of steel.
5. Determine transverse steel and detail the sections suitably.

7.6.3 Column Subjected to Axial Load and Biaxial Bending

As mentioned in Chapter 5, the trial cross-sectional dimensions are generally fixed in advance during the stage of analysis for design forces. The cross-sectional dimensions are either based on the architectural considerations or on designer's judgement and experience. However, following points may help in selection of suitable trial section:

1. For economy, $\dfrac{d_y}{d_x} \approx \dfrac{M_{ux}}{M_{uy}}$

   where d_x and d_y are the effective depths in y- and x- directions, respectively.

2. From the point of view of good design practice,

$$\frac{D}{b} \le 2.0$$

After trial dimensions of cross-section of the column are selected, design of biaxially loaded columns reduces to determination of suitable reinforcement detailing. The trial percentage of reinforcement can be determined by designing the column section as uniaxially loaded column for the moment,

$$M_u = \sqrt{M_{ux}^2 + M_{uy}^2}$$

This moment is considered to act about the axis which provides maximum value of ratio moment-to-corresponding effective depth.The adequacy of the section is checked by using the interaction formula for biaxial moment suggested by IS:456:

$$\left(\frac{M_{ux}}{M_{uxi}}\right)^{\alpha_n} \pm \left(\frac{M_{uy}}{M_{uyi}}\right)^{\alpha_n} \le 1.0 \qquad (7.12)$$

where

M_{ux}, M_{uy} = Moments about x- and y-axes, respectively, due to the design load P_u

M_{uxi}, M_{uyi} = Maximum uniaxial moment capacities of the trial section for bending about x- and y-axes, respectively, along with the axial load P_{uz}

α_n = Exponent, given in Table 7.4

Here

$$P_{uz} = 0.447 f_{ck} A_c + 0.75 f_y A_s \qquad (7.13)$$

Table 7.4 Relationship of P_u/P_{uz} to Exponent α_n

P_u/P_{uz}	< 0.20	0.40	0.60	≥ 0.80
α_n	1.0	1.33	1.67	2.0

7.6.4 Slender Columns

For the columns with slenderness ratios l_{ex}/D and/or $l_{ey}/b > 12$, the additional moments M_{ax} and M_{ay} in appropriate directions shall be taken into account while designing the column section:

$$M_{ax} = \left(\frac{P_u D}{2000}\right)\left(\frac{l_{ex}}{D}\right)^2$$

and

$$M_{ay} = \left(\frac{P_u b}{2000}\right)\left(\frac{l_{ey}}{b}\right)^2 \qquad (7.14)$$

where

P_u = Axial load on the member

l_{ex}, l_{ey} = Effective lengths in respect of major and minor axes, respectively

D = Depth at right angles to the major axis

b = Width of the section

These additional moments are to be reduced by the factor k given by

$$k = \frac{P_{uz} - P_u}{P_{uz} - P_{ub}} \leq 1 \tag{7.15}$$

where P_{ub} is the axial load corresponding to the condition of maximum compressive strain of 0.0035 in concrete and tensile strain of 0.002 in the outermost layer of tensile steel.

7.6.5 Design Column Interaction Diagrams

Typical column design interaction diagrams presented in Figs. 5.12 to 5.17 are for the use for uniaxial bending of column of rectangular section and are valid when reinforcement is to be provided on two sides of the column section, whereas those presented in Figs. 5.18 to 5.20 are for steel on four sides and are used for biaxial bending. For circular sections refer to design curve of Figs. 5.21 and 5.22. The readers should refer to the special publication of BIS, SP:16—Design Aids for Reinforced Concrete to IS:456, and appendix-D for more details. The following examples will illustrate the procedure and types of problems encountered in practice.

Example 7.10 Design a column: (i) to carry an axial service load of 1225 kN. The grades of concrete and steel to be used are M25 and Fe415, respectively. The effective length of the column is 3.25 m. (ii) Re-design the column if one of its cross-sectional dimensions is restricted to 300 mm due to architectural considerations.

Solution For M25 concrete and Fe415 steel:

$$f_{ck} = 25 \text{ MPa and } f_y = 415 \text{ MPa}$$

$$\text{Factored load } P_u = 1.5 \times 1225 = 1837.5 \text{ kN}$$

Longitudinal steel. Consider 2.0 per cent longitudinal steel, i.e., $A_s = 0.02A_g$ and $A_c = 0.98A_g$.

$$\text{Load equation } P_u = 0.40 f_{ck} A_c + 0.67 f_y A_s$$

Therefore,

$$1837.5 \times 10^3 = 0.4 \times 25 \times 0.98A_g + 0.67 \times 415 \times 0.02A_g$$

or

$$A_g = 119621.12 \text{ mm}^2$$

Hence,

$$A_s = 0.02A_g = 2392.42 \text{ mm}^2.$$

CASE I Adopt a square cross-section of size 350 × 350 mm. Minimum design eccentricity is:

$$e_{\min} = \frac{L}{500} + \frac{D}{30}$$

$$= \frac{3.25 \times 10^3}{500} + \frac{350}{30} = 18.17 \text{ mm} < 20 \text{ mm}$$

Thus, the eccentricity to be considered is $e = e_{\min} = 20$ mm.

Minimum eccentricity permitted is given by

$$e_{\min,\text{ permitted}} = 0.05D = 0.05 \times 350 = 17.5 \text{ mm} < e_{\min}$$

Hence the adequacy of the section is required to be checked for the following forces:

$$P_u = 1837.5 \text{ kN}$$

$$\therefore \quad M_u = P_u \times \text{e}$$

$$= 1837.5 \times 0.02 = 36.75 \text{ kNm}$$

Consider the column interaction diagram for the section with steel on two sides.

$$\frac{P_u}{f_{ck}bD} = \frac{1837.5 \times 10^3}{25 \times 350 \times 350} = 0.60$$

$$\frac{M_u}{f_{ck}bD^2} = \frac{36.75 \times 10^6}{25 \times 350 \times 350^2} = 0.0343$$

Consider the cover to longitudinal steel $d' = 50$ mm. Thus,

$$\frac{d'}{D} = \frac{50}{350} = 0.143 \approx 0.14$$

Refer to column interaction diagram of Fig. 5.15 for $f_y = 415$ MPa and $d'/D = 0.15$. Then

$$\frac{p}{f_{ck}} = 0.076$$

Therefore,

$$p = 0.076 f_{ck} = 0.076 \times 25 = 1.9 \text{ per cent}$$

Hence 2.0 per cent steel provided is adequate. Adopt four bars of 25 mm ϕ and four bars of 16 mm ϕ ($A_s = 2768$ mm^2), as shown in Fig. 7.11(a).

Lateral ties. Minimum diameter of the bar for ties

$$\phi_t = \frac{\text{Diameter of the longitudinal bar}}{4} (= 6.25 \text{ mm}) \text{ or 5 mm, whichever is greater.}$$

Use 8 mm ϕ bar ties at a spacing which is the least of the following:

1. The least lateral dimension, i.e., 350 mm
2. 16 × ϕ of longitudinal bar, i.e., 256 mm
3. 300 mm

Provide 8 mm ϕ mild steel ties @ 250 mm c/c.

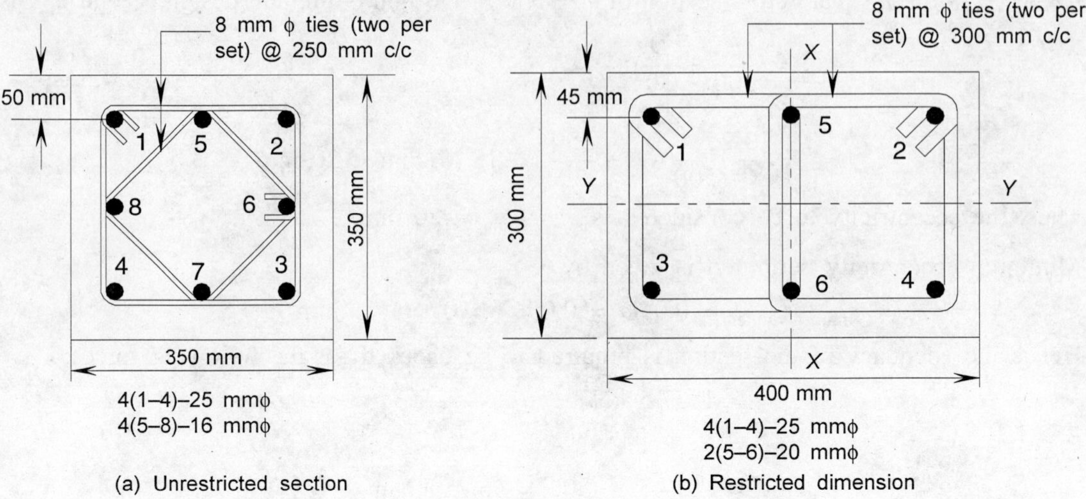

Fig. 7.11 Reinforcement details of the column of Example 7.10.

CASE II If one side of column section is restricted to 300 mm, the other side of the section is:

$$D = \frac{119621.12}{300} = 398.74 \text{ mm}$$

Adopt a trial section of size 300 × 400 mm.

(i) Minimum design eccentricity along the *y*-axis is expressed as:

$$e_{\min} = \frac{3.25 \times 10^3}{500} + \frac{400}{30} = 19.83 < 20 \text{ mm}$$

$$e_{\min,\text{ permited}} = 0.05D = 0.05 \times 400 = 20 \text{ mm} > e_{\min}$$

Hence no check for adequacy of the section need to be considered in this direction.

(ii) Minimum design eccentricity along the *x*-axis is obtained as:

$$e_{\min} = \frac{3.25 \times 10^3}{500} + \frac{300}{30} = 16.5 \text{ mm} < 20 \text{ mm}$$

Thus, the design eccentricity to be considered, $e = e_{\min} = 20$ mm, and

$$e_{\min \text{ permitted}} = 0.05 \times 300 = 15 \text{ mm} < e_{\min}$$

Therefore, the adequacy check is required in this direction for the following forces:

$$P_u = 1837.5 \text{ kN}$$

$$M_{uy} = P_{u.}e = 1837.5 \times 0.02 = 36.75 \text{ kNm}$$

Consider the effective cover d' in this direction to be 45 mm. Here, b = 400 mm and D = 300 mm. Therefore, d'/D = 45/300 = 0.15, and

$$\frac{P_u}{f_{ck}bD} = \frac{1837.5 \times 10^3}{25 \times 400 \times 300} = 0.6125$$

$$\frac{M_u}{f_{ck}bD^2} = \frac{36.75 \times 10^6}{25 \times 400 \times (300)^2} = 0.041$$

From the column interaction diagram of Fig. 5.15 for f_y = 415 MPa and d'/D = 0.15,

$$\frac{p}{f_{ck}} = 0.086$$

Longitudinal steel ratio required p = 0.086 × 25 = 2.15 per cent

Therefore, $$A_s = \frac{pbD}{100} = \frac{2.15 \times 300 \times 400}{100} = 2580 \text{ mm}^2$$

Provide 4 bars of 25 mm ϕ and 2 bars of 20 mm ϕ (A_s = 2592 mm^2) along with 8 mm ϕ mild steel ties @ 300 mm c/c. The arrangement of reinforcement is shown in Fig. 7.11(b).

Example 7.11 Design a column having an effective length of 4.75 m to support a factored load of 1650 kN. Consider the reinforcement ratio p to be in the range 1.5 to 2.0 per cent and the effective cover to longitudinal steel of 55 mm. The materials to be used are: M25 grade concrete and HYSD steel bars of grade Fe415.

Solution For the given materials:

$$f_{ck} = 25 \text{ MPa and } f_y = 415 \text{ MPa}$$

Factored load P_u = 1650 kN and

$$P_u = 0.40 f_{ck} A_c + 0.67 f_y A_s$$

Consider a gross reinforcement ratio of 1.5 per cent, i.e. $A_s = 0.015A_g$ and

$$A_c = A_g - 0.015A_g = 0.985A_g$$

Therefore,

$$1650 \times 10^3 = 0.4 \times 25 \times 0.985A_g + 0.67 \times 415 \times 0.015A_g$$

or

$$A_g = 117682.72 \text{ mm}^2$$

Consider 350 × 350 mm trial section with 1.5 per cent steel. The minimum design eccentricity is given by

$$e_{\min} = \frac{4.75 \times 10^3}{500} + \frac{350}{30} = 21.2 \text{ mm} > 20 \text{ mm (minimum)}$$

Permitted value.

$$e_{min,permitted} = 0.05 \times 350 = 17.5 < e_{min}$$

Hence, the adequacy of the section needs to be checked for the following forces:

$$P_u = 1650 \text{ kN and } M_u = 1650 \times 0.0212 = 34.98 \text{ kNm}$$

Consider the steel to be distributed on two sides:

$$\frac{P_u}{f_{ck}bD} = \frac{1650 \times 10^3}{25 \times 350 \times 350} = 0.539$$

$$\frac{M_u}{f_{ck}bD^2} = \frac{34.98 \times 10^6}{25 \times 350 \times (350)^2} = 0.0265$$

For the ratio, $d'/D = 55/350 = 0.157$ (say 0.15), refer to the column interaction diagram of Fig. 5.15 for $f_y = 415$ and $d'/D = 0.15$, giving $p/f_{ck} = 0.053$. Therefore, the longitudinal steel required is given as:

$$p = 0.053 f_{ck} = 1.325 \text{ per cent}$$

Hence the assumed area of 1.5 per cent is adequate.
Check for slenderness

$$\frac{l_{ex}}{D} = \frac{l_{ey}}{b} = \frac{4.75 \times 10^3}{350} = 13.57 > 12$$

The column is slender with respect to both the axes. Additional moments due to slenderness to be considered in the design are:

$$M_{ax} = M_{ay} = \left(\frac{P_u D}{2000}\right)\left(\frac{l_{ex}}{D}\right)^2 = \frac{(1650 \times 10^3) \times 350}{2000} \times (13.57)^2 \times 10^{-6} = 53.172 \text{ kNm}$$

These additional moments are to be reduced by a factor k given by

$$k = \frac{P_{uz} - P_u}{P_{uz} - P_{ub}}$$

Here,

Factored axial load $P_u = 1650$ kN

Capacity of the column section under pure axial load

$$\begin{aligned} P_{uz} &= 0.447 f_{ck} A_c + 0.75 f_y A_s \\ &= 0.447 \times 25 \times (0.985 \times 350^2) + 0.75 \times 415 \times (0.015 \times 350^2) \\ &= 1920.33 \times 10^3 \text{ N} = 1920.33 \text{ kN} \end{aligned}$$

P_{ub} is the axial load corresponding to the balanced condition of the maximum compressive strain in concrete of 0.0035 occurring simultaneously with a maximum tensile strain in reinforcement equal to 0.0020. To compute P_{ub}, consider the balanced strain diagram shown in Fig. 7.12(a).

$$\frac{0.0035}{x_u} = \frac{0.0020}{(D - d') - x_u} \quad \text{or} \quad x_u = \frac{0.0035}{0.0055}(D - d')$$

Here, D = 350 mm, d' = 55 mm, and hence

$$x_u = \frac{0.0035}{0.0055} \times (350 - 55) = 187.73 \text{ mm}$$

$$\text{Strain in compression steel } \varepsilon_{sc} = 0.0035\left(\frac{x_u - d'}{x_u}\right) = 0.00247$$

Therefore,

$$P_{ub} = 0.362 f_{ck} b x_u + f_{sc} A'_{sc} - f_{st} A'_{st}$$

Since $A'_{st} = A'_{sc}$, and $f_{sc} = f_{st}$, as the strains in compression and tension steels are more than 0.0020. Therefore

$$P_{ub} = 0.362 f_{ck} b x_u = 0.362 \times 25 \times 350 \times 187.73$$

$$= 594.63 \times 10^3 \text{ N} = 594.63 \text{ kN}$$

Therefore, the reduction factor:

$$k = \frac{1920.33 - 1650}{1920.33 - 594.63} = 0.204$$

Net additional moments due to slenderness is:

$$kM_{max} = 0.204 \times 53.172 = 10.85 \text{ kNm}$$

$$\text{Total design moment } M_u = 34.98 + 10.85 = 45.83 \text{ kNm}$$

The modified values of dimensionless parameters for column interaction curves are:

$$\frac{P_u}{f_{ck} bD} = \frac{1650 \times 10^3}{25 \times 350 \times 350} = 0.539$$

$$\frac{M_u}{f_{ck} bD^2} = \frac{45.83 \times 10^6}{25 \times 350 \times 350^2} = 0.043$$

and

$$\frac{d'}{D} = \frac{55}{350} = 0.157$$

Refer to the column interaction diagram of Fig. 5.17 corresponding to f_y = 415 MPa and d'/D = 0.15. For $(P_u/f_{ck}bD)$ = 0.539 and $(M_u/f_{ck}bD^2)$ = 0.043, the parameter p/f_{ck} = 0.065. Thus,

$$p = 0.065 f_{ck} = 0.065 \times 25 = 1.625 \text{ per cent}$$

Therefore,

$$A_s = \frac{pbD}{100} = \frac{1.625 \times 350 \times 350}{100} = 1990.63 \text{ mm}^2$$

Adopt 8 mm ϕ mild steel ties at the spacing which is least of the following:

(i) 350 mm, (ii) 16 ϕ (16 × 18 = 288 mm) and (iii) 300 mm

Provide 8 bars of 18 mm ϕ (A_s = 2035 mm^2) with 8 mm ϕ mild steel ties at 280 mm c/c. The reinforcement details are shown in Fig. 7.12(b).

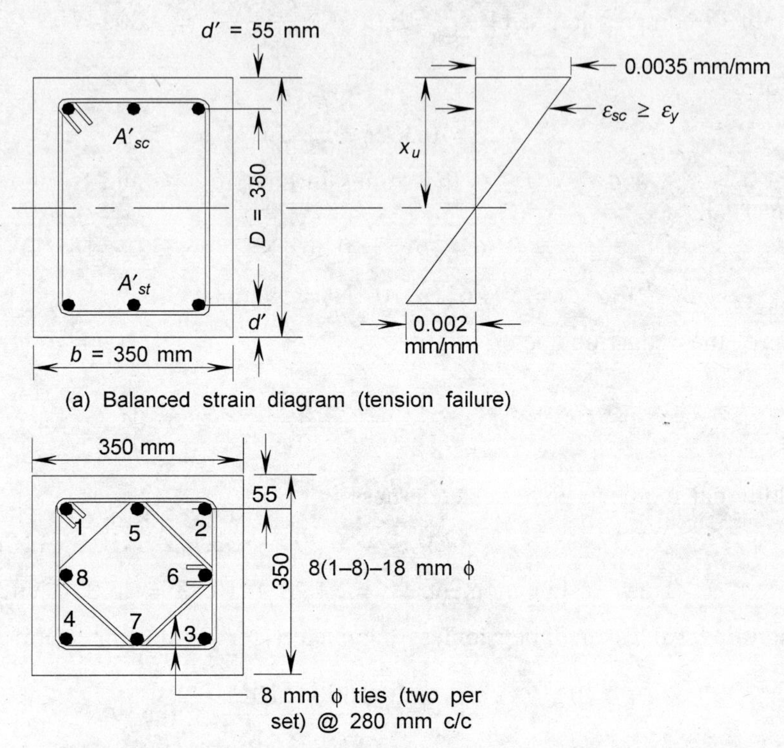

Fig. 7.12 *Reinforcement details of the column of Example 7.11.*

Example 7.12 A braced reinforced concrete column of circular cross-section of 500 mm diameter is to support a factored axial load of 2250 kN alongwith a factored moment of 160 kNm. The unsupported length of the column is 6.3 m with effective length of 5.5 m. Design the column when it is to be provided with: (a) lateral ties and (b) spiral reinforcement. The grades of concrete and steel used are M25 and Fe415, respectively.

Solution For M25 concrete and Fe415 steel:

f_{ck} = 25 MPa and f_y = 415 MPa,

D = 500 mm, P_u = 2250 kN and M_u = 160 kNm

$$\text{Slenderness ratio } \frac{l_{\text{eff}}}{D} = \frac{5.5 \times 10^3}{500} = 11 < 12$$

Therefore it is a short column. The minimum design eccentricity is

$$e_{\min} = \frac{6.3 \times 10^3}{500} + \frac{500}{30} = 29.27 \text{ mm} > 20 \text{ mm (minimum)}$$

and

$$e_{\text{min,permitted}} = 0.05 \times 500 = 25.0 < e_{\min}$$

Therefore, the moment due to minimum eccentricity is:

$$M'_u = P_u e_{\min} = 2250 \times \frac{29.27}{1000} = 65.85 \text{ kNm} < M_u$$

Consider 25 mm ϕ bars at nominal cover of 40 mm, thus effective cover is given by

$$d' = 40 + \frac{25}{2} = 52.5 \text{ and } \frac{d'}{D} = \frac{52.5}{500} = 0.105 \approx 0.10$$

CASE ***(a) Column with ties:*** The dimensionless load parameters are:

$$\frac{P_u}{f_{ck}D^2} = \frac{2250 \times 10^3}{25 \times 500^2} = 0.36$$

and

$$\frac{M_u}{f_{ck}D^3} = \frac{160 \times 10^6}{25 \times 500^3} = 0.0512$$

From Fig. 5.18 for Fe415 and $d'/D = 0.10$, $p/f_{ck} = 0.073$ whence $p = 1.825$ per cent.

Therefore, $$A_s = \frac{p\pi D^2}{4 \times 100} = 3583.38 \text{ mm}^2$$

Provide 8 bars of 25 mm ϕ ($A_s = 3927 \text{ mm}^2$).

Transverse reinforcement. Adopt 8 mm ϕ mild steel ties at the spacing which is least of the following:

(i) 500 mm, (ii) 16 ϕ (16 × 25 = 400 mm) and (iii) 300 mm.

Provide 8 mm ϕ mild steel ties at 300 mm c/c. The reinforcement details are shown in Fig. 7.13(a).

CASE ***(b) Column with helical reinforcement:*** The dimensionless load parameters are:

$$\frac{P_u}{f_{ck}D^2} = \frac{(2250 \times 10^3)/1.05}{25 \times 500^2} = 0.343$$

and

$$\frac{M_u}{f_{ck}D^3} = \frac{(160 \times 10^6)/1.05}{25 \times 500^3} = 0.0488$$

From Fig. 5.18 for f_y = 415 MPa and d'/D = 0.10, p/f_{ck} = 0.06 whence p = 1.50 per cent.

Therefore,
$$A_s = \frac{p\pi D^2}{4 \times 100} = 2945.24 \text{ mm}^2$$

Provide 6 bars of 25 mm ϕ (A_s = 2945 mm²).

Helical reinforcement. Adopt 6 mm ϕ mild steel bar for helix. Therefore, the diameter of the core is:

$$D_c = 500 - 40 - 40 + 12 = 432 \text{ mm}$$

The pitch of helix is:

$$s_v \le \left(\frac{11.11 a_{sp} D_c}{D^2 - D_c^2}\right)\left(\frac{f_{sp}}{f_{ck}}\right) = \left(\frac{11.11 \times (\pi \times 6^2/4) \times 432}{500^2 - 432^2}\right)\left(\frac{250}{25}\right) = 21.41 \text{ mm}$$

However, IS:456 has imposed restrictions on the pitch as follows:

$$s_v > \begin{cases} 25 \text{ mm} \\ 3\phi_{sp} = 18 \text{ mm} \end{cases}$$

$$s_v < \begin{cases} 75 \text{ mm} \\ 432/6 = 72 \text{ mm} \end{cases}$$

Provide 6 mm ϕ spiral of 25 mm pitch. The reinforcement details are shown in Fig. 7.13(b).

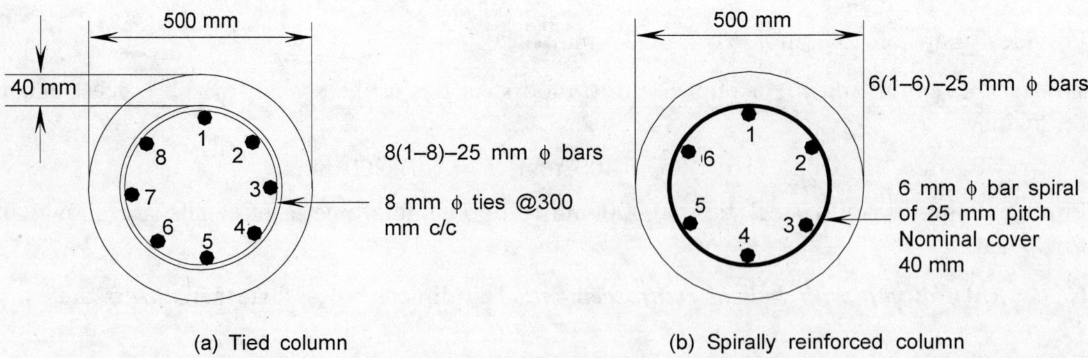

Fig. 7.13 *Reinforcement details of circular column of Example 7.12.*

Example 7.13 Design a column section to carry an ultimate axial load of 2250 kN, and factored design moments of 150 kNm and 100 kNm about major and minor axes, respectively. One of the dimensions of the column section is restricted to 300 mm. The materials to be used are: concrete of grade M25 and HYSD steel bars of grade Fe415. Consider effective cover of 50 mm.

Solution From the problem stipulations,

$$f_{ck} = 25 \text{ MPa and } f_y = 415 \text{ MPa}, b = 300 \text{ mm}$$

$$P_u = 2250 \text{ kN}, M_{ux} = 150 \text{ kNm and } M_{uy} = 100 \text{ kNm}$$

Preliminary design. To predetermine a trial section, consider an average reinforcement ratio of 2.5 per cent as an axially loaded column.

$$P_u = 0.40 f_{ck} A_c + 0.67 f_y A_s$$

$$2250 \times 10^3 = 0.40 \times 25 \times 0.975 A_g + 0.67 \times 415 \times 0.025 A_g$$

Therefore,

$$A_g = 134{,}720.56 \text{ mm}^2$$

One side of the section is limited to 300 mm. Therefore,

$$\text{Other side} = \frac{134{,}720.56}{300} = 449.07 \text{ mm}$$

As the section is also subjected to biaxial moments, consider a bigger trial section of size 300 × 500 mm. The trial percentage of steel can be obtained by designing the trial section for equivalent uniaxial bending moment:

$$M_u = \sqrt{M_{ux}^2 + M_{uy}^2} = \sqrt{(150)^2 + (100)^2} = 180.28 \text{ kNm}$$

acting about the minor axis.
Dimensionless load parameters are:

$$\frac{P_u}{f_{ck} bD} = \frac{2250 \times 10^3}{25 \times 300 \times 500} = 0.60$$

$$\frac{M_{ux}}{f_{ck} b^2 D} = \frac{180.28 \times 10^6}{25 \times 300^2 \times 500} = 0.16$$

From the *P-M* curve of Fig. 5.16 corresponding to $d'/D = 0.10$ and Fe415,

$$\frac{p}{f_{ck}} = 0.185, \text{ therefore, } p = 0.185 \times 25 = 4.625 \text{ per cent}$$

Consider 300 × 500 mm section with 4.5 per cent longitudinal reinforcement uniformly distributed along the four sides.

Check for adequacy of the trial section. The moments due to minimum eccentricities are less than the applied moments.

Uniaxial moment capacity about the major axis M_{uxi} is determined as follows:

As we know, $b = 300$ mm, $D = 500$ mm, $\frac{p}{f_{ck}} = \frac{4.5}{25} = 0.18$, $\frac{d'}{D} = \frac{50}{500} = 0.1$

and
$$\frac{P_u}{f_{ck}bD} = \frac{2250 \times 10^3}{25 \times 300 \times 500} = 0.60$$

Refer to the column interaction diagram of Fig. 5.16 corresponding to $f_y = 415$, and $d'/D = 0.10$. For $P_u/(f_{ck}bD) = 0.60$ and $p/f_{ck} = 0.18$, the moment dimensionless parameter is,

$$\frac{M_u}{f_{ck}bD^2} = 0.15$$

Therefore,

$$M_{uxi} = 0.15 f_{ck}bD^2 = 0.15 \times 25 \times 300 \times 500^2$$
$$= 281.25 \times 10^6 \text{ Nmm} = 281.25 \text{ kNm}$$

Uniaxial moment capacity about the minor axis M_{uyi} is given by

$$b = 500 \text{ mm}, D = 300 \text{ mm}, \frac{d'}{D} = \frac{50}{300} = 0.167 \text{ and } \frac{P_u}{f_{ck}bD} = 0.60$$

Refer to the column interaction diagram of Fig. 5.17 corresponding to $f_y = 415$ MPa and $d'/D = 0.15$. For $P_u/(f_{ck}bD) = 0.60$ and $p/f_{ck} = 0.18$, the moment dimensionless parameter is:

$$\frac{M_u}{f_{ck}bD^2} = 0.13$$

Therefore,

$$M_{uyi} = 0.13 f_{ck}bD^2 = 0.13 \times 25 \times 500 \times 300^2$$
$$= 146.25 \times 10^6 \text{ Nmm} = 146.25 \text{ kNm}$$

The pure axial load capacity of the column is expressed as:

$$P_{uz} = 0.447 f_{ck}A_c + 0.75 f_y A_s$$
$$= 0.447 \times 25 \times (0.955 \times 300 \times 500) + 0.75 \times 415 \times (0.045 \times 300 \times 500)$$
$$= 3701.76 \times 10^3 \text{ N} = 3701.76 \text{ kN}$$

For the ratio,

$$\frac{P_u}{P_{uz}} = \frac{2250}{3701.76} = 0.61$$

The exponent

$$\alpha_n = \left(\frac{0.61 - 0.2}{0.8 - 0.2} + 1\right) = 1.68$$

For checking the adequacy of the trial section, substitute the above values in the interaction formula:

$$\left(\frac{M_{ux}}{M_{uxi}}\right)^{\alpha_n} \pm \left(\frac{M_{uy}}{M_{uyi}}\right)^{\alpha_n} = \left(\frac{150}{281.25}\right)^{1.68} + \left(\frac{100}{146.25}\right)^{1.68} = 0.876 < 1.0$$

The section is overdesigned with 4.5 per cent steel.

Second trial. Decrease the amount of steel to 4 per cent, equally distributed on four sides and recheck the adequacy of the section with the parameter, $p/f_{ck} = 4/25 = 0.16$.

Refer to the column interaction diagram of Fig. 5.16 corresponding to $f_y = 415$ MPa and $d'/D = 0.10$. The value of $M_u/(f_{ck}bD^2)$ corresponding to $P_u/(f_{ck}bD) = 0.6$ and $p/f_{ck} = 0.16$ is 0.134. Therefore,

$$M_{uxi} = 0.134 f_{ck}bD^2 = 0.134 \times 25 \times 300 \times 500^2 \times 10^{-6} = 251.25 \text{ kNm}$$

Similarly, from the column interaction diagram of Fig. 5.17 corresponding to $f_y = 415$ MPa and $d'/D \approx 0.15$, the value of $M_u/(f_{ck}bD^2)$ for $P_u/(f_{ck}bD) = 0.60$ and $p/f_{ck} = 0.16$ is 0.117. Therefore,

$$M_{uyi} = 0.117 f_{ck}Db^2$$

$$= 0.117 \times 25 \times 500 \times 300^2 \times 10^{-6} = 131.63 \text{ kNm}$$

and

$$P_{uz} = 0.447 \times 25 \times (0.96 \times 300 \times 500) + 0.75 \times 415 \times (0.04 \times 300 \times 500)$$

$$= 3476.7 \times 10^3 \text{ N} = 3476.7 \text{ kN}$$

For the ratio, $P_u/P_{uz} = 0.647$

The exponent

$$\alpha_n = \left(\frac{0.647 - 0.2}{0.8 - 0.2} + 1\right) = 1.745$$

Therefore,

$$\left(\frac{M_{ux}}{M_{uxi}}\right)^{\alpha_n} \pm \left(\frac{M_{uy}}{M_{uyi}}\right)^{\alpha_n} = \left(\frac{150}{251.25}\right)^{1.745} + \left(\frac{100}{131.63}\right)^{1745} = 1.0256 > 1.0$$

A little increase in steel will make the section adequate. Hence adopt a column section of size 300 × 500 mm.

Longitudinal reinforcement.

$$A_s = \frac{pbD}{100} = \frac{4 \times 300 \times 500}{100} = 6000 \text{ mm}^2$$

Transverse reinforcement. Adopt 8 mm ϕ mild steel ties at the spacing which is least of the following:

(i) 300 mm, (ii) 16 ϕ (16 × 25 = 400 mm) and (iii) 300 mm

Provide 4 bars of 28 mm ϕ and 8 bars of 25 mm ϕ (A_{sc} = 6390 mm^2) with 8 mm ϕ bar ties at 300 mm c/c. The reinforcement details are shown in Fig. 7.14.

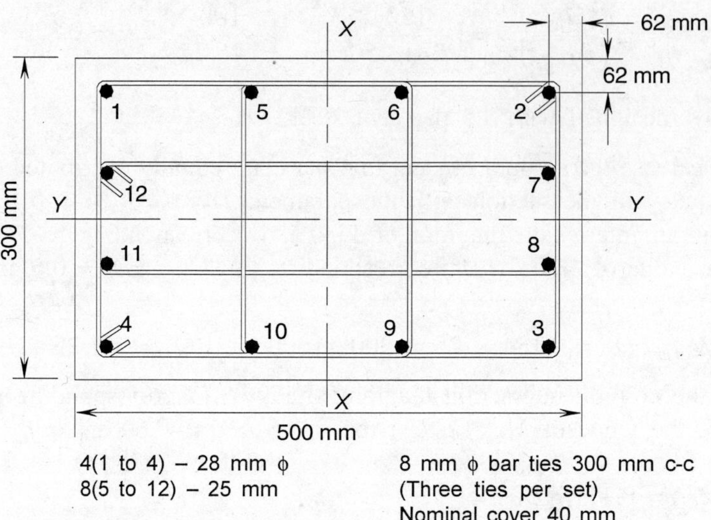

Fig.7.14 *Reinforcement details of the column of Example 7.13.*

7.7 CONCRETE WALLS CARRYING VERTICAL LOADS

A vertical load carrying member having breadth more than four times its thickness is called a **wall**. The reinforced concrete walls in buildings generally carry gravity loads. Sometimes they also carry lateral loads acting normal to the surface of the wall. If the lateral loads act in the plane of the wall, the wall is classified as shear wall. Shear walls which are commonly used in tall buildings are very effective in resisting earthquake and wind forces.

A wall with percentage of steel more than 0.40 is considered to be reinforced concrete wall, and the one having lesser steel is called **plain concrete wall**. In reinforced concrete walls the reinforcement is considered in calculating its load carrying capacity. IS:456 deals briefly with the design of ordinary reinforced concrete walls carrying vertical loads only, whereas ordinary plain walls are to be designed as per IS:1905. However, ordinary walls primarily designed for vertical loads should also be able to transmit the usual lateral or horizontal wind load to its peripheral members.

7.7.1 Classification

As in the case of columns, the reinforced concrete walls are classified as *braced* or *unbraced* according to the provisions made for carrying the lateral loads. The wall is considered as braced if all the lateral forces acting on it are resisted by the monolithically built cross walls. When less than 25 per cent of lateral load is resisted by cross-walls it is considered unbraced.

Like columns a reinforced wall is classified as *short wall*, if the ratio of effective height to thickness is less than 12, otherwise it is termed as *long* or *slender wall*. In a braced wall, slenderness is considered with respect to height only without reference to its length and slenderness effect should be considered in its design. On the other hand in case of plain and unbraced reinforced concrete walls with $p < 0.4$ per cent, the slenderness ratio is limited to 30, whereas in case of reinforced concrete walls with $(0.4 \le p < 1.0)$ and $(1.0 \le p < 4.0)$, the limiting slenderness ratios are 40 and 45, respectively.

7.7.2 Reinforcement Detailing

As per code recommendations:

1. A wall shall have minimum thickness of 100 mm.
2. The minimum amount of vertical reinforcement in plain walls is 0.12 per cent of the gross concrete area with HYSD steel bars or welded wire fabric not exceeding 16 mm ϕ.
   In case of mild steel reinforcement, minimum steel shall be 0.15 per cent of gross-cross-sectional area of concrete.
3. The vertical reinforcement may be provided in single layer in thin walls, and in two layers in the walls of thickness greater than 170.
4. The basic aim of transverse horizontal reinforcement is to restrain vertical bars against buckling.The minimum amount of horizontal reinforcement shall be 0.20 per cent of gross concrete area for HYSD bars or welded wire fabric not larger than 16 mm ϕ. For mild steel the minimum percentage of steel is 0.25.
5. The spacing of vertical and horizontal bars should not exceed 3 times the wall thickness or 450 mm whichever is less.
6. To maintain continuity of stress path it is desirable to provide two 16 mm ϕ bars around the openings. These bars are extended for the length L_d beyond the corners.
7. When vertical steel resisting compression exceeds 2 per cent, links of atleast 6 mm or one-quarter the size of largest compression bar shall be provided throughout the thickness of the wall. Spacing of these links should not exceed twice the wall thickness in either the horizontal or vertical directions and in vertical direction should not exceed 16 times the bar size. Any vertical compression bar not enclosed by a link should be within 20 mm of restrained bar. As the arrangement of links cause practical difficulties, it is recommended that the vertical reinforcement be limited to 2 per cent of the cross-section.

7.7.3 Design of Reinforced Concrete Walls

Due to similarity of its behaviour with column, a wall is designed in accordance with the recommendations given for columns. A wall subjected to axial load and moment is designed as a column subjected to uniaxial bending. In case of long or slender walls, the additional moments due to slenderness should be accounted for in the design. The design may be accomplished using P-M interaction curves for columns.

The design is based on the provisions of BS:8110. The load carrying capacity of reinforced concrete wall is calculated as in the case of columns. In the design of a concrete wall the strength of wall is increased by an amount given in Table 7.5. The increase in strength of the wall depends upon the *aspect ratio of the wall*, i.e., ratio of its height to length. The length of the wall is its overall length or in case of openings it is the length between adjacent openings. It can be seen from the table that *longer the wall compared to its height, larger is the increase in strength.*

TABLE 7.5 Increase in the Strength of Wall

Aspect ratio, storey height/length	≥ 1.5	1.0	≤ 0.5
Increase in the strength, per cent	nil	10	20

The axial load in a wall is calculated by treating the beam and slabs transmitting the load as simply supported. The load carrying capacity of various types of walls are:

Short braced reinforced concrete wall carrying axial load only.

$$P_u = 0.40 f_{ck} A_c + 0.75 A_s f_y$$

Short unbraced reinforced concrete wall carrying axial load only.

$$P_u = 0.35 f_{ck} A_c + 0.67 A_s f_y$$

Slender braced walls. A wall with slenderness ratio L_e/D limited to 40 to 45, is designed as slender compression member carrying additional moments,

$$M_{ax} = \frac{P_u D}{2000}\left(\frac{L_e}{D}\right)^2$$

Trial and modification procedure can be used for its design. For a wall reinforced with one central layer of steel, the additional moments are doubled.

Short unbraced walls. It is designed as a column carrying P_u and M_u, the minimum moment being due to an eccentricity of D/20(=0.05D) or 20 mm.

Slender unbraced wall. The slenderness ratio should not exceed 30. It is designed as in slender braced walls.

Procedure for design:

1. Determine the effective length and slenderness ratio.
2. Design the wall using the recommendations given for the columns.
3. Detail the steel.

The arrangement of reinforcement is shown in Fig. 7.15.

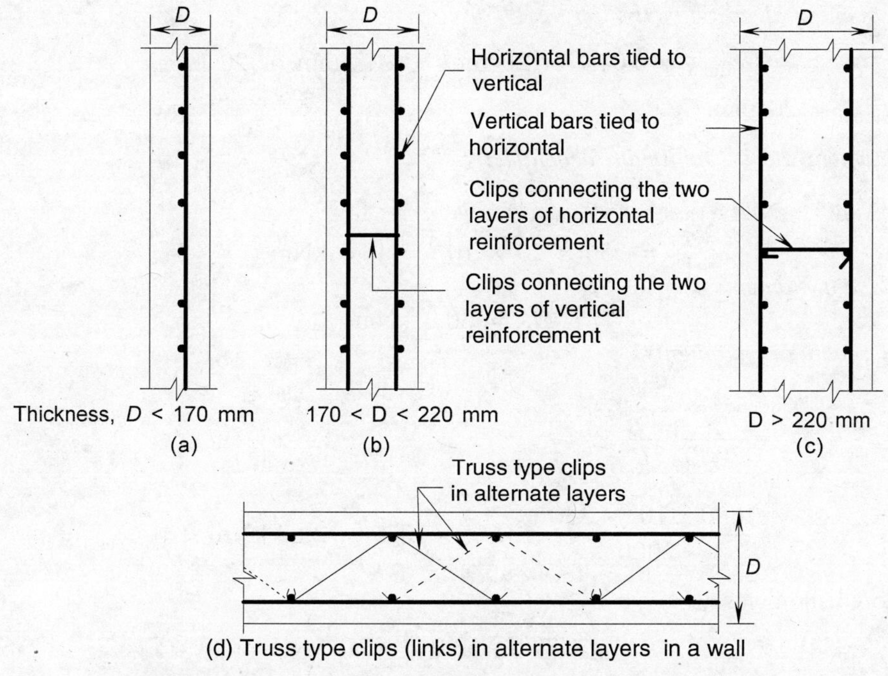

Fig. 7.15 Detailing of reinforcement in concrete walls.

Example 7.14 Design a reinforced concrete wall of 4.6 m height to support a factored load of 650 kN/m length and factored moment of 30 kNm/m at right angles to the length of wall. The distance between cross walls is 4.0 m. The materials to be used in construction are: M25 grade concrete mix and HYSD steel bars of grade Fe415.

Solution From the given data, f_{ck} = 25 MPa and f_y = 415 MPa, L_o = 4.6 m, P_u = 650 kN/m and M_u = 30 kNm/m.

Thickness of wall. Consider a moderate height-to-thickness ratio of 20, therefore,

$$D = \frac{4.6 \times 10^3}{20} = 230 \text{ mm} > 100 \text{ mm}$$

Consider

$$L_e = 0.65L_o = 0.65 \times (4.6 \times 10^3) = 2990 \text{ mm}$$

Then Slenderness ratio $\dfrac{L_e}{D} = \dfrac{2990}{230} = 13 > 12(< 40)$

The wall is slender.

Minimum accidental eccentricity.

$$e_{x,\min} = 0.05D \text{ or } 20 \text{ mm} = 11.5 \text{ mm or } 20 \text{ mm}$$

Adopt e_x, $e_{\min}$ = 20 mm.

Primary moment due to minimum eccentricity.

$$M_{ux,e\min} = P_u e_{x,\min}$$
$$= 650 \times 20 \times 10^{-3} = 13.0 \text{ kNm}$$

Primary design moments.

$$M_{ux} = 30.0 \text{ kNm}$$

Additional secondary moments.

$$M_{xa} = \frac{P_u D}{2000}\left(\frac{L_{ex}}{D}\right)^2$$

$$= \frac{650 \times 10^3 \times 230}{2000}(13.0)^2 \times 10^{-6} = 12.63 \text{ kNm} < M_{ux,e\min}$$

Total factored moments is given by,

$$M_{ux} = 30.0 + 13.0 = 43.0 \text{ kNm}$$

Consider 12 mm diameter bars uniformly spaced at nominal cover of 40 mm. Therefore,

$$d' = 40 + \left(\frac{12}{2}\right) = 46 \text{ mm and } \frac{d'}{D} = 46/230 = 0.20$$

Vertical steel. The dimensionless load parameters are:

$$\frac{P_u}{f_{ck}bD} = \frac{650 \times 10^3}{25 \times 1000 \times 230} = 0.113$$

and
$$\frac{M_{ux}}{f_{ck}bD^2} = \frac{43.0 \times 10^6}{25 \times 1000 \times 230^2} = 0.031$$

From the P-M curve for d'/D = 0.20 and Fe 415 given in SP:16:

For $\dfrac{P_u}{f_{ck}bD} = 0.113$ and $\dfrac{M_{ux}}{f_{ck}bD^2} = 0.031$, $\dfrac{p}{f_{ck}} = 0.00$

Therefore, provide minimum steel of 0.40 per cent.

$$A_s = \frac{pbD}{100} = \frac{0.40 \times 1000 \times 230}{100} = 920 \text{ mm}^2$$

Adopt 12 mm diameter bars uniformly distributed at spacing of

$$s = \frac{1000 \times 113}{0.5 \times 920} = 245.65 \text{ mm}$$

Thus, provide vertical steel in the form of 12 mm diameter bars at 245 mm c/c on both the faces. This satisfies maximum spacing requirements, i.e., less than $3D$ or 450 mm.

Horizontal steel. Provide minimum horizontal steel of 0.20 per cent in the form of 12 mm diameter bars at 450 mm c/c on both the faces. This satisfies maximum spacing requirements i.e., less than $3D$ or 450 mm.

Links. No links are required as the steel is less than 2.0 per cent.

7.8 DESIGN OF TENSION MEMBERS

In reinforced concrete members subjected to direct tension, the tension is carried primarily by the steel and concrete is considered only to provide protective cover to the steel reinforcement. In such structures the values of allowable stresses in steel and concrete are restricted so that the strains in steel and concrete are not high, consequently crack widths are limited. This provision minimizes the danger of corrosion of steel. The common example of members subjected to direct tension are hanger and ties of bow girder bridges and circular water tanks. On the other hand, in tanks with fixed bases, the walls are subjected to bending tension. The members under direct tension are designed by elastic theory and those subjected to bending tension are designed either by elastic or limit states theory.

7.8.1 Member in Direct Tension

As mentioned earlier in the absence of accurate method for calculating maximum widths of tensile cracks, the design procedure can be based on elastic theory and modular ratio which assumes uncracked section at working or characteristic loads, i.e., both steel and concrete are elastic and whole section is effective in resisting the direct tension. The design procedure consists in arriving at size of concrete member which does not crack under service loads. The reinforcement is calculated assuming that the whole tension is carried by the steel only. Though the method is not very sound and tend to give the section larger in size with higher margin of safety, but it has been found to give satisfactory performance in the field. The following provisions relate to the thin sections like walls of water tanks.

Minimum steel: A minimum amount of steel shall always be provided in two principal directions perpendicular to each other to take care of shrinkage, temperature, etc.The minimum amount is 0.3 per cent of the gross cross-sectional area of concrete for HYSD bars and 0.5 per cent for plain mild steel bars.

Spacing of steel: For member up to 170 mm thick, the reinforcement may be provided in one layer and in members over 200 mm thick, the steel should be equally divided on both faces. The maximum spacing of steel is restricted to 300 mm.

Cover to reinforcement: The cover to be provided on the surfaces not exposed to water is based on environmental conditions as stipulated in the code, but minimum cover should be 40 mm for the surfaces in contact with water. BS:5337, the british code on water retaining structures, has classified environmental exposure into the following three categories:

Class A. Exposed to wetting and drying (allowed crack width, 0.10 mm)

Class B. Exposed to continuous contact with water, e.g. walls of liquid retaining structures (allowed crack width, 0.20 mm)

Class C. Exposed only to outside air, i.e., to normal condition (allowed crack width, 0.30 mm).

TABLE 7.6 Permissible Stress in Steel Under Direct Tension and Lap Length (BS: 5337)

Type of stress	*Exposure category*	*Permissible stress, MPa (Lap length)*	
	(Grade of concrete)	*Plain bars, Fe250*	*Deformed bars, Fe415*
Flexural tension and shear	A (M25)	85 (24d)	100 (20d)
	B (M25)	115 (32d)	130 (26d)
	C (M30)	125 (35d)	140 (28d)
Compression	A to C	125	140

TABLE 7.7 Permissible Stress in Concrete in Direct Tension

Cracking condition	*Permissible direct tension, MPa*			
	M20	*M25*	*M30*	*M35*
Without cracking of concrete	—	1.31	1.44	—
With cracking of concrete, IS:456	2.8	3.2	3.6	4.0

Thickness of wall: Thickness of concrete wall in tension which does not crack under service loads can be obtained as follows:

$$\text{Equivalent area of concrete } A_e = A_c + (m - 1)A_s \tag{7.16}$$

Area of concrete required to carry the tension without cracking of concrete $A_c = T/f_{ct}$ and area of steel required to carry the tension

$$A_s = T/f_s \tag{7.17}$$

Substituting these values in Eq. (7.16), the thickness t can be obtained from:

$$A_e = bt = T\left[\frac{1}{f_{ct}} + \frac{m-1}{f_s}\right] \quad \text{or} \quad t = \left(\frac{T}{b}\right)\left[\frac{1}{f_{ct}} + \frac{m-1}{f_s}\right] \tag{7.18}$$

Procedure for design: Procedure for the design of reinforced concrete section subjected to direct tension is as follows:

1. Based on the exposure condition and grade of reinforcement to be used determine f_s and calculate the area of steel.

$$A_s = \frac{T}{f_s} \nless 0.3 \text{ or } 0.5 \text{ per cent depending upon the type of steel}$$

2. Determine size and spacing of bars, $s \leq 300$ mm.
3. Determine the dimensions of the section depending upon the stress condition in the concrete, i.e., whether the concrete is allowed to crack or the section is uncracked.

$$A_c = A_e - (m - 1)A_s = \left(\frac{T}{f_{ct}}\right) - (m - 1)A_s$$

4. Check the stress in concrete in the section selected earlier.

$$f_{ct} = \frac{T}{A_c + (m-1)A_s}$$

5. Compute the minimum secondary steel to be provided in the section,

$$A_{sd} = 0.003\, A_c \quad \text{for deformed bars}$$
$$= 0.005\, A_c \quad \text{for plain bars}$$

6. Select the size of the bar and determine its number and clear distance between the bars ($\ngtr$ 300 mm to limit the crack width).
7. Depending upon the exposure condition select the nominal or clear cover to the reinforcement.

Example 7.15 Design a reinforcement concrete section for a circular water tank wall subjected to a hoop tension of 125 kN per metre height. The materials to be used are: M30 grade concrete mix and Fe250 grade plain steel bars.

Solution

For exposure category of Class B, f_s = 115 MPa
For no cracking of concrete, f_{ct} = 1.44 MPa and m = 15

Thickness of concrete section.

$$t = \left(\frac{T}{b}\right)\left[\frac{1}{f_{ct}} + \frac{m-1}{f_s}\right] = \left(\frac{125 \times 10^3}{1000}\right)\left[\frac{1}{1.44} + \frac{15-1}{115}\right] = 102.02 \text{ mm}$$

and

$$\text{Area of steel } A_s = \frac{125 \times 10^3}{115} = 1086.96 \text{ mm}^2$$

Consider 12 mm ϕ bars at a spacing,

$$s = \frac{1000a_s}{A_s} = \frac{1000 \times 113}{1086.96} = 103.96 \text{ mm}$$

Provide 125 mm thick concrete section (minimum thickness for water retaining structure) with 12 mm ϕ rings at 100 mm c/c vertically (A_{st} = 1130 mm^2 and s < 300 mm).

Check for stress in concrete.

$$A_e = (125 \times 1000) + (15 - 1) \times (1130) = 140820 \text{ mm}^2$$

Therefore,

$$f_{ct} = \frac{125 \times 10^3}{140820} = 0.888 \text{ MPa} < 1.44 \text{ MPa}$$

Area of secondary steel. Minimium vertical steel is 0.5 per cent of cross-sectional area.

Therefore, $$A_{sd} = \left(\frac{0.50}{100}\right) \times (125 \times 1000) = 625 \text{ mm}^2$$

Provide 10 mm ϕ bars at 125 mm c/c horizontally (A_{sd} = 629 mm^2) at nominal cover of 40 mm (minimum).

7.8.2 Member Subjected to Bending and Direct Tension

Consider a reinforced concrete section of thickness t subjected to moment M and tension T acting at the mid-point of the section. Transfer the tension T to the level of steel which results in an additional moment, $T(d - 0.5t)$ and tension T at the level of steel. The resultant moment on the section is:

$$M_e = M - T(d - 0.5t)$$

or

$$\frac{M_e}{T} = \frac{M}{T} - (d - 0.5t) = e - (d - 0.5t) \tag{7.19}$$

where $(d - 0.5t)$ is the distance of steel from the mid point of the section, and $e = (M/T)$. It should be noted that:

1. When $e = (d - 0.5t)$, $M_e = 0$ and T only need be considered at the level of steel for design, i.e., $A_s = T/f_s$.
2. When $e < (d - 0.5t)$, M_e is negative, i.e., T predominates and entire section is in tension. The design of tension reinforcement on both the faces is based on allowable stresses. The reinforcement on the two faces are:

$$A_{s1} = \left(\frac{T}{2f_s}\right)\left[1 + \frac{e}{(d - 0.5t)}\right] \quad \text{and} \quad A_{s2} = \left(\frac{T}{2f_s}\right)\left[1 - \frac{e}{(d - 0.5t)}\right] \tag{7.20}$$

3. When $e > (d - 0.5t)$, the resulting moment M_e is positive and the bending effect predominates, i.e., both tension and compression occur in the section. The presence of tension reduces the original moment from M to M_e, and hence the depth of neutral axis from x to say x_1. Iterative procedure can be used to obtain the correct value of x and hence of p. Since the moment is fairly proportional to p for the given values of b, t and f_s, first trial value p_1 can be obtained by scaling down p in proportion to (M_e/M), i.e.,

$$p_1 = p\left[\frac{[e + (t/2) - d]}{e + (t/2)}\right] = p\left[1 - \frac{d}{e + (t/2)}\right] \tag{7.21}$$

and the corresponding depth of neutral axis is given by

$$x_1 = d\left(-mp + \sqrt{2mp + (mp)^2}\right) \tag{7.22}$$

and for second trial value of p,

$$p_2 = p_1\left[1 - \frac{d - (x/3)}{[e + (t/2) - (x/3)]}\right] \tag{7.23}$$

The new value of x_2 can be determined from the modified value of p_2. The process is repeated till the value of x converges, i.e., difference in values of x in two consecutive iterations is small. Once the converged value of x is reached, serviceability check on crack width can be applied.

Area of reinforcement can be obtained from:

$$A_s = pbd$$

or alternatively from

$$A_s = \frac{M_e}{f_s[d - (x/3)]} \tag{7.24}$$

The shear stress is given by

$$\tau_v = \frac{V}{bz} \quad \text{where } z = \text{Lever arm} = d - \frac{x}{3}$$

However, the method using elastic analysis and considering the predetermined section to be uncracked with restricted tensile stress in concrete gives satisfactory results. In this method, the design is based on the assumption that the maximum crack width will not exeed 0.30 mm, if allowable tension in concrete is limited to values given in Table 7.8.

TABLE 7.8 Permissible Tension in Bending for Crack Control (IS:456)

Grade of concrete	*Tension, MPa*		
	Direct	*Bending*	*Shear strength, MPa*
M20	1.2	1.7	1.7
M25	1.3	1.8	1.9
M30	1.5	2.0	2.2

Equation (7.24) can be written as:

$$\frac{M_e}{bd^2} = \left(\frac{A_s}{bd}\right) f_s \left(1 - \frac{x}{3d}\right)$$

or

$$\frac{100M_e}{bd^2} = \left(\frac{100A_s}{bd}\right) f_s \left(1 - \frac{x}{3d}\right) = pf_s \left(1 - \frac{x}{3d}\right) \tag{7.25}$$

where p is percentage of steel. From Eq. (7.25), it is evident that (M/bd^2) is fairly proportional to p.

Design procedure:

1. Calculate the depth of neutral axis of uncracked section of total depth t as follows:

$$x = \frac{t + 2(m-1)[A_s d/(bt)]}{2(m-1)[A_s/(bt) + 2]} \tag{7.26}$$

2. Calculate the maximum tensile stress f_{ct} from the following expression:

$$M = f_{ct} bh \left[\left(\frac{t-x}{3}\right) + (m-1)\left(\frac{A_s}{bt}\right)\left(\frac{d-x}{t-x}\right)\left(d - \frac{x}{3}\right)\right] \tag{7.27}$$

3. f_{ct} shall not exceed allowable tensile stress in concrete. If this condition is not satisfied, either increase the depth of the section or the area of steel.
4. Check shear stresses.

Design aids: For limit state of failure under pure tension,

$$T_u = (0.87 f_y)\left(\frac{pbt}{100}\right) \quad \text{or} \quad \frac{T_u}{f_{ck} bt} = \left(\frac{p}{f_{ck}}\right)\left(\frac{0.87 f_y}{100}\right) \tag{7.28}$$

As in case of compression members interaction diagrams of rectangular section subjected to combined bending and direct tension can be constructed by plotting $T_u/(f_{ck}bd)$ on y-axis and $M_u/(f_{ck}bd^2)$ on x-axis, with steel distributed on two or four sides. It should be noted that the charts given in SP:16 do not take cracking into account. However, these interaction diagrams simplify the computation of steel considerably.

Example 7.16 The section of a reinforced concrete wall of a rectangular tank is subjected to direct tension of 75 kN/m and moment of: (i) 90 kNm/m and (ii) 5.5 kNm/m. Design the section when the M25 grade concrete mix and HYSD steel bars of grade Fe415 are to be used under Class B exposure conditions.

Solution For the given material and exposure conditions:

f_{st} = 190 MPa, f_{cb} = 8.5 MPa, f_{ct} = 1.8 MPa and m = 11 (uncracked section)

The depth of neutral axis $x = kd$ where

$$k = \frac{1}{1 + 190/(11 \times 8.5)} = 0.33, \qquad j = 1 - \left(\frac{k}{3}\right) = 0.89$$

CASE I Eccentricity $e = M/T = (90 \times 10^6)/(75 \times 10^3) = 1200$ mm

Consider thickness of wall as 350 mm with 20 mm ϕ steel bars at a nominal cover of 40 mm. Then

$$\text{Effective depth } d = 350 - 40 - (20/2) = 300 \text{ mm}$$

To ascertain the type of section, the parameter is:

$$d - 0.5t = 300 - 0.5 \times 350 = 125 \text{ mm} < e \ (= 1200 \text{ mm})$$

Therefore, the moment dominates, and both tension and compression occur in the section.

$$\text{Equivalent moment } M_e = M - T(d - 0.5t)$$

$$= 90 - 75\ (0.300 - 0.175\ ) = 80.625 \text{ kNm/m}$$

$$\text{Flexural steel } A_{s1} = \frac{M}{jdf_s}$$

$$= \frac{80.625 \times 10^6}{0.89 \times 300 \times 190} = 1589.30 \text{ mm}^2$$

$$\text{Tension steel } A_{s2} = \frac{T}{f_s}$$

$$= \frac{75 \times 10^3}{190} = 394.74 \text{ mm}^2$$

$$\text{Total steel } A_s = A_{s1} + A_{s2}$$

$$= 1589.30 + 394.74 = 1984.04 \text{ mm}^2$$

Provide 20 mm ϕ @ 150 mm c/c ($A_s = 2094$ mm^2) on tension face. The percentage of steel is:

$$p = \frac{2094 \times 100}{1000 \times 300} = 0.698 \text{ per cent}$$

On the other face minimum stipulated steel is provided.

CASE II Consider thickness of wall as 250 mm with 20 mm ϕ bars at nominal cover of 40 mm. Then

$$\text{Effective thickness } d = 250 - 40 - \frac{20}{2} = 200 \text{ mm}$$

$$\text{Eccentricity } e = \frac{5.5 \times 10^6}{75 \times 10^3} = 73.33 \text{ mm}$$

and

$$\text{Parameter} = d - 0.5t = 200 - 0.5 \times 250 = 75 \text{ mm} > e \ (=73.33 \text{ mm})$$

Tension predominates and entire section is in tension. The section is designed as uncracked section.

Tension steel on both faces.

$$A_s = \left(\frac{1}{2f_s}\right)\left[T \pm \frac{M}{(d - 0.5t)}\right]$$

$$A_{s1} = \left(\frac{1}{2 \times 190}\right)\left[75 \times 10^3 + \frac{5.5 \times 10^6}{75}\right] = 390.35 \text{ mm}^2$$

$$A'_{s1} = \left(\frac{1}{2 \times 190}\right)\left[75 \times 10^3 - \frac{5.5 \times 10^6}{75}\right] = 4.386 \text{ mm}^2$$

Provide minimum steel on both the faces. Since the section is designed as uncracked section, check maximum tension for safety against cracking.

7.9 DESIGN OF FOUNDATIONS

The *foundation* or *substructure* is that part of the structure which is in direct contact with the soil, generally below the ground level, and transfers the loads from superstructure to the underlying strata or subsoil safely. The foundation is generally enlarged at the base to distribute the load transferred from the structure over a larger area such that the pressure on the soil does not exceed its safe or permissible bearing capacity.

The safe bearing capacity of soil is based on its strength and settlement characteristics. It is the pressure at the base of the foundation that the soil can sustain without the risk of shear failure or undergoing settlement more than the permissible value for the structure. It is obtained by using the principles of soil mechanics by dividing ultimate capacity of soil by a suitable factor of safety thus represents a serviceability condition. Therefore, the combinations of design service loads acting on footing are: (a) 1.0(*DL* + *LL*), (b) 1.0(*DL* + *WL*) and (c) 1.0*DL* + 0.8(*LL* + *WL*/*SL*). The safe bearing capacity multiplied by the area of foundation gives the *safe bearing load.*

The horizontal loads are resisted by friction, cohesion of soil and passive pressure of soil on the vertical face of foundation in contact with the soil. However, in computation of frictional resistance 0.90 times of dead load is considered. In the framed buildings with columns fixed with foundations, the horizontal load is transferred to the subsoil through fixing moments. The area of footing A_F is obtained as follows:

$$A_F = \frac{\text{Service gravity load from column or wall}}{\text{Safe bearing capacity of soil underneath}}$$

However, it should be noted that the consolidation of the soil causes the supported structure to settle. The settlement should be restricted to be within reasonable limits. Differential or unequal settlement under different parts of the structure can be minimized by providing foundation components such that the intensity of pressure on the soil is equal under different parts of the footing. Generally, the foundations are so proportioned that the centre of gravity of external load system coincides with the center of gravity of loaded area to obtain uniform pressure on the base of the foundation. A non-uniform base pressure can lead to differential settlement which may result in tilting of column.

In the foundation subjected to vertical load P and moment M, i.e., P acting at an eccentricity e (= M/P), the base pressure under the footing is non-uniform. In case of small eccentricity, the centre of gravity of foundation itself may be offset from the line of action of P for uniform pressure on the base. An eccentricity in the range of $L/10$ to $L/12$ may be considered to be small.

The types of footings commonly used in general practice are illustrated in Fig. 7.16. *Spread footings* are designed to distribute a heavy load over a large area of soil in order to reduce the intensity of load at the base of the footing so that the soil will safely support the structure. These footings, when provided for independent columns, are called *isolated footings,* as illustrated in Figs.7.16(a) and 7.16(b). The isolated footing may be circular, square or rectangular in plan. The foundation for a structure may be composed of many isolated footings.

Combined footing or *multiple column footing* is a single footing provided for two or more columns, and is proportioned in such a way that the centroid of the column loads coincides with the centroid of footing slab. Depending upon the shape in plan, they are called *rectangular, trapezoidal and strap footings.*

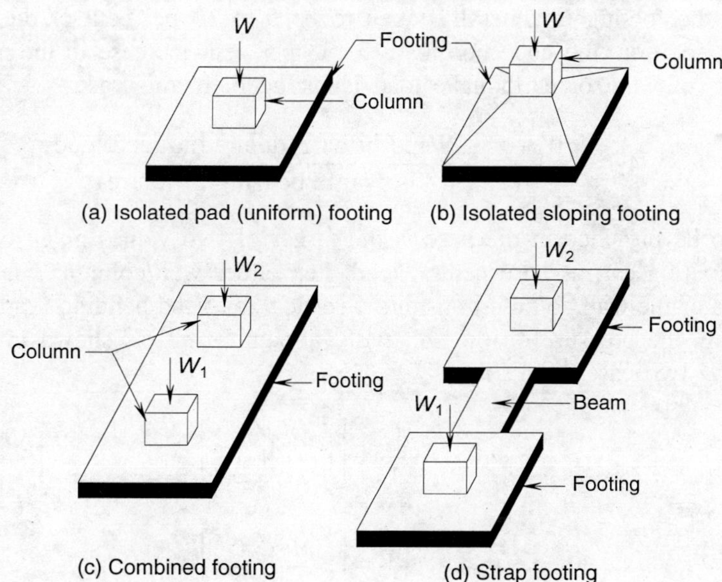

Fig. 7.16 Common types building footing.

Distribution of load to the soil: It is well established that the pressure distribution beneath the footing is not uniform even under symmetrically loaded footing, as shown in Fig. 7.17. The

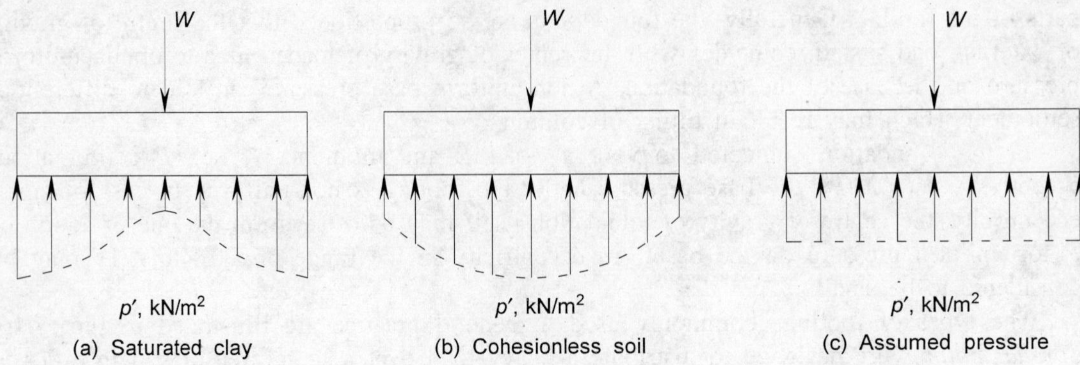

Fig. 7.17 Effect of type of soil on pressure distribution on the base of rigid footing.

pressure distribution on the soil under footing depends on types of soil, footing and applied load. However, it is customary to disregard non-uniformities as the numerical values of pressure are uncertain and highly variable. For a concentrically loaded column, the footing area A_F required is calculated from:

$$A_F = \frac{\text{Load carried by the column } W + \text{Self-weight of the footing } w}{\text{Allowable bearing pressure } p_o} \tag{7.29}$$

Weight of the footing is generally taken to be 5 to 10 per cent of the load imposed by the column. In addition, the code permits a 33.33 per cent increase in the allowable pressure when the effect of wind or earthquake load is included. In this case,

$$A_F = \frac{W + w + \text{Wind or earthquake induced load}}{(1.33)(\text{Allowable bearing pressure})} \tag{7.30}$$

Footing area to be provided at the base is the larger of two values given by Eqs. (7.29) and (7.30). If the footing carries an eccentric load, i.e., either wall/column is not concentric with the footing area or the wall/column transmits a vertical load and bending moment at its junction with the footing, the pressure on the soil will vary uniformly as shown in Fig.7.18 and one of the following two cases will arise:

$$p_{\max} = \frac{W_t}{A} + \left[\left(\frac{M}{I}\right)\left(\frac{L}{2}\right)\right]$$

$$p_{\min} = \frac{W_t}{A} - \left[\left(\frac{M}{I}\right)\left(\frac{L}{2}\right)\right] \tag{7.31}$$

where W_t (= $W + w$) and M are the total vertical load and imposed moment, respectively, and L is the length of the footing along the direction of eccentricity.

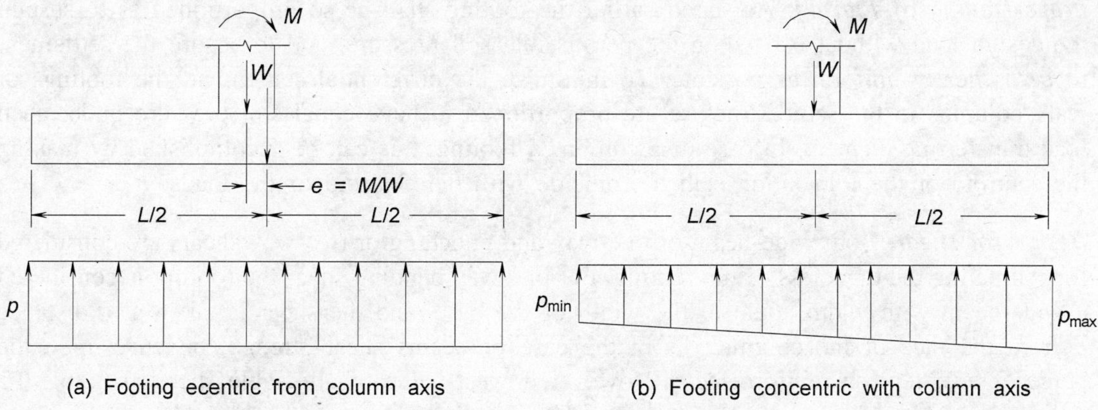

Fig. 7.18 Effect of eccentricity of load on pressure distribution on the base of rigid footing.

7.9.1 General Design Considerations

1. The thickness at the edge of footing shall not be less than 150 mm for footings on soil to minimize corrosion that can be caused by ground salts and 300 mm for the footing slab above the top of piles or the footing on piles.
2. Since the column transfers the load on top of footing by bearing, the bearing pressure shall not exceed $0.45f_{ck}$ when bearing or supporting area is of same size as loaded area. However, when supporting area, i.e., area of footing A_1 is larger than loading area, i.e., area of column base A_2, the above bearing pressure shall be increased by $\sqrt{(A_1/A_2)} \not> 2.0$. The area A_1 is taken as that of lower base of the largest frustrum of a pyramid or cone contained wholly within the footing with area A_2 on the top with a side slope 1 in 2. If these stresses are exceeded, the transfer of forces should be accomplished by the reinforcement bars extended into the footing or by the providing dowel (starter) bars extending into column a distance of L_d (47ϕ for M20 grade concrete and HYSD steel of grade Fe415). If the depth required to provide a distance L_d is very large, it is economical to provide pedestal or sloping footing.
3. Area of extended longitudinal bars or dowels (minimum of four bars with $\phi >(\phi + 3$ mm) shall be greater than 0.5 per cent of cross-sectional area of supporting column or pedestal.
4. The rectangular columns should preferably be provided with rectangular footings with base projecting equal distance beyond column faces.
5. The steel shall not be less than the minimum specified for slabs, i.e., 0.12 per cent for Fe415 and 0.15 per cent for Fe250 grade steels.
6. The structural design is carried out for factored load using limit states design method. The design basically consists of determining the thickness of footing and its reinforcement. The thickness should be sufficient to resist shear force without shear reinforcement and bending without compression reinforcement.

Proportioning of footing: As stated earlier the footing shall be so proportioned as to sustain the design load without exceeding the permissible soil pressure, and to ensure the settlement to be as nearly uniform as possible. To minimize the differential settlement, the footings of walls/columns in the same structure are proportioned to have equal soil pressure under each load transferring element. In case of a combined footing, this can be accomplished by making the centroid of the foundation slab to coincide with the centroid of the loads.

Design for shear: Both wide-beam or one-way and punching or two-way shears are considered for estimating the thickness of the footing. For one-way bending shear the footing is considered a wide-beam with width equal to the width of the base, and the shear is taken at a distance d from the face of the column. As in the case of beams shear strength depends upon the percentage of tension reinforcement. However, lowest value of allowable shear of 0.36 MPa may be considered for one-way shear action for preliminary design. Punching or two-way shear indicates the tendency of column to punch through the footing slab. The punching shear is resisted along the surface of a truncated prism around the column load called **critical perimeter** at distance $d/2$ from the column face. The maximum value of shear stress as stipulated by IS:456 is $0.25f_{ck}$.

Design for bending moment: The design bending moment is computed at:

1. The face of column for the footings supporting reinforced concrete columns.
2. Half-way between centre line and edge of wall for footings under masonry walls.

The moments are considered in both x- and y- directions. Since, for designing a footing is considered as a slab, the recommendations for minimum reinforcement in slabs shall apply. The depth of footing should be adequate for: (a) one-way and two-way shears without shear reinforcement, (b) bending moment without compression reinforcement and (c) developing transfer bond length by main reinforcement or dowel bars.

The reinforcements required in x- and y- directions shall be distributed across the cross-section of footing as follows: (a) In one-way and wall footings, the main steel is distributed uniformly across the full width, (b) In two-way square footings, steel is distributed across the full width in both directions (c) In two-way rectangular footings, the steel in long direction is placed uniformly over the full width, while in short direction major portion of steel is placed at equal spacing in the column band equal to the width of footing. The remaining steel is placed at uniform spacing outside the column band. The amount of reinforcement placed in column band is given by

$$A_{cb} = \frac{2A_{st}}{(L/B)+1}$$

where A_{st} is the total reinforcement in the short direction and L/B is the ratio of length to breath of footing.

The sizes of bars are so chosen that the full development length can be provided in the available dimension of the footing, i.e., L_d should be less than one-half of lengths of footing in the two directions. The bars in the footing can be bend at 90° at the end.

7.9.2 Design Procedure

The following procedure may be adopted for the design of column footings:

1. For the given materials, calculate the design constants.
2. Consider the self-weight of the footing in the range of 5 to 10 per cent of the load carried by the column. Determine the area of the footing from the service loads transferred from the column and self weight of footing, and safe bearing capacity of soil. Decide a suitable layout and determine dimensions of the required type of footing.
3. Calculate the upward factored pressure on the footing exerted by the underlying soil.
4. Estimate the thickness of footing slab required from: (a) bending moment considerations, (b) shear considerations: (i) by treating the footing as a wide-beam and (ii) by assuming two-way action of footing.

   Adopt the largest value of the depth obtained in steps (a), b(i) and b(ii).
5. Check for self weight of footing.
6. Calculate reinforcements in *x*- and *y*- directions from bending moment considerations. These steels shall not be less than the minimum specified for slabs i.e., 0.12 per cent for Fe415 and 0.15 per cent for Fe250 grade steels.
7. Choose diameter of bars for the required development length and distribute steel as specified in the code.
8. Draw reinforcement details.

7.9.3 Spread Footings for Walls

The spread footings provided under masonry or concrete walls carrying direct vertical loads may be designed either of plain concrete or of reinforced concrete. A wall footing essentially deflects upward in transverse direction as cantilever (one-way action) and a strip of unit width along its length is considered for its design. The concrete walls of high rigidity develop flexural tension cracks under the face of the wall as shown in Fig.7.19(a). Whereas for masonry walls of smaller rigidity, maximum moment is computed at the critical section half-way between the middle and edge of the wall as shown in Fig.7.19(a). A wall footing is designed for bending moment, shear and development length. The design bending moment and shear force for spread footing per unit length of wall are determined as follows:

1. Concrete wall $M_u = \dfrac{p'_u(B - b_w)^2}{8}$ (7.32)

2. Masonry wall $M_u = \dfrac{p'_u(0.5B - 0.25b_w)^2}{2} = \dfrac{p'_u(B - 0.5\, b_w)^2}{8}$ (7.33)

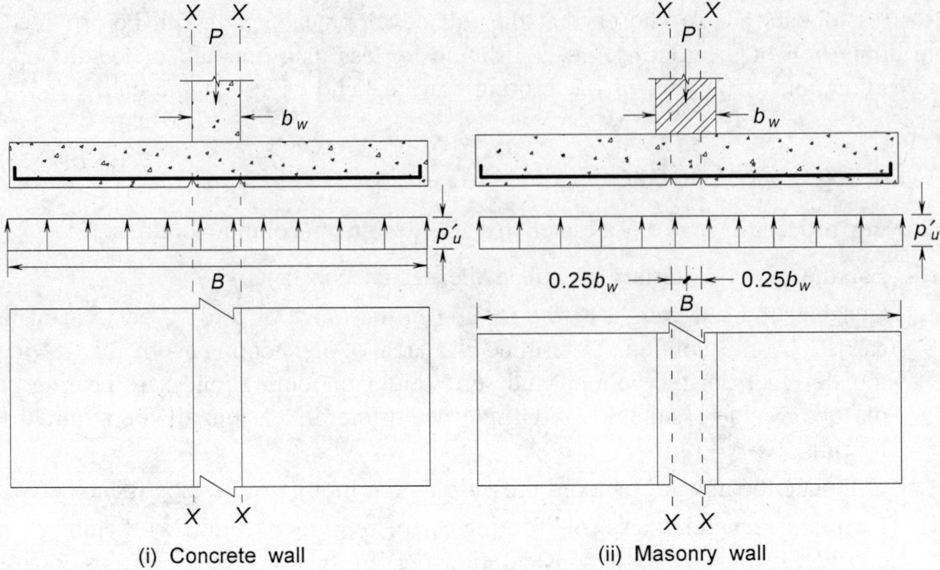

(i) Concrete wall (ii) Masonry wall

(a) Critical section x-x for maximum moment in wall footing.

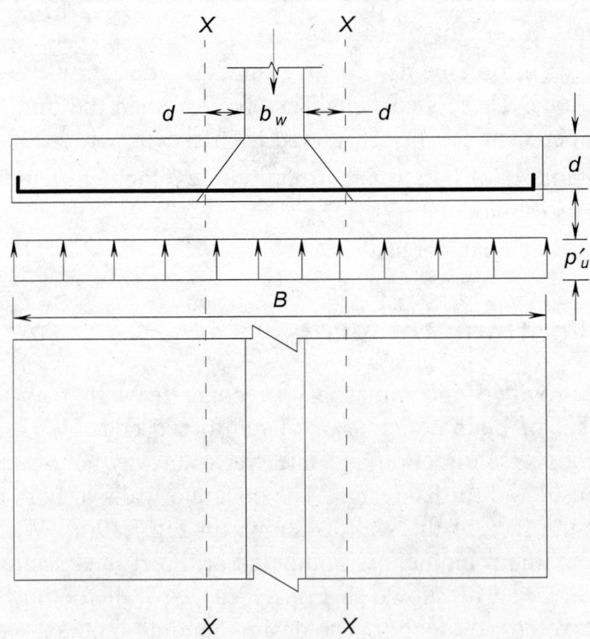

(b) Critical section for shear in wall footing.

Fig. 7.19 *Critical sections for computation of design moment and shear in a wall footing.*

The critical section for shear is located at a distance d, the effective depth of footing slab, from the face of the wall as shown in Fig. 7.19(b). The design shear force is computed as:

$$V_u = p'_u\left[\frac{(B-b_w)}{2} - d\right] \tag{7.34}$$

where B and b_w are the width of footing slab and thickness of the wall, respectively, and p'_u is the intensity of net upward soil pressure at the critical section. The following example will illustrate the procedure for the design of reinforced concrete wall footings:

Example 7.17 A one brick, i.e., 230 mm thick masonry wall is to be provided with a reinforced concrete footing on a site having soil with safe bearing capacity, unit weight and angle of repose of 125 kN/m^2, 17.5 kN/m^3 and 30°, respectively. The construction material specified are the concrete of grade M20 and HYSD steel of grade Fe415. Design the footing when the wall supports at service state: (a) a load of 150 kN/m length and (b) a load of 140 kN/m and a moment of 13.5 kNm/m.

Solution The design stipulations are:

b_w = 230 mm, p_o = 125 kN/m^2, γ = 17.5 kN/m^3, ϕ = 30°, f_{ck} = 20 MPa and f_y = 415 MPa

Depth of footing.

$$h = \left(\frac{p_o}{\gamma}\right)\left(\frac{1-\sin\phi}{1+\sin\phi}\right)^2$$

$$= \left(\frac{125}{17.5}\right)\left(\frac{1-\sin 30°}{1+\sin 30°}\right)^2 = 0.794 \approx 1.0\ m$$

Adopt the depth of footing of 1.0 m.

CASE I ***Wall carrying vertical load only:***

Width of footing. Consider one metre length of the wall. The width of footing is given by

$$B = \frac{\text{Load on the wall } w + \text{Self weight of footing @ 10\% of } w}{p_o}$$

$$= \frac{(1.0+0.1)w}{p_o} = \frac{1.1\times150}{125} = 1.32 \text{ m} \approx 1.40 \text{ m}$$

Provide 1.4 m wide footing. Net upward factored soil pressure is:

$$p'_u = \frac{1.5\times150}{1.4} = 160.71 \text{ kN/m}^2$$

Thickness of footing. The thickness of footing is governed by the magnitude of bending moment and shear force acting on it. The critical section for bending moment shall be located half-way between middle and edge of the wall.

$$M_u = \frac{p'_u(0.5B - 0.25b_w)^2}{2}$$

$$= \frac{160.71 \times (0.5 \times 1.4 - 0.25 \times 0.23)^2}{2} = 33.17 \text{ kNm/m}$$

Consider a balanced section:

$$M_u = M_{u,\text{lim}} = 0.1388 f_{ck} b d^2 = 33.17 \times 10^6$$

Therefore,

$$\text{Depth for flexure } d = \sqrt{\frac{33.17 \times 10^6}{0.1388 \times 20 \times 1000}} = 109.31 \text{ mm}$$

The factored shear V_u at the critical section at a distance d from the face of the wall is considered to be resisted without shear reinforcement. Thus,

$$V_u = p'_u \left[\frac{B - b_w}{2} - d \right] = \tau_{uc} b d \tag{7.35}$$

where τ_{uc} is the shear strength of concrete in the footing slab which may be taken as its minimum value of 0.36 MPa (for $p_t \leq 0.25$ per cent) since flexural steel is normally small due to large depth of footing slab based on the shear consideration. Consider the multiplying factor k_s to be 1.2 for the assumed total depth of 200 mm. Thus, $k_s \tau_{uc} = 1.2 \times 0.36 = 0.432$ MPa. Therefore, Eq. (7.35) can be written as:

$$160.71 \times 10^3 \times \left[\frac{1.4 - 0.23}{2} - \frac{d}{1000} \right] = 0.432 \times 1000d$$

or

$$d = 158.62 \text{ mm}$$

Consider 12 mm ϕ bars at a nominal or clear cover of 40 mm, the overall depth D is given by

$$D = 158.62 + 40 + \frac{12}{2} = 204.62 \text{ mm}$$

Adopt 210 mm thick footing slab with d = 164 mm.

Reinforcement. The flexural reinforcement parameter is:

$$R_u = \left(\frac{4.6 M_u}{f_{ck} b d^2} \right)$$

$$= \left(\frac{4.6 \times 33.17 \times 10^6}{20 \times 1000 \times 164^2} \right) = 0.2837$$

and

$$A_{st} = \left(\frac{20 \times 1000 \times 164}{2 \times 415}\right) \times \left[1 - \sqrt{1 - 0.2837}\right] = 606.79 \text{ mm}^2/\text{m}$$

Minimum steel required is:

$$A_{st,\min} = \left(\frac{0.85}{f_y}\right) bd = \left(\frac{0.85}{415}\right) (1000 \times 164) = 335.9 \text{ mm}^2/\text{m} < A_{st}$$

Provide 12 mm ϕ bars @ 180 mm c/c (≤ $3d$ or 450 mm whichever is smaller) with A_{st} = 628 mm^2/m. With side face nominal cover of 25 mm, the length available for embedment shall not be less than L_d (= 47.0ϕ), i.e.,

$$0.5B - 0.25b_w - 25 \geq L_d$$

$$0.5 \times 1400 - 0.25 \times 230 - 25 \; (= 617.5 \text{ mm}) \not< 47 \times 12 \; (564 \text{ mm})$$

Hence, the adequate length is available for embedment. However, 90° bends may be provided at the end of the bars. Distribution steel is:

$$A_{st} = \frac{0.12}{100} \times 1000 \times 210 = 252 \text{ mm}^2/\text{m}$$

Provide 8 mm ϕ @ 190 mm c/c (A_{st} = 265 mm^2/m).

Check for shear.

$$p_t = \frac{100A_{st}}{bd}$$

$$= \frac{100 \times 628}{1000 \times 164} = 0.383 \text{ per cent}$$

For p_t = 0.383 per cent, τ_{uc} = 0.429 MPa and for D = 210 mm, k_s = 1.18. Thus,

$$k_s\tau_{uc} = 1.18 \times 0.429 = 0.506 \text{ MPa}$$

Shear at the critical section V_u = 160.71 × [0.5 (1.4 – 0.23) – 0.164] = 67.66 kN and

$$\text{Nominal shear } \tau_v = \frac{V_u}{bd}$$

$$= \frac{67.66 \times 10^3}{1000 \times 164}$$

$$= 0.4126 \text{ MPa} < k_s\tau_{uc} (= 0.506 \text{ MPa})$$

Therefore, no shear reinforcement is required. The reinforcement details are shown in Fig. 7.18(a).

CASE II *Wall carrying axial load and moment.* In this case the footing may be proportioned such that the resultant of axial load and moment passes through the centre of footing at the base of wall resulting in uniform soil pressure. On the other hand, if the centre of footing coincides with the axial load, it results in non-uniform soil pressure.

(a) When resultant of applied forces passes through centre of footing

As in the previous case the base of footing can be located at 1.0 m depth below ground level. The eccentricity of the resultant of applied forces, $e = M/W = 13.5/140 = 0.096$ m.

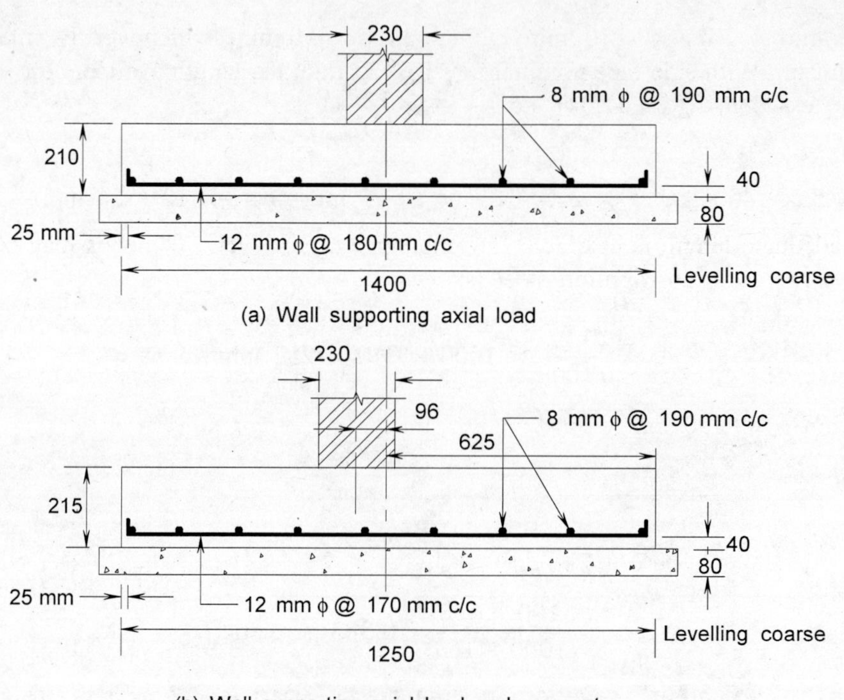

Fig. 7.20 *Detailing of concrete footing for the brick wall of Example 7.17.*

From Eqn. (7.29).

$$\text{Width of footing } B = \frac{1.1W}{p_o}$$

$$= \frac{1.1 \times 140}{125} = 1.232 \text{ m}$$

Consider 1.25 m wide footing. Net upward factored soil pressure is:

$$p'_u = \frac{1.5 \times 140}{1.25} = 168 \text{ kN/m}^2$$

Thickness of the footing. The bending moment at critical section is:

$$M_u = \frac{p_u'(0.5B - 0.25b_w + e)^2}{2}$$

$$= \frac{168 \times (0.5 \times 1.25 - 0.25 \times 0.23 + 0.096)^2}{2} = 36.98 \text{ kNm/m}$$

Consider a balanced section, $M_u = M_{u,\text{lim}} = 0.1388 f_{ck} bd^2$
Depth from flexural considerations is:

$$d = \sqrt{\frac{36.98 \times 10^6}{0.1388 \times 20 \times 1000}} = 115.42 \text{ mm}$$

As computed earlier,

$$k_s \tau_{uc} = 1.2 \times 0.36 = 0.432 \text{ MPa}$$

Equating the critical shear force to the shear resisting capacity of concrete, we get

$$V_u = p_u' \left[\left(\frac{1}{2}\right)(B - b_w) + e - d \right] = k_s \tau_{uc} bd$$

or

$$168 \times \left[\frac{1.25 - 0.23}{2} + 0.096 - \frac{d}{1000} \right] \times 10^3 = 0.432 \times 1000 \times d$$

Therefore, $d = 169.68$ mm.
Consider 12 mm ϕ bars at a nominal cover of 40 mm. The overall depth D is given by

$$D = 169.68 + 40 + (12/2) = 215.68 \text{ mm}$$

Adopt 215 mm thick footing slab with $d = 169$ mm.
Reinforcement. The flexural reinforcement parameter is:

$$R_u = \left(\frac{4.6 M_u}{f_{ck} bd^2} \right)$$

$$= \frac{4.6 \times 36.98 \times 10^6}{20 \times 1000 \times (169)^2} = 0.2978$$

Then

$$A_{st} = \left(\frac{20 \times 1000 \times 169}{2 \times 415} \right) \times \left[1 - \sqrt{1 - 0.2978} \right] = 659.82 \text{ mm}^2\text{/m}$$

$$A_{st} \not< A_{st,\min} = \left(\frac{0.85}{f_y} \right) bd = \left(\frac{0.85}{415} \right) (1000 \times 169) = 346.145 \text{ mm}^2$$

Provide 12 mm ϕ bars @ 170 mm c/c (A_{st} = 665 mm^2). With side face nominal cover of 25 mm, the available embedment length from the section of maximum moment i.e., face of wall is:

$$(0.5B - 0.25b_w + e) - 25 = (0.5 \times 1250 - 0.25 \times 230 + 96) - 25$$

$$= 638.5 \text{ mm} > L_d (= 47\ \phi = 47 \times 12 = 564 \text{ mm})$$

Hence adequate embedment length is available, i.e., slab is safe in bond.

$$\text{Distribution steel } A_{st} = 0.12 \times \frac{1000 \times 215}{100} = 258 \text{ mm}^2$$

Provide 8 mm ϕ bars @ 190 mm c/c (A_{st} = 265 mm^2).
Check for shear.

$$p_t = \frac{100A_{st}}{bd} = \frac{100 \times 665}{1000 \times 169} = 0.393 \text{ per cent}$$

For p_t = 0.393 per cent, τ_{uc} = 0.433 MPa and for D = 215, k_s = 1.17. Therefore,

$$k_s \tau_{uc} = 1.17 \times 0.433 = 0.5066 \text{ MPa}$$

Shear capacity of concrete is obtained as:

$$V_{uc} = (0.5066 \times 1000 \times 169) \times 10^{-3} = 85..62 \text{ kN/m}$$

Shear at critical section is:

$$V_u = 168 \times \left[\frac{1.25 - 0.23}{2} + 0.096 - \frac{169}{1000}\right] = 73.416 \text{ kN/m}$$

Since $V_u < V_{uc}$, shear reinforcement is not required. The reinforcement details are shown in Fig. 7.20(b).
(b) When centre of footing coincides with the axial load

Width of footing. The width of footing in this case can be obtained by equating maximum soil pressure to the bearing capacity of soil, i.e.,

$$\frac{1.1W}{1.0 \times B} + \frac{6M}{1.0 \times B^2} = p_o$$

or

$$\frac{1.1 \times 140}{B} + \frac{6 \times 13.5}{B^2} = 125$$

or

$$B^2 - 1.232B - 0.648 = 0 \text{ or } B = 1.63 \text{ m}$$

Consider width of footings B to be 1.65 m. Intensities of net upward factored soil pressure are:

$$p_{u,\max},\ p_{u,\min} = 1.5\left(\frac{W}{B} \pm \frac{6M}{B^2}\right)$$

$$= 1.5 \times \left[\frac{140}{1.65} \pm \frac{6 \times 13.5}{(1.65)^2}\right]$$

$$= 171.9 \text{ kN/m}^2 \text{ and } 82.64 \text{ kN/m}^2$$

Intensity of soil pressure at the critical section located at the point half-way between middle and edge of the wall is:

$$p'_{u,b} = p'_{u,\max} - (p'_{u,\max} - p'_{u,\min}) \times \frac{0.5B - 0.25b_w}{B}$$

$$= 130.38 \text{ kN/m}^2$$

Thickness of footing. Design moment

$$M_u = \frac{p'_{u,b}(0.5B - 0.25b_w)^2}{2} + (p'_{u,\max} - p'_{u,b})\ \frac{(0.5B - 0.25\,b_w)^2}{3}$$

$$= 130.38 \times \frac{(0.5 \times 1.65 - 0.25 \times 0.23)^2}{2}$$

$$+ (171.9 - 130.38) \times \frac{(0.5 \times 1.65 - 0.25 \times 0.23)^2}{3}$$

$$= 46.55 \text{ kNm}$$

Effective depth from flexural considerations is obtained as:

$$d = \sqrt{\frac{46.55 \times 10^6}{0.1388 \times 20 \times 1000}} = 129.49 \text{ mm}$$

and

$$V_u = \frac{1}{2}(p'_{u,\max} + p'_{u,s})\ [0.5(B - b_w) - d]$$

where $p'_{u,s}$ is the factored soil pressure at critical section for the shear which is given by

$$p'_{u,s} = p'_{u,\max} - (p'_{u,\max} - p'_{u,\min})\ \frac{0.5(B - b_w) - d}{B}$$

$$= 171.9 - (171.9 - 82.64) \times \frac{0.5(1.65 - 0.23) - d}{1.65}$$

$$= 133.49 + 54.097d$$

Therefore,

$$V_u = 0.5[171.9 + 133.49 + 54.097d](0.71 - d)$$

$$= (152.7 + 27.049d)(0.71 - d) = -27.049d^2 - 133.495d + 108.417$$

Equate shear capacity of concrete, $k_s\tau_{uc}(bd)$ to the critical shear force V_u. Consider $k_s\tau_{uc}$ = 0.432 MPa as in the previous case, i.e.,

$$-27.049d^2 - 133.459d + 108.417 = 0.432 \times 1000 \times d$$

or

$$-27.049d^2 - 565.459d + 108.417 = 0$$

or

$$d^2 + 20.9d - 4.0 = 0 \text{ or } d = 0.1897 \text{ m} = 189.7 \text{ mm}$$

$\therefore$ Overall depth $D = 189.7 + 40 + \left(\frac{12}{2}\right) = 235.7$ mm

Adopt D = 240 mm with effective depth of 194 mm.

Reinforcement. The flexural steel parameter is given by

$$R_u = \left(\frac{4.6M_u}{f_{ck}bd^2}\right)$$

$$= \frac{4.6 \times 46.55 \times 10^6}{20 \times 1000 \times 194^2} = 0.2845$$

and

$$A_{st} = \left(\frac{f_{ck}bd}{2f_y}\right)\left[1 - \sqrt{1 - R_u}\right]$$

$$= \left(\frac{20 \times (1000 \times 194)}{2 \times 415}\right) \times \left[1 - \sqrt{1 - 0.2845}\right]$$

$$= 720.5 \text{ mm}^2\text{/m} \not< \left(\frac{0.85}{f_y}\right)(bd) \quad [= 397.35 \text{ mm}^2\text{/m}]$$

Provide 12 mm ϕ bars @ 150 mm c/c (A_{st} = 754 mm^2).

$$\text{Embedment length} = 0.5 \times 1650 - 0.25 \times 230 - 25 = 742.5 \text{ mm} > L_d \ (=47\phi = 564 \text{ mm})$$

Hence adequate embedment length is available, i.e., slab is safe in bond. Distribution steel is:

$$A_{st} = 0.12 \times \frac{1000 \times 240}{100} = 288 \text{ mm}^2\text{/m}$$

Provide 8 mm ϕ bars @ 170 mm c/c (A_{st} = 296 mm^2).

Check for shear.

$$p_t = \frac{754 \times 100}{1000 \times 194} = 0.389 \text{ per cent}$$

For $p_t = 0.389$ per cent, τ_{uc} is 0.432 MPa and k_s for $D = 240$ mm is 1.12, therefore,

$$V_{uc} = (1.12 \times 0.432) \times (1000 \times 194) \times 10^{-3} = 93.86 \text{ kN/m}$$

and

$$V_u = -\ 27.049d^2 - 133.495d + 108.417 = 81.5 \text{ kN/m} < V_{uc}$$

Hence the footing is safe in shear. It may be noted that the footing designed by coinciding its centre with the resultant of axial load and moment is economical.

7.9.4 Design of Isolated or Independent Footing

The following example will illustrate the procedure for the design of an isolated footing.

Example 7.18 Design an isolated pad footing for a reinforced concrete column of size 300 × 350 mm carrying an axial load of 750 kN. The safe bearing capacity of the soil at site is 175 kN/m^2 with angle of repose of 30°. The materials to be used are: concrete of grade M20 and HYSD steel bars of grade Fe415. The unit weight of soil at the site is 19.5 kN/m^3.

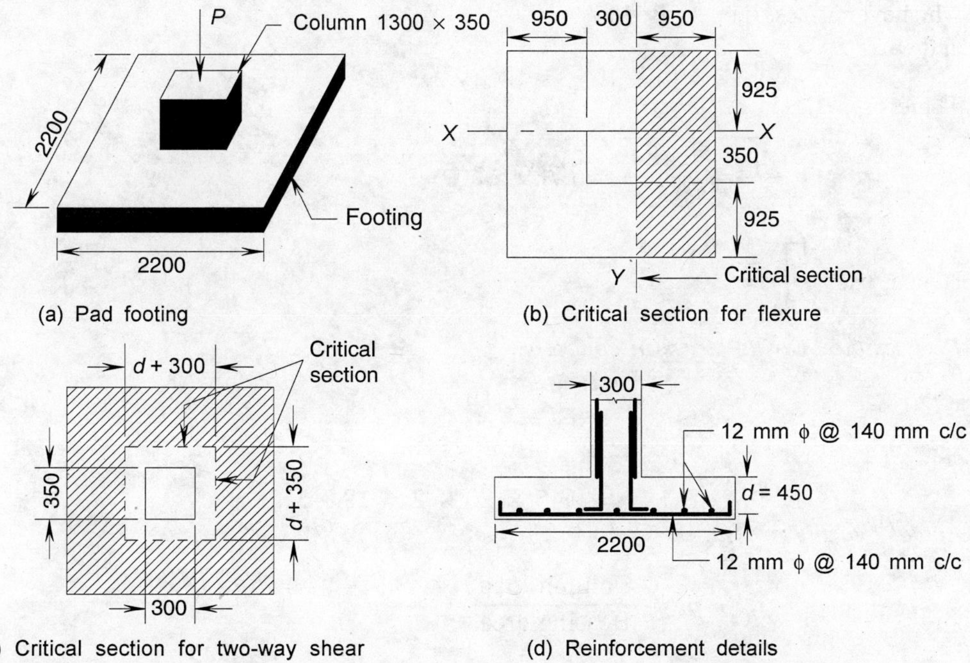

Fig. 7.21 Design and detailing of the isolated footing of Example 7.18.

Solution Design constants for M20 concrete and Fe415 steel are:

$$f_{ck} = 20 \text{ MPa}, f_y = 415 \text{ MPa}, x_{u,\max} = 0.4791d$$

$$p_{t,\lim} = 0.961 \text{ per cent and } M_{u,\lim} = 0.1388 f_{ck} bd^2$$

Depth of footing. The depth of footing h is given by

$$h = \left(\frac{p_o}{\gamma}\right)\left(\frac{1-\sin\phi}{1+\sin\phi}\right)^2 = \frac{175}{19.5} \times \left(\frac{1-\sin 30^\circ}{1+\sin 30^\circ}\right)^2 = 0.997 \text{ m} \approx 1.0 \text{ m}$$

Size of footing.

Axial load on the column $W' = 750$ kN

Dead load of footing (consider @ 10 per cent of column load) $w = 75$ kN

Total load on the soil $W = W' + w = 825$ kN

Therefore,

$$\text{Area of footing required } A_F = \frac{825}{175} = 4.714 \text{ m}^2$$

Provide square footing of size 2.2 × 2.2 m ($A = 4.84$ m^2).

Transfer of load at the base of the column. Consider the thickness of the footing to be 500 mm.

$$A_1 = (2.2 \times 2.2) = 4.84 \text{ m}^2$$

$$\text{or } (0.30 + 4 \times 0.5) \times (0.35 + 4 \times 0.5) = 5.405 \text{ m}^2$$

whichever is less i.e. A_1 is 4.84 m^2

and

$$A_2 = 0.30 \times 0.35 = 0.105 \text{ m}^2$$

Therefore,

$$\sqrt{\frac{A_l}{A_2}} = \sqrt{\frac{4.84}{0.105}} = 6.789 > 2$$

Hence, use $\sqrt{\frac{A_1}{A_2}} = 2.0$.

Permissible bearing stress in concrete.

$$p_c = 0.45 f_{ck} \sqrt{\frac{A_1}{A_2}}$$

$$= 0.45 \times 20 \times 2 = 18.0 \text{ MPa}$$

Actual bearing pressure.

$$p_a = \frac{\text{Column load}}{\text{Bearing area}}$$

$$= \frac{750 \times 10^3}{300 \times 350} = 7.14 \text{ MPa} < 18.0 \text{ MPa}$$

Upward soil pressure. Net upward factored pressure is given by

$$p'_u = \frac{\text{Factored load on column}}{\text{Area of footing}}$$

$$= \frac{1.5 \times 750}{2.2 \times 2.2} = 232.44 \text{ kN/m}^2$$

Depth of footing from shear considerations. The permissible shear stress in the concrete of footing slab is given by $k_s \tau_{uc}$.
where

$$k_s = 0.5 + \beta_c = 0.5 + \frac{\text{Short side of column}}{\text{Long side of column}}$$

$$= 0.5 + \frac{300}{350} = 1.357 > 1.0$$

Therefore, $k_s = 1.0$ and

$$\tau_{uc} = 0.25\sqrt{f_{ck}}$$

$$= 0.25 \times \sqrt{20} = 1.118 \text{ MPa } (= 1118 \text{ kN/m}^2)$$

Thus, Permissible shear stress $k_s \tau_{uc} = 1.118$ MPa $= 1118$ kN/m^2
The effective depth of the footing is to be so adopted that $\tau_v \le k_s \tau_{uc}$.

One-way shear (wide beam action). Consider d to be the effective depth of footing in metres and the critical section is at distance d from the face of the column on the side giving a greater shear area. Then

$$\text{Shear force } V_u = 2.2 \times (0.95 - d) \times 232.44$$

Equating nominal shear stress $\tau_v = V_u/bd$ to the permissible shear stress $k_s \tau_{uc}$, i.e.,

$$2.2\ (0.95 - d) \times 232.44 = 1118 \times (\text{d} \times 2.2)$$

Therefore,

$$d = 0.164 \text{ m} = 164 \text{ mm}$$

Two-way shear (two-way action): Shear force is:

$$V_u = [2.2^2 - (0.3 + d) \times (0.35 + d)] \times 232.44$$

and

$$\text{Shear force resisted by concrete } V_{uc} = \tau_{uc} bd$$

where b is the periphery of the section under consideration which is given by

$$b = 2[(0.3 + d) + (0.35 + d)]$$

Thus, for V_u to be equal to V_{uc},

$$[(2.2)^2 - (0.105 + 0.65d + d^2)] \times 232.44 = 1118 \times 2 \times [(0.3 + d) + (0.35 + d)\} \times d$$

or

$$d^2 + 0.34d - 0.234 = 0$$

Therefore,

$$d = 0.343 \text{ m} = 343 \text{ mm}$$

Consider 12 mm ϕ bars in both directions at a nominal cover of 50 mm. Overall depth is:

$$D = d + \text{Cover} + \phi + \frac{\phi}{2} = 343 + 50 + 12 + 6 = 411 \text{ mm}$$

Adopt a 450 mm thick foundation slab with an effective depth of 380 mm.

Bending moments. The net cantilever projections in x-and y-directions are:

$$l_x = \frac{2200 - 300}{2} = 950 \text{ mm}$$

and

$$l_y = \frac{2200 - 350}{2} = 925 \text{ mm}$$

Bending moments per metre width of the footing are:

$$M_{u,y} = \frac{p_u l_x^2}{2} = \frac{232.44 \times 0.950^2}{2} = 104.89 \text{ kNm}$$

and

$$\text{M}_{\text{u,x}} = \frac{p_u l_y^2}{2} = \frac{232.44 \times 0.925^2}{2} = 99.44 \text{ kNm}$$

Thus,

$$M_u = 104.89 \text{ kNm}$$

To compute the effective depth, equate M_u to $M_{u,\text{lim}}$:

$$0.1388 f_{ck} b d^2 = 104.89 \times 10^6$$

or

$$d = \left[\frac{104.89 \times 10^6}{0.1388 \times 20 \times 1000}\right]^{\frac{1}{2}} = 194.47 \text{ mm} < 380 \text{ mm (provided)}$$

Reinforcement. Provide reinforcement for maximum bending moments in both the directions. Dimensionless parameter for flexural reinforcement is:

$$R_u = \frac{4.6 \times 104.89 \times 10^6}{20 \times 1000 \times 380^2} = 0.16707$$

Therefore,

$$A_{st} = \left[\frac{20 \times 1000 \times 380}{2 \times 415}\right] [1 - \sqrt{1 - 0.16707}\ ] = 799.83 \text{ mm}^2$$

Since the footing behaves essentially as a wide beam, the minimum reinforcement shall not be less than that given by

$$\frac{A_{st}}{bd} = \frac{0.85}{f_y} \quad \text{or} \quad \frac{100A_{st}}{bd} = \frac{85}{f_y} = p_t$$

Therefore, $$A_{st} = 0.85 \times \frac{1000 \times 380}{415} = 778.31 \text{ mm}^2$$

Provide 12 mm ϕ bars at 140 mm c/c (A_{st} = 807.78 mm^2).

Check for development length. Development length is given by

$$L_d = \frac{0.87 f_y \phi}{4\tau_{bd}} = \frac{0.87 \times 415\phi}{4 \times 1.92} = 47.0\phi$$

For 12 mm ϕ bars, $L_d = 47.0 \times 12 = 564$ mm
Available minimum anchorage length is:

$$= l_y - \text{Side cover} - \text{Bend allowance } (= 3\phi) + 8\phi$$

$$= 925 - 40 + 5 \times 12 = 945 \text{ mm} > 564 \text{ mm}$$

Check for cracking.

Clear distance between bars = 140 – 12 = 128 mm < 300 mm

Hence the reinforcement arrangement is satisfactory. Reinforcement details are shown in Fig. 7.21(d).

Design of pedestal. In reinforced concrete construction, an enlargement of column at its base called **pedestal** results in a larger bearing area to foundation and provides enough development length for reinforcement. In such cases pedestal and footing act together as a stepped footing. The pedestal is reinforced with longitudinal bars and lateral ties just like a column.

7.10 BASEMENT WALL

A high basement wall behaves as a vertical slab in flexure under the action of lateral pressure of outside soil. The design procedure depends upon the method of construction which determines the support constraints at the top and bottom of the basement wall. A four-way support action may be obtained from column and pilasters, and basement and ground floor slabs, whereas two-way support action will occur if the slabs alone support the wall as shown in Fig.7.22(a). The basement wall should ordinarily be designed as a simply supported member with a bending moment diagram as shown in Fig. 7.22(a). The effect of vertical load may be ignored because its beneficial effect is small as compared to the uncertainties involved in application of loads. Similarly, the economical effect of end restraints of basement wall is not of much consequence.

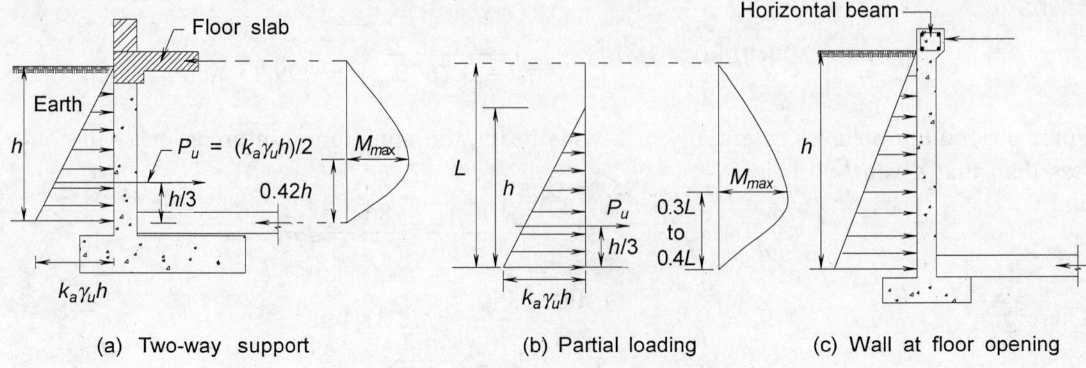

(a) Two-way support (b) Partial loading (c) Wall at floor opening

Fig. 7.22 Pressure distribution on basement slabs.

When a large part of the height of basement wall is above the ground, the bending moment due to lateral pressure may be assumed as shown in Fig. 7.22(b). In case of two-storey basement walls, slab can be designed as a two-span continuous vertical slab subjected to uniformly varying lateral load. Long walls may be designed as one-way slab in the vertical direction. In case of short walls (i.e., aspect ratio less than 2) and at corners of long walls, the wall has two-way action. Unless the horizontal bending moment is very large, the wall is generally designed as one-way slab and the minimum wall reinforcement will be adequate to resist earth pressure.

Other important points about the basement walls are:

1. To insure lateral support at the top, the ground floor slab should be cast before the backfill is placed against the wall.
2. To ensure dry building spaces, the outside walls must be made water-proof and proper drainage should be provided.
3. The horizontal reaction at the bottom of basement wall is resisted by the friction of the footing on base soil. If the basement slab is cast right against the wall, the slab can resist the horizontal reaction by friction against the soil or by bearing against another wall with an opposing horizontal reaction.
4. If the water table is very low, the hydrostatic pressure may be ignored, and if within one metre of the surface of the ground, it may be assumed to be at the surface.

The preliminary dimensions of different components of basement wall may be determined as in the case of cantilever retaining wall. The following example will illustrate the procedure.

Example 7.19 A basement wall of height 3.75 m above the basement floor slab to soffit of ground floor beam retains soil to its full height. The exterior face of the wall is on the property line. The unit weight of soil and its angle of repose are 19.5 kN/m^3 and 30°, respectively. The safe bearing capacity of soil at the site is 150 kN/m^2. Design the basement wall using M20 concrete and HYSD steel of grade Fe415.

Solution For M20 concrete and HYSD steel of grade Fe415:

$$f_{ck} = 20 \text{ MPa}, f_y = 415 \text{ MPa},$$

$$M_{u,\lim} = 0.1388 f_{ck} b d^2 \text{ and } p_{t,\lim} = 0.961 \text{ per cent}$$

For the soil at the site:

$$\text{Factored unit weight of soil } \gamma_u = 1.5 \times 19.5 = 29.25 \text{ kN/m}^3$$

and

$$\text{Coefficient of active earth pressure } k_a = \frac{1 - \sin\varphi}{1 + \sin\varphi}$$

$$= \frac{1 - \sin 30°}{1 + \sin 30°} = \frac{1}{3}$$

Consider 1 m length of the wall of effective height of 3.75 m. Factored total horizontal pressure of earth against wall is given by

$$P_u = \frac{1}{2} k_a \gamma_u h^2$$

$$= \frac{1}{2} \times \frac{1}{3} \times 29.25 \times (3.75)^2 = 68.55 \text{ kNm}$$

Neglecting the restraint of the floor systems at the top and bottom of the basement wall, the maximum bending moment occurs at $0.423h$ from the bottom.

$$M_u = \left[\frac{2}{9\sqrt{3}}\right] P_u h = 0.1283 \times 68.55 \times 3.75 = 32.981 \text{ kNm}$$

Equate M_u to $M_{u,\lim}$.

$$0.1388 f_{ck} b d^2 = 32.981 \times 10^6$$

$$d = 109.00 \text{ mm}$$

Consider overall thickness of 200 mm with 12 mm ϕ bars.
Therefore,

$$\text{Effective depth } d = 200 - 40 - \left(\frac{12}{2}\right) = 154 \text{ mm}$$

Reinforcement. The parameter for flexural steel is:

$$R_u = \frac{4.6 \times 32.981 \times 10^6}{20 \times 1000 \times 154^2} = 0.3198$$

Therefore,

$$A_{st} = \frac{20 \times 1000 \times 154}{2 \times 415} \times [1 - \sqrt{1 - 0.3198}] = 650.35 \text{ mm}^2$$

Provide 12 mm ϕ vertical bars @ 170 mm c/c ($A_{st} = 665$ mm^2) and 10 mm ϕ horizontal bars @ 200 mm c/c ($A_{st} = 393$ mm^2) on the interior face of the wall as distribution steel.

Design of footing. As the footing can not be extended beyond the exterior face of the wall, the wall footing forms L-shape cross-section. Consider the thickness of base D as 200 mm with effective depth,

$$d = 200 - (50 + 6) = 144 \text{ mm}$$

Adopt base width (generally presumed $h/4$ to $h/5$) B = 850 mm. Then

$$\text{Projection of footing beyond the wall} = 850 - 200 = 650 \text{ mm}$$

Bending moment on the footing:

$$M_{u,\max} = \frac{(1.5 \times 150)(0.65)^2}{2} = 47.53 \text{ kNm}$$

Therefore,
$$d = \sqrt{\frac{47.53 \times 10^6}{0.1388 \times 20 \times 1000}} = 130.85 \text{ mm} < 144 \text{ mm}$$

The thickness is adequate.

Reinforcement. Flexural reinforcement parameter is:

$$R_u = \frac{4.6 \times 47.53 \times 10^6}{20 \times 1000 \times 144^2} = 0.5272$$

Therefore,

$$A_{st} = \frac{20 \times 1000 \times 144}{2 \times 415} \times [1 - \sqrt{1 - 0.5272}] = 1083.96 \text{ mm}^2$$

Provide 16 mm ϕ bars @ 185 mm c/c (A_{st} = 1087 mm^2) at the bottom side of the base slab.

$$\text{Minimum steel} = \frac{0.12}{100}(154 \times 1000) = 184.8 \text{ mm}^2$$

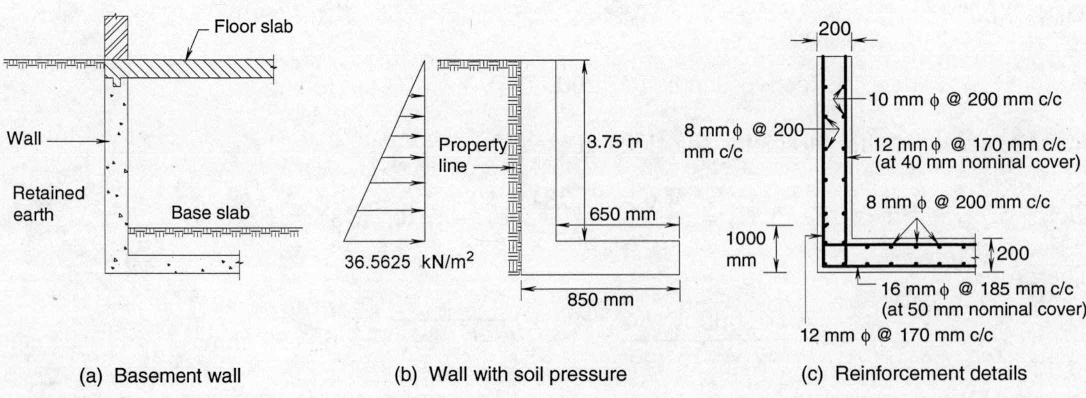

Fig. 7.23 *Geometry, design forces and detailing of the basement slabs of Example 7.19.*

Provide 8 mm ϕ bars @ 200 mm c/c (A_{st} = 250 mm^2) both horizontally and vertically on the exterior face of wall and horizontally in two perpendicular directions on the top side of base slab. Also provide horizontal bars on the bottom face of base slab along the length of the wall. The reinforcement details are shown in Fig. 7.23(c).

Review Questions

7.1 Explain the difference in the behaviour of one-way and two-way slabs.

7.2 What are the considerations that govern thickness of: (a) one-way and (b) two-way slabs?

7.3 What type of construction will enable two-way slabs to be classified as: (a) simply supported on four sides and (b) restrained on all the sides? What is the difference in the behaviour of these slabs?

7.4 Distinguish the action of slabs continuous over their supports from those that are cast monolithically with their supporting beams.

7.5 Why is it essential to provide corner reinforcement in two-way rectangular slabs whose corners are prevented from lifting up?

7.6 How are deflections checked and crack width controlled in two-way slabs?

7.7 Is there any difference between the action of shear in slabs and shear in beams? Compare the allowable shear strength in slabs to that in beams.

7.8 Mark the edge and middle strip portions of a two-way slab and explain the differences in their action. List IS:456 recommendations for reinforcing these strips.

7.9 What is a lintel? How is its thickness decided?

7.10 What are the conditions under which the load transferred to the lintel is of triangular shape?

7.11 When will be the base angle of triangular load 60°?

7.12 Illustrate, with the help of diagrams, the terms: rise, tread and going.

7.13 What is meant by a dog-legged staircase? Sketch a layout of such a staircase and indicate the spans for design.

7.14 Sketch the effective span and the loading to be used to calculate the BM and SF in a step of a staircase cantilevering from the wall.

7.15 Explain the difference in structural behaviour of stair slabs spanning longitudinally and transversely.

7.16 Is the live load specified in IS code in kN/m^2 taken along the length of the slab or on projected horizontal length of the slab?

7.17 Sketch the layout of a slab staircase spanning longitudinally and supported on beams.

7.18 Detail the longitudinal bars in longitudinally spanning waist-slab type stairs at the junction of the stab with: (a) upper landing slab and (b) lower landing slab.

7.19 Discuss the load effects produced by gravity loads on tread-riser type stairs spanning longitudinally.

7.20 What is meant by moment redistribution and what is its significance in design?

7.21 Why does the code not permit moment redistribution when bending moments in continuous beams are based on the moment coefficients specified in the code?

7.22 State the differences between: (a) column and wall, (b) plain and reinforced concrete walls and (c) braced and unbraced walls.

7.23 State why are the provisions of IS:456 for walls not applicable to design of shear walls.

7.24 How are the effective heights and slenderness ratios of plain and reinforced concrete walls calculated?

7.25 What are the limits of slenderness ratios laid down for plain and reinforced concrete walls?

7.26 What is the influence of the length-to-height ratio on the behaviour of reinforced concrete walls as stipulated in IS:456?

7.27 What is the minimum reinforcement to be placed in concrete walls to control shrinkage and temperature cracking? Does this also apply to plain walls?

7.28 What are the stipulations regarding detailing of transverse steel in reinforced concrete walls? When are the links provided through the thickness in reinforced concrete walls and what functions do they perform?

7.29 List two examples where the reinforced concrete member is subjected to: (a) direct tension and (b) direct tension combined with bending.

7.30 What is the basic assumption used for design of reinforced concrete member subjected to direct tension?

7.31 Discuss the classifications of exposure conditions used in IS:456 and BS:5337. What are the maximum crack widths allowed in each of the exposure conditions for durability?

7.32 How can the crack width in the reinforced concrete members subjected to tension be controlled?

7.33 Explain how the *P-M* design charts given in SP:16 are used for members subjected to direct tension combined with bending. Can these charts be used for the design of water tanks?

7.34 Explain the detailing of reinforcement in the reinforced concrete members subjected to direct tension.

7.35 Describe the distribution of secondary reinforcement in members subjected to (a) direct tension and (b) direct tension and bending.

7.36 What are the main requirements of the footing of a structure?

7.37 In limit states design, to determine the base area of an isolated footing for a given column load and safe bearing capacity of soil, which of the factored load and the service load is used as the basis of calculation?

7.38 Having determined the base area of the reinforced concrete footing, should we use the factored load or the service load for design of the footing?

7.39 When is a foundation considered as rigid? What is the difference in the action if the footing is flexible?

7.40 What are one-way and two-way shears in footings? What are the corresponding permissible values?

7.41 Sketch the reinforcement details for: (a) square footing and (b) rectangular footing.

7.42 What is the amount of minimum steel prescribed for footings according to IS:456?

7.43 How is the diameter of the reinforcement bars used in isolated footings selected? How can the development length of these bars be increased?

7.44 Is it necessary to continue all the steel in a reinforced concrete column into the footing?

7.45 Why is a footing so proportioned that the bearing pressure underneath the footing is more or less of same order of magnitude as the bearing capacity of soil?

7.46 Describe briefly the situations where combined footings are preferred to isolated footings.

7.47 What is meant by eccentric load on a footing and when does this occur?

7.48 Why is it desirable that a footing be so proportioned that eccentricity of loading is eliminated?

7.49 Why is a pedestal provided to a column?

7.50 What are the conditions in which transfer of forces at the interface of column (or pedestal) and footing is achieved without any reinforcement?

7.51 What is the purpose of a basement wall?

7.52 When will the two-way and four-way actions occur in basement wall?

7.53 Why is basement wall ordinarily designed as simply supported vertical member?

Tutorial Problems

T.7.1 The roof of a hall of internal dimensions of 27.5 × 10 m consists of continuous one-way slab supported on 250 mm wide beams spaced 4 m, centre to centre, with the ends of the slab being supported on walls. The centres of the end beams are 3.75 m from the face of the walls. The slab carries a service superimposed load of 1.5 kN/m^2 and a dead load due to mud phuska, tiles and plaster of 1.0 kN/m^2, in addition to the self-weight of the slab. Design and detail the continuous one-way slab, if the grades of concrete and steel to be used are M20 and Fe415, respectively.

T.7.2 A rectangular isolated reinforced concrete simply supported beam has a clear span of 9.25 m and carries a service uniformly distributed load of intensity 22 kN/m. Design and detail the beam to be constructed with concrete of grade M20 and steel of grade Fe415. Assume $b = 0.5d$.

T.7.3 A rectangular reinforced concrete beam is supported on two 460 mm thick brick walls 6.75 m (clear) apart. The beam carries a service superimposed load of 40 kN/m. The section of the beam is restricted to 300 × 625 mm from architectural considerations. Design the appropriate reinforcements at the tension and compression faces to carry the required load. The materials to be used are: concrete of grade M20 and HYSD steel bars of grade Fe415.

T.7.4 A laboratory floor consists of a monolithic one-way slab system on beams spaced 3.0 m centre to centre. The clear span of the beam is 10 m. The floor supports a uniformly distributed service load of 5 kN/m^2. The floor finish may be taken as 0.75 kN/m^2. Design and detail the slab, and the edge spandrel L-beam. The grades of concrete and steel to be used are M25 and Fe415, respectively. Take $b_w = 0.55d$.

T.7.5 A reinforced concrete canopy slab is to be monolithically cast with a lintel spanning an opening of 3.5 m in a 460 mm thick wall. The height of the wall above the lintel is 4.5 m. The canopy slab projects 2.5 m from the face of the wall and is provided with 50 mm thick and 350 mm high reinforced concrete fascia. The weight of the roof finish and water-proofing treatment may be taken 0.25 kN/m^2. The roof of the canopy is inaccessible, i.e., service live load may be taken 0.75 kN/m^2.

The weight of the masonry wall above the lintel may be assumed to be adequate for the stability of the cantilever canopy. The unit weight of the masonry is 19.2 kN/m^3. Design and detail the canopy slab and the lintel if the materials to be used are: M25 grade concrete and Fe250 grade steel.

T.7.6 Design and detail a reinforced concrete slab for a room measuring 4 × 6 m from face to face of walls. The slab is to carry a service live load of 2.0 kN/m^2 and a dead load due to the floor finish and partitions of 1.5 kN/m^2. The slab can be considered to be simply supported on all the four edges with the corners free to lift. The grades of concrete and steel to be used are M25 and Fe415, respectively. Redesign the slab if corners are held down.

T.7.7 Design and detail an interior panel of size 4.5 × 6.0 m of a reinforced concrete floor slab of a building. The floor is to carry a service live load of 4.0 kN/m^2 and a load due to movable partitions of 1.0 kN/m^2. The weight due to the floor finish and ceiling plaster may be taken as 1.0 kN/m^2. The materials to be used are: concrete of grade M20 and mild steel of grade Fe250.

T.7.8 A straight staircase flight with rise, tread, and width of 150 mm, 290 mm and 1.45 m, respectively, is to be provided in a building which is not likely to be overcrowded. Design the staircase when: (a) The stair consists of structurally independent tread slabs, projecting from the face of a reinforced concrete wall. (b) The stairs have a tread-riser arrangement, (c) the isolated tread slabs are to be supported on two stringer beams, each 230 mm wide with clear spacing between the beams as 1.45 m and (d) a waist slab type arrangement. The materials are M20 grade concrete and HYSD steel of grade Fe415.

T.7.9 Design and detail a reinforced concrete square isolated footing for a column of size 700 × 500 mm carrying an axial service load of 3000 kN inclusive of its own weight. The test borings and plate load test indicate that soil is composed of medium compacted and gravelly sands with the safe bearing capacity at the level of the footing base as 250 kN/m^2. The grades of concrete and steel to be used are M25 and Fe415, respectively.

T.7.10 Design and detail an isolated column footing for the column of Problem 7.9 when one of the plan dimensions of footing is spatially restricted to 3.0 m, i.e., the footing is rectangular in plane.

T.7.11 Design and detail a tied, non-slender reinforced concrete column to carry a service axial load of 800 kN and a service moment of 300 kNm. Assume the total reinforcement ratio between 2 and 3 per cent. The materials to be used are: concrete of grade M25 and HYSD steel of grade Fe415.

T.7.12 A short column is subjected to a factored axial load of 1600 kN and a factored moment of 175 kNm. Consider that the gross reinforcement ratio lies between 2.0 and 3.5 per cent, and the effective cover to the centre of longitudinal steel (d') is 60 mm. The grades of concrete and steel to be used are M25 and Fe415, respectively. Design and detail the column when: (a) the cross-sectional dimensions are not restricted and (b) one side is restricted to 275 mm.

T.7.13 Design and detail the following rigid frame columns using M30 grade concrete and Fe415 grade steel when one of the sides is restricted from architectural considerations to 300 mm when:

(a) P_u = 1750 kN and M_{ux} = 250 kNm and

(b) P_u = 1750 kN and M_{uy} = 125 kNm

where M_{ux} and M_{uy} are the factored bending moments about the major and minor axes, respectively, and P_u is the factored axial force.

T.7.14 Design a reinforced concrete member of a rigid frame subjected to a factored axial load of 1500 kN, and factored moments of 175 kNm and 125 kNm about its major and minor axes, respectively. Use concrete of grade M25 and steel of grade Fe415.

T.7.15 Design a slender reinforced concrete member of a rigid plane frame to support a factored axial load of 2000 kN, and factored bending moments about minor axis of $M_{uy,1}$ = 80 kNm and $M_{uy,2}$ = 120 kNm at the top and bottom ends, respectively. The grades of concrete and steel to be used are M30 and Fe415, respectively.

T.7.16 A reinforced concrete slender braced circular column of 325 mm diameter has an unsupported length of 7.5 m and effective length of 5.25 m. Design the column which bends in single curvature to support a factored axial load of 450 kN, and factored moments of 40 kNm and 22.5 kNm at the top and bottom, respectively. The materials to be used are M25 grade concrete mix and Fe415 grade HYSD steel bars.

T.7.17 A reinforced concrete rectangular column of size 425 × 575 mm bent in single curvature has an unsupported length of 7.5 m and the effective lengths of 6.0 m and 6.5 m about the major and minor axes, respectively. Design the column using M25 grade concrete and Fe415 grade steel for the following service load system:

Axial load: P =1350 kN

Moments at the top: M_{xx} = 165 kNm, M_{yy} = 70 kNm

Moments at the bottom: M_{xx} = 230 kNm, M_{yy} =130 kN

T.7.18 A reinforced concrete braced slender column of size 350 × 475 mm is to support a factored axial load of 1750 kN alongwith factored moments of:

At the top: M_{xx} = 135 kNm, M_{yy} = 88 kNm

At the bottom: M_{xx} = 205 kNm, M_{yy} = 105 kNm

Design longitudinal and transverse reinforcements using M30 grade concrete mix and HYSD steels bars of grade Fe415.

T.7.19 The wall of a reinforced concrete circular water tank is subjected to a hoop tension of 132 kN per metre height. Design suitable section of wall using M30 concrete and HYSD steel bars of grade Fe415.

T.7.20 The wall of a reinforced concrete rectangular water tank is subjected to a bending moment of 26.5 kNm per metre length and a pull (tension) of 43.0 kN from the cross wall. Design a suitable section for the wall using M30 grade concrete mix and HYSD steel bars of grade Fe415.

T.7.21 A 125 mm thick concrete wall of height 3.3 m with a length of 4.8 m between cross walls is to support a factored load of 725 kN per metre length. Design the wall using M20 grade concrete mix and HYSD bars of grade Fe415.

T.7.22 Design a reinforced concrete wall of height 6.6 m for a staircase to support a factored load of 525 kN per metre length and factored moment of 21 kNm/m at right angles to the plane of the wall. The wall is restrained in position and direction at the bottom, and in position at the top. The materials to be used are M20 grade concrete mix and HYSD steel bars of grade Fe415.

T.7.23 A basement wall of height 3.3 m from basement floor to the soffit of ground floor beam retains soil up to 2.8 m height. The exterior face of the wall is on the property line. The unit weight of soil and its angle of repose are 18.5 kN/m^3 and 30°, respectively. The safe load bearing capacity of soil at the site is 165 kN/m^2. Design and detail the basement wall using M25 concrete mix and HYSD steel of grade Fe415.

CHAPTER

8

Detailing the Reinforcement

8.1 INTRODUCTION

As discussed earlier, the current practice is to design the structures mainly by the Limit States Method with partial safety factors lower than the factors of safety inherent in the WSM. Furthermore, with the availability of more sophisticated and accurate methods of analysis and design, the conservatism in-built in approximate methods is no longer available. Modern structures are comparatively slender, flexible (in terms of deflections) and are more crack prone. Thus, the serviceability criteria assume far greater importance in modern structures.

In the preceding chapters, the attention was mainly focused on the design of critical sections of structural components such as slabs, beams, columns, etc, which entailed a proposition for proportioning the structural section and the provision of necessary reinforcement, followed by some form of analysis which either proved or disproved the original proposition and suggested subsequent refinements or amendments in the original proposition. However, not much attention has been given to combining these structural components together to ensure an integral action particularly to the joint details. Special attention need to be placed on: termination, extension and bending of bars; anchorage and development length; stirrup anchorage; splices; construction details at connections, provision for continuity or discontinuity at junctions of members; construction sequencing and reinforcement placement; deflection control; crack control; special needs of the structures such as inverted beams, edge and spandrel beams, cantilever members; cover, bar support and reinforcement protection; durability; recognition of an allowance for the long-term effects of creep, shrinkage and temperature; details of construction/expansion joints; structures (tanks, etc.) needing special procedures.

On completion of the structural design, the design ideas normally need to be communicated for construction at the site, by translating them into detailed structural drawings. In fact, an elaborate analysis becomes worthless if the computations are not translated into successful structures. Such may be the case when the structure is represented by a set of poorly detailed drawings.

8.2 DETAILED STRUCTURAL DRAWINGS

In addition to graphic delineation of a structure a drawing acts as a guide and defines the order to perform certain operations on the site in a specified manner. The drawing also serves as a record of some of the important assumptions made in the design. Engineering drawings prepared by the designer should specify grades of concrete and steel, live load, dimensions, reinforcement, lap lengths, concrete cover and all other information needed for detailing the reinforcement, building the forms, fabricating the reinforcement and placing the concrete.

A simple, clear and comprehensive drawing can minimize confusion and helps in avoiding costly or dangerous errors due to misunderstanding or misinterpretation. Notes and statements should be clear and unambiguous.

Different methods of detailing and drawing have been developed to suit different types of works. The reinforced concrete detailing is aimed at:

1. providing an outline of the concrete to give necessary information regarding formwork to provide finished sections of the desired size.
2. providing enough reinforcement details for fabrication of the steel skeleton and its placement in the desired position.

The detailing also incorporates the whole thought process by which the designer enables each part of the structure to perform safely. It is to be realized that the poor detailing can lead to disaster just as surely as a defective design; in fact, it is much more the frequent cause of trouble.

To reinforce a concrete structure effectively, a designer must possess a thorough understanding of its structural behaviour based on the extensive knowledge of the properties of materials, type of structural system employed, failure modes, overall deformation patterns, compatibility conditions, load transmission paths to supports, stress concentration points and possible weak links. Furthermore, the designer should also be able to identify the critical and primary load carrying components, and to perform an approximate manual check on the design resistance of such members. For example, an approximate estimation of the ultimate moment of resistance of a beam/slab section can be obtained as $(0.87 f_y A_{st})(0.9d)$.

The purpose of reinforcement should be kept in mind while detailing the reinforcement. The reinforcing steel is provided in concrete to accomplish the following:

1. Resist the internal tensile forces as obtained from analysis. The reinforcement must ensure that a structure possesses adequate strength at service and limit states.
2. Ensure that crack widths under service conditions do not exceed the permissible values.

3. Prevent excessive cracking due to shrinkage and temperature changes when the structural elements are restrained.
4. Resist the compressive force when the concrete alone is not capable of resisting the internal compressive force.
5. Prevent buckling by restraining compression bars against lateral movement, and provide confinement in highly stressed areas of compression zones of columns, beams and joints.
6. Provide temporary support to the reinforcing system during construction, and provide protection against spalling of the fire protective cover in composite sections.

Some of the important guiding principles in the art of detailing are described in this chapter.

8.3 REINFORCEMENT LAYOUT FOR FLEXURAL MEMBERS

It is recognized that the good detailing leads to improved structural behaviour, and adequate strength is attained by efficient interaction of concrete and steel.

8.3.1 Detailing of Slabs

In the structures like one-way or cantilever slabs where design reinforcement is placed only in one direction cracks could develop across the perpendicular direction due to contraction and shrinkage, and it is necessary to provide some reinforcement in the direction perpendicular to the main reinforcement. This reinforcement is usually called the **secondary** or **distribution** or **temperature reinforcement**. In case of the members exposed to the sun such as sunshades, canopies, balconies, roof slabs without adequate insulation cover, etc., the minimum reinforcement specified by the Code should be increased by 50 to 100 per cent depending upon the severity of exposure, size of member and local conditions.

8.3.2 Detailing of Beams

Tension reinforcement: The minimum area of tension reinforcement shall not be less than $0.85bd/f_y$, where b and d are the breadth of the beam or web in case of the T-beam and effective depth, respectively. The maximum area of tension reinforcement shall not exceed $0.04bD$.

Compression reinforcement: The maximum area of compression reinforcement shall not exceed $0.04bD$. Compression steel shall be enclosed by stirrups for the effective lateral restraint.

Anchorage: It is common practice to have flexural bars terminated in the compression zone where the compressive stress acts parallel to the bar. However, the compressive stress acting transversely to an anchored bar has been found to be more beneficial. Such a situation is encountered when the reaction acts at the tension face of a beam. For example, the bottom bars in the end span of a continuous beam show better anchorage at the simply supported end than in the vicinity of the point of contra-flexure where they enter a compression zone. Thus, it is better to seek areas of normal pressure in preference to compression zones for anchorage of flexural reinforcement. The hooked anchorage at the simply supported end of a beam can be much improved if the hooks are tilted or lie in a near horizontal position. The bars can also be terminated in the tension zone of the beam which contains sufficient transverse web reinforcement.

Concentrated and multilayered arrangement of negative moment (top) reinforcement deteriorates the bond, resulting in an increase of the crack widths. The negative moment steel can be spread into the adjoining monolithic slab preferably by using smaller diameter bars, as shown in Fig. 8.1. The arrangement will give a slightly larger lever arm and provide better access for vibrators in a usually overcrowded beam-column joint. However, the major part of the flexural reinforcement must be within the multi-legged cage of stirrups.

Shear reinforcement: The longitudinal bars passing through the corners of the stirrups are

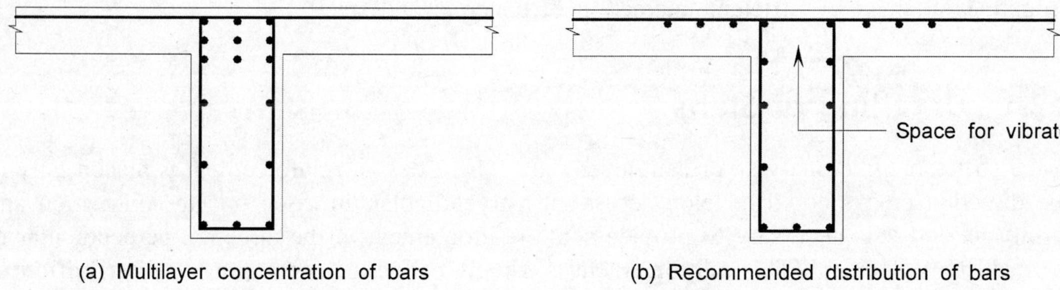

Fig. 8.1 *Recommended arrangement of top reinforcement in the flanged beams.*

essential because they must distribute the concentrated bearing received from stirrups in tension. The stirrups should fit tightly and be in contact with the longitudinal bars that they surround. The normal practice is to bend stirrups around longitudinal bars with an angle of 135°. Some codes permit a 90° turn for stirrups.

When shearing forces are present and more than two bars are used to resist flexure, it is desirable to form a truss at each of the longitudinal bars by using multi-legged stirrups. In the absence of vertical stirrup legs, the centre bars are incapable of resisting vertical forces, and thus are inefficient in carrying the bond forces.

Curtailment of flexural steel: The curtailment of flexural bars in the tension zone creates a discontinuity which may result in a sudden increase in tension strain, causing cracks. It is, therefore, essential that the shear strength of such areas of a beam be supplemented by the web reinforcement in the form of additional stirrups in the vicinity of cut-off points of flexural reinforcement in the tension zone.

Contrary to general belief, the bent-up bars from the flexural reinforcement are often responsible for inferior performance.

Side reinforcement: When the depth of the web in a beam exceeds 750 mm, side face reinforcement shall be provided along the two faces. The total area of such reinforcement shall not be less than 0.10 per cent of the web area and shall be distributed equally on the two faces at a spacing not exceeding 300 mm or the web width, whichever is less as shown in the Fig. 8.2. To guard against the bar yielding locally at crack, in BS:8110, the spacing of the side reinforcement is limited to 250 mm and the diameter of the bars is restricted to that given by the following expression:

$$\phi = \sqrt{\frac{S_b b}{f_y}}$$

where

S_b = Vertical spacing (mm)

b = Width of the beam (mm)

f_y = Characteristic strength (MPa)

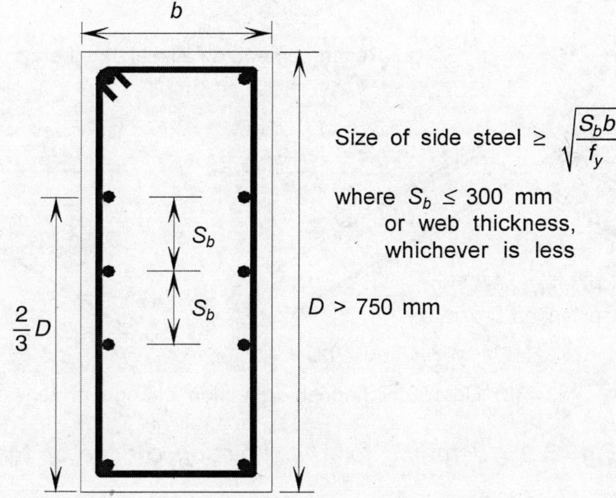

Fig. 8.2 *Spacing of side reinforcement.*

8.3.3 Member with a Change in Direction

Whenever, a loaded concrete member is not straight or its dimensions are abruptly changed or the direction of main compression or tension reinforcement changes, internal resultant radial force F is induced at the location of kink as shown in Fig. 8.3(a), tending to split the concrete.

When the angular change is small (say < 15°), the resultant radial forces F is small and can be carried and transferred to the compression zone by providing adequate number of stirrups at the location of the kink on either side at close spacing as shown in Fig. 8.3(a). When the angular change is relatively large, the reinforcement from either side should be continued straight and anchored so that no transverse force is generated at the kink and steel develops the full design stress. In the situations where a bar cannot get a straight length of L_d beyond the location of the kink in the beam face, it is continued on the compression face and anchored there as shown in Fig. 8.3(b), so that no outward splitting force is developed due to the bend in this bar. The common examples of these situations are: junction of stairs and landing, inside corners of rigid frames and where the soffit of beam forms an angle as in a gable bent. Rigid corner connections of beams to columns often require closed stirrups or ties around the bend.

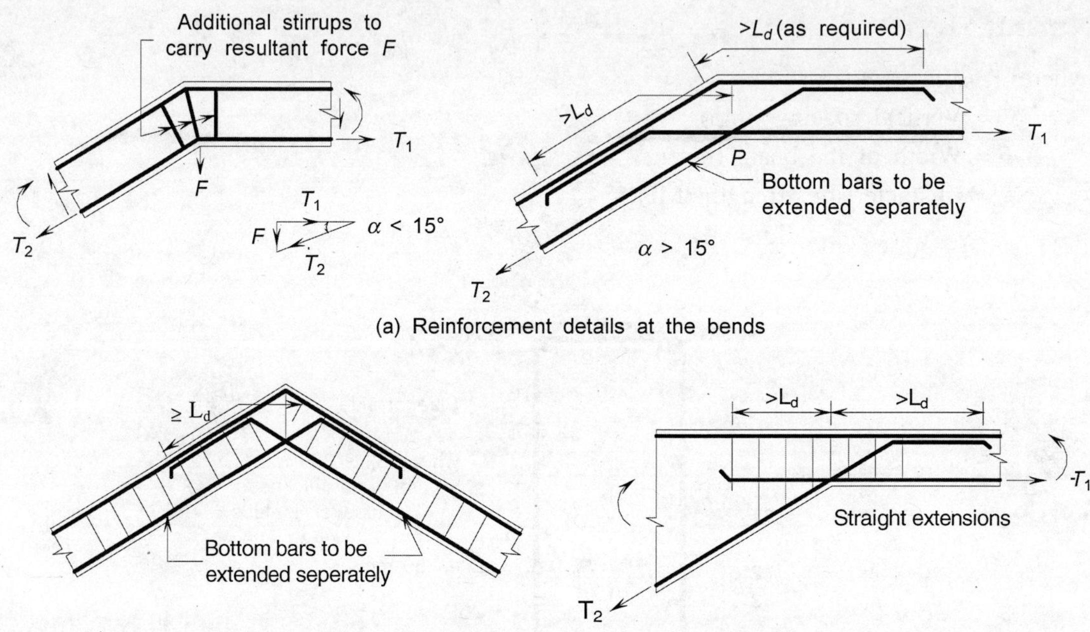

Fig. 8.3 Detailing for the direction change of tension force.

8.3.4 Edge Beams

In edge and spandrel beams, stirrups must be of the closed type and at least one longitudinal bar should be located at each corner of the beam section. Sometimes two-piece stirrups forming closed ties are used as shown in Fig. 8.4(d).

8.3.5 Detailing of Support Points

A typical connection of a beam supporting a cantilever slab is shown in Fig. 8.4(a). When beams of equal depth meet, the bottom steel of the secondary beam must lie above the bottom reinforcement of the supporting girder. Suspender stirrup reinforcement should be provided in the supporting girder in addition to usual stirrups, for supporting the whole shear or reaction force from the secondary beam, as shown in Fig. 8.4(b).

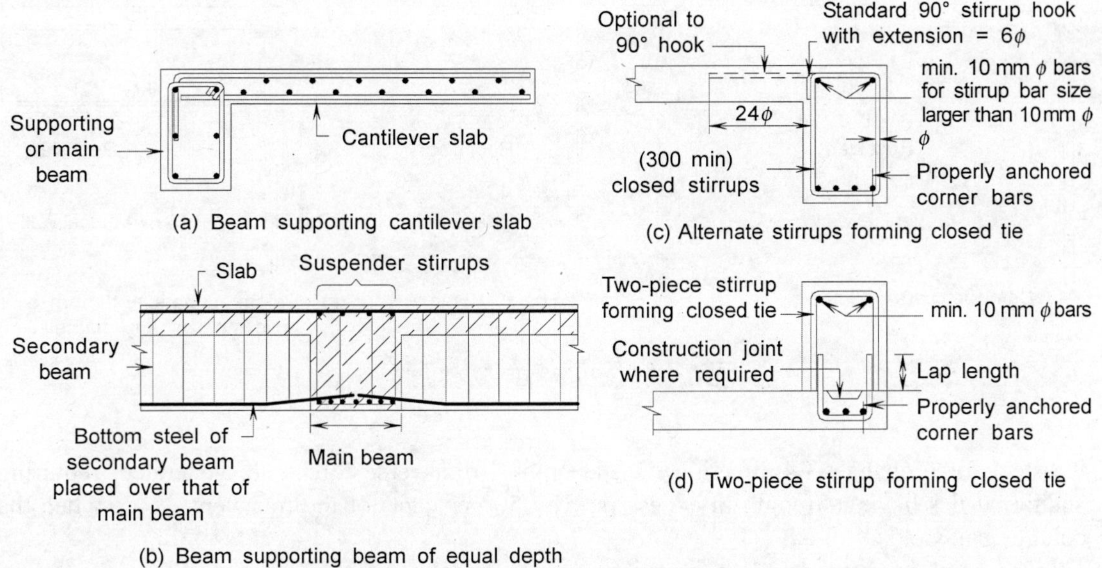

Fig. 8.4 Detailing at supporting points.

8.3.6 Corners of Walls

In concrete walls, horizontal reinforcement may be required to resist tension, moment, shear, temperature and shrinkage effects. All such bars on both faces of wall must be sufficiently extended past the corner or intersection to satisfy development length requirements.

There are situations where the embedment length available for end anchorage of bars is insufficient to develop the design stress in the bar through bond, e.g. corbels, deep beams, small size footings, pre-cast beams, etc. In such cases, special devices such as welded cross bars and end plates must be provided. The detailing of the joint of concrete wall with a slab is shown in Fig 8.5.

8.4 DETAILING OF COMPRESSION MEMBERS

Bond and anchorage conditions are more favourable in the compression members. Consequently, fewer difficulties arise with respect to the detailing of longitudinal bars. Longitudinal bars are usually located close to the periphery (for better flexural resistance), but may be placed in the interior of the column when eccentricities in loading are minimal. When a large number of bars are to be accommodated, they may be bundled or, alternatively, grouped. A few points related to splicing of column bars deserve special attention. Wherever

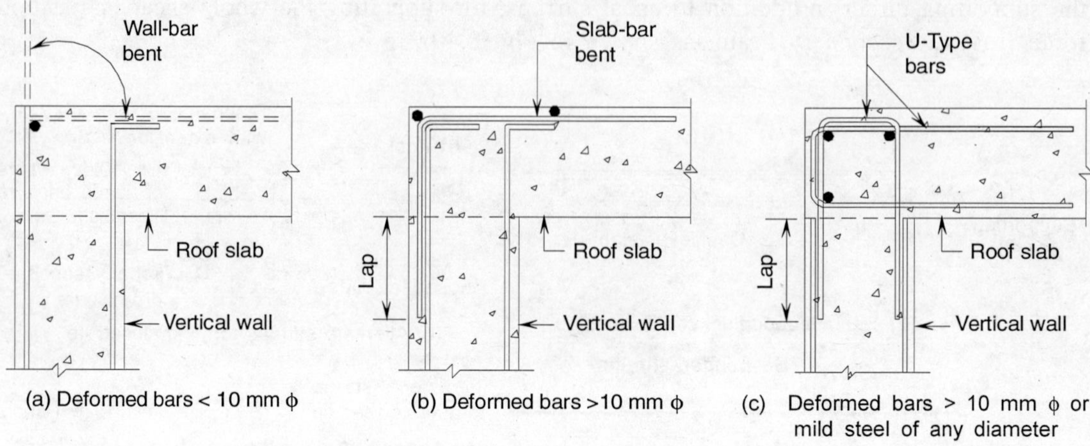

Fig. 8.5 Splices at top of walls.

the steel force changes its direction, i.e., at cranks, transverse forces are generated, requiring additional ties having strength in excess (say by 50 per cent) of the transverse force when the column bar yields.

The reinforcement detailing provisions for a column are semi-empirical in nature. The recommendations made by IS:456 are as follows:

8.4.1 Longitudinal Reinforcement

Minimum steel: The minimum area of longitudinal steel shall always be provided to resist bending moment that may be induced even though the section is acted upon by the concentric loads, and to reduce the effect of shrinkage and creep under sustained load. IS:456 has recommended a minimum area of 0.8 per cent of the gross area of concrete. In any column that has a larger cross-sectional area than required to support the load, the minimum percentage of steel shall be based upon the area of concrete required to resist the direct stress and not upon the actual area.

Maximum steel: The maximum area of longitudinal steel is limited to 6 per cent of the gross area of the concrete section from the economic consideration, and the requirements of placing and compaction of concrete. However, at laps, the cross-sectional area shall not exceed 4 per cent. In other words, at laps, the total area of steel shall not exceed 8 per cent.

Number of bars: The minimum number of longitudinal bars shall be four in a rectangular column and six in a circular or helically reinforced column.

Diameter of bars: The diameter of the bars shall not be less than 12 mm.

Spacing of bars: Spacing of longitudinal bars along the periphery of the column shall not exceed 300 mm to ensure confinement of the concrete core and to restrict the width of cracks.

Splices: The splices in column reinforcement are usually made by lapping of bars. This causes congestion and is not practicable for the bars of larger diameter. Therefore, the built splices are becoming popular. It may be welded type, sleeved type with filled metal alloy.

Usually, it is better to place splices in a low stress area (e.g. mid-height of a column), if such a place exists. In an earthquake-resistant structure, a column may be subjected to very large bending moments, therefore, possible hinges may form at the top and bottom ends. For this reason, splicing should be near mid-height.

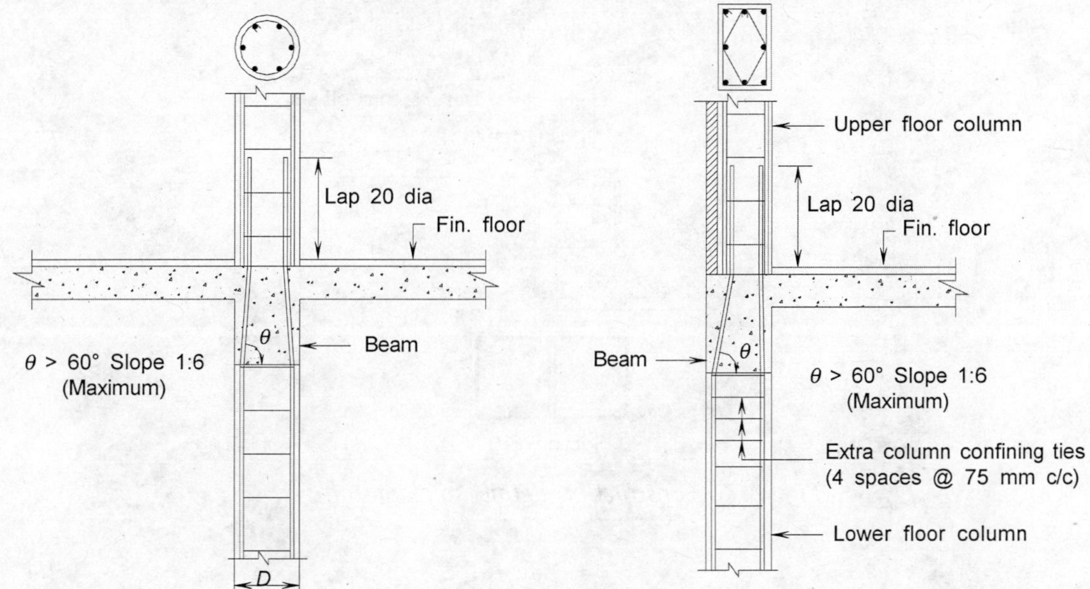

Fig. 8.6 *Detailing at offsets in columns and splicing the bars.*

Offset columns: When the cross-section of a column is reduced in the upper storey, the longitudinal bars from lower column are offset to come with in the column above or dowel splices may be used. The slope of the inclined portion of such bars with the axis of the column shall not exceed 1 in 6 and the portion of the bar above and below the offset shall be parallel to the axis of the column, as shown in Fig. 8.6. Adequate horizontal support at the offset bends shall be provided by the additional ties or spirals. The ties shall be designed to resist horizontal thrust equal to 1.5 times the horizontal force generated due to the change of direction at the bend.

When the offset between column faces exceeds 75 mm, the vertical bars in the column below shall be terminated at the floor slab, and splicing of column bars by dowels may be necessary as shown in Fig. 8.7. Dowels may also be necessary when the construction of a part of the structure is delayed, and also between various units of structures [such as footings and columns as shown in Fig. 8.8(a)]. Dowels should, in general, be of the same size and grade as the bars joined, and should be of sufficient length to splice with the main bars. However, in case of foundations the load transfer from the column to foundation can be achieved by extending the column bars into the foundation as shown in the Fig. 8.8(b).

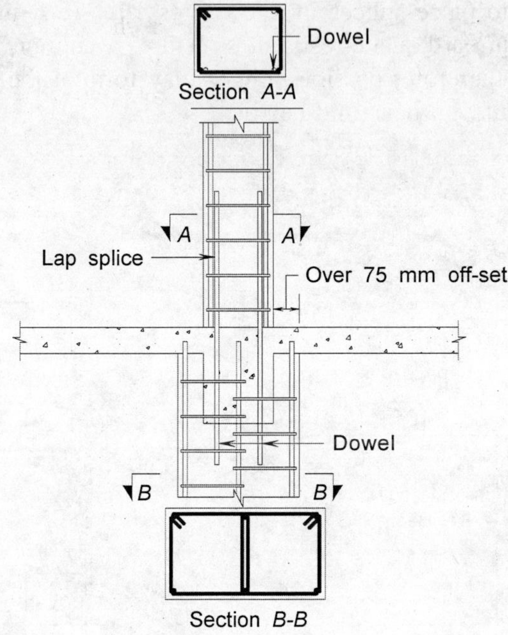

Fig. 8.7 Construction details for large off-set.

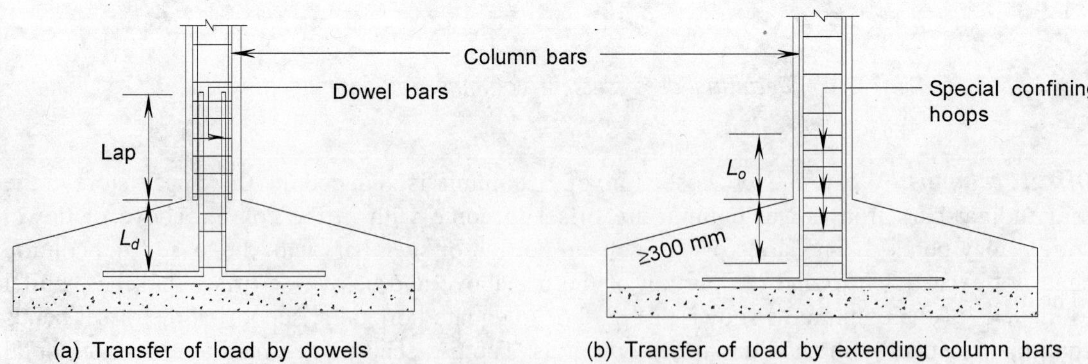

Fig. 8.8 Detailing in columns for load transfer.

When column bars are spliced, additional ties shall be provided at and near the ends of spliced bars, to provide confinement to the highly stressed concrete in the regions of the bar ends.

Cover to reinforcement: Adequate cover to reinforcement is required for its durability and to prevent bond failure of deformed bars.

1. For a longitudinal reinforcing bar in a column, the cover shall not be less than 40 mm or the diameter of the bar, whichever is more. In the case of columns of minimum dimension of 200 mm or less, whose reinforcing bars do not exceed 12 mm ϕ; a cover of 25 mm may be provided.
2. Increased cover thickness may be provided when surfaces of the concrete members are exposed to the action of harmful chemicals, acid vapour, saline atmosphere, sulfurous smoke (as in the case of steam-operated railways). Such an increase in cover may be between 15 mm and 50 mm in addition to that recommended for normal conditions. For reinforced concrete members totally immersed in sea water, the additional cover shall be 40 mm, whereas for the members periodically immersed in sea water or subjected to sea spray, the additional cover shall be 50 mm. When the concrete of grade M25 and above is used, the additional cover specified earlier is reduced to half. In all such cases the cover shall not exceed 75 mm.

8.4.2 Transverse Reinforcement

The purpose of transverse reinforcement in the column is to provide adequate lateral support to longitudinal bars carrying a compressive load to prevent instability due to outward buckling. It is provided either in the form of circular rings capable of taking up circumferential tension or by polygonal links (lateral ties) with internal angles not exceeding 135°. To develop the full strength of ties, the ends of transverse reinforcing bars shall be anchored, satisfying either of the following specifications:

1. The ends of the tie should be bent through an angle of at least 90° around a bar of at least its own diameter and continued beyond the end of the curve for a length of at least eight times the bar diameter.
2. The ends of the bar should be bent through an angle 135° around a bar of at least its own diameter and continued beyond the end of the curve for a length of at least six times the bar diameter.
3. The ends of the bar should be bent through an angle of 180° around a bar of at least its own diameter and continued beyond the end of the curve for a length of at least four times the bar diameter.

The preceding cases of anchorage are illustrated in Fig. 8.9.

Arrangements of transverse steel: IS:456 has recommended the following arrangements:

1. If the longitudinal bars are not spaced more than 75 mm, transverse reinforcement needs only to go around the corner and alternate bars for the purpose of providing effective lateral supports, as shown in Fig.8.10(a).

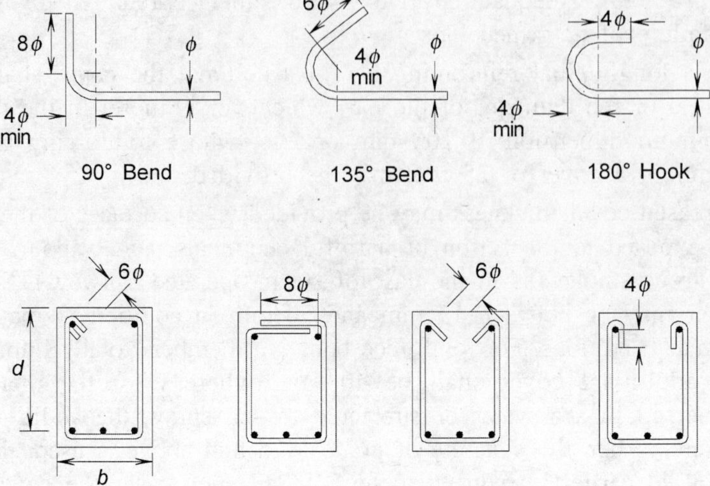

Fig. 8.9 *Types of hooks and bends for stirrups or ties.*

2. If the longitudinal bars are spaced at a distance not exceeding 48 times the diameter of the tie bar they are effectively tied in two directions, additional longitudinal bars in between these bars need to be tied in one direction by open ties, as shown in Fig. 8.10(b).
3. When the longitudinal reinforcing bars in a compression member are placed in more than one row, effective lateral support to the longitudinal bars in the inner rows may be assumed to have been provided if the transverse reinforcement is provided for the outermost row and no bar of the inner row is closer to the nearest compression face than three times the largest diameter of the bars in the inner row, as shown in Fig. 8.10(c).
4. When the longitudinal bars in a compression member are grouped (not bundled) and each group is adequately tied with transverse reinforcement, the transverse reinforcement for a compression member as a whole may be provided on the assumption that each group is a single longitudinal bar for the purpose of determining the diameter and spacing of the transverse reinforcement. The diameter of such a transverse reinforcement need not exceed 20 mm, as shown in Fig. 8.10(d).

For rectangular (not covered in type-1 above), circular and L-shaped typical columns the detailing arrangements are shown in Figs. 8.10(e), (f) and (g), respectively.

Diameter and pitch of lateral ties: The diameter and pitch of the transverse reinforcement shall be as follows:

1. The diameter of the steel bar for the polygon links or lateral ties and helical reinforcement shall not be less than one-fourth of the diameter of the largest longitudinal bar, and in no case less than 6 mm.

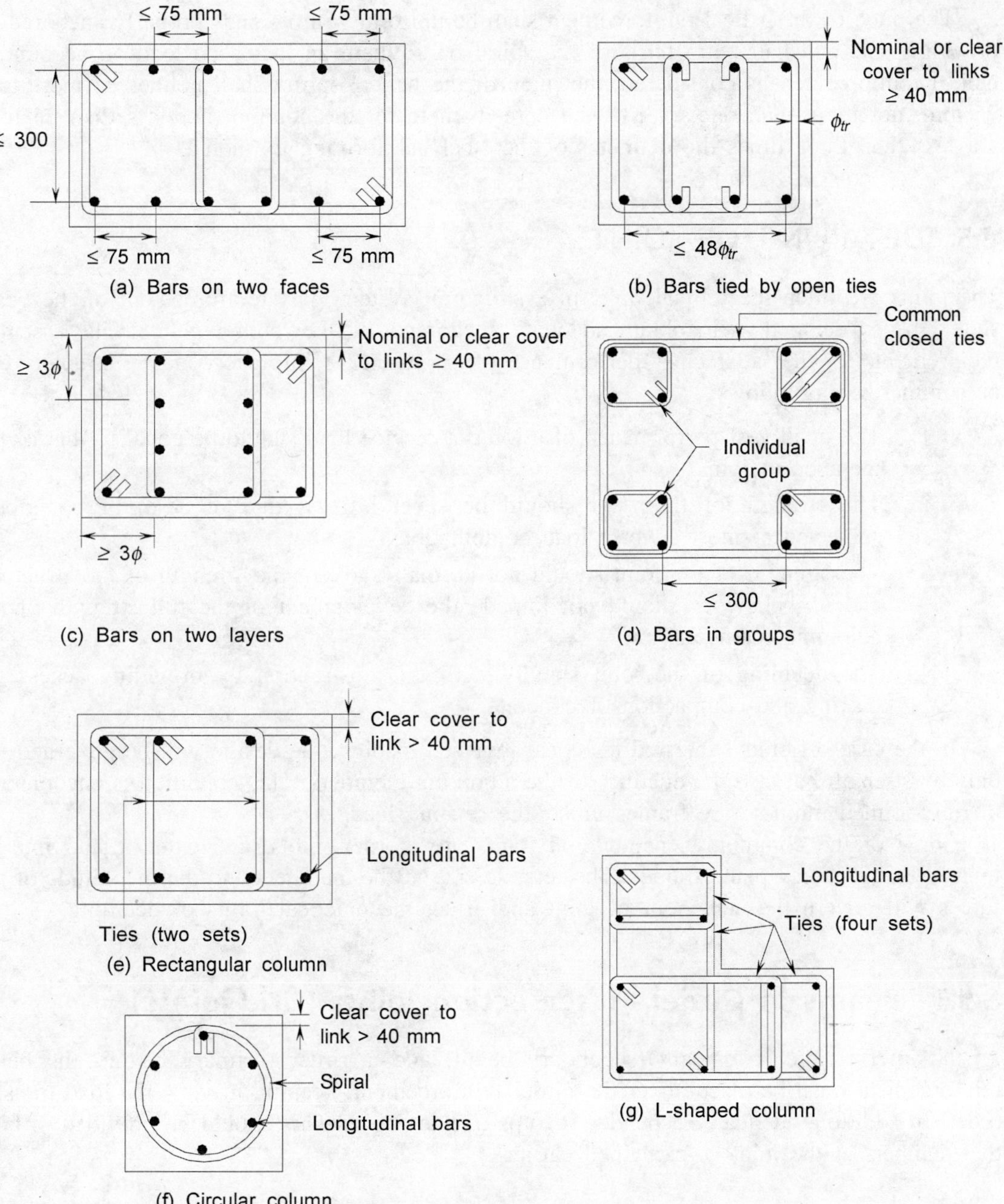

Fig. 8.10 *Typical code recommendations for detailing in columns.*

2. The pitch of the ties shall not be more than the least of the following: (a) the least lateral dimension of the compression member, (b) sixteen times the smallest diameter of the longitudinal reinforcement bar to be tied and (c) 300 mm.

The pitch of the helical reinforcement shall be uniform with its ends properly anchored by providing one and half extra turns of the spiral bar. Where an increased load on account of helical reinforcement is considered, the pitch of the helical spring shall neither be more than 75 mm, nor more than one-sixth of the core diameter of the column, nor less than 25 mm, nor less than three times the diameter of the steel bar forming the helix.

8.5 DETAILING OF JOINTS

The joints are often the critical links in a structural system. The reinforcement of the beam should be so detailed as to obtain adequate anchorage, as shown in Fig. 8.8. The essential requirements for the satisfactory performance of a joint in a reinforced concrete structure can be summarized as follows:

1. The quality of performance of a joint at service load should be equal to that of the members it joins.
2. The strength of the joint should be several times that of adjoining members subjected to most adverse load combinations.
3. The strength of the joint should not normally govern the strength of the structure, and its behaviour should not impede the development of the full strength of the adjoining members.
4. The detailing of the joint should facilitate construction by providing access for placing and compaction of concrete.

In the case of joints subjected to repeat reversed loading, the design will be governed not only by strength but also the ductility of the adjoining members. This condition is encountered in rigid-jointed multi-storey frames under the seismic load.

Some of the commonly encountered joints are: corner joints, and exterior and interior joints of multi-storey plane frames. The relative size of the members and the magnitude of the forces will govern the behaviour of joint and affect the practical limits of detailing.

8.5.1 Beams or Girders Intersection Joints (Grid-joints)

At the intersection of a beam (secondary beam) and a girder (primary beam), the beam reinforcement must be placed over the girder reinforcement, keeping in view the load transfer order. In addition, adequate suspender stirrups or hanging up bars should be provided in both the members at the joint as explained earlier.

8.5.2 Beam and Column Joints (Rigid-frame joints)

Joints of beams and columns in rigid jointed frames should have adequate strength to enable the development of the full design strengths under the most adverse loading combination, without distress in the joint itself. Joint is usually subjected to axial and shearing forces, in addition to the bending moment. A rigid joint must have full continuity between various members meeting at the joint. During deformation, the main flexural reinforcement in the joint zone undergoes a change in direction and as a consequence transverse forces are developed. There can be three types of stress patterns: (a) hogging moments tending to close the joint, i.e., causing tension in the outer fibres, (b) sagging moments tending to open the joint and (c) joints subjected to reversal of moments as in the case of seismic loading. For satisfactory performance in this case, the outer tension bars should be continuous around the corner.

The case of a knee joint subjected to opening moment i.e., moment tending to open the knee joint, is more critical than the case of moment tending to close the knee joint. Suggested reinforcing details for large size joints of this type are shown in Fig. 8.11.

When the joint is subject to high intensity reversed loading for several cycles, it is recommended that the resistance offered by the concrete should not be taken into account.

8.5.3 Corner Joints

The outer bars are generally continuous. Thus, they have sufficient anchorage and prevent splitting failure because of high bearing within the bend. In case of the closing joint, no reliance is placed on the inner compression steel; it does not seem to matter how the inner bars are anchored. The detailing recommended is shown in Fig. 8.11(c). Secondary reinforcement in the form of stirrups is required to preserve the integrity of the concrete within the joint.

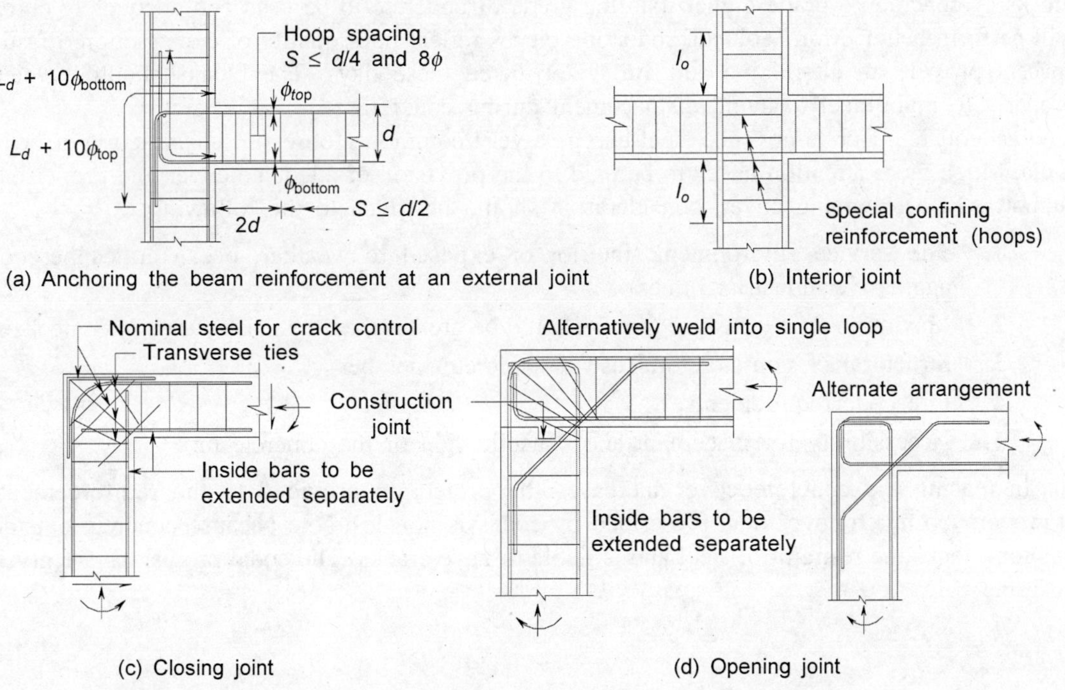

(a) Anchoring the beam reinforcement at an external joint

(b) Interior joint

(c) Closing joint

(d) Opening joint

Fig. 8.11 Detailing the beam-column joints.

The right angled corner joint subjected to moments, tending to open this angle, is more severely affected than the closing joint. The type of detailing recommended for such a joint is shown in Fig. 8.11(d).

8.5.4 Exterior and Interior Joints

The suggested detailing for exterior and interior joints is shown in Figs. 8.11(a) and 8.11(b), respectively.

8.6 BAR SUPPORTS AND COVER

It is essential to have the reinforcing steel accurately located in the forms and firmly held in place before and during the placing of concrete by means of *supports* and *spacers*. Such supports should be adequate to prevent displacement during the course of construction and to keep the bars at the proper distance (cover) from the forms.

Although common practice is to use pre-cast cement sand mortar blocks (of thickness equal to the specified cover) for bottom bars and individual (locally fabricated) metal chairs for top/bent bars, but actual site practices vary considerably. The mortar blocks are very porous and provide passage to moisture, oxygen, chlorides and other deleterious materials which cause deterioration of concrete and steel. Cover blocks should be strong and impermeable. Cover blocks of concrete; a grade higher than the grade of concrete to be used and adequately cured will perform better. Mortar blocks and stone pieces reflect poor quality of construction. Plastic spacers provide an alternative and are widely used these days. The blocks should be tied properly to main steel to avoid displacement during concreting.

Inadequate concrete cover to steel bars is a very commonly observed construction error in India. Much more attention needs to be paid to the provision of adequate cover and proper bar supports. In determining cover, consideration should be given to the following:

1. The service environment (interior or exposed to weather, ocean atmosphere or aggressive industrial fumes)
2. Maximum aggregate size (cover should be greater than the maximum aggregate size)
3. Structural or non-structural use of concrete member
4. Fire code requirements
5. Accessibility for placement and consolidation of the concrete mix

Increasing the concrete cover increases the protection provided to the reinforcement. However, too much cover than prescribed by codes is not desirable because concrete outside this limit lacks the restraint of steel and is liable to have cracks. The codal provisions are given in Table 8.1.

TABLE 8.1 Nominal Cover for Reinforcement in Cast-in-place Concrete

IS:456		*ACI:318/318R-68*	
Exposure class	*Minimum cover (mm)*	*Site condition*	*Minimum cover (mm)*
Mild	20	(a) Concrete cast against and permanently exposed to earth	75
Moderate	30	(b) Concrete exposed to weather	
Severe	45	(i) Main bars 20 mm and above	50
Very severe	50	(ii) Main bars 16 mm and below	40
Extreme	75	(c) Concrete not exposed to weather	
Notes:		**Slabs, Walls, Joists**	
(a) For footing minimum cover 50 mm		(i) Main bars 40 mm and above	40
(b) For columns, 40 mm or equal to main bar dia whichever is more.		(ii) Main bars 32 mm and below	20
(c) For mild exposure and 12 mm main bars, cover can be reduced by 5 mm		**Beams, Columns**	
(d) Tolerance + 10 mm, – 0 mm		(i) Primary reinforcement, ties, stirrups, spirals	40
(e) For 'severe' and 'very severe' exposure condition, reduction of 5 mm may be made, where concrete grade is M35 and above.		(ii) Shells, folded plates	20

8.7 DEFLECTION CONTROL

In Section 6.2, it is explained that modern designs result in relatively slender members with associated larger deflections. Creep and shrinkage cause deflections to increase with age, and such increases may be as much as 2 to 3 times the initial elastic deflections.

8.8 SYMBOLS FOR DETAILING

For one particular job uniform symbols and abbreviations should be used. The readers should refer to IS:5525, Recommendations for Detailing of Reinforcement in Reinforced Concrete Works, for more details.
The following symbols are suggested:

For reinforcement:

Bt	Bent bar
Ct	Column tie
Sp	Spiral
St	Straight bar

Stp	Stirrups
ϕ	Diameter of plain round bar
Φ	Diameter of HYSD (or Tor steel) bar
@	Spacing centre to centre
↔	Direction of spanning

For drawings:

Dimension lines (0.2 mm)
Concrete outlines, ties and stirrups (0.3 mm)
Reinforcement except as above (0.5 mm to 0.8 mm)
Column in plan (fully dark)
Beam in plan (0.3 mm)

The following scales are recommended in drawings:

Plan	1:100	1:5
Elevation	1:5	1:30
Section	1:50, 1:30	1:25, 1:15, 1:10

Based on the sketches prepared by the designer and the architectural drawings, structural drawings are prepared by a draughtsman under the constant supervision of the designer. Structural drawings should be prepared such that all the information regarding the job to be executed is covered. Such drawings should be clear and easy to understand.

Much of the general information regarding the job can be covered by general notes in the first drawing. For example, for a reinforced concrete building of moderate size, the typical general information may be provided in the form of general notes as follows:

1. All structural drawings should be read in conjunction with appropriate architectural drawings. In case of discrepancy between the architectural and structural drawings, the structural engineer and the architect should be consulted.
2. Unless otherwise specified the concrete shall be of grade M20.
3. Under no circumstances should concreting be done until the reinforcement is checked, approved and certified by the structural engineer.
4. It should be noted that ϕ indicates mild steel bars and Φ indicates HYSD bars.
5. The mild steel and Fe250 grade 1 bars shall conform to IS:432.
6. All HYSD bars shall conform to IS:786 for grade Fe415.
7. The reinforcement shall be bent and fixed in accordance with the procedure specified in IS:2502.
8. Lap in reinforcement shall be provided nearer to the support in the case of simply supported members and staggered. For a continuous beam, the lap shall be nearer to the support for bottom bars and no lap shall be preferred for extra top bars. For columns, the lap shall start 100 mm above the floor level.

9. Lap length for reinforcing bars shall be as follows:

Column, walls, etc.	54 ϕ for mild steel
	57 for HYSD bars
Slabs, beams, etc.	58 ϕ for mild steel
	69 ϕ for HYSD bars

10. All bent-up bars shall be at 45°, unless otherwise specified.
11. Twenty-five mm spacer bars shall be used at 600 mm c/c to separate the layers of beam bars.
12. Concrete cover refers to minimum clear distance from the reinforcement to the face of concrete. The minimum prescribed cover is measured to the outer edge of stirrups, ties, or spirals if the transverse reinforcement is used; to the outer layer of bars if more than one layer is used without stirrups. Nominal or clear cover to main reinforcement shall be as follows:

Footing	50 mm
Columns	40 mm
Stairs, lintel, etc.	15 mm
Beams	25 mm
Slabs	15 mm

13. Chairs of approved types shall be used to support the slab steel where necessary. At least one chair shall be used for each 1.5 m^2 of floor area.
14. Formwork for slabs must be kept for a minimum of 7 days; for beams: 14 days up to 6 m span, and 21 days for more than 6 m span; and that for walls, column, etc., for 48 hours. The formwork of cantilever beams should be stripped on the instruction of the site engineer.
15. The safe bearing capacity of soil, is p kN/m^2 at a depth h m with a permissible settlement s mm as per the soil-testing report.
16. In footings, the reinforcement parallel to the longer side should be placed in the bottom layer.
17. All dimensions are in millimetres.
18. The cement-mortar ratio (1: r) for all brick masonry work should comprise of 1 part cement and r parts sand and watered well for 7 days.

In Notes 15 and 18, the values of parameters p, h, s and r should be specified as per the design stipulations.

Depending upon the type of job the preceding general notes may be modified. In structural drawings, these instructions shall be listed at one place.

8.9 DESIGN AND DETAILING FOR DUCTILITY

For acceptable performance during an earthquake a structure should have appropriate dynamic and structural characteristics. To satisfy the performance criteria as enunciated in IS:1893 a structure must have adequate strength and ductility. If the lateral force resisting structural element loaded to its full strength during an earthquake is brittle, it will fail throwing its share of load on to the other elements. On the other hand, if it is sufficiently ductile, it would continue to participate in resisting the lateral force up to its full strength after it yields, redistributing only the excess load to the remaining elements.

The term ductility implies the ability of a member to sustain significant inelastic deformations prior to collapse. A brittle material fails suddenly upon attaining its maximum load whereas a ductile material undergoes larger strains while resisting loads. Typical generalized force-deformation relationships for brittle and ductile elements are shown in Fig. 8.12. The force may be the load, moment or stress, while the deformation could be elongation, curvature, rotation or strain.

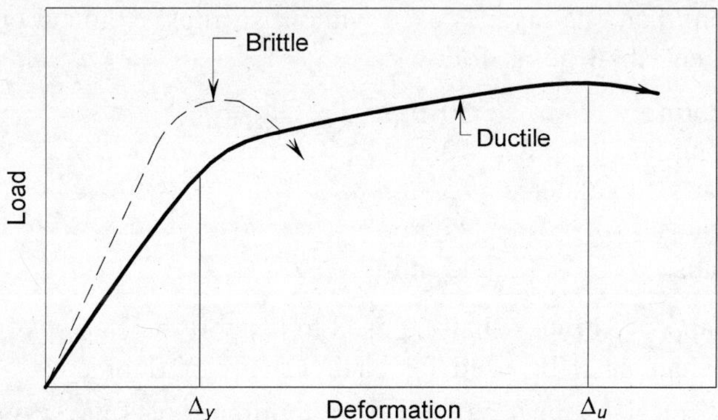

Fig. 8.12 Typical force-deformation relationships for brittle and ductile materials.

8.9.1 Significance of Ductility

A ductile structure can be expected to adapt to unexpected overloads, load reversals, impact and structural movements due to foundation settlement and volume changes. These factors are generally difficult to quantify and are ignored in the analysis and design but presumed to have been taken care of by presence of some ductility in the structure. A ductile structure provides sufficient warning to the users of its impending failure and thus reducing probability of loss of life in the event of collapse.

The limit states design procedure for reinforced concrete structures assumes that all the critical sections in the structure will reach their maximum capacities at design load for the structure. For this to happen all joints and sections must be able to withstand forces and deformations corresponding to the yielding of the reinforcement.

The structural ductility μ is generally defined as *the ratio of absolute maximum deformation (strain, rotations, curvatures or deflections)* Δ_u *to the deformation* Δ_y*, corresponding to yielding of reinforcement in the cross-section or to a major deviation from linear force-deformation curve for the structure.* The strain based ductility definition depends entirely on the material properties, while rotation or curvature based definition also includes the effects of shape and size of the cross-section. On the other hand, a deflection based definition takes into account entire configuration of structure and the loading.

8.9.2 Ductility of Beams

For simplicity curvature ductility of reinforced concrete beams can be based on the behaviour of its cross-section.

Singly reinforced beam: The yield curvature of a singly reinforced beam can be computed by using elastic theory as follows:

$$\phi_y = \frac{\varepsilon_y}{(d - kd)} \tag{8.1}$$

where yield strain of tension steel $\varepsilon_y = f_y/E_s$ and $k = [\,-mp + \sqrt{(mp)^2 + 2mp}\,]$.

The ultimate curvature at the ultimate strain of concrete at crushing, i.e., ε_{cu} (= 0.0035), is computed as follows:

$$\phi_u = \frac{\varepsilon_{cu}}{x_u} \tag{8.2}$$

where

$$x_u = \frac{0.87 f_y A_{st}}{0.362 f_{ck} b} = \frac{0.87 f_y pd}{0.362 f_{ck}} \leq x_{u,\max}$$

Therefore, ductility μ is given by

$$\mu = \frac{\varepsilon_{cu}}{\varepsilon_y}\left(\frac{d - kd}{x_u}\right) = \frac{\varepsilon_{cu}}{(f_y / E_s)}\left[1 + mp - \sqrt{(mp)^2 + 2mp}\right]\left(\frac{1}{x_u / d}\right) \tag{8.3}$$

Doubly reinforced beam: The presence of compression steel has relatively little effect on its yield curvature. However, there is a large increase in the ultimate curvature. For a doubly reinforced concrete section,

$$0.362 f_{ck} x_u b + A_{sc} f_{sc} = 0.87 f_y A_{st}$$

or

$$\frac{x_u}{d} = \left[0.87 p_t - \left(\frac{f_{sc}}{f_y}\right) p_c\right]\left(\frac{f_y}{0.362 f_{ck}}\right) \tag{8.4}$$

where $f_{sc} = 0.87 f_y$, then Eq. (8.4) reduces to

$$\frac{x_u}{d} = 2.403(p_t - p_c)\left(\frac{f_y}{f_{ck}}\right) \leq \frac{x_{u,\max}}{d} \tag{8.5}$$

From linear strain variation,

$$\frac{\varepsilon_{ym}}{d - x_u} = \frac{\varepsilon_{cu}}{x_u} \quad \text{or} \quad \frac{d - x_u}{x_u} = \frac{\varepsilon_{ym}}{\varepsilon_{cu}} \quad \text{or} \quad \frac{x_u}{d} = \frac{\varepsilon_{cu}}{\varepsilon_{cu} + \varepsilon_{ym}} \tag{8.6}$$

Therefore, from Eqs. (8.5) and (8.6),

$$(p_t - p_c) \leq 0.416\left[\frac{\varepsilon_{cu}}{\varepsilon_{cu} + \mu_s \varepsilon_{ym}}\right]\left(\frac{f_{ck}}{f_y}\right) \tag{8.7}$$

where ε_{ym} is the maximum strain in tensile steel and μ_s is the strain ductility in the steel.

It should be noted that

1. The ductility of a beam cross-section increases as the steel ratio p_t or $(p_t - p_c)$ decreases. If high percentage of reinforcement is provided, the concrete will crush before the yielding of steel, leading to a *brittle failure* corresponding to $\mu = 1.0$. Thus, a beam should be designed as an *under-reinforced section*. The other factors affecting ductility are ε_{cu}, f_{ck} and f_y. The parameter ε_{cu} is mainly a function of f_{ck}, rate of loading and transverse (shear) reinforcement. For a given value of ε_{cu} (= 0.0035), the ductility increases with the increase in f_{ck} and decreases with the increase in f_y. Since the ductility is inversely proportional to f_y, Fe415 grade steel is preferable to Fe500 grade steel from ductility considerations.
2. As mentioned earlier, the ductility increases with the decrease in $(p_t - p_c)$ value, i.e., ductility increases with increase in compression steel.
3. In a T-beam, the enlarged compression flange reduces depth of compression zone at collapse and thus increases the ductility. If the neutral axis falls in the flange, the ductility can be computed by using Eq. (8.3).
4. The lateral reinforcement tends to improve ductility by preventing premature shear failures, restraining compression reinforcement against buckling and by confining the compression zone, thereby increasing deformation capability of reinforced concrete beam.

    Therefore, adequate ductility in a flexural member can be ensured by restricting amount of tensile reinforcement and providing more compression reinforcement enclosed by stirrups. The shear reinforcement should be adequate to ensure that strength in shear exceeds that in flexure and thus, prevent a non-ductile shear failure before full flexural strength of member has been developed.

    As per IS:13920, the shear resistance of a flexural member shall be the maximum of the following:

    (i) Factored shear force as per analysis.

    (ii) Shear force due to formation of plastic hinges at both the ends plus due to the factored gravity load $w_u = 1.2\ (DL + LL)$ on the clear span L_c. For reversible seismic loading, the shears at the left and right supports 1 and 2 are given by

$$V_l = \frac{M_{p1} + M_{p2}}{L_c} \pm 0.5 w_u L_c \tag{8.8}$$

$$V_r = \frac{M_{p1} + M_{p2}}{L_c} \mp 0.5 w_u L_c \tag{8.9}$$

where M_{p1} and M_{p2} are probable hogging or sagging and sagging or hogging moments capacity at the left and right ends of the beam, respectively. The length of the yield regions is related to the relative magnitude of ultimate and yield strengths. *The larger is the ratio of ultimate to yield moment, the longer is the yield region.* Assuming the ratio of actual ultimate tensile stress in longitudinal steel to the actual tensile yield strength of steel to be not less than 1.25, the ultimate moment capacity M_p is obtained by multiplying yield moment capacity by a factor of 1.4 ($\approx$ 1.25/0.87). The design shear at each end shall be the absolute maximum of the corresponding two values of V_l and V_r.

5. A reinforced concrete framed structure should be so designed that the inelasticity is confined to the beams only and columns remain elastic. This is termed as **weak girder-strong column requirement**. To ensure this at a beam-column joint

$$\sum M_{\text{column}} > 1.2 \sum M_{\text{beam}}$$

The flexural resistance should be summed up such that the column moments oppose the beam moments. The offsets of beams to columns or offsets of columns from floor to floor should be avoided. The change in stiffness should be gradual from floor to floor.

Closed stirrups or spirals should be provided to confine concrete at the sections of maximum moment to increase the ductility of members. Such sections include upper and lower ends of columns, and within beam-column joint which do not have beams on all sides. If the axial load exceeds 0.4 times the balanced axial load, a spiral column is recommended.

6. Splices and bar anchorages must be adequate to prevent bond failures.
7. The beam-column connections should be made monolithic.

8.9.3 Detailing for Ductility

For the enhanced ductility following recommendations based on the provisions of IS:4326, IS:13920 and ACI:318 should be followed:

Design of beam sections:

1. The sagging (bottom) and hogging (top) reinforcement ratios p_t and p_c, respectively, should be greater than

$$p_{t,\min} > \frac{0.24\sqrt{f_{ck}}}{f_y} \qquad \text{(IS:13920)}$$

$$p_{t,\min} > \frac{1.4}{f_y} \qquad \text{(ACI:318)}$$

where $p = A_s/(bd)$ for the rectangular sections and $A_s/(b_w d)$ for the flanged sections. A_s is the area of reinforcement on either face. However, reinforcement ratio p should not exceed 0.025.

2. At least two bars should continue at both top and bottom throughout.
3. At the face of joint the positive-moment of resistance should not be less than one-half of the negative-moment of resistance.
4. Neither the negative nor the positive moment of resistance at any section along the member length should be less than one-fourth of the maximum moment of the resistance provided at the face of either of the joints.
5. When a beam frames into a column, both the top and bottom bars of beam should be anchored into the column so as to develop their full strength in bond beyond the section of beam at the face of the column. When the beams exist on both sides of column, bars at both faces of beams must be taken continuously through the column.
6. The spacing of designed vertical stirrups should not exceed $0.25d$ in a length equal to $2d$ near each end of beam and $0.5d$ in the remaining length.

Design of columns:

1. If the average axial stress P_u/A_g on a column subjected to seismic loading is less than $0.10f_{ck}$, the column reinforcement is designed according to the beam requirements discussed earlier. On the other hand, for the case $P_u/A_g \geq 0.10f_{ck}$, special confining reinforcement is required.
   (a) Cross-section area of bar forming circular hoops or spirals is given by

$$a_{sp} = 0.09SD_c\left(\frac{A_g}{A_c} - 1\right)\left(\frac{f_{ck}}{f_{y,sp}}\right) \tag{8.10}$$

   where S is pitch of spiral or spacing of ties and other symbols have usual meaning.
   (b) In the case of rectangular closed stirrups or ties used in the rectangular sections, the area of bars is given by

$$a_{sp} = 0.18Sh\left(\frac{A_g}{A_c} - 1\right)\left(\frac{f_{ck}}{f_{y,sp}}\right) \tag{8.11}$$

   where h is the longer dimension of the rectangular hoop to its outer face.
2. Special confining steel where required must be provided above and below the beam connection in a length of column at each end which is largest of the: (a) 1/6 of clear height of the column, (b) larger lateral dimension of the column and (c) 450 mm.
   The pitch of lateral ties should neither exceed 1/4th of minimum dimension of the member nor 100 mm.
3. Shear reinforcement must be provided in columns to resist nominal shear developed at the limit state of collapse of frame. In the presence of axial compression, an increased value of permissible shear strength of concrete τ_c is stipulated in IS:456.

$$\tau_c' = \delta\tau_c$$

where

$$\delta = \left(1 + \frac{3P_u}{A_g f_{ck}}\right) \tag{8.12}$$

where P_u and A_g are axial load in N, and gross area of concrete section in mm^2, respectively. The spacing of shear reinforcement should not exceed $0.5d$ where d is the effective depth of column measured from compression fibre to tension steel.

4. Spiral column should be used wherever possible especially if $P_u > 0.4P_b$ where P_b is the balanced axial load.

Design of beam-column connection: The beam-column joints are generally the critical links in a structure. To avoid frame failure due to inadequate joints, the joint details must be carefully considered as described earlier. However, special attention must be paid to the anchorage of beam reinforcement in the joint. The ties as required at the end of column must be provided through the joint. However, if the connection is confined by beams from all the four sides, the amount of this reinforcement is reduced to half the value. This reinforcement is generally known as **joint hoops**.

Example 8.1 In a multi-storey RCC frame building, a typical floor beam with 140 mm thick slab carries service negative bending moment and shear force of 500 kNm and 360 kN, respectively, at the face of beam-column joint due to gravity and seismic loads. The size of the beam web has been fixed at 345 × 500 mm from architectural considerations. Design the beam section for adequate ductility. The materials used are M20 concrete and HYSD steel of grade Fe415. The effective cover to tension steel is 55 mm.

Solution The overall cross-section of the beam is 345 × 640 mm deep.

$$\text{Effective depth } d = 640 - 55 = 585 \text{ mm}$$

$$\text{Factored bending moment } M_u = 1.2 \times 500 = 600 \text{ kNm}$$

and

$$\text{The moment parameter } \frac{M_u}{bd^2} = \frac{600 \times 10^6}{345 \times 585^2} = 5.08 \text{ MPa}$$

The section is designed as a doubly reinforced section with the effective cover to compression steel of 50 mm. The ratio d'/d = 0.855 (≈ 0.10). From Table C.4, Tension steel p_t = 1.670 per cent and compression steel p_c = 0.746 per cent.

$$\text{Minimum tension steel } p_{t,\min} = \frac{0.24\sqrt{f_{ck}}}{f_y}$$

$$= \frac{0.24 \times \sqrt{20}}{415} = 0.002586 \text{ or } 0.26 \text{ per cent}$$

and

$$\text{Maximum tension steel } p_{t,\max} = 2.5 \text{ per cent} > 1.670 \text{ per cent}$$

The positive moment capacity M_{up} (= $2.285bd^2$) as a singly reinforced beam with p_t = 0.746 per cent as obtained from Table C.1 is less than 50 per cent of the negative moment capacity M_{un} (=5.08 bd^2) at the same face and hence violates the ductility provisions of

IS:13920. This can be corrected by designing the section such that the compression steel is atleast 50 per cent of the tension steel at the section. Thus, with d'/d = 0.10 from Table C.2 for M_{up} = 2.54bd^2 (=M_{un}/2), p_t = p_c = 0.856 per cent.

Therefore, p_t = 1.670 per cent (=3368 mm^2) and p_c = 0.856 per cent (= 1726 mm^2). Provide 7 –25 mm ϕ bars at the top (A_{st} = 3436 mm^2) and 3–28 mm ϕ bars at the bottom face (A_{sc} = 1847 mm^2).

Shear reinforcement.

$$\text{Factored shear force } V_u = 1.2 \times 360 = 432 \text{ kN}$$

$$\text{Nominal shear stress } \tau_v = \frac{432 \times 10^3}{345 \times 585} = 2.14 \text{ MPa} < 2.8 \text{ MPa}$$

For p_t =1.670 per cent, the shear strength of concrete τ_c = 0.74 MPa. Since $\tau_c < \tau_v < \tau_{c,\max}$, the section is acceptable with shear reinforcement.

Consider 8 mm ϕ – 4 legged stirrups. The spacing is given by

$$S_v = \frac{0.87 f_y A_{sv}}{(\tau_v - \tau_c)b}$$

$$= \frac{0.87 \times 415 \times (4 \times 50)}{(2.14 - 0.74) \times 345} = 149.5 \text{ mm}$$

$$\ngtr \frac{d}{4} = \frac{585}{4} = 146.25 \text{ mm}$$

$$\ngtr 8 \times 25 = 200 \text{ mm}$$

$$\nless 60 \text{ mm (suggested for better compaction of concrete)}$$

Thus, provide 8 mm ϕ – 4 legged stirrups @ 140 mm c/c in a length equal to 1170 mm (= 2d) from the face of the beam-column joint.

Example 8.2 In a multi-storey reinforced concrete frame building a typical column of 3.36 m clear height carries an axial force of 3200 kN and a bending moment of 750 kNm under gravity and seismic load conditions. Design the column section with adequate ductility. The materials used are: M25 grade concrete mix and HYSD steel of grade Fe415.

Solution Consider square column of size 650 mm. The design loads are:

$$\text{Factored axial force } P_u = 1.2 \times 3200 = 3840 \text{ kN}$$

and

$$\text{Factored bending moment } M_u = 1.2 \times 750 = 900 \text{ kNm}$$

Longitudinal steel.

The dimensionless parameters are:

$$\frac{P_u}{f_{ck}bD} = \frac{3840 \times 10^3}{25 \times 650 \times 650} = 0.364$$

$$\frac{M_u}{f_{ck}bD^2} = \frac{900 \times 10^6}{25 \times 650 \times 650^2} = 0.131$$

From Fig. 5.16 for a column reinforced on all the four faces with d'/D = 0.10, the factor p/f_{ck} = 0.102 giving p = 2.55 per cent < 4.0 per cent.

Therefore, Area of steel A_s = 0.0255 × 650 × 650 = 10773.75 mm^2

Provide 16-32 mm ϕ bars (A_s = 12868 mm^2).

Lateral steel. Consider 8 mm ϕ bar ties at nominal or clear cover of 40 mm.

$$\text{Confining reinforcement } a_{sp} = 0.18Sh\left(\frac{A_g}{A_c} - 1\right)\left(\frac{f_{ck}}{f_{y,sp}}\right)$$

where

$$h = 650 - 40 - 40 = 570 \text{ mm} > 300 \text{ mm}$$

Therefore, provide cross tie around the middle longitudinal bar.

$$\text{Effective } h = \frac{570}{2} = 285 \text{ mm} < 300 \text{ mm}$$

Gross area of concrete A_g = 650 × 650 = 422500 mm^2

Area of concrete core A_c = (650 – 40 – 40)^2 = 324900 mm^2

Therefore,

$$\text{Ratio } \frac{A_g}{A_c} = 1.30$$

Cross-sectional area of stirrup a_{sp} = 50 mm^2

$$\text{and Spacing of stirrups } S = \frac{a_{sp}(f_{y,sp}/f_{ck})}{0.18h\left[\frac{A_g}{A_c} - 1\right]}$$

$$= \frac{50 \times (415/25)}{0.18 \times 285 \times 0.30} = 53.93 \text{ mm}$$

$$\not> \frac{D}{4}\left(= \frac{650}{4} = 162.5 \text{ mm}\right)$$

$$\not> 100 \text{ mm}$$

$$< 75 \text{ mm}$$

Special confinement is to be provided over a length which is larger of the following:

(i) $L/6 = (3.36 \times 10/6) = 560$ mm

(ii) $D = 650$ mm

(iii) 450 mm

Provide 8 mm ϕ bar stirrups @ 50 mm c/c in a distance of 650 mm from the face of the joint. In the remaining portion, the spacing is minimum of (a) 650 mm, (b) $16 \times 32 = 512$ mm and (c) 300 mm. Adopt spacing of 8 mm ϕ stirrups of 300 mm.

Example 8.3 The behaviour of a multi-storey reinforced concrete frame building subjected to lateral loads is modelled by that of plane-frames spaced 3.5 m centre-to-centre with storey height of 3.3 m. An interior beam-column joint in the ground floor of a typical frame is to be checked for its adequacy from ductility considerations. The materials used are: M25 concrete and HYSD steel of grade Fe415. The unit weights of concrete, finish material and brickwork are 25, 22 and 20 kN/m^3, respectively. The design stipulations are:

Thicknesses of slab and floor finish = 130 and 40 mm, respectively

Thickness of wall supported on the beam = 230 mm

Live load = 2.5 kN/m^2

Axial load in the column at the joint = 1100 kN

Clear span of beams on two sides of joint = 4.75 and 4.2 m

Size of beams = 230×600 mm

Size of the column = 230×550 mm

Reinforcement in the column = 8 – 25 mm ϕ (A_s = 3927 mm^2, p = 3.1 per cent)

Tension and compression steels in the beam on either side of the joint are (6-22 mm ϕ, p_t = 1.8 per cent) and (4–20 mm ϕ, p_c = 0.994 per cent), respectively.

Solution Gravity loads on the beams are:

(i) Dead load = $(0.130 \times 25 + 0.040 \times 22) \times 3.5$

$+ 0.230 \times (3.3 - 0.6) \times 20 + 0.230 \times (0.6 - 0.13) \times 25$

$= 29.58$ ($\approx$ 30 kN/m)

(ii) Live load on the beam = $2.5 \times 3.5 = 8.75$ kN/m

Factored gravity load $w_u = 1.2 \times (30 + 8.75) = 46.5$ kN/m

Check in flexure.

For a doubly reinforced section with p_t = 1.8 per cent, p_c = 0.994 per cent and $d'/d = 0.10$ from Table C.4, we have

$$\frac{M_u}{bd^2} = 5.42$$

Therefore

Hogging moment capacity $M_{un} = 5.42 \times 230 \times 550^2 \times 10^{-6} = 377.10$ kNm

For sagging moment capacity as a singly reinforced section with p_t = 0.994 per cent, from Table C.1, we get

$$\frac{M_u}{bd^2} = 3.0$$

Thus

$$M_{up} = 3.0 \times 230 \times 550^2 \times 10^{-6} = 208.73 \text{ kNm}$$

Axial load dimensionless parameters are:

$$\frac{P_u}{f_{ck}bD} = \frac{1.2 \times 1100 \times 10^3}{25 \times 230 \times 550} = 0.417$$

and

$$\frac{p}{f_{ck}} = \frac{3.1}{25} = 0.124$$

Therefore, from Fig. 5.16 for the above parameters, $M_u/f_{ck}bD^2 = 0.142$
Thus,

$$M_u = 0.142 \times 25 \times 230 \times 550^2 \times 10^{-6} = 247.0 \text{ kNm}$$

Now,

$$\Sigma M_g = 377.10 + 208.73 = 585.83 \text{ kNm}$$

$$\Sigma M_c = 2 \times 247.0 = 494.0 \text{ kNm} < 1.2\ \Sigma M_g$$

Thus, the design of beam-column joint does not satisfy the weak girder-strong column criteria. Hence, here is a need for an increase in the dimensions of the column such that the above criteria is satisfied.

Check in shear.

(i) Beam

Factored shear due to gravity load $V_{u,g} = 1.2 \times 46.5 \times \dfrac{4.75}{2} = 110.44$ kN

Shear due to formation of plastic hinge in the beam is given as:

$$V_{u,p} = 1.4 \times \frac{585.83}{4.75} = 172.67 \text{ kN}$$

$$\text{Shear force } V_u = 110.44 + 172.67 = 283.11 \text{ kN}$$

$$\text{Nominal shear stress } \tau_u = \frac{283.11 \times 10^3}{230 \times 550} = 2.238 \text{ MPa}$$

Shear strength of concrete with p_t = 1.8 per cent, τ_c = 0.79 MPa
Provide 10 mm ϕ bar – 2 legged stirrups @ 170 mm c/c.

(ii) Column

Factored shear in the column $V_u = 1.4 \times \dfrac{585.83}{3.30} = 248.53$ kN

Nominal shear stress $\tau_v = \dfrac{248.53 \times 10^3}{230 \times 550} = 1.96$ MPa

For M25 concrete with p_t (on one face) = 1.55 per cent, $\tau_c = 0.75$ MPa.

Coefficient of increase in shear strength $\delta = 1 + \dfrac{3P_u}{A_g f_{ck}} = 1 + \dfrac{3 \times 1.2 \times 1100 \times 10^3}{(230 \times 550) \times 25} = 2.252$

The enhanced permissible shear strength of concrete is:

$$\tau'_c = \delta\tau_c = 2.252 \times 0.75 = 1.69 \text{ MPa} < 1.96 \text{ MPa}$$

Hence column-section need to be increased.

Review Questions

8.1 What are the basic objectives of detailed structural drawings?

8.2 List the salient general notes for a reinforced concrete building of moderate size.

8.3 Explain the functions of distribution or temperature steel.

8.4 Draw reinforcement details for: (a) flexural member with change in direction, (b) corner of concrete wall, (c) grid-joint, and (d) rigid plane frame-joint.

8.5 Write short notes on bar support and cover.

8.6 What is meant by structural ductility? Explain briefly quantitative measures of ductility in reinforced concrete.

8.7 What are the measures taken for improving the ductility of reinforced concrete structures?

8.8 Describe the objectives of special detailing provisions of IS:13920.

8.9 Why are limits placed on tensile reinforcement ratios in beams in earthquake-resistant design?

8.10 How are the design shears estimated in beams of ductile frame?

8.11 What are the design requirements of beam-column joint in earthquake-resistant design?

8.12 Is it desirable to have: (a) high strength steel and (b) high strength concrete in earthquake resistant design of reinforced concrete structures?

8.13 What is meant by special confining reinforcement in a column of a frame designed for ductility?

8.14 What are the reasons that inclined stirrups and bent-up bars not preferred as shear reinforcement in earthquake-resistant design?

8.15 Draw reinforcement details for offset columns.

Tutorial Problems

T.8.1 The behaviour of a multi-storey reinforced concrete frame building subjected to lateral loads is modelled by that of plane-frames spaced 3.75 m centre-to-centre with storey height of 3.4 m. An interior beam-column joint in an intermediate floor of a typical frame is to be checked for adequacy of ductility. The materials used are: concrete of grade of M25 and HYSD steel of grade Fe415. The unit weights of concrete, finish and brickwork may be taken as 25, 21 and 19.5 kN/m^3.The design stipulations are:

Thicknesses of slab and floor finish	= 125 and 35 mm
Thickness of brick wall supported on the beam	= 230 mm
Live load	= 3.5 kN/m^2
Axial load in the column at the joint	= 1250 kN
Clear span of beams on two sides of joint	= 4.85 and 4.25 m
Size of beams	= 230 × 600 mm
Tension and compression steels in the beam on either side of the joint	= (6 – 20 mm ϕ) and (4 – 22 mm ϕ)
Size of the column	= 230 × 530 mm
Reinforcement in the column	= 8 – 25 mm ϕ

T.8.2 In a multi-storey reinforced concrete frame building a typical column of clear height of 3.4 m carries an axial load of 3500 kN and a bending moment of 780 kNm under gravity and seismic load conditions. Design the column section with adequate ductility. The materials are: M25 grade concrete mix and HYSD steel of grade Fe415.

T.8.3 In a multi-storey reinforced concrete frame building a typical floor beam with 125 mm thick slab carries a negative bending moment of 480 kNm and a shear force of 330 kN at the face of beam-column joint due to gravity and seismic loads. The size of the beam web has been fixed at 345 × 520 mm from architectural considerations. Design the beam section for adequate ductility. The materials are M25 concrete and HYSD steel of grade Fe415. The effective cover to tension steel is 50 mm.

Appendix

A

Working Stress Design Method

A.1 INTRODUCTION

Working stress design method is based on the behaviour of a member subjected to the loads expected to be encountered by it during its service period. It ensures satisfactory behaviour under service load and is assumed to possess adequate safety against collapse. Both the concrete and steel are assumed to behave perfectly elastically, i.e., the stress is proportional to the strain. The distribution of strain across the section is assumed to be linear, i.e., the plane sections that are plane before bending remain plane after bending. Thus, the strain and hence the stress at any point are proportional to the distance of the point from the neutral axis. With this a triangular stress distribution in concrete is obtained, varying from zero at the neutral axis to maximum at the extreme fibre of the section as shown in Fig. A.1. The allowable stresses are taken to be fixed proportions of the ultimate strength of the materials. The factor of safety for concrete in compression due to bending, and that of steel in tension are 3.0 and 1.78, respectively. The corresponding allowable stresses are given in Tables A.1 and A.2, respectively. This elastic method fails to give the maximum load that the structure can support and an idea of reserve strength of the member. The actual stresses developed in the structure at the working loads often differ considerably from the theoretical values. The reason for this difference is the assumption that the concrete is a perfectly elastic material, which is not correct; on the contrary it is a highly complex heterogeneous material.

In the working stress design approach, members are proportioned so as to sustain the anticipated service or working loads without the stresses in the concrete or reinforcement exceeding the permissible values for the individual materials. In this approach the entire margin of safety is provided for by the fact that the calculated stresses in the members under the action

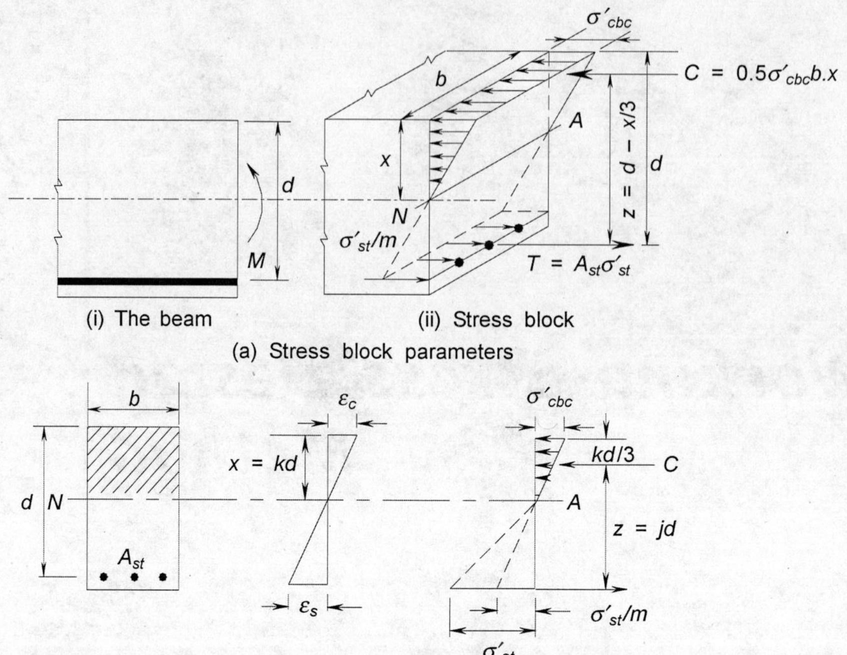

Fig. A.1 Notation, stress and strain distributions.

of working loads are such that they are considerably below the ultimate or yield stress of constituent materials.

Where the stresses due to wind load (or earthquake forces) are combined with those due to dead load, superimposed and impact loads, IS:456 permits the members to be designed for stresses in concrete and steel which are one-third greater than the permissible values. The wind and seismic forces need not be considered as acting simultaneously.

A.2 FLEXURE

The application of the working stress design procedure is based on the following basic assumptions required for the derivation of working stress design expressions:

1. Plane sections before bending remain plane after bending, i.e., the strain is proportional to the distance from the neutral axis.
2. Both concrete and reinforcing steel obey Hooke's law, i.e., stress-strain relations are straight lines under working loads.
3. The tensile strength of concrete is neglected.

4. There is a perfect bond or adhesion between the concrete and reinforcing steel so that there is no slippage between the two materials.
5. The modular ratio m has the value $280/(3\sigma_{cbc})$ (rounded off to the nearest integer value in accordance with IS:2-1960), where σ_{cbc} is the permissible stress (in MPa) in concrete in compression due to bending.
6. The steel area is assumed to be concentrated at the centroid of the steel.
7. The other basic assumptions concerning deformation and flexure of homogeneous members are valid.

The external forces are assumed to be resisted by internal compressive forces C developed in concrete and tensile forces T in steel. The moment at a section due to external loads is resisted by the internal couple formed due to the above two forces in concrete acting through the centroid of triangular distribution of compressive stress and tension acting at the centroid of steel. The distance between the lines of action of resultant resistive forces is known as **lever arm** denoted by $z(=jd)$, as shown in Fig. A.1.

When a straight line stress distribution is assumed, the neutral axis passes through the centroid of the effective cross-section for the condition $C = T$. The word effective is used here for the reason that contribution of concrete in tension is neglected and thus the effective area of concrete is the area of concrete above the neutral axis.

If a member is so proportioned that the maximum stresses in both the concrete and the reinforcing steel reach their permissible values at the same time then the section is termed as **balanced section**. The neutral axis of this section is called the **balanced** or **critical neutral axis** and its depth below the top compression fibre is denoted by x_c as shown in Fig. A.2(a). The balanced design means that there is exactly enough reinforcement to develop the permissible compressive stress in concrete. If there is lesser amount of steel then compressive strength of concrete cannot be developed, i.e., concrete is stressed below the permissible limit and the section is under-reinforced. Under such a condition the failure of the beam will be initiated due to overstress of steel. The steel being ductile gives sufficient warning before the final collapse of the structure. For this case the actual neutral axis lies above the critical neutral axis as shown in Fig. A.2(b).

If there is more steel than that required for the balanced section, the section is over-reinforced, i.e., the steel is not fully stressed to its permissible value while the maximum compressive stress in concrete reaches its permissible limit. The failure will be initiated due to overstress in concrete. As the concrete failure is brittle, the structure fails suddenly without any warning. The actual neutral axis lies below the critical neutral axis as shown in Fig. A.2(c).

In the balanced design, though both the materials are stressed to their permissible values, it is not necessarily the most economical design. In fact the greatest economy in most of the cases is attained using under-reinforced sections.

A.2.1 Rectangular Beams Reinforced in Tension

This type of beam is commonly known as **singly reinforced beam**. Figure A.1 depicts the

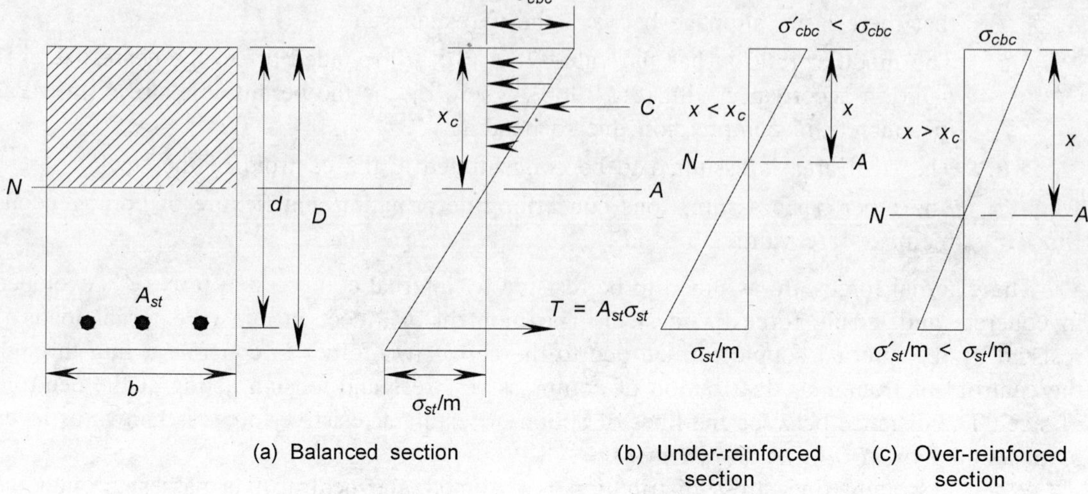

Fig. A.2 *Types of singly reinforced sections.*

assumed distribution of strain and stress across the rectangular beam section. Following symbols are used in the derivation of expressions:

σ'_{cbc} = Actual compressive stress in concrete under bending

σ'_{st} = Actual stress in steel in tension

ε_c = Strain in concrete = σ'_{cbc}/E_c

ε_s = Strain in steel = σ'_{st}/E_s

E_c, E_s = Modulii of elasticity for concrete and steel, respectively

m = Modular ratio = E_s/E_c

b = Width of beam

d = Effective depth from extreme compression fibre to the centre of tensile reinforcement

x = Depth of neutral axis from extreme compression fibre

z = Lever arm, the distance between centroid of compression force to the centroid of tensile force.

From the strain distribution curve,

$$\frac{\varepsilon_c}{\varepsilon_s} = \left(\frac{\sigma'_{cbc} / E_c}{\sigma'_{st} / E_s}\right) = \frac{x}{d - x}$$

or

$$m\left(\frac{\sigma'_{cbc}}{\sigma'_{st}}\right) = \frac{x}{d - x}$$

or

$$\sigma'_{st} = m\sigma'_{cbc}\left(\frac{d - x}{x}\right) \tag{A.1}$$

Giving,

$$x = \frac{d}{[1 + \sigma'_{st}/(m\sigma'_{cbc})]} = kd \tag{A.2}$$

where $k = 1/[1 + \sigma'_{st}/(m\sigma'_{cbc})]$, and is called the **neutral axis coefficient.**
Since the centroid of compressive force is at a distance of $(x/3)$ from the top edge of the section, the lever arm is given by:

$$z = d - \left(\frac{x}{3}\right) = \left(d - \frac{kd}{3}\right) = d\left(1 - \frac{k}{3}\right) = jd \tag{A.3}$$

where

$j = (1 - k/3)$, and is termed as the **lever arm constant**.

Total compressive force on the section $C = 0.5\sigma'_{cbc}bx$
Total tensile force $T = \sigma'_{st}A_{st}$
For the equilibrium of the section $C = T$.

Moment of resistance: The moment of resistance of a reinforced concrete section M_s is equal to the moment of the couple formed by the total compressive force C acting at the centroid of the compressive stress block, and total tensile force T in steel acting at the centroid of reinforcement. Thus, the moment of resistance of section with respect to the compressive force is:

$$M_r = Cz = 0.5\sigma'_{cbc}bx\left(d - \frac{x}{3}\right) \tag{A.4}$$

The moment of resistance of the section with respect to the tensile force is given as:

$$M_r = Tz = (\sigma'_{st}A_{st})\left(d - \frac{x}{3}\right) \tag{A.5}$$

The moment of resistance capacity of a concrete section may be computed from the permissible stresses in both concrete and steel, permissible stress in steel or permissible stress in concrete depending on whether the section is balanced, under-reinforced or over-reinforced, respectively. A section is termed as **balanced** when maximum stresses in concrete and steel reach their permissible values at the same time, i.e., $\sigma'_{cbc} = \sigma_{cbc}$ and $\sigma'_{st} = \sigma_{st}$. The neutral axis for this case is called **critical neutral axis** and is denoted by x_c, as illustrated in Fig. A.2. Thus, from Eq. (A.2),

$$x_c = \frac{d}{[1 + \sigma_{st}/(m\sigma_{cbc})]} = k_c d \tag{A.6}$$

The corresponding value of the moment of resistance with respect to compression is obtained from Eq. (A.4) as:

$$M_r = Cz = 0.5\sigma_{cbc} b(k_c d)\left(d - \frac{k_c d}{3}\right)$$

$$= 0.5\sigma_{cbc} bd^2 k_c\left(1 - \frac{k_c}{3}\right) = (0.5\sigma_{cbc} k_c j)bd^2 = Qbd^2 \quad \text{(A.7)}$$

Q is called the **moment of resistance factor** for a balanced rectangular section. The moment of resistance with respect to tensile force is obtained as:

$$M_r = Tz = (\sigma_{st} A_{st})\left(d - \frac{k_c d}{3}\right) = (\sigma_{st} A_{st})jd \quad \text{(A.8)}$$

Area of reinforcement can be obtained from the condition, $T = C$.
Thus, for a balanced section,

$$A_{st}\sigma_{st} = 0.5\sigma_{cbc}(bx) = 0.5\sigma_{cbc} bk_c d$$

or

$$A_{st} = \frac{\sigma_{cbc} k_c bd}{2\sigma_{st}} \quad \text{(A.9)}$$

The percentage of reinforcing steel is expressed as:

$$p_t = \frac{A_{st}}{bd} \times 100$$

Thus, for the balanced section from Eq. (A.9),

$$p_{t,\text{bal}} = \frac{100A_{st}}{bd} = \frac{50k_c\sigma_{cbc}}{\sigma_{st}} \quad \text{(A.10)}$$

Design constants for a balanced section: For M20 grade concrete and Fe250 grade mild steel, σ_{cbc} = 7.0 MPa, σ_{st} =140 MPa and m = 280/(3σ_{cbc}) = 13.33 (= 13, rounded off to the nearest integer value). The neutral axis coefficient is given by

$$\text{k}_\text{c} = \frac{1}{\left(1 + \dfrac{\sigma_{st}}{m\sigma_{cbc}}\right)} = \frac{1}{\left(1 + \dfrac{140}{13 \times 7}\right)} = 0.3939$$

and thus,

$$j = 1 - \frac{k_c}{3} = 1 - \frac{0.394}{3} = 0.869$$

$$Q = 0.5\sigma_{cbc} k_c j = 0.5 \times 7 \times 0.394 \times 0.869 = 1.198 \text{ MPa}$$

and

$$p_{t,\text{bal}} = \frac{50k_c\sigma_{cbc}}{\sigma_{st}} = \frac{50 \times 0.394 \times 7}{140} = 0.985$$

The design constants for some of the material combinations are given in Table A.1. If the amount of steel provided is less than that required for the balanced section, i.e., $p_t < p_{t,\text{bal}}$, the steel is stressed to its permissible value while the concrete is under-stressed. The actual neutral axis moves upward ($x < x_c$) and such a section is called **under-reinforced section**, as shown in Fig. A.2. For an under-reinforced section,

$$\sigma'_{st} = \sigma_{st},\ \sigma'_{cbc} < \sigma_{cbc},\ \text{and } x < x_c$$

and allowable

$$M_r = Tz = A_{st}\sigma_{st}\left(d - \frac{x}{3}\right)$$

and

$$\sigma'_{cbc} = \frac{A_{st}\sigma_{st}}{0.5bx}$$

Table A.1 Design Constants for the Balanced Sections

Grade of concrete	*Allowable stress in flexure σ_{cbc}, MPa*	*Modular ratio, m*	*Fe250, σ_{st} = 140 MPa*				*Fe415, σ_{st} = 230 MPa*			
			k_c	*j*	*Q, MPa*	$p_{t,bal}$	k_c	*j*	*Q, MPa*	$p_{t,bal}$
M20	7.0	13	0.394	0.869	1.198	0.985	0.283	0.906	0.898	0.431
M25	8.5	11	0.400	0.867	1.475	1.214	0.289	0.904	1.110	0.534
M30	10.0	9	0.391	0.870	1.701	1.396	0.281	0.906	1.273	0.611

On the other hand, if the area of steel provided is more than that required for a balanced section, the section is termed as **over-reinforced**. In this case, steel is not fully stressed to its permissible value while compressive stress in extreme compression fibre reaches its permissible value. For an over-reinforced section,

$$\sigma'_{cbc} = \sigma_{cbc},\ \sigma'_{st} < \sigma_{st},\ \text{and } x > x_c$$

and allowable

$$M_r = Cz = 0.5\sigma_{cbc}bx\left(d - \frac{x}{3}\right)$$

and

$$\sigma_{st} = \frac{0.5\sigma_{cbc}bx}{A_{st}}$$

For a section with given properties as shown in Fig. A.3, i.e., *b*, *d* and A_{st} being known, the position of the neutral axis *x* (= *kd*) can be obtained by equating total compressive and total tensile forces on the section; *C* = *T*, i.e.,

$$0.5\sigma_{cbc}bx = A_{st}\sigma_{st} = A_{st}m\left[\frac{\sigma_{cbc}(d - x)}{x}\right]$$

or

$$bx^2 + 2mA_{st}x - 2mA_{st}d = 0$$

or

$$x = -\frac{mA_{st}}{b} \pm \left[\left(\frac{mA_{st}}{b}\right)^2 + \frac{2mA_{st}d}{b}\right]^{1/2}$$

Substituting $A_{st}/bd = p$, we get

$$k = -m\frac{A_{st}}{bd} \pm \left[\left(\frac{mA_{st}}{bd}\right)^2 + 2m\frac{A_{st}}{bd}\right]^{1/2}$$

$$= -mp + (m^2p^2 + 2mp)^{1/2} \qquad \text{(A.11)}$$

(+ve sign is taken since k cannot be negative.)

Following types of problems are generally encountered in the analysis of concrete beams reinforced in tension:

Type 1 Determination of moment of resistance and load-carrying capacity of a beam section: In the analysis problems the dimensions of the beam section (b, d), area of reinforcement (A_{st}), and permissible stresses in the material (σ_{cbc} and σ_{st}) are given. The steps involved are:

1. Find the position of actual neutral axis x from the known values of b, d and A_{st} using Eq. (A.11).
2. Find the position of critical neutral axis x_c for the known permissible stresses in concrete and steel from Eq. (A.6).
3. Compare x with x_c to determine the type of section:
   (a) If $x > x_c$, the section is over-reinforced.
   (b) If $x < x_c$, the section is under-reinforced.
4. Calculate the moment of resistance for the appropriate type of section:

$$M_r = 0.5\sigma_{cbc}bx\left(d - \frac{x}{3}\right) \qquad \text{for the over-reinforced section}$$

or

$$M_r = \sigma_{st}A_{st}\left(d - \frac{x}{3}\right) \qquad \text{for the under-reinforced section}$$

If the effective span and support conditions of the beam are known, the load-carrying capacity can also be computed.

Example A.1 A reinforced concrete beam section of size 300 × 700 mm effective depth is reinforced with 3 bars of 20 mm ϕ in tension. Determine the moment of resistance and the maximum stresses induced in the materials, i.e., in concrete and steel. The concrete mix and HYSD steel reinforcement used are of grades M20 and Fe415, respectively.

If the effective span of the beam is 5.0 m, determine the maximum superimposed uniformly distributed load the beam can carry safely. The unit weight of concrete is 25 kN/m^3.

Solution For the given materials:

$$\sigma_{cbc} = 7 \text{ MPa}, \sigma_{st} = 230 \text{ MPa}$$

and

$$\text{Modular ratio } m = \frac{280}{3\sigma_{cbc}} = \frac{280}{3 \times 7} = 13.33 \text{ (say 13)}$$

Area of tension steel,

$$A_{st} = 3 \times \frac{\pi}{4}(20)^2 = 942 \text{ mm}^2 \text{ (alternatively from Table E.1, Appendix E)}$$

Thus

$$p = \frac{A_{st}}{bd} = \frac{942}{300 \times 700} = 0.004486,$$

and

$$mp = 0.05832.$$

The coefficient of depth of the actual neutral axis is: $k = (m^2p^2 + 2mp)^{1/2} - mp = 0.288$
The coefficient of depth of the critical neutral axis is:

$$k_c = \frac{1}{[1 + \sigma_{st}/(m\sigma_{cbc})]} = \frac{1}{[1 + 230/(13 \times 7)]} = 0.2835$$

Since $k > k_c$, the section is over-reinforced and the concrete will reach its maximum or permissible stress first, i.e., $\sigma'_{cbc} = \sigma_{cbc} = 7$ MPa.
The depth of the actual neutral axis is:

$$x = kd$$
$$= 0.288 \times 700 = 201.60 \text{ mm}$$

Thus,

$$\text{Lever arm } z = \left(d - \frac{x}{3}\right) = 700 - \left(\frac{201.6}{3}\right) = 632.80 \text{ mm}$$

The moment of resistance is given by

$$M_r = Cz = (0.5\sigma_{cbc}bx)z$$
$$= 0.5 \times 7 \times 300 \times 201.6 \times 632.80$$
$$= 133.951 \times 10^6 \text{ Nmm} = 133.951 \text{ kNm}$$

Stresses in materials are:

$$\sigma'_{cbc} = 7 \text{ MPa}$$

$$\sigma'_{st} = \frac{C}{A_{st}} = \frac{0.5\sigma_{cbc}bx}{A_{st}}$$

$$= \frac{0.5 \times 7 \times 300 \times 201.1}{942} = 224.71 \text{ MPa} < 230 \text{ MPa } (=\sigma_{st})$$

Load-carrying capacity. Consider w to be the distributed load on the beam in kN per metre run. The maximum bending moment is:

$$M = \frac{wl^2}{8} = \frac{w \times (5)^2}{8}$$

Equate M to M_t. Thus,

$$\frac{25w}{8} = 133.951$$

Giving

$$w = 42.86 \text{ kN/m}$$

Let the effective cover to reinforcement be 50 mm then the total depth of beam is given as:

$$D = d + 50 = 700 + 50 = 750 \text{ mm}$$

$$\text{Self-weight of beam } w_d = (0.300 \times 0.750 \times 1) \times 25 = 5.625 \text{ kN/m}$$

Superimposed load the beam can support

$$w_s = w - w_d = 42.86 - 5.625 = 37.235 \text{ kN/m (say 37 kN/m)}$$

Type II Determination of stresses developed in concrete and steel: In this type of analysis problems the sectional properties (b, d and A_{st}) and applied moment M are given. The steps involved are:

1. Find the position of the actual neutral axis using the stipulated sectional properties. Take moments of effective transformed areas about the neutral axis as shown in Fig. A.3.

$$bx\left(\frac{x}{2}\right) = mA_{st}(d - x)$$

or alternatively by using Eq. (A.11).

2. Determine the lever arm, $z = d - (x/3)$.

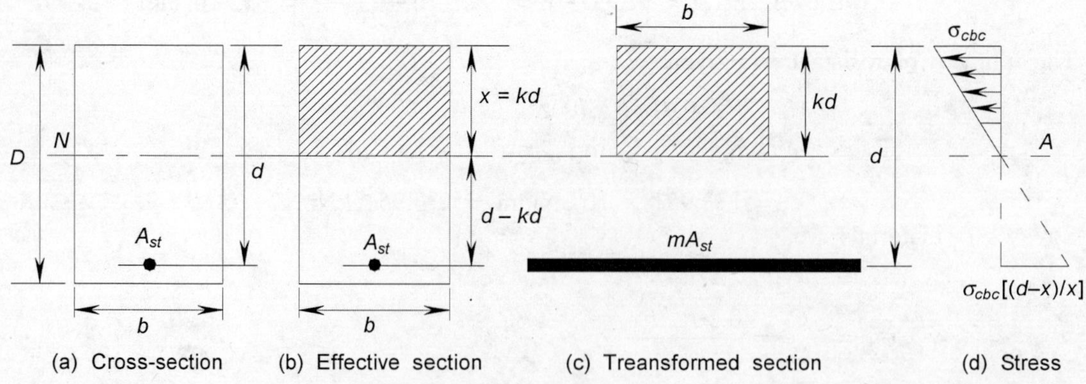

Fig. A.3 *Transformed area of a reinforced concrete section under flexure after cracking.*

3. Find the stresses induced in the materials due to applied moment M:

$$\text{Stress in steel } \sigma'_{st} = \frac{M}{A_{st}z}$$

$$\text{Stress in concrete, } \sigma'_{cbc} = \frac{M}{0.5bxz}$$

or alternatively from the stress diagram:

$$\sigma'_{cbc} = \left(\frac{\sigma'_{st}}{m}\right)\frac{x}{d-x}$$

A.2.2 Design of Rectangular Beams

The following types of problems are generally encountered.

Type I Determination of tensile steel: The cross-sectional dimensions are predetermined from architectural or other considerations. The materials to be used and the moment to be carried are known. The steps involved are:

1. Compute the design constants m, k and Q for the given materials.
2. Determine the allowable moment of resistance or moment-carrying capacity of the section

$$M_r = Qbd^2$$

3. Compare the moment to be resisted with M_r obtained in Step (2). The following cases may arise:

CASE I If $M < M_r$, the section is to be designed as an under-reinforced section. Take moment of transformed areas about the neutral axis as shown in the Fig. A.3

$$bx\left(\frac{x}{2}\right) = mA_{st}(d-x) \quad \text{or} \quad A_{st} = \frac{bx^2}{2m(d-x)} \tag{A.12}$$

Since the section is under-reinforced,

$$M = M_r = \sigma_{st}A_{st}\left(d - \frac{x}{3}\right)$$

or

$$M = \sigma_{st}\left(\frac{bx^2}{2m(d-x)}\right)\left(d - \frac{x}{3}\right) = \frac{\sigma_{st}bx^2\left(3 - \frac{x}{d}\right)}{6m\left(1 - \frac{x}{d}\right)}$$

or

$$\left(\frac{6Mm}{\sigma_{st}bd^2}\right) = \frac{(3-k)k^2}{(1-k)} \quad \text{or} \quad k^3 - 3k^2 - Rk + R = 0 \tag{A.13}$$

where

$$R = \left(\frac{6Mm}{\sigma_{st}bd^2}\right) \quad \text{and} \quad k = \left(\frac{x}{d}\right)$$

Equation (A.13) is solved for k and A_{st} is calculated from Eq. (A.12).

CASE II If $M > M_r$, the section is over-reinforced then

$$M = M_r = 0.5\sigma_{cbc}bx\left[d - \frac{x}{3}\right]$$

First determine x and then A_{st} from the equation obtained by equating the moment of transformed areas about the neutral axis:

$$mA_{st}(d - x) = bx\left(\frac{x}{2}\right) \quad \text{or} \quad A_{st} = \frac{bx^2}{2m(d - x)}$$

However, IS:456 does not permit use of the over-reinforced section. In such cases, it is preferable to design it as a doubly reinforced beam where the reinforcement is also provided in compression to give additional strength to the concrete.

Example A.2 A rectangular beam of size 250 × 500 mm overall is to carry a design moment of 40 kNm. Determine the area of tension reinforcement to be provided at an effective cover of 50 mm. The materials to be used are M20 grade concrete and high yield strength deformed steel of grade Fe415.

If the design moment is 50 kNm instead of 40 kNm, determine the amount of steel required.

Solution For M20 grade concrete and Fe415 grade steel:

$$\sigma_{cbc} = 7.0 \text{ MPa}, \ \sigma_{st} = 230 \text{ MPa}, \ m = 13, \ k_c = 0.283 \text{ and } Q = 0.898 \text{ MPa},$$

$$b = 250 \text{ mm}, \ d = 500 - 50 = 450 \text{ mm}$$

Critical moment of resistance is:

$$M_r = Qbd^2 = 0.898 \times 250 \times 450^2 = 45.46 \times 10^6 \text{ Nmm} = 45.46 \text{ kNm}$$

(i) When $M = 40$ kNm.

Since $M_r > M$, the section is to be designed as the under-reinforced section. From Eq. (A.13),

$$R = \left(\frac{6Mm}{\sigma_{st}bd^2}\right) = \frac{6 \times (40 \times 10^6) \times 13}{230 \times 250 \times 450^2} = 0.268$$

$$k^3 - 3k^2 - Rk + R = k^3 - 3k^2 - 0.268k + 0.268 = 0$$

Giving,

$$k = 0.28 < k_c (= 0.283)$$

Therefore,

$$x = kd = 0.268 \times 450 = 120.6 \text{ mm}$$

Substituting this value of x in Eq. (A.12),

$$A_{st} = \frac{250 \times (125.54)^2}{2 \times 13 \times (450 - 125.54)} = 424.56 \text{ mm}^2$$

Provide 2 × 12 mm + 1 × 16 mm ϕ bars (A_{st} = 427.3 mm^2).

(ii) When M = 50 kNm.

Since $M_r < M$, the section will be treated as an over-reinforced section. Hence,

$$M_r = 0.5\sigma_{cbc} bx\left(d - \frac{x}{3}\right)$$

or

$$50 \times 10^6 = 0.5 \times 7 \times 250x\left(450 - \frac{x}{3}\right)$$

or

$$x^2 - 1350x + 171428.57 = 0$$

Giving

$$x = 141.9 \text{ mm}$$

Take moment of transformed areas about the neutral axis,

$$bx\left(\frac{x}{2}\right) = mA_{st}(d - x)$$

or

$$A_{st} = \frac{bx^2}{2m(d - x)} = \frac{250 \times 141.9^2}{2 \times 13 \times (450 - 141.9)} = 628.41 \text{ mm}^2$$

Provide two bars of 20 mm ϕ (A_{st} = 628.4 mm^2).

Type II Design or proportioning a section to support the given loads: In this type of problems, cross-sectional dimensions and amount of tension reinforcement to be provided are computed by assuming the section to be a balanced section. The steps involved are:

1. Calculate the maximum bending moment M due to applied loads.
2. Determine the design constants k, j and Q for balanced design.
3. Choose a depth-to-breadth ratio of the beam section preferably between 2 and 3.
4. Determine the sectional dimensions b and d from the relation, $M = Qbd^2$.

5. Obtain the overall depth D by adding nominal concrete cover to steel to the effective depth d.
6. Compute the area of tension reinforcement using the expression $A_{st} = M/[\sigma_{st} jd]$.

Example A.3 Design a rectangular section for a simply supported reinforced concrete beam of effective span of 5 m carrying a concentrated load of 40 kN at its mid-span. The concrete to be used is of grade M20 and the reinforcement consists of HYSD steel bars with σ_{st} = 230 MPa.

Solution For M20 grade concrete and HYSD steel of grade Fe415:

$$\sigma_{cbc} = 7 \text{ MPa}, \ \sigma_{st} = 230 \text{ MPa and } m = 13$$

Coefficient for the critical neutral axis is:

$$k_c = \frac{1}{(1 + \sigma_{st}/m\sigma_{cbc})} = 0.283$$

The corresponding lever arm coefficient is given by

$$j = 1 - \frac{k_c}{3} = 0.906$$

The critical or balanced moment of resistance coefficient is obtained as:

$$Q = 0.5\sigma_{cbc}k_c j$$

$$= 0.5 \times 7 \times 0.283 \times 0.906 = 0.898 \text{ MPa}$$

Thus, the critical moment of resistance of a section is given by

$$M_r = 0.898bd^2$$

Consider $b = 0.5d$, hence $M_r = 0.449d^3$.

CASE I When the self-weight of the beam is ignored, the maximum bending moment is given by

$$M = \frac{WL}{4} = 40 \times \frac{5}{4} = 50 \text{ kNm}$$

Therefore,

$$M_r = 0.449d^3 = 50 \times 10^6$$

Thus, d = 481.11 mm (say 500 mm) and $b = 0.5\ d$ = 250 mm

Now,

$$A_{st}\sigma_{st} jd = M \quad \text{or} \quad A_{st} = \frac{M}{\sigma_{st}\, jd}$$

Therefore

$$A_{st} = \frac{50 \times 10^6}{230 \times 0.906 \times 500} = 479.89 \text{ mm}^2$$

Hence, provide 3 bars of 16 mm ϕ (A_{st} = 603 mm^2) at an effective cover of 50 mm. Thus, the overall dimensions of the section are 250 × 550 mm.

CASE II The self-weight of the beam can be taken into consideration by revising the section arrived at in Case I. Consider an enhanced section of size 300 × 600 mm overall as the trial section. Self-weight of the beam $w_d = 0.300 \times 0.600 \times 25 = 4.5$ kN/m.

Maximum bending moment due to uniformly distributed self-weight is:

$$\frac{w_d L^2}{8} = \frac{4.5 \times 5^2}{8} = 14.06 \text{ kNm}$$

Total design moment $M = 50 + 14.06 = 64.06$ kNm
Equate M_r to M.

$$0.449d^3 = 64.06 \times 10^6$$

Therefore,

$$d = 522.53 \text{ mm}$$

Adopt 16 mm ϕ bars at nominal cover of 30 mm.

$$\text{Overall depth } D = d + (\text{Half diameter of bar}) + \text{Clear cover}$$

$$= 522.53 + \left(\frac{16}{2}\right) + 30 = 560.53 \text{ mm (say 570 mm)}$$

and

$$\text{Effective depth } d = 570 - 8 - 30 = 532 \text{ mm}$$

Thus

$$b = 0.5d = 0.5 \times 532 = 266 \text{ mm (say 270 mm)}$$

Hence a section of size 270 × 570 mm overall will be adequate. The amount of tension steel required is given by

$$A_{st} = \frac{64.06 \times 10^6}{230 \times 0.906 \times 532} = 577.86 \text{ mm}^2$$

Provide 3 bars of 16 mm ϕ (A_{st} = 603 mm^2) at a nominal cover of 30 mm.

A.2.3 Rectangular Beams Reinforced in Tension and Compression

A section reinforced both in tension and compression is known as a **doubly reinforced section**. Doubly reinforced sections are provided in the cases where cross-section is limited or where reversal of stresses may take place, e.g. in case of wind and earthquake loads. As compared to singly reinforced sections the doubly reinforced sections are uneconomical. It is, therefore, advisable to restrict the depth of neutral axis *kd* to that corresponding to balanced section. The doubly reinforced sections are recommended only when they are absolutely necessary.

In reinforced concrete structures, the creep deformation of concrete produces an additional strain in compression steel and gradually increases the stress which can be accounted for by increasing the modular ratio. IS:456 stipulates that the permissible compressive stress in reinforcing bars in a beam or slab, when the compressive resistance of the concrete is taken into account, shall be taken as (1.5m) times the calculated compressive stress in the surrounding concrete or σ_{sc} whichever is lower, where σ_{sc} is the permissible stress in compression in bars. Consider a rectangular section reinforced on tension as well as on compression faces, as shown in Fig. A.4. The notations used in the figure are:

A_{st} = Total area of tension steel

A_{sc} = Area of compression steel

σ_{cbc} = Permissible stress in concrete

σ_{st} = Permissible stress in tension steel

σ_{sc} = Permissible stress in compression steel

σ'_{sc} = Stress in compression steel

From the strain distribution curve,

$$\frac{\varepsilon_c}{\varepsilon_{sc}} = \frac{\sigma'_{cbc}/E_c}{\sigma'_{sc}/E_s} = m\frac{\sigma'_{cbc}}{\sigma'_{sc}} = \frac{x}{x-d'}$$

or

$$\sigma'_{sc} = \frac{m\sigma'_{cbc}(x-d')}{x} = m \times \text{Stress in concrete at the level of compression steel}$$

However, as explained earlier, the effective modular ratio shall be taken as 1.5m for compression steel in flexure. Thus,

$$\sigma'_{sc} = 1.5m\sigma'_{cbc}\left(\frac{x-d'}{x}\right) \le \sigma_{sc} \tag{A.14}$$

Equivalent concrete area resisting compression is:

$$A_c = (bx - A_{sc}) + 1.5mA_{sc} = bx + (1.5m - 1)A_{sc} \tag{A.15}$$

as illustrated in Fig. A.4(a).

To determine depth of neutral axis take moment of effective areas about neutral axis,

$$bx\left(\frac{x}{2}\right) + (1.5m - 1)A_{sc}(x - d') = mA_{st}(d - x)$$

For an economical design, $x = k_c d$. Thus,

$$b\left[\frac{(k_c d)^2}{2}\right] + (1.5m - 1)A_{sc}(k_c d - d') = mA_{st}(d - k_c d) \tag{A.16}$$

Moment of resistance: C_c and C_s are the total compressive forces in concrete and steel, respectively. M_r is obtained by taking moment about tension steel:

$$M_r = C_c\left(d - \frac{x}{3}\right) + C_s(d - d')$$

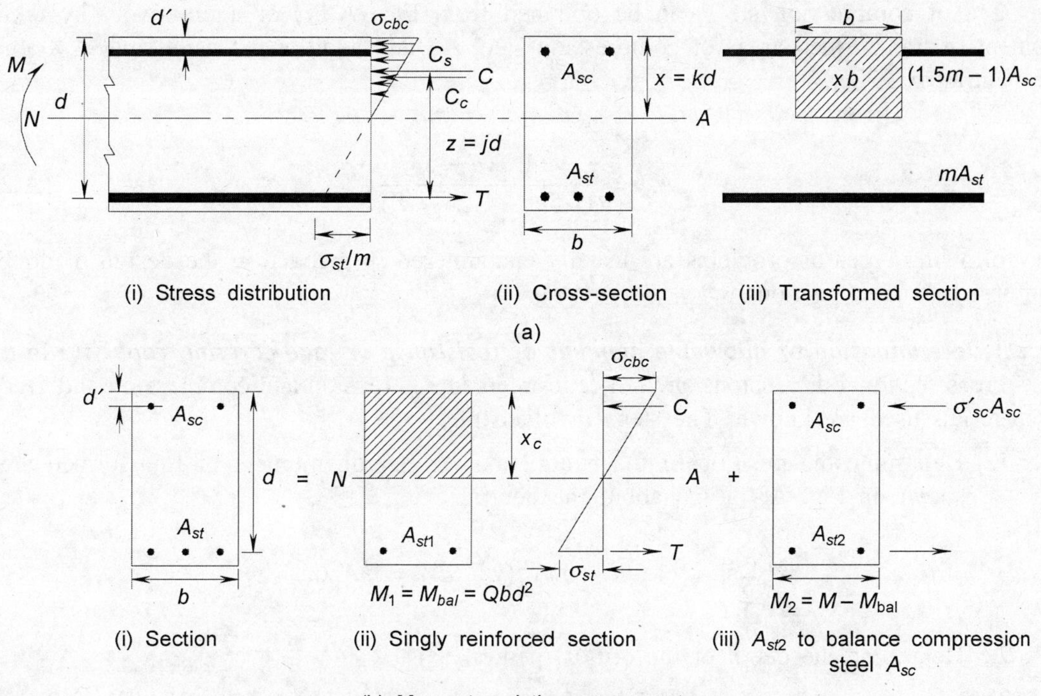

Fig. A.4 *Beams reinforced both in tension and compression.*

$$= bx(0.5\sigma_{cbc})\left(d - \frac{x}{3}\right) + \left[\frac{(1.5m - 1)A_{sc}\sigma_{cbc}(x - d')}{x}\right](d - d')$$

$$= M_1 + M_2 \tag{A.17}$$

Thus, in the design of doubly reinforced beams, the moment of resistance is considered to consist of two parts. First is the resisting moment M_1 of the beam section with only tension reinforcement A_{st1} for the balanced condition, as illustrated in Fig. A.4[b(ii)]. The second is the moment of resistance M_2 of an auxiliary section due to the couple formed by the additional compressive force in compression steel A_{sc} and the tensile force in a like amount of additional tension steel A_{st2}.

The corresponding values of tension steel are:

$$A_{st} = \frac{M_1}{\sigma_{st}\, jd} \tag{A.18}$$

and

$$A_{st2} = \frac{M_2}{\sigma_{st}(d - d')} \tag{A.19}$$

Therefore, total tension steel $A_t = A_{st1} + A_{st2}$ (A.20)

The area of compression steel can be obtained from Eq. (A.17) or alternatively by taking moment of transformed areas of compression steel A_{sc} and additional tensile steel A_{st2} about the neutral axis, i.e.,

$$(1.5m - 1)A_{sc}(x - d') = mA_{st2}(d - x)$$

$$A_{sc} = \frac{mA_{st2}(d - x)}{(1.5m - 1)(x - d')} \tag{A.21}$$

The following types of problems are usually encountered in practice in the design of doubly reinforced concrete sections.

Type I Determination of allowable moment of resistance or load-carrying capacity: In this case, cross-sectional dimensions, area of reinforcements in tension and compression and grades of materials used are known. The steps involved are:

1. Determine the position of the neutral axis by taking moment of transformed areas shown in Fig. A.4[a(iii)] about the neutral axis:

$$bx\left(\frac{x}{2}\right) + (1.5m - 1)A_{sc}(x - d') = mA_{st}(d - x)$$

2. Determine the depth of the critical neutral axis x_c.
3. Compare x with x_c to determine the stress level in the two materials. The following two cases arise:

   CASE I If $x > x_c$, the beam is over-reinforced and concrete reaches its permissible stress σ_{cbc}. M_r is given by

$$M_r = 0.5\sigma_{cbc}bx\left(d - \frac{x}{3}\right) + (1.5m - 1)A_{sc}\sigma_{cbc}\left(\frac{x - d'}{x}\right)(d - d')$$

   CASE II If $x < x_c$, steel reaches its permissible stress σ_{st} earlier, and

$$M_r = \left(\frac{bx}{2}\right)\left(\frac{\sigma_{st}}{m}\right)\left(\frac{x}{d - x}\right)\left(d - \frac{x}{3}\right) + (1.5m - 1)\,A_{sc}\left(\frac{\sigma_{st}}{m}\right)\left(\frac{x - d'}{d - x}\right)(d - d')$$

4. If the effective span and the support conditions of the beam are known, compute the load-carrying capacity.

Example A.4 A reinforced concrete simply supported beam of rectangular section of size 300 × 500 mm effective is reinforced with 4 × 20 mm ϕ bars at tension face, and 4 × 14 mm ϕ bars at compression face at an effective covers of 50 mm. Determine: (a) the moment of resistance of the section, (b) the safe uniformly distributed load the beam can support in addition to its self-weight over an effective span of 8 m and (c) the maximum stresses developed in concrete, tension and compression steels.

What will be the effect of modifying tension steel to 4 × 16 mm ϕ bars? The materials used are M20 grade concrete and HYSD reinforcement of grade Fe415.

Solution For M20 grade concrete and Fe415 grade steel:

$$\sigma_{cbc} = 7 \text{ MPa}, \ \sigma_{st} = 230 \text{ MPa and } m = \frac{280}{3\sigma_{cbc}} = 13.33 \text{ (say 13)}$$

CASE I $A_{st} = 4 \times 20 \text{ mm } \phi = 1256 \text{ mm}^2$
$A_{sc} = 4 \times 14 \text{ mm } \phi = 615 \text{ mm}^2$

Transformed area of compression steel = $(1.5m - 1)A_{sc} = (1.5 \times 13 - 1) \times 615 = 11377.5 \text{ mm}^2$

Transformed area of tension steel = $mA_{st} = 13 \times 1256 = 16328 \text{ mm}^2$
Consider x to be the depth of the neutral axis. Take moment of transformed areas shown in Fig. A.4(a) about neutral axis:

$$300x\left(\frac{x}{2}\right) + 11377.5(x - 50) = 16328\ (500 - x)$$

or

$$x^2 + 184.70x - 58219.17 = 0$$

Giving $x = 166.01$ mm and $k = \dfrac{x}{d} = 0.332$

The coefficient of critical depth of the neutral axis is:

$$k_c = \frac{1}{[1 + \sigma_{st}/m\sigma_{cbc}]} = \frac{1}{[1 + 230/(13 \times 7)]} = 0.283$$

Since $k > k_c$, the section is over-reinforced and concrete will reach its permissible value. Thus, $\sigma'_{cbc} = \sigma_{cbc} = 7$ MPa, and stress in compressive steel is:

$$\sigma_{sc} = 1.5m \times \left[\frac{\sigma_{cbc}(x - d')}{x}\right]$$

$$= 1.5 \times 13 \times 7 \times \frac{166.01 - 50}{166.01} = 95.39 \text{ MPa}$$

Stress in tension steel is given by

$$\sigma'_{st} = \frac{m\sigma_{cbc}(d - x)}{x}$$

$$= 13 \times 7 \times \frac{500 - 166.01}{166.01} = 183.08 \text{ MPa}$$

The allowable moment of resistance of the section is:

$$M_r = 0.5\sigma_{cbc}bx\left(d - \frac{x}{3}\right) + (1.5m - 1)A_{sc}\sigma_{cbc}\left(\frac{x - d'}{x}\right)(d - d')$$

$$= 0.5 \times 7 \times 300 \times 166.01 \times \left[500 - \frac{166.01}{3}\right]$$

$$+ (1.5 \times 13 - 1) \times 615 \times 7 \times \left[\frac{166.01 - 50}{166.01}\right](500 - 50)$$

$$= 77.51 \times 10^6 + 25.044 \times 10^6 = 102.554 \times 10^6 \text{ Nmm} = 102.554 \text{ kNm}$$

If the load on the beam is w kN/m then the maximum moment is:

$$M = w \times \frac{8^2}{8} = 102.554$$

Giving,

$$w = 12.819 \text{ kN/m}$$

Now,

$$\text{Self-weight of beam } w_d = 0.300 \times 0.550 \times 25 = 4.125 \text{ kN/m}$$

Superimposed uniformly distributed load w_s the beam can support safely is:

$$w_s = w - w_d = 12.819 - 4.125 = 8.694 \text{ kN/m}$$

CASE II

$$A_{st} = 4 \times 16 \text{ mm } \phi = 804 \text{ mm}^2$$
$$A_{sc} = 4 \times 14 \text{ mm } \phi = 615 \text{ mm}^2$$

For the position of the neutral axis, take moment of transformed areas about the neutral axis as:

$$bx\left(\frac{x}{2}\right) + (1.5m - 1)A_{sc}(x - d') = mA_{st}(d - x)$$

$$150x^2 + (1.5 \times 13 - 1) \times 615 \times (x - 50) = 13 \times 804 \times (500 - x)$$

or

$$x^2 + 145.53x - 38{,}632.5 = 0$$

Giving

$$x = 136.82 \text{ mm}$$

Therefore,

$$k = \frac{x}{d} = 0.274$$

The value of k_c as determined in Case I is 0.283.
Since $k < k_c$, the section is under-reinforced and stress in tension steel will reach its permissible value. Thus, $\sigma'_{st} = \sigma_{st} = 230$ MPa, and compressive stress in extreme fibre of concrete from the stress diagram:

$$\sigma'_{cbc} = \left(\frac{\sigma_{st}}{m}\right)\left(\frac{x}{d - x}\right)$$

$$= \left(\frac{230}{13}\right) \times \left(\frac{136.82}{500 - 136.82}\right) = 6.67 \text{ MPa}$$

and

$$\text{Stress in compression steel } \sigma'_{sc} = 1.5m\left(\frac{\sigma_{st}}{m}\right)\left(\frac{x-d'}{d-x}\right)$$

$$= 1.5 \times 13 \times \left(\frac{230}{13}\right) \times \left(\frac{136.82-50}{500-136.82}\right) = 82.47 \text{ MPa}$$

The allowable moment of resistance of the section is:

$$M_r = 0.5\sigma'_{cbc}bx\left(d-\frac{x}{3}\right)+\left(\frac{\sigma'_{sc}}{1.5m}\right)(1.5m-1)(A_{sc}(d-d')$$

$$= 0.5 \times 6.67 \times 300 \times 136.82 \times \left(500-\frac{136.82}{3}\right)$$

$$+\left[\frac{82.47}{1.5\times 13}\right] \times (1.5 \times 13 - 1) \times 615 \times (500 - 50)$$

$$= 83.854 \times 10^6 \text{ Nmm} = 83.854 \text{ kNm}$$

Type II Checking of adequacy of section: In this type of problems, the cross-sectional dimensions and areas of tension and compression steels are given. It is required to check the stresses developed in the materials under the action of applied moment. The steps involved are:

1. Find the position of the actual neutral axis.
2. Equate moment of resistance of the section in terms of compressive stress in the extreme fibre of concrete σ'_{cbc} to the applied moment M and determine σ'_{cbc}.
3. Compute stresses in tension and compression steels.

Example A.5 A reinforced concrete simply supported beam of rectangular cross-section of size 300 × 500 mm effective is reinforced with four 25 mm ϕ mild steel bars on tension side and four 20 mm ϕ mild steel bars on the compression side. The effective cover to compression steel is 40 mm from the top edge. The concrete mix used is of grade M20. Determine the stresses in steel and concrete, if the beam carries total uniformly distributed load of 10 kN/m over an effective span of 10 m.

Solution For the given beam section:

$$b = 300 \text{ mm},\ d = 500 \text{ mm},\ d' = 40 \text{ mm and } m = 13$$

$$A_{st} = (4 \times 25 \text{ mm } \phi) = 1963 \text{ mm}^2,\ A_{sc} = (4 \times 20 \text{ mm } \phi) = 1256 \text{ mm}^2$$

Transformed areas corresponding to:

$$\text{Tension steel} = mA_{st} = 13 \times 1963 = 25519 \text{ mm}^2$$

$$\text{Compression steel} = (1.5m - 1)A_{sc} = (1.5 \times 13 - 1) \times 1256 = 23236 \text{ mm}^2$$

$$\text{Maximum moment } M = \frac{wL^2}{8} = \frac{10\times 10^2}{8} = 125 \text{ kNm}$$

To locate the neutral axis take moment of transformed areas about the neutral axis as:

$$\frac{300x^2}{2} + 23236 \times (x - 40) = 25519 \times (500 - x)$$

$$x^2 + 325.03x - 91259.6 = 0$$

Giving

$$x = 180.52 \text{ mm}$$

Let σ'_{cbc} be the compressive stress in the extreme fibre of concrete in compression then stress in concrete at the level of compression steel is given by

$$\sigma''_{cbc} = \sigma'_{cbc} \times \frac{180.52 - 40}{180.52} = 0.7784\sigma'_{cbc} \text{ MPa}$$

Taking moment of compressive forces about the tensile steel, we have

$$M'_r = 0.5\ \sigma'_{cbc} bx \left(d - \frac{x}{3} \right) + (1.5m - 1)\ A_{sc}\sigma''_{cbc}(d - d')$$

$$= [0.5 \times 300 \times 180.52 \times \left(500 - \frac{180.52}{3} \right)$$

$$+ (1.5 \times 13 - 1) \times 1256 \times 0.7784 \times (500 - 40)]\sigma'_{cbc} = 20.23 \times 10^6\ \sigma'_{cbc}$$

Equating M'_r to M, we get

$$20.23 \times 10^6\ \sigma'_{cbc} = 125 \times 10^6$$

Giving

$$\sigma'_{cbc} = 6.179 \text{ MPa}$$

Stress in compressive steel is:

$$\sigma_{sc} = 1.5m\sigma''_{cbc}$$

$$= 1.5 \times 13 \times 0.7784 \times 6.179 = 93.79 \text{ MPa}$$

Stress in tension steel is:

$$\sigma'_{st} = m\sigma'_{cbc}\left(\frac{d - x}{x} \right) = 13 \times 6.179 \times \frac{500 - 180.52}{180.52}$$

$$= 142.16 \text{ MPa}$$

Type III Design of a section: Design of a section for a given moment and cross-sectional dimensions involve the determination of areas of tension and compression reinforcements. The steps involved are:

1. Compute design constants for the given materials.
2. Adopt suitable concrete cover to compression steel.

3. Determine allowable moment of resistance of the section as M_{bal} ($=M_1$) $= Qbd^2$ and comparing it with the design moment M. The following three cases may arise:

   (a) $M = M_{\text{bal}}$, (b) $M < M_{\text{bal}}$ and (c) $M > M_{\text{bal}}$

   For the first two cases, the section is designed as singly reinforced beam as discussed earlier. In the third case, the section is designed as doubly reinforced beam.

4. Compute the area of tension reinforcement A_{st} for resisting M_{bal}:

$$A_{st1} = \frac{M_{\text{bal}}}{\sigma_{st}\, jd}$$

5. Compute the additional moment to be resisted by internal couple system formed by additional tensile steel A_{st2} and compression steel A_{sc}:

$$M_2 = M - M_{\text{bal}}$$

6. Calculate additional tension steel:

$$A_{st2} = \frac{M - M_{\text{bal}}}{\sigma_{st}(d - d')}$$

   and

$$A_{st} = A_{st1} + A_{st2}$$

7. Compute area of compression steel from the equation:

$$(M - M_{\text{bal}}) = (1.5m - 1)A_{sc}\sigma_{cbc}\left[\frac{x_c - d'}{x_c}\right](d - d') \tag{A.22}$$

Example A.6 A reinforced concrete beam of rectangular section has to carry a uniformly distributed load of 8 kN/m over an effective span of 10 m. Design the section using M20 grade concrete and Fe415 grade HYSD steel, when: (a) there is no restriction on the size of the beam section and (b) the size of the section is restricted to 250 × 600 mm effective.

Solution For M20 grade concrete and Fe415 grade steel,

$$\sigma_{cbc} = 7 \text{ MPa},\ \sigma_{st} = 230 \text{ MPa and } m = 13,$$

$$k_c = 0.283,\ j = 0.906,\ x_c = k_c d = 169.8 \text{ mm and } Q = 0.898 \text{ MPa}$$

$$\text{Design moment } M = \frac{wL^2}{8} = \frac{8 \times 10^2}{8} = 100 \text{ kNm}$$

CASE I Design the beam such that a singly reinforced balanced section is obtained, i.e.,

$$M_{\text{bal}} = Qbd^2 = M \text{ or } 0.898bd^2 = 100 \times 10^6$$

Consider $d/b = 2.5$.

$$d^3 = 278.4 \times 10^6$$

Therefore,

$$d = 652.96 \text{ mm}, \; b = \frac{652.96}{2.5} = 261.19 \text{ mm}$$

An overall section of size 260 × 700 mm with the effective depth d = 655 mm will be adequate. Area of tension reinforcement for this singly reinforced beam is given by

$$A_{st} = \frac{M}{\sigma_{st} jd} = \frac{(100 \times 10^6)}{230 \times 0.906 \times 655} = 732.66 \text{ mm}^2$$

Provide 3 bars of 18 mm ϕ (A_{st} = 763 mm^2) for tension reinforcement.

CASE II The section is restricted to 250 × 600 mm effective. The allowable moment of resistance of the singly reinforced balanced section is:

$$M_{\text{bal}} = Qbd^2 = 0.898 \times 250 \times 600^2 \times 10^{-6} = 80.82 \text{ kNm}$$

Since $M > M_{\text{bal}}$, the beam section is to be doubly reinforced. The area of steel A_{st1} resisting the moment M_{bal} is obtained as:

$$A_{st1} = \frac{M_{\text{bal}}}{\sigma_{st} jd} = \frac{80.82 \times 10^6}{230 \times 0.906 \times 600} = 646.42 \text{ mm}^2$$

Additional moment $M_2(= M - M_{\text{bal}})$ is resisted by the additional tension steel A_{st2} and compression steel A_{sc}. Consider the effective cover to the compression steel d' to be 40 mm. Thus,

$$A_{st2} = \frac{M - M_{\text{bal}}}{\sigma_{st}(d - d')}$$

$$= \frac{(100 - 80.82) \times 10^6}{230 \times (600 - 40)} = 148.91 \text{ mm}^2$$

Then

$$\text{Total area of tension steel } A_{st} = A_{st1} + A_{st2} = 795.33 \text{ mm}^2$$

The compression steel can be computed from the equation obtained by equating moments of transformed areas of compression and tension steels about the neutral axis:

$$(1.5m - 1)A_{sc}(x_c - d') = mA_{st2}(d - x_c)$$

or

$$A_{sc} = \frac{13 \times 148.91 \times (600 - 169.8)}{(1.5 \times 13 - 1)(169.8 - 40)} = 346.81 \text{ mm}^2$$

Provide four 16 mm ϕ (A_{st} = 804 mm^2) bars as tension steel and two 16 mm ϕ (A_{st} = 402 mm^2) bars as compression steel.

A.2.4 Flanged Beams

As discussed in Chapter 3 that the reinforced concrete floor systems where concrete slab is cast monolithically with and, connected to, rectangular beams, a portion of the slab above the beam behaves structurally as a part of the beam in compression. The slab portion is called **flange** and beam the **web**. Such a beam with flange and web is termed as **flanged beam**. Flanged beams may be T-beams or L-beams, so called because of their resemblance with the letters T, and inverted L, respectively. The flanged beams are shown in Fig. A.5. The terms b_f, b_w, D_f and d represent the width of flange, breadth of the web, thickness of the flange (i.e., of the slab), and effective depth of the beam, respectively. The effective width of the flange b_f may be taken as stipulated by IS:456:

For T-beam: $$b_f = \left(\frac{l_o}{6}\right) + b_w + 6D_f$$

For L-beam: $$b_f = \left(\frac{l_o}{12}\right) + b_w + 3D_f \qquad \text{(A.23)}$$

where l_o is the distance between points of zero moments in the beams. This is equal to the effective span for the simply supported beams and is 0.7 times the effective span for the continuous beams. However, the effective flange width in no case should exceed the breadth of web plus half the sum of the clear distances to the adjacent beams on either side.

In case of isolated beams, the effective flange width obtained below shall, in no case, be greater than the actual width of the flange.

For T-beam: $$b_f = b_w + \frac{l_o}{(l_o/b) + 4}$$

For L-beam: $$b_f = b_w + \frac{0.5l_o}{(l_o/b) + 4} \qquad \text{(A.24)}$$

A.2.4.1 Flanged Beams with Tension Reinforcement

Neutral axis: As in the case of a rectangular beam the first step in the analysis of the flanged section is to determine the position of the neutral axis. If the material properties, i.e., σ_{cbc}, σ_{st} and m, are known, the critical neutral axis can be determined from the stress distribution as follows:

$$\frac{\sigma_{cbc}}{\sigma_{st}/m} = \frac{x_c}{d - x_c} \quad \text{or} \quad x_c = \frac{d}{1 + \sigma_{st}/(m\sigma_{cbc})} = k_c d$$

The neutral axis may lie either in the flange or in the web as shown in Fig. A.5 On the other hand, if the cross-sectional dimensions, i.e., b_f, D_f, b_w, d, and the area of steel A_{st} are known, the depth x of the actual neutral axis can be computed by equating moments of transformed areas of compression and tension sides about the neutral axis.

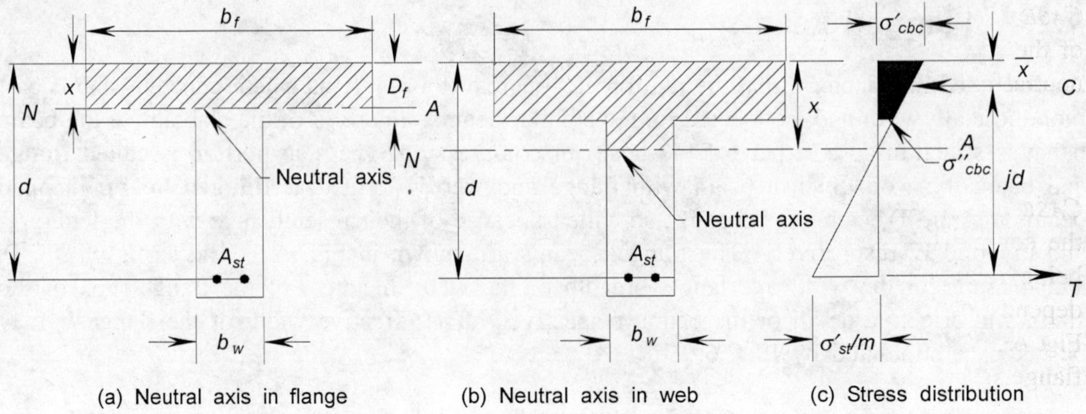

***Fig. A.5** Normal stress distribution in reinforced concrete T-beam under flexure after cracking.*

CASE I $x < D_f$: The neutral axis lies in the flange as shown in Fig. A.5(a). Equate the moments of transformed areas about the neutral axis as follows:

$$b_f x\left(\frac{x}{2}\right) = mA_{st}(d - x)$$

CASE II $x > D_f$: The neutral axis lies in the web as shown in Fig. A.5(b). Equate the moments of transformed areas about the neutral axis:

$$b_f D_f\left(x - \frac{D_f}{2}\right) + \frac{b_w(x - D_f)(x - D_f)}{2} = mA_{st}(d - x)$$

However, the compression taken by the web is very small as compared to that taken by the flange and hence may be neglected for preliminary design, i.e.,

$$b_f D_f\left(x - \frac{D_f}{2}\right) = mA_{st}(d - x)$$

Thus, while calculating the depth of the neutral axis, it is essential to assume for the first trial that either $x < D_f$ or $x > D_f$. To reduce the calculations for the first trial, consider that the neutral axis coincides with the underneath of the flange slab, and calculate moments M_c and M_s, respectively of transformed areas of concrete flange and transformed area of steel about the bottom of the flange. Then

1. If $M_c > M_s$, the neutral axis lies in the flange.
2. If $M_c = M_s$, the neutral axis coincides with the bottom of the flange.
3. If $M_c < M_s$, the neutral axis lies in the web.

Lever arm: Depending upon the position of the neutral axis, there will be two cases of lever arm.

CASE I: If the neutral axis of the section falls within the flange ($x \le D_f$), then the analysis of the section is the same as for a rectangular beam, because the effect of concrete below the neutral axis has been neglected:

$$z = jd = \left(d - \frac{x}{3}\right)$$

CASE II: If the neutral axis lies in the web ($x > D_f$), the total compressive force consists of the compressive force in the flange and that in the web. The compressive force in the web, being relatively very small, can be neglected in the analysis. The lever arm of the section depends on the position of the centroid of the trapezoidal stress diagram as illustrated in Fig. A.5(c). The distance of the centroid of the trapezoidal stress diagram from the top of the flange is given by

$$\bar{x} = \left(\frac{\sigma'_{cbc} + 2\sigma''_{cbc}}{\sigma'_{cbc} + \sigma''_{cbc}}\right)\left(\frac{D_f}{3}\right) \tag{A.25a}$$

where σ'_{cbc} and σ''_{cbc} are the stresses at the top and bottom of the flange, respectively. From the stress distribution diagram,

$$\sigma''_{cbc} = \sigma'_{cbc}\left(\frac{x - D_f}{x}\right) \text{ and } \sigma'_{st} = \frac{m\sigma'_{cbc}(d - x)}{x}$$

Thus,

$$\bar{x} = \frac{3x - 2D_f}{2x - D_f}\left(\frac{D_f}{3}\right) \tag{A.25b}$$

Now,

$$\text{Lever arm } z(=jd) = d - \bar{x}$$

Moment of resistance: If the neutral axis falls in the flange, the moment of resistance of the section is:

$$M_r = Cz = b_f x\left(\frac{\sigma'_{cbc}}{2}\right)\left(d - \frac{x}{3}\right)$$

or

$$M_r = Tz = A_{st}\sigma'_{st}\left(d - \frac{x}{3}\right)$$

For the case when the neutral axis falls in the web, the moment of resistance with respect to compression is:

$$M_r = b_f D_f\left(\frac{\sigma'_{cbc} + \sigma''_{cbc}}{2}\right)(z)$$

where $z = d - \bar{x}$

and that with respect to tension is:

$$M_r = (A_{st}\sigma'_{st})(z)$$

Types of problems:

Type I Analysis of the section. If the dimensions of the section, area of tension steel and permissible stress in the materials are given, the following procedure may be adopted:

1. Determine the depths of the critical neutral axis x_c and the actual neutral axis x. Compare the two to ascertain whether the section is balanced, under- or over-reinforced. If $x < x_c$ the beam is under-reinforced, and if $x > x_c$, the beam is over-reinforced.
2. Based on the controlling permissible stress compute the stress in the other materials.
3. Compute the lever arm.
4. Determine the moment of resistance.

Example A.7 A reinforced concrete T-beam section having an effective flange width of 1500 mm, flange thickness of 100 mm, width of web as 250 mm and effective depth of 450 mm, is reinforced with (a) 6 × 22 mm ϕ bars, (b) 6 × 25 mm ϕ bars, on tension side. Calculate the moment of resistance and stresses induced in the extreme compression fibre of concrete and in steel. The materials used are: M20 grade concrete and HYSD steel of grade Fe415. Determine the uniformly distributed load inclusive of self-weight the beam can support safely over an effective span of 6.5 m.

Solution For M20 concrete and Fe415 grade steel reinforcement:

$$\sigma_{cbc} = 7 \text{ MPa}, \ \sigma_{st} = 230 \text{ MPa and } m = 13$$

$$k_c = 0.283, \ j = 0.906, \text{ and } Q = 0.898 \text{ MPa}$$

The sectional properties are:

$$b_f = 1500 \text{ mm}, \ d = 450 \text{ mm}, \ D_f = 100 \text{ mm}, \ b_w = 250 \text{ mm}$$

$$A_{st} = 6 \times 22 \text{ mm } \phi = 2280 \text{ mm}^2$$

$$A_{st} = 6 \times 25 \text{ mm } \phi = 2945 \text{ mm}^2$$

The position of the critical neutral axis is obtained from:

$$x_c = k_c d = 0.283 \times 450 = 127.35 \text{ mm}$$

CASE *I* Determine moments M_c and M_s, respectively, of the transformed areas about the bottom of the flange.

$$M_c = b_f D_f \left(\frac{D_f}{2} \right) = 1500 \times 100 \times 50 = 7.5 \times 10^6 \text{ mm}^3$$

and

$$M_s = mA_{st}(d - D_f) = 13 \times 2280 \times 350 = 10.37 \times 10^6 \text{ mm}^3$$

Since $M_s > M_c$, the neutral axis lies in the web ($x > D_f$). Equate moments about the neutral axis:

$$b_f D_f \left(x - \frac{D_f}{2} \right) = mA_{st}(d - x)$$

or

$$1500 \times 100 \times (x - 50) = 13 \times 2280 \times (450 - x)$$

Giving

$$x = 116.00 \text{ mm}$$

Since $x < x_c$, the section is under-reinforced with $\sigma'_{st} = \sigma_{st} = 230$ MPa. The corresponding stresses in concrete at the top and bottom of the flange are:

$$\sigma'_{cbc} = \left(\frac{\sigma'_{st}}{m}\right)\frac{x}{(d-x)}$$

$$= \frac{230 \times 116.00}{13 \times (450 - 116.00)} = 6.14 \text{ MPa}$$

$$\sigma''_{cbc} = \left(\frac{\sigma'_{st}}{m}\right)\frac{(x - D_f)}{(d-x)}$$

$$= \frac{230 \times (116.00 - 100)}{13 \times (450 - 116.00)} = 0.85 \text{ MPa}$$

Depth of the centroid of compressive from the top of the flange is:

$$\bar{x} = \left(\frac{\sigma'_{cbc} + 2\sigma''_{cbc}}{\sigma'_{cbc} + \sigma''_{cbc}}\right) \times \left(\frac{D_f}{3}\right) = 37.39 \text{ mm}$$

Alternatively,

$$\bar{x} = \frac{3x - 2D_f}{2x - D_f} \times \frac{D_f}{3}$$

$$= \frac{3 \times 116.00 - 2 \times 100}{2 \times 116.00 - 100} \times \frac{100}{3} = 37.37 \text{ mm}$$

Thus,

$$\text{Lever arm } z = 450 - 37.37 = 412.63 \text{ mm}$$

Allowable moment of resistance $M_r = Tz$

$$= (A_{st}\sigma_{st})z$$

$$= (2280 \times 230) \times 412.63$$

$$= 216.38 \times 10^6 \text{ Nmm} = 216.38 \text{ kNm}$$

Let w be the uniformly distributed load inclusive of self-weight in kN/m the beam can carry safely. Equate maximum moment M to M_r:

$$w \times \frac{(6.5)^2}{8} = 216.38$$

Therefore,

$$w = 40.97 \text{ kN/m (say 40.9 kN/m)}$$

Case II $A_{st} = 6 \times 25$ mm $\phi = 2945$ mm^2
Moments of transformed areas about the bottom of the flange are:

$$M_c = 1500 \times 100 \times 50 = 7.5 \times 10^6 \text{ mm}^3$$

$$M_s = 13 \times 2945 \times 350 = 13.40 \times 10^6 \text{ mm}^3$$

Since $M_s > M_c$, the neutral axis falls in the web ($x > D_f$). Thus,

$$b_f D_f \left(x - \frac{D_f}{2} \right) = m A_{st}(d - x)$$

or

$$1500 \times 100 \times (x - 50) = 13 \times 2945 \times (450 - x)$$

Giving

$$x = 131.33 \text{ mm}$$

Since $x > x_c$, the section is over-reinforced with $\sigma'_{cbc} = \sigma_{cbc} = 7$ MPa. The corresponding stresses in concrete at the bottom of the flange and tension steel are:

$$\sigma''_{cbc} = \frac{\sigma'_{cbc}(x - D_f)}{x} = 7 \times \frac{131.33 - 100}{131.33} = 1.67 \text{ MPa}$$

$$\sigma'_{st} = \frac{m\sigma'_{cbc}(d - x)}{x} = 13 \times 7 \times \frac{450 - 131.33}{131.33} = 220.81 \text{ MPa}$$

Depth of the centroid of compressive from the top of the flange is:

$$\bar{x} = \left[\frac{\sigma'_{cbc} + 2\sigma''_{cbc}}{\sigma'_{cbc} + \sigma''_{cbc}} \right] \times \frac{D_f}{3} = 39.75 \text{ mm}$$

$$\text{Alternatively } \bar{x} = \left[\frac{3x - 2D_f}{2x - D_f} \right] \times \frac{D_f}{3}$$

$$= \left[\frac{3 \times 131.33 - 2 \times 100}{2 \times 131.33 - 100} \times \frac{100}{3} \right] = 39.75 \text{ mm}$$

Thus,

Lever arm $z = 450 - 39.75 = 410.25$ mm

$$\text{Now,} \quad M_r = Cz = \left[\frac{b_f D_f (\sigma'_{cbc} + \sigma''_{cbc})}{2} \right] z$$

$$= \frac{1500 \times 100 \times (7 + 1.67) \times 410.25}{2} = 266.77 \times 10^6 \text{ Nmm} = 266.77 \text{ kNm}$$

Alternatively moment of resistance can be computed w.r.t. tension steel as:

$$M_r = Tz$$

$$= (A_{st}\sigma_{st})z = (2945 \times 220.81) \times 410.25$$

$$= 266.78 \times 10^6 \text{ Nmm} = 266.78 \text{ kNm}$$

Let w be the safe distributed load inclusive of self-weight in kN/m. Equate maximum moment M to allowable moment of resistance M_r:

$$w \times \frac{(6.5)^2}{8} = 266.78$$

Therefore,

$$w = 50.51 \text{ kN/m} \qquad \text{(say 50.5 kN/m)}$$

Type II Check of adequacy. The cross-sectional dimensions b_f, D_f, b_w, d, and area of tension reinforcement are given. It is required to compute the stresses developed in the materials under the action of given loading. The steps involved are:

1. Determine the depth of the actual neutral axis as explained in Type I problems.
2. Compute the lever arm.
3. Determine M_r in terms of unknown tensile stress σ_{st} and equating it to the moment due to external load, and determine σ_{st}.
4. Calculate σ_{cbc} using the stress diagram.

Example A.8 A singly reinforced concrete beam of T-shaped section has an effective flange of width 1500 mm and thickness of 125 mm. The tension reinforcement is provided at an effective depth of 600 mm. The beam supports a uniformly distributed load 45 kN/m (inclusive of self-weight) over an effective span of 6.5 m. The grade of concrete mix used is M20. Determine the stresses induced in the top compression fibre of concrete and in tension steel for the following cases of tension steel: (a) 6 × 20 mm ϕ and (b) 6 × 25 mm ϕ.

Solution The properties of the T-section are:

$$b_f = 1500 \text{ mm}, \; D_f = 125 \text{ mm and } d = 600 \text{ mm}$$

For M20 grade concrete, $\sigma_{cbc} = 7$ MPa, $m = 13$
Then

$$\text{Maximum moment } M = \frac{wl^2}{8}$$

$$= 45 \times \frac{(6.5)^2}{8} = 237.66 \text{ kNm}$$

CASE I $A_{st} = 6 \times 20 \text{ mm } \phi = 1885 \text{ mm}^2$
Moments of transformed areas about the bottom of the flange are:

$$M_c = 1500 \times 125 \times \left(\frac{125}{2}\right) = 11.72 \times 10^6 \text{ mm}^3$$

$$M_s = 13 \times 1885 \times (600 - 125) = 11.64 \times 10^6 \text{ mm}^3$$

Since $M_c > M_s$, the neutral axis lies in the flange. Take moment about the neutral axis,

$$b_f x\left(\frac{x}{2}\right) = mA_{st}(d - x)$$

or

$$\frac{1500x^2}{2} = 13 \times 1885 \times (600 - x)$$

or

$$x^2 + 32.67x - 19{,}604 = 0$$

Giving

$$x = 124.63 \text{ mm} < D_f$$

The section behaves as a rectangular section.
Now,

$$\text{Lever arm } z = d - \left(\frac{x}{3}\right)$$

$$= 600 - \left(\frac{124.63}{3}\right) = 558.46 \text{ mm}$$

$$\text{Stress in steel } \sigma'_{st} = \frac{M}{A_{st}z} = \frac{237.66 \times 10^6}{1885 \times 558.46} = 225.76 \text{ MPa}$$

$$\text{Stress in concrete } \sigma'_{cbc} = \left(\frac{\sigma'_{st}}{m}\right)\frac{x}{(d-x)} = \frac{225.76}{13} \times \frac{124.63}{(600 - 124.63)} = 4.553 \text{ MPa}$$

CASE II $A_{st} = 6 \times 25 \text{ mm } \phi = 2944 \text{ mm}^2$
Moments of transformed areas about the bottom of the flange are:

$$M_c = 1500 \times 125 \times \left(\frac{125}{2}\right) = 11.72 \times 10^6 \text{ mm}^3$$

$$M_s = 13 \times 2944 \times (600 - 125) = 18.18 \times 10^6 \text{ mm}^3$$

Since $M_c < M_s$, the neutral axis falls in the web ($x > D_f$). Take moment about the neutral axis,

$$b_f D_f\left(x - \frac{D_f}{2}\right) = mA_{st}(d - x)$$

$$1500 \times 125\,(x - 62.5) = 13 \times 2944 \times (600 - x)$$

Giving

$$x = 153.61 \text{ mm}$$

The depth of the centroid of compression force from the top of the flange is given by

$$\bar{x} = \left[\frac{3x - 2D_f}{2x - D_f}\right]\left(\frac{D_f}{3}\right)$$

$$= \left[\frac{3 \times 153.61 - 2 \times 125}{2 \times 153.61 - 125}\right] \times \left(\frac{125}{3}\right) = 48.21 \text{ mm}$$

Therefore, lever arm $z = d - x$

$$= 600 - 48.21 = 551.79 \text{ mm}$$

$$\text{Stress in steel } \sigma'_{st} = \frac{237.66 \times 10^6}{2944 \times 551.79} = 146.30 \text{ MPa}$$

$$\text{Stress in concrete } \sigma'_{cbc} = \left(\frac{\sigma'_{st}}{m}\right)\frac{x}{(d-x)} = \frac{146.30 \times 153.61}{13 \times (600 - 153.61)} = 3.87 \text{ MPa}$$

Alternatively, let σ'_{cbc} be the stress at the extreme compression fibre then the stress at the bottom fibre of the flange is:

$$\sigma''_{cbc} = \frac{\sigma'_{cbc}(x - D_f)}{x} = \frac{\sigma'_{cbc}(153.61 \times 125)}{153.61} = 0.186\sigma'_{cbc}$$

Total compressive force in flange is expressed as:

$$C = \sigma'_{cbc} \times (1 + 0.186) \times 1500 \times \frac{125}{2} = 74.125\sigma'_{cbc}$$

Therefore,

$$M_r = Cz = 74.125\sigma'_{cbc} \times 551.79$$

Equating M_r to M, we have

$$\sigma'_{cbc} = \frac{237.66 \times 10^6}{74.125 \times 551.79} = 3.87 \text{ MPa}$$

Corresponding stress in tension steel is given as:

$$\sigma'_{st} = \frac{\sigma'_{cbc}(d - x)}{x} = 3.87 \times \frac{600 - 153.61}{153.61} = 146.20 \text{ MPa}$$

Type III Design of the flanged beam. The thickness of the flange is fixed by the design of the slab, and the effective width of the flange is that part of the slab which deflects monolithically with the beam and is under compression. The width of the web is based on shear and reinforcement placement considerations. Thus, the design of the T-or L-beam section requires the determination of the effective depth of the beam and the area of reinforcement. The steps involved are:

1. Estimate the depth of the flanged beam for preliminary computations as follows:
   (a) 1/12th of the span for heavy loads
   (b) 1/12th to 1/15th of the span for medium loads
   (c) 1/15th to 1/20th for light loads
2. Estimate the rib width b_w which is approximately one-half of the rib projection below the slab flange.
3. Determine the maximum bending moment M to be carried by the beam.

4. Check the depth from economic considerations:

$$d = \frac{D_f}{2} + \left\{ \frac{Mr}{\sigma_{st} b_w} \right\}^{1/2} \tag{A.26}$$

where r is the cost ratio of steel to concrete.

5. Ascertain the type of section, i.e., whether it is balanced, under-reinforced or over-reinforced.
6. Calculate the sectional area of reinforcement based on z_c or on assumed approximate value,

$$z = 0.90d \text{ or } \left(d - \frac{D_f}{2} \right) \tag{A.27}$$

7. Based on the sectional area of reinforcement obtained in Step 6, determine the actual position of the neutral axis and hence the actual value of the lever arm.
8. Recalculate the sectional area of reinforcement.
9. Check the resulting maximum stresses in concrete and steel.

Example A.9 A reinforced concrete floor system consists of a 100 mm thick slab cast monolithically with simply supported beams having an effective span of 6.5 m and spaced 3 m centre-to-centre. The super-imposed load on floor is 6.0 kN/m^2. Design a typical intermediate beam, if the materials to be used are M20 grade concrete and mild steel reinforcement. The unit weight of concrete is 25 kN/m^3.

Solution For M20 concrete and mild steel reinforcement,

$$\sigma_{cbc} = 7 \text{ MPa}, \ \sigma_{st} = 140 \text{ MPa}, \ m = 13 \text{ and } k = 0.394$$

For preliminary design consider a total depth of approximately 1/12th of the span, i.e., say 550 mm, and the width of the rib,

$$b_w = \frac{550 - 100}{2} = 225 \text{ mm, say } 250 \text{ mm}$$

The effective width of the flange b_f is the minimum of the following:

(i) $l_o/6 + b_w + 6D_f = 6500/6 + 250 + (6 \times 100) = 1933$
(ii) 3m, i.e., 3000 mm

Adopt a flange width b_f = 1933 mm

Load on beam.

Self-weight of slab = 0.10 × 25	= 2.5 kN/m^2
Superimposed load	= 6.0 kN/m^2
Total load on slab	= 8.5 N/m^2
Load per metre length of beam = 8.5 × 3	= 25.5 kN/m
Self-weight of beam = 0.25 × (0.55 – 0.1) × 25	= 2.813 kN/m
Total uniformly distributed load on the beam w	= 28.313 kN/m

Then,

$$\text{Maximum bending moment} = \frac{wl^2}{8} = \frac{28.313 \times 6.5^2}{8} = 149.53 \text{ kNm}$$

$$\text{Economic depth } d = \left(\frac{D_f}{2}\right) + \left(\frac{Mr}{\sigma_{st} b_w}\right)^{1/2}$$

Consider cost ratio, $r = 60$, then

$$d = \left(\frac{100}{2}\right) + \left[\frac{149.53 \times 10^6 \times 60}{140 \times 250}\right]^{1/2} = 556.3 \text{ mm}$$

Thus, adopt a total depth of 550 mm with $d = 500$ mm.
Assume the section to be balanced:

$$x_c = 0.394d = 197 \text{ mm} > D_f \text{ (=100 mm)}$$

Hence, the critical neutral axis falls in the web, and the stress underneath the flange is obtained as:

$$\sigma''_{cbc} = \frac{\sigma'_{cbc}(x_c - D_f)}{x_c} = 0.492\ \sigma'_{cbc}$$

The depth of the resultant compressive force in concrete from the top of the flange is given by

$$\bar{x} = \left[\frac{3x_c - 2D_f}{2x_c - D_f}\right]\left(\frac{D_f}{3}\right) = 44.33 \text{ mm}$$

and

$$\text{Lever arm } z_c = (d - \bar{x})$$
$$= 500 - 44.33 = 455.67 \text{ mm}$$

For the balanced moment of resistance,

$$\sigma'_{cbc} = \sigma_{cbc} = 7 \text{ MPa}$$

Therefore, $M_{\text{bal}} = 0.5 \times (\sigma'_{cbc} + \sigma''_{cbc}) b_f D_f z_c$

$$= 0.5 \times 7 \times (1 + 0.492) \times 1933 \times 100 \times 455.67$$
$$= 459.96 \times 10^6 \text{ Nmm} = 459.96 \text{ kNm}$$

Since $M_{\text{bal}} > M$, it is an under-reinforced section. Thus

$$\sigma'_{st} = \sigma_{st} = 140 \text{ MPa}$$

and

$$M = \sigma'_{st} A_{st} z_c$$

Giving

$$A_{st} = \frac{M}{\sigma'_{st} z_c} = \frac{149.53 \times 10^6}{140 \times 455.67} = 2344 \text{ mm}^2$$

To determine the position of the actual neutral axis, take moment of transformed areas about the neutral axis as:

$$b_f D_f\left(x - \frac{D_f}{2}\right) = mA_{st}(d - x)$$

$$1933 \times 100\ (x - 50) = 13 \times 2344 \times (500 - x)$$

Giving

$$x = 126.74 \text{ mm}$$

and

$$\bar{x} = \left[\frac{3x - 2D_f}{2x - D_f}\right]\left[\frac{D_f}{3}\right] = 39.32 \text{ mm}$$

and

$$\text{Lever arm } z = 460.68 \text{ mm}$$

The exact area of tension steel is:

$$A_{st} = \frac{149.53 \times 10^6}{140 \times 460.68} = 2318 \text{ mm}^2$$

The stresses in the materials are: σ_{st} = 140 MPa, and from the stress distribution diagram

$$\sigma'_{cbc} = \left(\frac{140}{13}\right)\left(\frac{126.74}{500 - 126.74}\right) = 3.66 \text{ MPa}$$

A.2.4.2 Flange Beams with Tension and Compression Reinforcements

When a flange section is reinforced on the compression side also, it is known as a **doubly reinforced flange section**. It can be designed by using a procedure similar to that for a doubly reinforced rectangular beam. To ascertain the position of neutral axis consider that the neutral axis coincides with the underneath of the slab flange, and compute moments M_c and M_t of transformed areas on compression and tension sides, respectively, about the bottom of the flange:

$$M_c = b_f D_f\left(\frac{D_f}{2}\right) + (1.5m - 1)A_{sc}(D_f - d') \quad \text{(A.28)}$$

$$M_s = mA_{st}(d - D_f) \quad \text{(A.29)}$$

If $M_c > M_s$, the neutral axis lies in the flange. On the other hand, if $M_c < M_s$, the neutral axis lies in the web.

CASE I: In the first case when the neutral axis lies in the flange ($x < D_f$), its depth can be computed from:

$$b_f x\left(\frac{x}{2}\right) + (1.5m - 1)A_{sc}(x - d') = mA_{st}(d - x) \quad \text{(A.30)}$$

The analysis is exactly similar to that for a rectangular section.

CASE II: When the neutral axis lies in the web ($x > D_f$), the section is analyzed as a T-beam section. Since the compressive force in the web is relatively very small in comparison with the compressive force in the flange, it may be neglected in the analysis. The depth of the neutral axis can be computed from:

$$b_f D_f\left(x - \frac{D_f}{2}\right) + (1.5m - 1)A_{st}(x - d') = mA_{st}(d - x) \tag{A.31}$$

The lever arm depends on the position of the resultant compressive force. The stresses at various levels in terms of the stress induced in the extreme compression fibre of concrete σ_{cbc} can be obtained from the stress distribution diagram:

$$\frac{\sigma'_{cbc}}{x} = \frac{\sigma'_{cs}}{x - d'} = \frac{\sigma''_{cbc}}{x - D_f} = \frac{\sigma'_{st}}{m(d - x)}$$

Therefore,

$$\sigma'_{cs} = \frac{\sigma'_{cbc}(x - d')}{x},\ \sigma''_{cbc} = \frac{\sigma'_{cbc}(x - D_f)}{x}$$

and

$$\sigma'_{st} = \frac{m\sigma'_{cbc}(d - x)}{x}$$

The position of the centroid of compressive force in concrete C_c from the top of the flange is given by

$$\bar{x}_c = \left[\frac{\sigma'_{cbc} + 2\sigma''_{cbc}}{\sigma'_{cbc} + \sigma''_{cbc}}\right]\left(\frac{D_f}{3}\right)$$

The distance of the force in compression steel C_s from the top of the flange is d'. Thus,

$$\bar{x} = \frac{C_c\bar{x}_c + C_s d'}{C_c + C_s}$$

and

$$\text{Lever arm } z(= jd) = d - \bar{x}.$$

The moment of resistance of the section is given by

1. With respect to the compressive force

$$M_r = [0.5(\sigma'_{cbc} + \sigma''_{cbc})b_f D_f + (1.5m - 1)A_{sc}\sigma'_{cs}]z$$

or

$$M_r = 0.5(\sigma''_{cbc} + \sigma''_{cbc})b_f D_f(d - \bar{x}) + (1.5m - 1)A_{sc}\sigma'_{cs}(d - d')$$

2. With respect to the tensile force,

$$M_r = (\sigma'_{st}A_{st})z$$

The nature of the problems and procedure for analysis are illustrated in the following examples:

Example A.10 The section of a T-beam has a flange of effective width and thickness of 1500 mm and 125 mm, respectively. The effective depth of the beam and the width of the web are 600 mm and 300 mm, respectively. The beam is reinforced with four bars of 25 mm ϕ on the tension side and three bars of 22 mm ϕ on the compression side. The effective cover to compression steel is 50 mm. The materials used are M20 grade concrete and HYSD steel of grade Fe415. Determine: (a) allowable moment of resistance, (b) the uniformly distributed load the beam can support safely over an effective simply supported span of 8 m and (c) stresses induced in the top compression fibre of concrete and tension steel when the beam is subjected to a bending moment of 200 kNm.

Solution For M20 concrete and Fe415 steel,

$$\sigma_{cbc} = 7 \text{ MPa}, \ \sigma_{st} = 230 \text{ MPa and } m = 13$$

$$b_f = 1500 \text{ mm}, \ D_f = 125 \text{ mm}, \ b_w = 300 \text{ mm}, \ d = 600 \text{ mm and } d' = 50 \text{ mm}$$

Areas of reinforcing steels are:

$$A_{st}(4 \times 25 \text{ mm } \phi) = 1963 \text{ mm}^2$$

$$A_{sc}(3 \times 22 \text{ mm } \phi) = 1140 \text{ mm}^2$$

To ascertain the position of the neutral axis, take moments of transformed areas on compression and tension sides about the bottom of the flange as:

$$M_c = b_f D_f \left(\frac{D_f}{2}\right) + (1.5m - 1)A_{sc}(D_f - d')$$

$$= 1500 \times 125 \times 62.5 + (1.5 \times 13 - 1) \times 1140 \times (125 - 50) = 13.30 \times 10^6 \text{ mm}^3$$

$$M_s = mA_{st}(d - D_f)$$

$$= 13 \times 1963 \times (600 - 125) = 12.12 \times 10^6 \text{ mm}^3$$

Since $M_c > M_t$, the neutral axis lies in the flange. The exact position of the neutral axis can be determined by taking moments of transformed areas about the neutral axis:

$$b_f x \left(\frac{x}{2}\right) + (1.5m - 1)A_{sc}(x - d') = mA_{st}(d - x)$$

$$\frac{1500x^2}{2} + (1.5 \times 13 - 1) \times 1140 \times (x - 50) = 13 \times 1963 \times (600 - x)$$

$$x^2 - 62.15x - 21821.2 = 0$$

Giving,

$$x = 119.88 \text{ mm } (< D_f)$$

The position of the balanced neutral axis is given by

$$x_c = \frac{d}{1 + (\sigma_{st}/m\sigma_{cbc})} = \frac{600}{1 + 230/(13 \times 7)} = 170.09 \text{ mm}$$

(i) Since $x < x_c$, the section is under-reinforced and hence $\sigma'_{st} = \sigma_{st} = 230$ MPa. The stresses in the extreme compression fibre of concrete and in compression steel can be obtained as:

$$\sigma'_{cbc} = \left(\frac{\sigma'_{st}}{m}\right)\left(\frac{x}{d - x}\right)$$

$$= \left(\frac{230}{13}\right)\left(\frac{119.88}{600 - 119.88}\right) = 4.42 \text{ MPa}$$

$$\sigma'_{sc} = \frac{(1.5m)\sigma'_{cbc}(x - d')}{x}$$

$$= 1.5 \times 13 \times 4.42 \times \frac{119.88 - 50}{119.88} = 50.22 \text{ MPa}$$

To determine the allowable moment of resistance of the section, take moment about tension steel as:

$$M_r = b_f x(0.5\sigma'_{cbc})\left(d - \frac{x}{3}\right) + A_{sc}\sigma'_{sc}(d - d')$$

$$= 1500 \times 119.88 \times (0.5 \times 4.42) \times \left(600 - \frac{119.88}{3}\right) + 1140 \times 50.22 \times (600 - 50)$$

$$= 254.05 \times 10^6 \text{ Nmm} = 254.05 \text{ kNm}$$

(ii) Let w kN/m be the total uniformly distributed load (inclusive of self-weight) the beam can support safely. The maximum bending moment is given as:

$$M = \frac{w \times 8^2}{8} = 8 \text{ w kNm}$$

Equating M to M_r, $8w = 254.05$

Giving $$w = 31.76 \text{ kN/m}$$

(iii) Since the beam is under-reinforced, equate allowable moment of resistance with respect to tensile force to that with respect to compressive force obtained earlier:

$$M_r = \sigma_{st}A_{st}z = 254.05 \times 10^6$$

Therefore

$$\text{Lever arm } z = \frac{254.05 \times 10^6}{230 \times 1963} = 562.69 \text{ mm.}$$

The stresses in the materials due to a moment of 200 kNm are:

$$\sigma'_{st} = \frac{200 \times 10^6}{1963 \times 562.69} = 181.07 \text{ MPa}$$

From the stress distribution diagram,

$$\sigma'_{cbc} = \left(\frac{\sigma'_{st}}{m}\right)\left(\frac{x}{d - x}\right)$$

$$= \frac{181.07}{13} \times \frac{119.88}{600 - 119.88} = 3.478 \text{ MPa}$$

$$\sigma'_{sc} = 1.5m\frac{(\sigma'_{st}/m)(x - d')}{(d - x)}$$

$$= \frac{(1.5 \times 181.07)(119.88 - 50)}{(600 - 119.88)} = 39.53 \text{ MPa}$$

Alternatively, the stresses induced in materials due to applied moment can also be computed by multiplying the stresses obtained in Part (i) by the ratio (M/M_r).

$$\sigma'_{cbc} = 4.42 \times \frac{200}{254.05} = 3.48 \text{ MPa}$$

$$\sigma'_{sc} = 50.22 \times \frac{200}{254.05} = 39.54 \text{ MPa}$$

$$\sigma'_{st} = 230 \times \frac{200}{254.05} = 181.07 \text{ MPa.}$$

Example A.11 A typical beam of a floor system behaves as a T-beam section having a flange of size 1600 × 125 mm. The width of the web adopted is 300 mm. Design the T-beam if it is subjected to a moment of 300 kNm for the following cases: (a) there is no restriction on the depth of the beam and (b) the effective depth of beam is to be 500 mm. Use M20 grade concrete mix and HYSD steel of grade Fe415 as reinforcement.

Solution For M20 grade concrete and Fe415 grade steel,

$$\sigma_{cbc} = 7 \text{ MPa}, \ \sigma_{st} = 230 \text{ MPa and } m = 13$$

$$k_c = \frac{1}{[1 + 230/(13 \times 7)]} = 0.283, \ j = 1 - \frac{k_c}{3} = 0.906 \text{ and } x_c = 0.283d$$

The known cross-sectional dimensions are:

$$b_f = 1600 \text{ mm}, \ D_f = 125 \text{ mm and } b_w = 300 \text{ mm}$$

CASE I The beam may be designed as a balanced section with the neutral axis falling in the flange. For this case, the section will behave as a rectangular section with $b = b_f$, and

$$M_{\text{bal}} = 0.5\sigma_{cbc} b_f x_c \left(d - \frac{x_c}{3} \right)$$

$$= 0.5 \times 7 \times 1600 \times 0.2835d \times \left(1 - \frac{0.2835}{3} \right) d = 1437.572 \ d^2 \text{ Nmm}$$

Equating M_{bal} to the applied moment M,

$$1437.572d^2 = 300 \times 10^6$$

Giving

$$d = 456.82 \text{ mm}$$

For the neutral axis to lie in the flange $x_c < D_f$, i.e., $0.283\ d \leq 125$ or $d \leq 441.70$ mm. Adopt $d = 435$ mm, the neutral axis will lie in the flange. The allowable moment of resistance of the singly reinforced rectangular balanced section is:

$$M_{\text{bal}} = 1437.572 \times (435)^2 \times 10^{-6} = 272.02 \text{ kNm}$$

Since $M > M_{bal}$, it is to be analyzed as a doubly reinforced beam. The area of steel A_{st1} for resisting M_{bal} is given by

$$A_{st1} = \frac{M_{bal}}{\sigma_{st} jd} = \frac{272.02 \times 10^6}{230 \times 0.906 \times 435} = 3000.93 \text{ mm}^2$$

Additional moment $(M - M_{bal})$ is resisted by the additional tension steel A_{st2} and compression steel A_{sc}. Let the effective cover to compression steel be $d' = 35$ mm. Then

$$A_{st2} = \frac{(300 - 272.02) \times 10^6}{230 \times (435 - 35)} = 304.13 \text{ mm}^2$$

Thus,

$$\text{Total area of tension steel } A_{st} = A_{st1} + A_{sc2} = 3305.06 \text{ mm}^2$$

The compression steel may be obtained by taking moment of compressive force in compression steel about the centroid of tension steel with

$$x_c = 0.283 \times 435 = 123.11 \text{ mm}$$

Therefore,

$$M - M_{bal} = (1.5m - 1)A_{sc}\sigma_{cbc}\left[\frac{(x_c - d')}{x_c}\right](d - d')$$

$$= (1.5 \times 13 - 1)A_{sc} \times 7 \times \left[\frac{123.11 - 35}{123.11}\right](435 - 35)$$

$$= 37073.33A_{sc}$$

Therefore,

$$A_{sc} = \frac{(300 - 272.02) \times 10^6}{37073.33} = 754.72 \text{ mm}^2$$

CASE II The effective depth in this case is 500 mm. The design of the section requires the determination of the area of reinforcement. To ascertain whether the section is to be singly reinforced, balanced, under-reinforced or doubly reinforced, assume the section to be the balanced one:

$$x_c = 0.283 \times 500 = 141.5 \text{ mm} > D_t$$

Hence the critical neutral axis falls in the web. The compressive stress at the underneath of the flange slab is given as:

$$\sigma''_{cbc} = \frac{\sigma'_{cbc}(x_c - D_f)}{x_c} = \frac{\sigma'_{cbc}(141.5 - 125)}{141.5} = 0.1166\, \sigma'_{cbc}$$

Depth of the resultant compressive force in concrete from the top of the flange is given by

$$\bar{x} = \frac{(\sigma'_{cbc} + 2\sigma''_{cbc})}{(\sigma'_{cbc} + \sigma''_{cbc})}\frac{D_f}{3} = 46.02 \text{ mm}$$

Alternatively $\bar{x} = \left[\frac{3x_c - 2D_f}{2x_c - D_f}\right] \times \frac{D_f}{3} = 46.02$ mm

Therefore, Lever arm $z_c = d - \bar{x} = 500 - 46.02 = 453.98$ mm

For the moment of resistance as a balanced section, $\sigma'_{cbc} = 7$ MPa. Therefore,

$$M_{bal} = b_f D_f \left[\frac{\sigma'_{cbc} + \sigma''_{cbc}}{2}\right] z_c$$

$$= 1600 \times 125 \times 7 \times \left[\frac{1 + 0.1166}{2}\right] \times 453.98$$

$$= 354.84 \times 10^6 \text{ Nmm} = 354.84 \text{ kNm}$$

Since $M_{bal} > M$, it is an under-reinforced section. Thus,

$$\sigma'_{st} = \sigma_{st} = 230 \text{ MPa}$$

To ascertain the position of the neutral axis, let it coincides with the bottom of the flange, i.e., $x = D_f$ and

$$\sigma'_{cbc} = \left(\frac{230}{13}\right)\left(\frac{125}{500 - 125}\right) = 5.897 \text{ MPa}$$

The moment of resistance for this case is:

$$M_c = 0.5\sigma'_{cbc} b_f D_f \left(d - \frac{D_f}{3}\right)$$

$$= 0.5 \times 5.897 \times 1600 \times 125 \times \left(500 - \frac{125}{3}\right) \times 10^{-6} = 270.28 \text{ kNm}$$

Since $M_c < M$, the neutral axis lies in the web. For the under-reinforced section,

$$M_s = Tz = \sigma_{st} A_{st} z_c$$

Equate M_s to M, i.e.,

$$230 \times A_{st} \times 453.98 = 300 \times 10^6$$

Therefore

$$A_{st} = \frac{300 \times 10^6}{230 \times 453.98} = 2873.1 \text{ mm}^2$$

Provide 6 bars of 25 mm ϕ (A_{st} = 2945 mm^2).

A.3 SHEAR

The majority of structural members subjected to bending are also subjected to shear forces. The value of shear strength lies between the tensile and compressive strengths of concrete. The reinforced concrete beams fail in shear abruptly without sufficient advance warning and the diagonal cracks are considerably wider than the flexural cracks. Shear failures which are brittle in nature are essentially diagonal tension failures. This necessitates a provision of reinforcement for diagonal tension. Most of the design recommendations are based on empirical relations derived from test results. Shear in a flexural member is given by the expression:

$$\tau = \frac{V}{bjd}$$

For convenience of computation, IS: 456 has adopted a simple equation:

$$\tau_v = \frac{V}{bd}$$

where

τ_v = Nominal shear stress

V = Design shear force

b = Width of section (= b_f in case of flanged beams)

d = Effective depth

In case of beams with varying depths, the shear stress is given by

$$\tau_v = \frac{V \pm (M/d) \tan \beta}{bd}$$

where

M = Bending moment at a section

β = Angle between top and bottom edges of the beam

The negative sign in the expression applies when bending moment M increases numerically in the same direction in which the effective depth d increases; the positive sign is used when the moment decreases numerically in this direction.

A.3.1 Shear Strength of Concrete without Shear Reinforcement τ_c

Shear strength of concrete varies according to the grade of concrete, and the crack control parameter expressed as $100 \times A_{st}/bd$. In this expression A_{st} is the area of reinforcement which extends at least one effective depth beyond the section under consideration.

TABLE A.2 Permissible Shear Stress in Concrete for Various Values of Crack Control Parameters

$\frac{A_{st}}{bd} \times 100$	*Permissible shear stress in concrete τ_c, MPa* — *Concrete grade*				
	M20	*M25*	*M30*	*M35*	*M40*
0.25	0.22	0.23	0.23	0.23	0.23
0.50	0.30	0.31	0.31	0.31	0.32
0.75	0.35	0.36	0.37	0.37	0.38
1.00	0.39	0.40	0.41	0.42	0.42
1.25	0.42	0.44	0.45	0.45	0.46
1.50	0.45	0.46	0.48	0.49	0.49
1.75	0.47	0.49	0.50	0.52	0.52
2.00	0.49	0.51	0.55	0.54	0.55
2.25	0.51	0.53	0.55	0.56	0.57
2.50	0.51	0.55	0.57	0.58	0.60
2.75	0.51	0.56	0.58	0.60	0.62
3.00	0.51	0.57	0.60	0.62	0.63

In solid slabs, the shear strength of concrete shall be $k\tau_c$, where k has the value given in Table A.3.

TABLE A.3 Value of k for Different Thicknesses of Slab

Overall depth of slab mm	*330 or more*	*275*	*250*	*225*	*220*	*175*	*150 or less*
k	1.00	1.05	1.10	1.15	1.20	1.25	1.30

A.3.2 Maximum Shear Stress $\tau_{c,max}$

Where nominal shear stress τ_v exceeds the shear strength of concrete τ_c, suitable shear reinforcement is provided. However, τ_v shall not exceed $\tau_{c,max}$ the maximum value given in Table A.4. By this provision the failure of the member by diagonal compression is prevented. If at any section the nominal shear stress exceeds the maximum permissible value, the section shall be redesigned by using higher grade concrete mix or by providing more depth.

TABLE A.4 Maximum Shear Stress in Concrete

Concrete grade	*M15*	*M20*	*M25*	*M30*	*M35*
$\tau_{c,max}$ (MPa)	1.6	1.8	1.9	2.2	2.3

A.3.3 Shear Reinforcement

When τ_v exceeds the shear strength of concrete τ_c, the shear reinforcement shall be provided to carry the shear force:

$$V_s = V - \tau_c bd$$

where V is the shear force due to externally applied design loads. The shear reinforcement is provided in the form of (a) vertical stirrups, (b) inclined stirrups, (c) bent-up or inclined bars, (d) combination of (a) and (c).

Vertical stirrups. The most commonly used form is generally two-legged. However, four-legged, six-legged, and so on, stirrups can also be formed. For vertical stirrups, the spacing S_v is given by

$$S_v = \frac{\sigma_{sv} A_{sv} d}{V_s}$$

where

A_{sv} = Total cross-sectional area of stirrup legs
d = Effective depth of beam
σ_{sv} = Permissible tensile stress in shear reinforcement $\ngtr$ 230 MPa

If a series of inclined stirrups are provided near the support crossing a potential diagonal crack then S_v in this case is given by

$$S_v = \frac{\sigma_{sv} A_{sv} (\sin \alpha + \cos \alpha) d}{V_s}$$

where α is the inclination of stirrups to the longitudinal axis of the beam.

Bent-up bars. The shear taken up by a single or group of bars, all bent up at the same section, is calculated as:

$$V_b = \sigma_{sv} A_{sv} \sin \alpha$$

where

A_{sv} = Area of bent-up bars
α = Angle between bent-up bar and axis of the member $\nless 45°$
σ_{sv} = Permissible tensile stress in the bar $\ngtr$ 230 MPa

Usually, the bars are bent at 45° or 60° to the longitudinal axis of the beam.

Minimum shear reinforcement: When the nominal shear stress τ_v is less than the shear strength of concrete τ_c, no shear reinforcement is required to be designed. However, under such circumstances, a minimum shear reinforcement in the form of vertical stirrups shall be provided such that

$$\frac{A_{sv}}{bS_v} \geq \frac{0.4}{f_y}$$

where

A_{sv} = Total cross-sectional area of stirrup legs effective in shear

S_v = Spacing of stirrups along the length of the member

b = Breadth of the member

f_y = Characteristic strength of shear reinforcement in MPa ($\ngtr$ 415 MPa)

The characteristic strengths for various types of steel are given in Table A.5.

The shear reinforcement is provided to prevent the shear cracks. It is seen that the horizontal distance between the two successive cracks is approximately equal to the effective depth d. The maximum spacing of shear reinforcement measured along the axis of the member shall not exceed $0.75d$ for the vertical stirrups and d for the inclined stirrups. In no case shall the spacing exceed 450 mm.

TABLE A.5 Characteristic Strengths for Various Grades of Steel

Types and grades of steel	*Characteristic strength* f_y (MPa)	*Permissible stress in steel* σ_{sv} (MPa)
Mild steel grade I	250	140
Medium tensile deformed bars	360	190
HYSD steel of grade Fe415	415	230
HYSD steel of grade Fe500	415	275

A.3.4 Design Procedure for Shear Reinforcement

1. Determine the maximum shear force and the corresponding nominal shear stress τ_v.
2. Compute steel ratio ($100A_{st}/bd$) and corresponding shear strength of concrete τ_c.
3. Compare τ_c with with τ_v.

   (a) If $\tau_v < \tau_c$, no shear design is required. Provide nominal vertical stirrups throughout the beam such that

$$\frac{A_{sv}}{bS_v} \geq \frac{0.4}{f_y} \quad \text{or} \quad S_v \leq \frac{A_{sv} f_y}{0.4b}$$

   where A_{sv} is the total cross-sectional area of stirrup legs.

   (b) If bent-up bars are not provided, go to Step 5. If the tension steel is available for shear, bend one or more bars from tension steel at 45° at a distance d from the supports. Compute the shear force taken by bent-up bars as follows:

$$V_b = \sigma_{sv} A_{sv} \sin \alpha \left(\ngtr \frac{1}{2} V_s \right)$$

   Here A_{sv} is the total cross-sectional area of bent-up bars.

4. Design the vertical stirrups for the balance shear force:

$$V_d = V_s - V_b$$

5. If bent-up bars are not provided then $V_d = V_s$.
Choose the diameter of the stirrup bar (usually 6 mm, 8 mm and 10 mm) and compute spacing S_v from

$$S_v = \frac{\sigma_{sv} A_{sv} d}{V_d}$$

6. Check the above spacing for the following:
   (a) $S_v \not> 0.75d$ or 450 mm, whichever is less
   (b) $S_v \not> \dfrac{A_{sv} f_y}{0.4b}$
7. Find the distance from the support up to which the designed stirrups are required. For the rest of the portion provide minimum shear reinforcement.

Example A.12 The reinforced concrete cantilever beam of span 3.5 m, shown in Fig. A.6, has a section 300 mm wide and 550 mm deep (effective) at the face of the support. The depth is reduced uniformly to 250 mm (effective) at the free end. The beam is reinforced on the tension face with 6 bars of 20 mm ϕ at the support. Two of these tension bars are curtailed

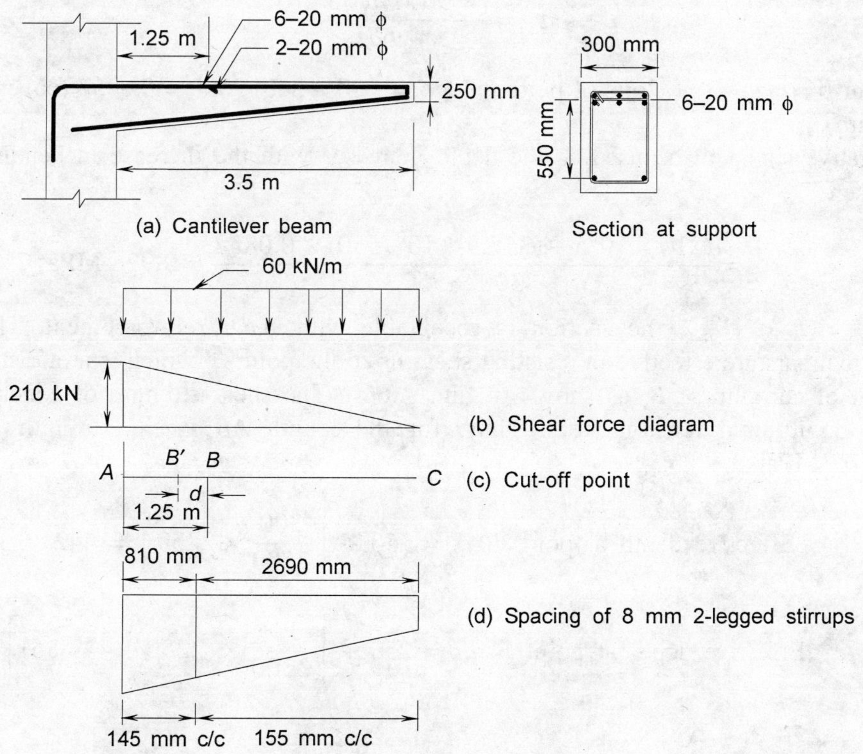

Fig. A.6 Design of cantilever beam of Example A.11.

at a distance of 1.25 m from the support. Design the shear reinforcement for the beam when it carries a uniformly distributed service load of 60 kN/m. The materials used are M20 grade concrete and HYSD steel of grade Fe415.

Solution For M20 grade concrete $\tau_{c,\max}$ = 1.8 MPa.
At the support, b = 300 mm, d = 550 mm, A_{st}(6 × 20 mm φ) = 1885 mm²

Therefore,
$$p = \frac{100A_{st}}{bd} = \frac{100 \times 1885}{300 \times 550} = 1.142$$

and
$$\tau_c = 0.39 + \left[\frac{(0.42 - 0.39)(1.142 - 1.00)}{(1.25 - 1.0)}\right] = 0.407 \text{ MPa}$$

Maximum shear force and moment at the support:
$$V = wl = 60 \times 3.5 = 210 \text{ kN}$$

$$M = \frac{wl^2}{2} = \frac{60 \times (3.5)^2}{2} = 367.5 \text{ kNm}$$

Nominal shear stress is given by

$$\tau_v = \frac{[V \pm (M/d)\tan\beta]}{bd}$$

where tan β = tangent of slope of bottom edge = (600 – 300)/3500 = 0.0857 (cover is assumed as 50 mm).
The negative sign will be used as the depth increases with the increase in bending moment. Thus,

$$\tau_v = \frac{(210 \times 10^6) - (367.5 \times 10^6 / 550) \times 0.0857}{300 \times 550} = 0.926 \text{ MPa}$$

Since $\tau_c < \tau_v < \tau_{c,\max}$, the section is acceptable with shear reinforcement. The curtailed longitudinal bars are effective in resisting shear up to the point B' which is at a distance d from the point of curtailment B, as shown in Fig. A.6(c). The shear strength of concrete depends on the longitudinal tension steel. Thus, for the length AB' shear strength of concrete τ_c = 0.407 MPa:

$$\text{Effective depth at point } B\ d_b = 250 + \left[\frac{300}{3500} \times 2250\right] = 442.86 \text{ mm}$$

$$\text{Effective depth at point } B'\ d_{b'} = 250 + \left[\frac{300}{3500} \times (2250 + 442.86)\right]$$

$$= 480.82 \text{ mm}$$

At effective cut-off point B', $A_{st}(4 \times 20 \text{ mm } \phi) = 1256 \text{ mm}^2$

Therefore, $$p = \frac{100A_{st}}{bd} = \frac{100 \times 1256}{300 \times 480.82} = 0.871$$

Shear strength of concrete is:

$$\tau_c = 0.35 + \frac{(0.39 - 0.35)(0.871 - 0.75)}{(1.00 - 0.75)} = 0.369 \text{ MPa}$$

Nominal shear stress $\tau_v = \dfrac{V - (M/d) \tan \alpha}{bd}$

where

$$V = 60 \times (2.250 + 0.443) = 161.572 \text{ kN}$$

$$M = 60 \times \frac{(2.250 + 0.443)^2}{2} = 217.545 \text{ kNm}$$

Therefore,

$$\tau_v = \frac{161.572 \times 10^3 - (217.545 \times 10^6/480.82) \times 0.0857}{300 \times 480.82}$$

$$= 0.8513 \text{ MPa}$$

Since $\tau_v > \tau_c$ shear reinforcement is required. Use 8 mm ϕ 2-legged vertical HYSD steel stirrups ($A_{sv} = 100 \text{ mm}^2$). Spacing in length AB' is given by

$$S_v = \frac{\sigma_{sv} A_{sv}}{(\tau_v - \tau_c)b} = \frac{230 \times 100}{(0.926 - 0.407) \times 300} = 147.72 \text{ mm}$$

Maximum spacing based on minimum shear reinforcement is:

$$S_v \leq \frac{A_{sv} f_y}{0.4b} = \frac{100 \times 415}{0.4 \times 300} = 345.83 \text{ mm}$$

Spacing shall be the least of the following:

(i) $0.75d = 0.75 \times 480.82 = 360.6$ mm, (ii) 450 mm, (iii) 345.83 mm and (iv) 147.72 mm (required)

Provide 8 mm ϕ two-legged stirrups 145 mm c/c.

Spacing of stirrups in the length BC is:

$$S_v = \frac{230 \times 100}{(0.8513 - 0.369) \times 300} = 158.96 \text{ mm}$$

Spacing shall be the least of the following:

(i) $0.75d = 0.75 \times 250 = 187.5$ mm, (ii) 450 mm, (iii) 345.83 mm (minimum steel) and (iv) 158.96 mm (required)

Hence provide 8 mm ϕ two-legged stirrups @ 155 mm c/c. The shear reinforcement is shown in Fig. A.6.

A.4 BOND AND DEVELOPMENT LENGTH

For reinforcement concrete to behave effectively, it is essential that stresses be transferred from one material to the other. In other words, there should be no slippage or relative movement between concrete and the reinforcement. To prevent failure in bond in concrete, end anchorage is provided mostly in the form of hooks. The extended length of the bar required to transmit the bar force to concrete is known as **development length L_d**. It is obtained by equating the bond resistance of concrete to the strength of the reinforcing bar.

$$\tau_{bd}\pi\phi L_d = \frac{\sigma_s \pi \phi^2}{4} \qquad \text{or } L_d = \frac{\phi\sigma_s}{4\tau_{bd}}$$

where

L_d = Development length

τ_{bd} = Permissible average stress in bond

ϕ = Nominal diameter of the bar

σ_s = Stress in reinforcing bar at the section under consideration at the design load

The permissible average bond stresses for plain and deformed bars in tension and compression, as recommended by IS:456, are given in Table A.6. The development length can be expressed as:

$$L_d = \frac{\sigma_s A_{st} jd}{V} = \frac{M_1}{V}$$

where M_1 is the moment of resistance of the section assuming all tension reinforcement to be stressed to σ_{st}. If the average bond strength τ_{bd} is not to be exceeded, the ratio (M_1/V) must be equal to or larger than the required development length L_d. To preclude the possibility of bond strength being exceeded even locally, additional anchorage length L_o is provided. At a simple support, L_o is the anchorage length beyond the centre of support and the equivalent anchorage value of any hook or mechanical anchorage. At a point of inflection, L_o is limited to the effective depth of the member or 12ϕ, whichever is greater. Preceding equation is, therefore, modified as:

$$L_d \le \left(\frac{M_1}{V}\right) + L_o$$

In case the ends of reinforcement are confined by compressive reaction, the value of M_1/V is to be increased by 30 per cent, i.e.,

$$L_d \le 1.3\left(\frac{M_1}{V}\right) + L_o$$

TABLE A.6 Permissible Stresses in Bond for Different Grades of Concrete

	Concrete grade		*M20*	*M25*	*M30*	*M35*	*M40*
	Tension	Plain bars	0.80	0.90	1.00	1.10	1.20
τ_{bd}		Deformed bars	1.12	1.26	1.40	1.54	1.68
MPa	Compression	Plain bars	1.00	1.125	1.25	1.375	1.50
		Deformed bars	1.40	1.575	1.75	1.925	2.10

Notes: 1. τ_{bd} for compression bars is 25 per cent greater than that for tension bars.
2. For deformed bars the values of τ_{bd} are increased by 40 per cent.

Example A.13 A simply supported reinforced concrete rectangular beam of size 250 × 500 mm effective is reinforced with six bars of 20 mm ϕ at the bottom and two bars of 16 mm ϕ at the top of the beam. The beam carries uniformly distributed service load of 55 kN/m (inclusive of self-weight) over an effective span of 6 m. The beam rests on 300 mm wide supports. If the materials used are M20 grade concrete and HYSD steel of grade Fe415, determine the development lengths required for the bars on tension and compression sides, from the face of the support and also at mid-span.

Solution For M20 grade concrete and Fe415 grade steel,

$$\sigma_{cbc} = 7 \text{ MPa}, \ \sigma_{st} = 230 \text{ MPa}, \ m = 13 \text{ and } j = 0.906$$

$$\tau_{bd} = 0.8 \times (1 + 0.4) = 1.12 \text{ MPa}$$

For the given section, b = 250 mm, d = 500 mm, and reinforcements are

$$A_{st}(6 \times 20 \text{ mm } \phi) = 1885 \text{ mm}^2$$

$$A_{sc}(2 \times 16 \text{ mm } \phi) = 402 \text{ mm}^2$$

$$\text{Reaction at the support } V = \frac{wl}{2} = \frac{55 \times 6}{2} = 165 \text{ kN}$$

Bending moment at the face of the support is expressed as:

$$M_1 = \left(165 \times \frac{0.30}{2}\right) - \left(55 \times 0.30 \times \frac{0.30}{2}\right) = 22.275 \text{ kNm}$$

Development length on the tension side. At support,

$$\sigma'_{st} = \frac{M_1}{A_{st}jd} = \frac{22.275 \times 10^6}{1885 \times 0.906 \times 500} = 26.09 \text{ MPa}$$

$$L_d = \frac{\phi\sigma_s}{4\tau_{bd}} = \frac{20 \times 26.09}{4 \times 1.12} = 116.47 \text{ mm}$$

Development length on the compression side.
Assume the section to be balanced. Then

$$x_c = \frac{500}{1 + 230/(13 \times 7)} = 141.5 \text{ mm}$$

Consider cover to the top steel as 40 mm; then the stress in concrete at the level of the compression steel is given by

$$\frac{\sigma'_{cbc}}{(26.09/13)} = \left[\frac{141.5 - 40}{500 - 141.5}\right]$$

Giving

$$\sigma'_{cbc} = 0.568 \text{ MPa and } \sigma'_{sc} = 1.5m \times \sigma'_{cbc} = 11.08 \text{ MPa}$$

For M20 concrete with HYSD steel bars in compression τ_{bd} = 1.40 MPa.

Therefore, $$L_d = \frac{\phi\sigma_s}{4\tau_{bd}} = \frac{16 \times 11.08}{4 \times 1.40} = 31.66 \text{ mm}$$

To compute the development length on either side of the point of maximum moment, i.e., at the mid-span, assume that the tension steel is stressed to its design value, i.e., 230 MPa. Then

$$L_d = \frac{20 \times 230}{4 \times 1.12} = 1026.79 \text{ mm}$$

The mid-span being the point of maximum stress, the bars shall be anchored on both sides of this point by 1027 mm. The length available on either side is 3000 – 50 (= end cover) = 2950 mm, and hence adequate.

A.5 COMPRESSION MEMBERS

The structural elements which are primarily subjected to axial compressive loads are termed as **compression members**. When a compression member is vertical, it is called a **column**, and when inclined or horizontal, it is termed as **strut**. For various types of columns readers should refer to Chapter 5.

A.5.1 Analysis of Axially Loaded Columns

The allowable load-carrying capacity P of a short column with lateral ties based on the equilibrium method is given by

$$P = \sigma_{cc}A_c + \sigma_{sc}A_{sc} \tag{A.32}$$

where

σ_{cc} = Permissible stress in concrete in direct compression

A_c = Net cross-sectional area of concrete

σ_{sc} = Permissible compressive stress in steel

A_{sc} = Cross-sectional area of longitudinal steel

Equation (A.32) is used for determining the load capacity as well as for design of tied column sections. In practice, the columns rarely carry purely axial loads, therefore, compression members are designed for a specified minimum eccentricity of load. This stipulated eccentricity has been accounted for in the permissible stresses in the materials.

Columns with spiral reinforcement: The allowable load-carrying capacity of a column having longitudinal reinforcement tied with spirals can be taken as 1.05 times the allowable load for a similar column with lateral ties, provided the ratio of the volume of helical reinforcement V_h to the volume of the concrete core V_c, i.e., (V_h/V_c) is not less than

$$\frac{0.36(A_g/A_c - 1)\sigma_{ck}}{f_y} \tag{A.33}$$

where

A_g = Gross area of cross-section

A_c = Area of core measured to the outside diameter of the spiral

σ_{ck} = Characteristic compressive strength of concrete

f_y = Characteristic strength of helical reinforcement ($\not>$ 415 MPa)

Composite columns: The allowable load P on such a column is given by

$$P = \sigma_{cc}A_c + \sigma_{sc}A_{sc} + \sigma_{mc}A_m \tag{A.34}$$

where

A_c = Net area of concrete section

A_{sc} = Cross-sectional area of longitudinal steel

A_m = Cross-sectional area of metal core, $A_m > 0.20\ A_g$

σ_{mc} = Permissible stress in metal core $\not>$ 125 MPa for steel core or $\not>$ 70 MPa for a cast iron core

In long columns, due to buckling tendency the allowable load capacity is likely to be less than that in the case of short columns. The allowable load capacity of a long column is obtained by multiplying the allowable load of a short column having the same cross-section by a reduction coefficient C_r which is given by

$$C_r = 1.25 - \left(\frac{l_{\text{eff}}}{48b}\right) \tag{A.35}$$

where

l_{eff} = Effective length of column

b = Least lateral dimension (or core diameter in case of a column with helical reinforcement)

The effective length l_{eff} depends on the degree of restraint at the ends of the compression member, and the unsupported length L which is the clear distance between the end restraints. The effective length l_{eff} of a compression member shall be taken as follows:

1. For the column with both ends effectively held in position and restrained against rotation, $l_{\text{eff}} = 0.65L$.
2. For the column with both ends effectively held in position and one end restrained against rotation, $l_{eff} = 0.80L$.

3. For the column with both ends effectively held in position, but not restrained against rotation, $l_{\text{eff}} = L$.

Example A.14 A reinforced concrete column of 350 mm diameter reinforced with 7 longitudinal bars of 25 mm ϕ placed at a clear cover of 25 mm, is provided with spiral reinforcement of 6 mm diameter at a pitch of 50 mm. The material used are: M20 grade concrete and HYSD steel of grade Fe415. Determine the safe load that the column can support if the effective length is (a) 3.0 m and (b) 6 m.

Solution For M20 grade concrete and Fe415 grade steel,

$$\sigma_{cc} = 5.0 \text{ MPa}, \sigma_{sc} = 190 \text{ MPa}, \sigma_{ck} = 20 \text{ MPa}, f_y = 415 \text{ MPa}$$

For the given column section,

$$A_{sc} = 7 \times 25 \text{ mm } \phi = 3436 \text{ mm}^2$$

$$A_c = \left[\pi \times \frac{(350)^2}{4}\right] - 3436 = 92775.28 \text{ mm}^2$$

The allowable load as in the case of a tied column is given by

$$P' = \sigma_{cc}A_c + \sigma_{sc}A_{sc} = 5.0 \times 92775.28 + 190 \times 3436 = 1116.72 \text{ kN}$$

Volume of helical reinforcement is:

$$V_h = \pi \times \frac{(300 + 6)(28.274)}{60} = 453 \text{ mm}^3\text{/mm height}$$

Volume of core is given by

$$V_c = \left[\frac{\pi(300 + 12)^2}{4}\right] \times 1 = 76453.80 \text{ mm}^3\text{/mm}$$

Thus,

$$\text{Ratio } \frac{V_h}{V_c} = 0.0059$$

The helical reinforcement parameter is:

$$(0.36)\left[\frac{A_g}{A_c} - 1\right]\left(\frac{\sigma_{ck}}{f_y}\right) = 0.36 \times \left[\frac{\pi(350)^2/4}{\pi(312)^2/4} - 1\right] \times \left(\frac{20}{415}\right) = 0.0045 < 0.0059$$

1. Effective length of the column, $l_{\text{eff}} = 3 \times 10^3$ mm

Therefore, $$C_r = 1.25 - \frac{3000}{48 \times 312} = 1.05 > 1.0$$

Hence $C_r = 1.00$ and $P = 1.05P' = 1.05 \times 1116.72 = 1172.56$ kN

Thus, the safe load that the column can support as a short column is 1172.56 kN.

2. Effective length of the column $l_{\text{eff}} = 6 \times 10^3$ mm

Therefore,

$$C_r = 1.25 - \left[\frac{6000}{48 \times 312}\right] = 0.849$$

Safe load $P = 1.05 \times 0.849 \times 1116.72 = 995.50$ kN

A.5.2 Design of Axially Loaded Columns

The design of a column consists in proportioning the column section along with determination of areas of longitudinal steel, and transverse steel in the form of ties and their placement. A long column is designed by using reduced permissible stresses in the materials, as it fails at a load smaller than the failure load of the corresponding short column. Alternatively, design loads and moments supported by a long column are magnified and the member is designed as a normal short column. For IS:456 recommendations regarding longitudinal and transverse reinforcements a reader should refer to Chapter 5.

Procedure for design: The following procedure may be adopted for the design of a reinforced concrete column:

1. For the stipulated grades of concrete mix and steel, determine the permissible stresses in concrete, longitudinal steel and ties.
2. Consider some suitable percentage p of longitudinal reinforcement A_{sc} (normally between 1 and 2) in terms of gross cross-sectional area A_g of the column, i.e.,

$$A_{sc} = \frac{pA_g}{100} = 0.01pA_g$$

3. Apply the load-carrying capacity formula:

$$P = \sigma_{cc}(A_g - A_{sc}) + \sigma_{sc}A_{sc} = \sigma_{cc}A_g(1 - 0.01p) + \sigma_{sc}A_g(0.01p) \qquad \text{(A.36)}$$

   and calculate the gross area A_g of concrete.
4. Knowing A_g, proportion the dimensions of the column of the desired shape.
5. For the given end conditions, determine the effective length l_{eff} of the column. From the ratio, l_{eff}/b, where b is the least lateral dimension of the column, check whether the column is long or short, based on the criteria: (a) if the ratio l_{eff}/b is less than or equal to 12, the column is a short column for which dimensions have already been determined in Step 4, and (*b*) if the ratio $l_{\text{eff}}/b > 12$, the design will be as a long column, for which following two alternatives are available:
   (i) If there is no restriction on the size of the column, increase the dimensions of the column section so that it is a short column and determine the modified value of A_{sc}.

(ii) If the size of the column is restricted, determine the reduction factor C_r, as follows:

$$C_r = 1.25 - \frac{l_{\text{eff}}}{48b}$$

Calculate the increased value of the load P' for an equivalent short column as: $P' = P/C_r$.

Using the revised value of P', recalculate the area A_g and determine the size of the column and area of longitudinal steel A_{sc}. In the case of a column having spiral reinforcement, the value of b for computing C_r, is equal to the core diameter measured to the outside diameter of the helix steel.

6. The area of longitudinal steel required can be obtained as:

$$A_{sc} = \frac{pA_g}{100}$$

Choose the diameter and determine the number of bars to be adopted and distribute the bars suitably around the periphery with an appropriate cover.

7. Design the transverse steel.

The following examples illustrate the procedure and types of problems encountered in practice.

Example A.15 A reinforced concrete column is to support a service load of 950 kN inclusive of its own weight. The column is effectively held in position at both the ends and restrained against rotation at one of the ends. The unsupported length of the column is 5.5 m. The materials to be used are M25 grade concrete mix and HYSD steel of grade Fe415. Design the column section when its shape is: (a) square and (b) circular.

Solution For M25 grade concrete and Fe415 grade steel,

$$\sigma_{cc} = 6 \text{ MPa}, \sigma_{sc} = 190 \text{ MPa and } l_{\text{eff}} = 0.80L = 4.4 \text{ m}$$

Consider that the minimum eccentricity is less than 0.05 times the lateral dimensions, and the area of longitudinal steel is 2 per cent of the gross cross-sectional area, i.e., $p = 2$. Then from Eq. (A.36),

$$950 \times 10^3 = 6 \times 0.98A_g + (190 \times 0.02)A_g$$

whence,

$$A_g = \frac{950 \times 10^3}{9.68} = 98140.5 \text{ mm}^2$$

(a) Square section: The cross-sectional dimensions are:

$$b = D = (A_g)^{1/2} = 313.27 \text{ mm}$$

Consider column section of size 315 × 315 mm.
Then

$$\frac{l_{\text{eff}}}{b} = \frac{4400}{315} = 13.97$$

Since $l_{\text{eff}}/b > 12$, the column should be designed as a long column. There are two alternatives.

Alternative I. The section is modified such that

$$\frac{l_{\text{eff}}}{b} \le 12 \text{ or } b \ge \frac{4400}{12} \quad (= 366.67 \text{ mm})$$

Adopt a column section of size 375 × 375 mm. The longitudinal reinforcement required is obtained from

$$950 \times 10^3 = 6\,[(375 \times 375) - A_{sc}] + 190\,(A_{sc})$$

or

$$A_{sc} = 577.45 \text{ mm}^2$$

Minimum steel to be provided is 0.8 per cent of the gross cross-section area. Thus,

$$A_{sc,\min} = 0.8 \times \frac{375 \times 375}{100} = 1125 \text{ mm}^2$$

Hence adopt A_{sc} = 1125 mm^2 and not 577.45 mm^2. Provide 4 bars of 20 mm ϕ (A_{sc} = 1257 mm^2).

Alternative II. For illustration of procedure consider the section of the column to be restricted to 325 × 325 mm from architectural considerations.
This column is a long column with the reduction factor,

$$C_r = 1.25 - \frac{l_{\text{eff}}}{48b} = 1.25 - \frac{4400}{48 \times 325} = 0.968$$

Revised axial load P' for the equivalent short column is:

$$P' = \frac{950}{0.968} = 981.4 \text{ kN}$$

Area of reinforcement required is obtained from:

$$981.4 \times 10^3 = 6\,[(325 \times 325) - A_{sc}] + 190(A_{sc})$$

or

$$A_{sc} = 1889.4 \text{ mm}^2$$

Provide 4 bars of 25 mm ϕ (A_{sc} = 1964 mm^2 > 1889.4 mm^2).

Design of ties. Minimum diameter of tie rod is:

$$\phi_t = 5 \text{ mm or } \frac{\text{Diameter of longitudinal bars}}{4} = 6.25 \text{ mm}$$

Use 8 mm diameter bar for ties. Spacing should be the least of the following :

(i) the least lateral dimension, i.e., 325 mm
(ii) 16 times the diameter of the longitudinal bar, i.e., 16 × 25 = 400 mm
(iii) 300 mm

Provide 8 mm ϕ bar ties at a spacing of 300 mm c/c.

(b) Circular section with lateral ties: From Case (a), A_g = 98140.5 mm^2. Diameter of the circular column is given by

$$D = \left(\frac{4A_g}{\pi}\right)^{1/2} = \left(\frac{4 \times 98140.5}{\pi}\right)^{1/2} = 353.5 \text{ mm}$$

Consider D = 350 mm, the ratio $\dfrac{l_{\text{eff}}}{b} = \dfrac{4400}{350}$ = 12.57.

Since $l_{\text{eff}}/b > 12$, the column is a long column with the reduction factor,

$$C_r = 1.25 - \left[\frac{l_{\text{eff}}}{48b}\right] = 1.25 - \left[\frac{4400}{48 \times 350}\right] = 0.988$$

The area of steel is given by

$$\frac{950 \times 10^3}{0.988} = 6\left[\frac{\pi(350)^2}{4} - A_{sc}\right] + 190A_{sc}$$

Giving

$$A_{sc} = 2088.43 \text{ mm}^2$$

Provide 7 bars of 20 mm ϕ (A_{sc} = 2199 mm^2).

Design of ties. Use 6 mm ϕ lateral ties. Spacing of the ties shall be the least of the following: (i) the least lateral dimension, i.e., 350 mm, (ii) 16 times the diameter of the longitudinal bar, i.e., 16 × 20 = 320 mm and (iii) 300 mm.

Provide 6 mm ϕ bar ties at a spacing of 300 mm c/c.

(c) Circular section with spiral reinforcement: Adopt the diameter D of the column section as 350 mm, the same as for the column with lateral ties. Consider the clear cover to longitudinal reinforcement as 40 mm.

Core diameter b = 350 – (2 × Cover) + (2 × Diameter of helix bar)

= 350 – (2 × 40) + (2× 8) = 286 mm

∴ Slenderness ratio $\dfrac{l_{\text{eff}}}{b} = \dfrac{4400}{286}$ = 15.38

Since $l_{\text{eff}}/b > 12$, the column is to be designed as a long column with a reduction factor,

$$C_r = 1.25 - \frac{l_{\text{eff}}}{48b} = 0.929$$

The allowable axial load capacity of a column with spiral reinforcement is 1.05 times that for an identical column with lateral ties. The area of reinforcement is obtained as follows:

$$\frac{P}{1.05 \times 0.929} = \sigma_{cc}\left[\frac{\pi D^2}{4} - A_{sc}\right] + \sigma_{sc}\, A_{sc}$$

$$\frac{950 \times 10^3}{1.05 \times 0.929} = 6\left[\frac{\pi(350)^2}{4} - A_{sc}\right] + 190\ A_{sc}$$

whence, $A_{sc} = 2155.66\ \text{mm}^2$

Use 6 bars of 22 mm ϕ ($A_{sc} = 2280\ \text{mm}^2$).

Helical reinforcement. The volume of helical reinforcement per mm length of column may be obtained from:

$$V_h = 0.36\left[\frac{A_g}{A_c} - 1\right]\left(\frac{f_{ck}}{f_y}\right) \text{ (volume of concrete core)}$$

$$= (0.36)\left[\left\{\frac{\pi(350)^2/4}{\pi(286)^2/4}\right\} - 1\right]\left(\frac{25}{415}\right)\left[\left\{\frac{\pi(286)^2}{4} - 2280\right\}\right]$$

$$= 668.69\ \text{mm}^3/\text{mm}$$

Minimum diameter of the helical bar. As per code stipulations the minimum diameter is 5 mm or (1/4) (22) = 5.5 mm. Use 8 mm ϕ bar for helical reinforcement.

Pitch. Spacing of spiral should satisfy the following:

Maximum: 75 mm or $\left(\frac{1}{6}\right)(350) = 58.33$ mm

Minimum: 25 mm or 3(8) = 24 mm

From Eq. (5.10),

$$s_v \le \left(\frac{11.11 a_{sp} D_c}{D^2 - D_c^2}\right)\left(\frac{f_{ysp}}{f_{ck}}\right) = \left(\frac{11.11 \times (\pi \times 8^2/4) \times 286}{350^2 - 286^2}\right)\left(\frac{415}{25}\right) = 65.14\ \text{mm}$$

Adopt pitch of 55 mm c/c, i.e., wind the column with 8 mm ϕ helical steel at a pitch of 55 mm.

A.5.3 Section Subjected to Combined Direct Load and Uniaxial Bending

In a majority of the cases, the column elements are subjected to combined axial load and bending moments. In such a case, the stresses produced are no longer uniform and may result in an uncracked or cracked section depending upon the relative magnitude of moment as shown in Fig. A.7.

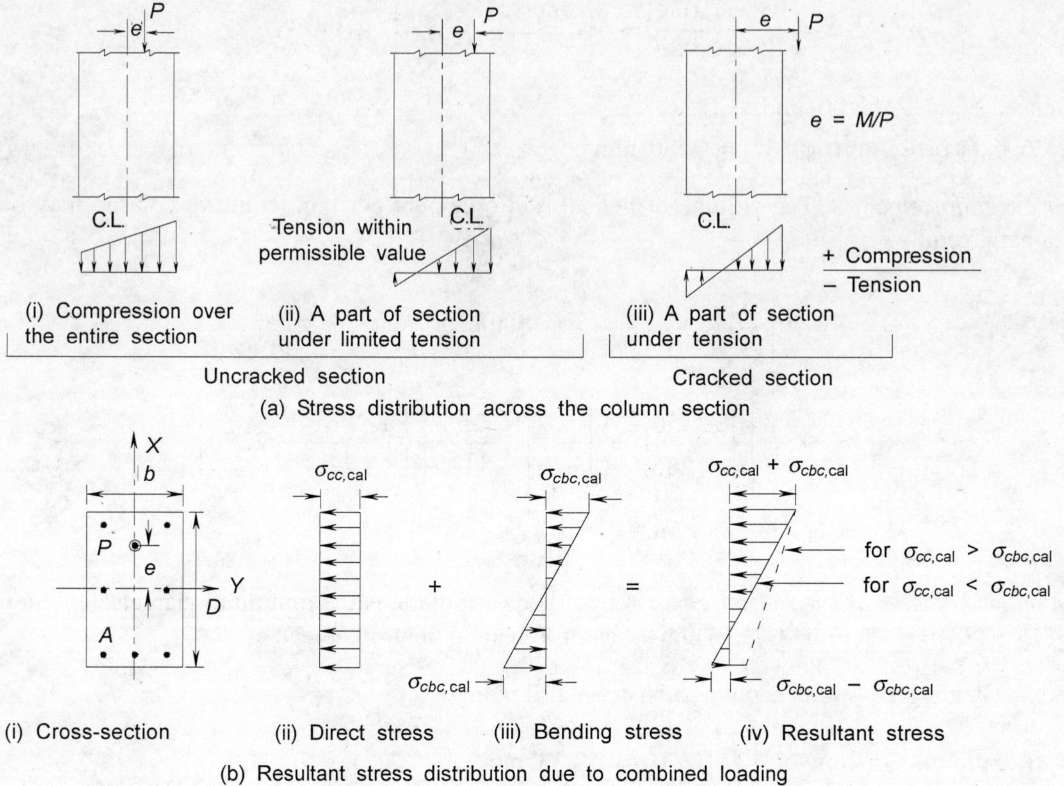

Fig. A.7 *Stress developed in a uniaxially eccentrically loaded column section.*

(a) Uncracked section: The analysis and design of the uncracked section is based on elastic theory. The stresses developed in a section subjected to an axial load P and uniaxial moment $M(=Pe)$ based on the transformed section and illustrated in Fig. A.7(b) are given by

$$\sigma_{cc,\text{cal}} = \frac{P}{A_e} \text{ and } \sigma_{cbc,\text{cal}} = \frac{M}{Z} \tag{A.37}$$

where.

$\sigma_{cc,\text{cal}}$ = Calculated direct compressive stress in concrete

$\sigma_{cbc,\text{cal}}$ = Calculated bending compressive stress in concrete

A_e = Area of the equivalent transformed section

Z = Modulus of the equivalent transformed section

For a rectangular section of size $b \times D$ symmetrically reinforced, the properties of the equivalent transformed section are:

$$A_e = A_c + 1.5mA_{sc} = bD + (1.5m - 1)A_{sc} \tag{A.38a}$$

$$l_{yy} = \left(\frac{bD^3}{12}\right) + (1.5m - 1)\Sigma A_{si} x^2_{si} \quad \text{(A.38b)}$$

$$Z = \frac{l_{yy}}{D/2} \quad \text{(A.38c)}$$

where

A_c = Net area of concrete (= $A_g - A_{sc}$)

l_{yy} = Moment of inertia of transformed area

At extreme fibres the maximum and minimum stresses are:

$$\sigma_{\max} = \sigma_{cc,\text{cal}} + \sigma_{cbc,\text{cal}} \quad \text{and} \quad \sigma_{\min} = \sigma_{cc,\text{cal}} - \sigma_{cbc,\text{cal}} \quad \text{(A.39)}$$

For the section to be safe and uncracked, the following conditions, as stipulated by IS:456, should be satisfied:

(i) For dead and live loads combination,

$$\left[\frac{\sigma_{cc,\text{cal}}}{\sigma_{cc}} + \frac{\sigma_{cbc,\text{cal}}}{\sigma_{cbc}}\right] \le 1 \quad \text{(A.40)}$$

(ii) For dead and wind or earthquake loads combination,

$$\left[\frac{\sigma_{cc,\text{cal}}}{\sigma_{cc}} + \frac{\sigma_{cbc,cal}}{\sigma_{cbc}}\right] \le 1.33 \quad \text{(A.41)}$$

(iii) Condition for tensile stress to be within permissible limits is:

$$\sigma_{ct} = (\sigma_{cbc,\text{cal}} - \sigma_{cc,\text{cal}}) \le 0.25(\sigma_{cbc,\text{cal}} + \sigma_{cc,\text{cal}})$$

$$\le 0.75 \times 7 \text{ days modulus of rupture of concrete} \quad \text{(A.42)}$$

The design of a section subjected to a combination of axial force P and uniaxial moment M consists in checking the adequacy of a pre-assigned section. The section is considered safe if the stresses developed in concrete and steel are within the permissible limits.

(b) Cracked section: A section which does not satisfy the checks as stipulated by IS:456 is to be considered a cracked section. The stresses in concrete and steel are calculated by ignoring the concrete in the tension zone, as shown in Fig. A.8(c). The line of no stress or neutral axis should satisfy the following conditions:

(i) The direct external load should be equal to the algebraic sum of the forces on concrete and steel.

(i) The moment of external loads about any reference line should be equal to the algebraic sum of the moments of forces in concrete and steel about the same line while ignoring the tensile force in concrete.

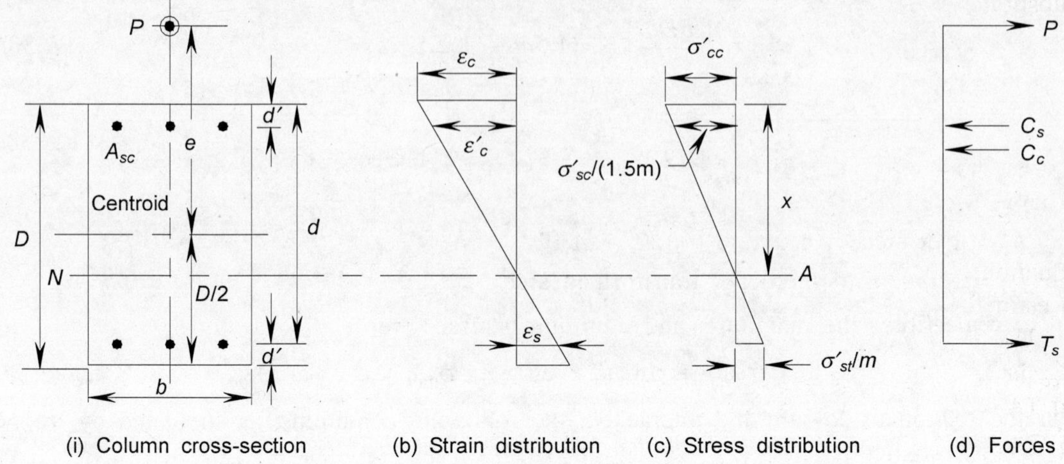

Fig. A.8 *Strain and stress distributions in a cracked section.*

For simplicity consider the case when reinforcement is provided only on compression and tension faces. The stresses induced in the section can be determined from the stress distribution diagram, as shown in Fig. A.8(c).

$$\frac{\sigma'_{cc}}{(\sigma'_{sc}/1.5m)} = \frac{x}{x-d'} \quad \text{or} \quad \sigma'_{sc} = 1.5m\sigma'_{cc}\left(\frac{x-d'}{x}\right) \tag{A.43}$$

$$\frac{\sigma'_{cc}}{(\sigma'_{st}/m)} = \frac{x}{d-x} \quad \text{or} \quad \sigma'_{st} = m\sigma'_{cc}\left(\frac{d-x}{x}\right) \tag{A.44}$$

For equilibrium of direct forces, $P = C_c + C_s - T_s$

or
$$P = \left(\frac{1}{2}\right)bx\sigma'_{cc} + \left[\frac{1.5m-1}{1.5m}\right]\sigma'_{sc}A_{sc} - \sigma'_{st}A_{st} \tag{A.45}$$

Taking moments of forces about tension reinforcement, we get

$$P\left[e + d - \left(\frac{D}{2}\right)\right] = \frac{1}{2}bx\sigma'_{cc}\left[d - \left(\frac{x}{3}\right)\right] + \left[\frac{1.5m-1}{1.5m}\right]\sigma'_{sc}A_{sc}(d-d') \tag{A.46}$$

Substitute values of σ'_{sc} and σ'_{st} from Eqs. (A.43) and (A.44), respectively, in Eq. (A.45),

$$P = \frac{1}{2}(bx\sigma'_{cc}) + [(1.5m-1)A_{sc}]\sigma'_{cc}\frac{(x-d')}{x} - mA_{st}\sigma'_{cc}\frac{(d-x)}{x} \tag{A.47}$$

The equation can be simplified as:

$$x^2 + 2\left[(1.5m-1)A_{sc} + mA_{st} - \left(\frac{P}{\sigma'_{cc}}\right)\right]\left(\frac{x}{b}\right) - \frac{2[(1.5m-1)A_{sc}d' + mA_{st}d]}{b} = 0 \tag{A.48}$$

Substitute the values of σ'_{sc} and σ'_{st} from Eqs. (A.43) and (A.44) in Eq. (A.46):

$$P\left[e + d - \left(\frac{D}{2}\right)\right] = \frac{1}{2} bx\sigma'_{cc}\left[d - \left(\frac{x}{3}\right)\right]$$

$$+ \ [(1.5m - 1)A_{sc}(d - d')] \left[\frac{\sigma'_{cc}(x - d')}{x}\right] \tag{A.49}$$

Equations (A.47) and (A.49) are solved iteratively for x and σ'_{cc}. Equation (A.48) can be used to estimate the value of x for a trial value of σ'_{cc} [= σ_{cbc} or $(P/bd) + (6Pe/bd^2)$].

Use this value of x to estimate a new value σ'_{cc} from Eq. (A.49). This improved value of σ'_{cc} is used to get a better approximation of x from Eq. (A.48) and hence of σ'_{cc} from Eq. (A.49). The process is repeated till no appreciable change in the value occurs. The values of σ'_{sc} and σ'_{st} are determined from Eqs. (A.43) and (A.44), respectively. The stresses should be within the permissible values for the section to be safe; otherwise the section is revised and rechecked.

A.5.4 Section Subjected to Combined Direct Load and Biaxial Bending

(i) Uncracked section: The design consists in checking the adequacy of pre-assigned section for the given axial load and moments by using elastic theory. The sections generally have equal reinforcements on all the four faces. The stresses developed are shown in Fig. A.9(a). Now,

Area of transformed section $A_e = A_c + 1.5mA_{sc}$

Moment of inertia about x-axis is:

$$l_{xx} = \left(\frac{b^3D}{12}\right) + (1.5m - 1)\sum A_{si}y_{si}^2$$

Moment of inertia about y-axis is:

$$l_{yy} = \left(\frac{bD^3}{12}\right) + (1.5\text{ m} - 1)\sum A_{si}x_{si}^2$$

Direct stress due to axial load $\sigma_{cc,\text{cal}} = \dfrac{P}{A_e}$

The bending stresses due to moments, M_x and M_y as illustrated in Fig. A.9(a) are:

$$\sigma_{cby,\text{cal}} = \left(\frac{M_y}{l_{yy}}\right)\left(\frac{D}{2}\right) = \frac{Pe_xD}{2l_{yy}}$$

$$\sigma_{cbx,\text{cal}} = \left(\frac{M_x}{l_{xx}}\right)\left(\frac{b}{2}\right) = \frac{Pe_yb}{2l_{xx}}$$

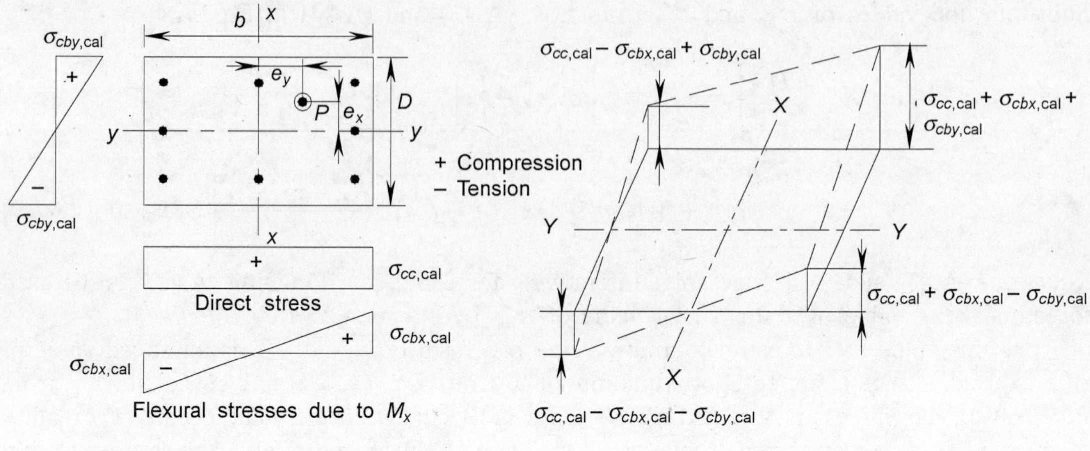

Fig. A.9 *Stress distribution in a biaxially accentrically loaded column section.*

For the section to be uncracked the following conditions should be satisfied:

(a) Interaction condition. For the section to be safe,

$$\frac{\sigma_{cc,\text{cal}}}{\sigma_{cc}} + \frac{\sigma_{cbx,\text{cal}} + \sigma_{cby,\text{cal}}}{\sigma_{cbc}} \le 1.0 \qquad \text{(for live and dead loads)}$$

$$\le 1.33 \qquad \text{(for dead and wind or earthquake loads)}$$

(b) Stress condition. For the tensile stress to be within permissible limits, the following conditions should be satisfied:

$$\sigma_{cbx,\text{cal}} + \sigma_{cby,\text{cal}} - \sigma_{cc,\text{cal}} \le 0.35(\sigma_{cbx,\text{cal}} + \sigma_{cby,\text{cal}} + \sigma_{cc,\text{cal}})$$

$$\le 0.75 \times 7 \text{ days modulus of rupture of concrete}$$

If the preceding conditions are not satisfied, the section is unsafe and cracked. The section is revised and rechecked for its adequacy.

(ii) Cracked section: The post-cracking behaviour is complicated and is beyond the scope of this book.

Example A.16 A reinforced concrete column section of size 300 × 550 mm is reinforced with four bars of 28 mm ϕ placed at corners with an effective cover of 50 mm. The materials used are M25 grade concrete and HYSD steel of grade Fe415. Determine the maximum and minimum stresses in concrete, if the column is subjected to a compressive load of 650 kN at an eccentricity of 110 mm along the larger dimension of the section. Also check the adequacy of the section as an uncracked section. What is the maximum eccentricity at which the load can be applied safely?

Solution For M25 grade concrete and HYSD steel of grade Fe415:

$$\sigma_{cc} = 6 \text{ MPa}, \; \sigma_{cbc} = 8.5 \text{ MPa and } m = 11$$

Properties of the column section are:

$$b = 300 \text{ mm}, \; D = 550 \text{ mm}, \; d' = 50 \text{ mm and } e = 110 \text{ mm}$$

$$A_{sc}(4 \times 28 \text{ mm } \phi) = 2463 \text{ mm}^2$$

Equivalent transformed area of the section is obtained as:

$$A_e = bD + (1.5m - 1)A_{sc}$$

$$= 300 \times 550 + (1.5 \times 11 - 1) \times 2463 = 203176.5 \text{ mm}^2$$

Moment of inertia of equivalent section about the axis of bending is given by:

$$I_{yy} = \frac{bD^3}{12} + (1.5m - 1)A_{sc}\left(\frac{D}{2} - d'\right)^2$$

$$= \frac{(300)(550)^3}{12} + [(1.5 \times 11) - 1]\,(2463)\,(275 - 50)^2 = 6.092 \times 10^9 \text{ mm}^4$$

Direct and bending stresses are as follows:

$$\sigma_{cc,\text{cal}} = \frac{650 \times 10^3}{203176.5} = 3.1992 \text{ MPa}$$

and

$$\sigma_{cbc,\text{cal}} = \left(\frac{M}{I}\right)\left(\frac{D}{2}\right) = \left(\frac{Pe}{I}\right)\left(\frac{D}{2}\right) = \left(\frac{650 \times 10^3 \times 110}{6.092 \times 10^9}\right)\left(\frac{550}{2}\right) = 3.2276 \text{ MPa}^2$$

Maximum stress = 3.1992 + 3.2276 = 6.4268 MPa (compression)
Minimum stress = 3.1992 – 3.2276 = – 0.0284 MPa (tension)

Part I. For the section to be safe and uncracked the following conditions should be satisfied:

(a) $$\frac{\sigma_{cc,\text{cal}}}{\sigma_{cc}} + \frac{\sigma_{cbc,\text{cal}}}{\sigma_{cbc}} = \frac{3.1992}{6.0} + \frac{3.2276}{8.5} = 0.9129 < 1.00$$

Thus, the interaction condition is satisfied.

(b) For tensile stress to be within permissible limits,

(i) $(\sigma_{cbc,\text{cal}} - \sigma_{cc,\text{cal}}) \le 0.25\,(\sigma_{cc,\text{cal}} + \sigma_{cbc,\text{cal}})$

$0.0284 \le 0.25(6.4268)\;(= 1.6067)$

Hence the section is uncracked.

(ii) $(\sigma_{cbc,\text{cal}} - \sigma_{cc,\text{cal}}) \le 0.75$ (7 days modulus of rupture of concrete

$$= 0.8 \times 0.7\sqrt{f_{ck}} = 0.56\sqrt{25} = 2.8 \text{ MPa})$$

Since 7 days modulus of rupture = 0.8 (modulus of rupture of concrete)
Thus, tension of 0.0284 MPa ≤ 0.75 × 2.8 (= 2.1 MPa)
Hence the section is adequate and uncracked.

Part II. The maximum eccentricity can be obtained from the following considerations:

(i) Interaction condition.

$$\frac{\sigma_{cc,cal}}{\sigma_{cc}} + \frac{\sigma_{cbc,cal}}{\sigma_{cbc}} = 1$$

or

$$\frac{3.1992}{6.0} + \frac{\sigma_{cbc,cal}}{8.5} = 1$$

whence

$$\sigma_{cbc,\text{cal}} = 3.9678 \text{ MPa}$$

(ii) For tensile stress to be within permissible limits.

$$\sigma_{cbc,\text{cal}} - 3.1992 = 0.25(\sigma_{cbc,\text{cal}} + 3.1992)$$

$$0.75\sigma_{cbc,\text{cal}} = 1.25 \times (3.1992)$$

Therefore,

$$\sigma_{cbc,\text{cal}} = 5.332 \text{ MPa}$$

Also,

$$\sigma_{cbc,\text{cal}} - 3.1992 = 0.75(2.8)$$

Whence

$$\sigma_{cbc,\text{cal}} = 5.2992 \text{ MPa}$$

$$\text{Critical value of } \sigma_{cbc,\text{cal}} = 3.9678 \text{ MPa}$$

From

$$\left[\left(\frac{Pe_{\text{max}}}{I}\right)\left(\frac{D}{2}\right)\right] = 3.9678$$

or

$$e_{\text{max}} = \frac{3.9678 \times (6.092 \times 10^9)}{(650 \times 10^3)275} = 135.227 \text{ mm}$$

Example A.17 Design a rectangular column section to support an axial load of 800 kN and a uniaxial bending moment of 85 kNm. Use M20 grade concrete and HYSD steel of grade Fe415.

Solution For M20 grade concrete and Fe415 grade steel,

$$\sigma_{cc} = 5 \text{ MPa}, \ \sigma_{cbc} = 7 \text{ MPa and } m = 13$$

$$P = 800 \text{ kN}, \ M = 85 \text{ kNm and } e = \frac{M}{P} = 106.25 \text{ mm}$$

Preliminary design. The section can be predetermined by using interaction condition,

$$\frac{\sigma_{cc,cal}}{\sigma_{cc}} + \frac{\sigma_{cbc,cal}}{\sigma_{cbc}} \le 1.0$$

where $\sigma_{cc,\text{cal}} = (P/A_e)$ and $\sigma_{cbc,\text{cal}} = (Pe/Z)$. Consider a section of size $b \times D$ having p per cent of longitudinal reinforcement equally distributed on two opposite faces at an effective cover of d'. Then

$$A_e = bD + (1.5m - 1)\ (0.010pbD)$$

$$= [1 + (1.5m - 1)(0.010p)]\ (bD)$$

and

$$Z = \frac{[(bD^3/12) + (1.5m - 1)(0.010pbD)(0.5D - d')^2]}{(D/2)}$$

$$= \left\{\left(\frac{1}{6}\right) + (1.5m - 1)(0.010p)\ \frac{[1 - (2d'/D)]^2}{2}\right\} bD^2$$

To start with let p = 1.0 per cent and $(d'/D) = 0.10$; then

$$A_e = 1.185bD \text{ and } Z = 0.2259bD^2$$

Substitute the values in the interaction formula:

$$\frac{P/(1.185bD)}{5} + \frac{Pe/(0.2259bD^2)}{7} \le 1.0$$

or

$$\frac{P}{(1.185)(5)} + \frac{P(e/D)}{(0.2259)(7)} \le bD$$

Consider $e/D = 0.20$, then

$$bD \ge 236203.67 \text{ mm}^2$$

Therefore

$$D = \frac{e}{0.2} = \frac{106.25}{0.2} = 531.25 \text{ mm (say 550 mm)}$$

and

$$b \ge (236203.67)/(550) = 429.46 \text{ mm (say 425 mm)}$$

Adopt a section of size 425 × 550 mm. Then

$$\text{Area of steel } A_{sc} = (0.010)\ (425)\ (550) = 2337.5 \text{ mm}^2$$

Consider 8 bars of 20 mm $\phi(A_{sc} = 2513 \text{ mm}^2)$ distributed equally on shorter faces at an effective cover of 50 mm. In addition provide 2 bars of 10 mm ϕ on each of the two longer faces such that the spacing of longitudinal bars is less than 150 mm.

Check for the adequacy of the section:

$$A_e = (425 \times 550) + [(1.5 \times 13 - 1)\ (2513)] = 280240.5\ \text{mm}^2$$

$$I_{yy} = \frac{425 \times 550^3}{12} + [(1.5 \times 13 - 1)\ (2513)\ (275 - 50)^2]$$

$$= 8246.03 \times 10^6\ \text{mm}^4$$

The stresses in the section are:

$$\sigma_{cc,\text{cal}} = \left(\frac{P}{A_e}\right) = \frac{800 \times 10^3}{280240.5} = 2.8547\ \text{MPa}$$

$$\sigma_{cbc,\text{cal}} = \left(\frac{M}{I}\right)\left(\frac{D}{2}\right) = \frac{85 \times 10^6}{8246.03 \times 10^6}\ (275)$$

$$= 2.8347\ \text{MPa}$$

Maximum stress $\sigma_{\text{max}} = \sigma_{cc,\text{cal}} + \sigma_{cbc,\text{cal}} = 5.6894$ MPa (compression)

Minimum stress $\sigma_{\text{min}} = \sigma_{cc,\text{cal}} - \sigma_{cbc,\text{cal}} = 0.02$ MPa (compression)

For the section to be safe and uncracked, the following conditions should be satisfied:

(i) $$\frac{\sigma_{cc,cal}}{\sigma_{cc}} + \frac{\sigma_{cbc,cal}}{\sigma_{cbc}} = \frac{2.8547}{5} + \frac{2.8347}{7} = 0.9759 < 1.0\ \ \text{(safe)}$$

(ii) $\sigma_{\text{min}} \leq 0.25\ \sigma_{\text{max}}$

$$0.02 \leq (0.25)\ (5.6894) = 1.4224\ \text{MPa (true)}$$

and $-\sigma_{\text{min}} \leq 0.75 \times$ (7 days modulus of rupture of concrete)

$$\leq 0.75 \times [(0.80) \times (0.70)(\sqrt{20})] = 1.8783\ \text{MPa (true)}$$

Hence section is uncracked and safe. Provide 6 mm ϕ bar ties at 250 mm c/c.

Example A.18 A reinforced concrete column section of size 350 × 500 mm is reinforced with 6 bar of 22 mm ϕ distributed equally on two opposite shorter sides at an effective cover of 50 mm. The materials to be used are M20 grade concrete mix and HYSD steel of grade Fe415. Check the adequacy of the section if the column is to support a direct compressive load of 450 kN, and moments of 35 kNm and 20 kNm about the axes parallel to the shorter and longer sides, respectively.

Solution For M20 grade concrete and Fe415 grade steel,

$\sigma_{cc} = 5.0$ MPa, $\sigma_{cbc} = 7.0$ MPa and $m = 13$

$b = 350$ mm, $D = 500$ mm, $d' = 50$ mm and $A_s(6 \times 22\ \text{mm}\ \phi) = 2280\ \text{mm}^2$

$A_e = (350 \times 500) + (1.5 \times 13 - 1)(2280) = 217180\ \text{mm}^2$

Moment of inertia about Y-axis,

$$l_{yy} = \left[\frac{(350)(500)^3}{12}\right] + [(1.5 \times 13 - 1)\ (2280)\ (250 - 50)^2]$$

$$= 5.333 \times 10^9 \text{ mm}^4$$

Moment of inertia about *X*-axis is:

$$l_{xx} = \left[\frac{(350)^3(500)}{12}\right] + [(1.5 \times 13 - 1)\ (1520)\ (175 - 50)^2]$$

$$= 2.225 \times 10^9 \text{ mm}^4$$

The stresses induced in the section are:
Direct stress.

$$\sigma_{cc,\text{cal}} = \frac{P}{A_e} = \frac{450 \times 10^3}{217180} = 2.072 \text{ MPa.}$$

Bending stresses.

$$\sigma_{cby,\text{cal}} = \left(\frac{M_y}{l_{yy}}\right)\left(\frac{D}{2}\right) = \left(\frac{35 \times 10^6}{5.333 \times 10^9}\right)\left(\frac{500}{2}\right)$$

$$= 1.641 \text{ MPa.}$$

$$\sigma_{cbx,\text{cal}} = \left(\frac{M_x}{l_{xx}}\right)\left(\frac{b}{2}\right) = \left(\frac{20 \times 10^6}{2.225 \times 10^9}\right)\left(\frac{350}{2}\right)$$

$$= 1.573 \text{ MPa.}$$

Now,

$$\text{Maximum stress } \sigma_{\max} = \sigma_{cc,\text{cal}} + \sigma_{cby,\text{cal}} + \sigma_{cbx,\text{cal}}$$

$$= 2.072 + 1.641 + 1.573 = 5.286 \text{ MPa (compression)}$$

and

$$\text{Minimum stress } \sigma_{\min} = \sigma_{cc,\text{cal}} - \sigma_{cbx,\text{cal}} - \sigma_{cby,\text{cal}} = 1.142 \text{ MPa (tension)}$$

Check for the adequacy of the section.

(a) *Apply the interaction condition for checking the adequacy.*

$$\frac{\sigma_{cc,\text{cal}}}{\sigma_{cc}} + \frac{\sigma_{cby,\text{cal}} + \sigma_{cbx,\text{cal}}}{\sigma_{cbc}}$$

$$= \frac{2.072}{5.0} + \frac{1.641 + 1.573}{7.0} = 0.8735 \leq 1.0 \text{ (safe)}$$

(b) *Conditions for the section to be uncracked.*

(i) $\sigma_{min} \leq 0.35\sigma_{max}$

$$1.142 \leq (0.35)(5.286) = 1.859 \text{ MPa (uncracked)}$$

(ii) $\sigma_{min} \leq 0.75$ (7 days modulus of rupture)

$$1.142 \leq (0.75) \times [(0.8) \times (0.7) (\sqrt{20})] = 1.878 \text{ MPa}$$

Conditions are satisfied and the section is uncracked. Therefore, the column section is uncracked and adequate.

Tutorial Problems

T.A.1 (a) Explain the working stress and limit state design methods for the design of reinforced cement concrete structures.

(b) State the assumptions made in theory of bending as applied to the design of reinforced concrete structures.

T.A.2 Explain briefly the following: (a) balanced section, (b) under-reinforced section, (c) over-reinforced section; explain why the over-reinforced design is not advisable, (d) modular ratio and its significance in the design of concrete structures and (e) transformed area.

T.A.3 A reinforced concrete beam of rectangular cross-section has a width of b mm and effective depth of d mm. Derive expressions for: (a) the position of the neutral axis, (b) the lever-arm, (c) the moment of resistance, and (d) the percentage of balanced steel, for various combinations of M20 grade concrete mixes, with mild steel and HYSD steel of grade Fe415.

[**Ans.** *M20 grade concrete and mild steel:*

$$m = 13,\ k = 0.394;\ j = 0.869;\ Q = 1.198 \text{ MPa},\ p_t = 0.985$$

M20 grade concrete and Fe415 grade steel:

$$m = 13,\ k = 0.283,\ j = 0.906,\ Q = 0.897 \text{ MPa},\ p_t = 0.4306]$$

T.A.4 A reinforced concrete beam of size 250 mm width × 550 mm overall depth is reinforced with 3 bars of 20 mm ϕ at an effective cover of 50 mm (i.e., bars are placed with their centres at 50 mm from the bottom face). Determine the moment of resistance, and the uniformly distributed superimposed load the beam can support over an effective span of 6 m. The materials used are M20 grade concrete (m = 13) and medium tensile steel (σ_{st} = 190 MPa). The unit weight of concrete is 25 kN/m^3.

[**Ans.** x = 177.74 mm, k_c = 0.324, M_r = 68.547 kNm, w_s = 11.795 kN/m]

T.A.5 A reinforced concrete cantilever beam of length 2.5 m and having a cross-section of size 250 × 450 mm effective is reinforced with 4 bars of 20 mm ϕ bars on the tension side at an effective cover of 50 mm. Determine the moment of resistance and the uniformly distributed superimposed load the beam can support without exceeding the permissible stresses in the materials. The materials used are M20 grade concrete and medium tensile steel (σ_{st} = 190 MPa). The unit weight of reinforced concrete is 25 kN/m^2.

[**Ans.** x = 185.81 mm, M_r = 63.09 kNm, w_s = 17.065 kN/m]

T.A.6 A reinforced concrete beam of rectangular cross-section of size 300 × 650 mm effective is reinforced with 4 bars of 20 mm ϕ on the tension side. Determine the maximum stresses developed in concrete and steel, if the beam carries a uniformly distributed load of 20 kN/m (inclusive of self-weight) over an effective span of 7 m with simply supported ends. The concrete used is of grade M20 and the steel is of grade Fe415.

[**Ans.** x = 217.13 mm, $m\sigma'_{cbc}$ = 6.51 MPa, σ'_{st} = 168.76 MPa]

T.A.7 A reinforced concrete simply supported beam of rectangular cross-section carries a uniformly distributed load of 12.5 kN/m inclusive of self-weight over an effective span of 8 m. From architectural considerations the width of the beam has been fixed at 300 mm. Determine the effective depth and area of tension steel required for balanced design. The materials used are M20 grade concrete and HYSD steel of grade Fe415.

T.A.8 Design a balanced section for a reinforced concrete beam subjected to a bending moment of 150 kNm. The material used are M20 grade concrete and mild steel reinforcement.

[**Ans.** M = 1.198bd^2. For a ratio d/b = 2; b = 315.1 mm, d = 630.31 mm. Provide 300 × 645.5 mm effective cross-section with A_{st} = 1910 mm^2]

T.A.9 A simply supported 125 mm thick roof slab of effective span of 3.5 m is reinforced with 12 mm ϕ bars with 100 mm c/c at an effective cover of 20 mm (i.e., the reinforcement is placed with its centre at 20 mm from the bottom). Determine the maximum allowable moment of resistance of 1.0 m wide strip of the slab, and uniformly distributed superimposed load the slab can support safely. The materials used are M25 grade concrete and HYSD steel reinforcement of grade Fe415.

[**Ans.** m = 11, k = 0.289, j = 0.904, A_{st} = 1130.97 mm^2/m width, d = 105 mm, M_r =15.64 kNm, and w_s = 7.09 kN/m^2]

T.A.10 A simply supported reinforced concrete rectangular beam carries a uniformly distributed load of 12.5 kN/m inclusive of its own weight, and also a concentrated load of 20 kN at the midpoint over an effective span of 6.5 m. Design a suitable balanced section for the flexural action taking (a) d = 2b and (b) d = 2.5b. The materials used are M20 grade concrete and HYSD steel of grade Fe415.

T.A.11 Calculate the moment of resistance of a reinforced concrete beam of rectangular cross-section of width b and effective depth d which is reinforced with 1.15 per cent steel both on tension and compression sides, the compression steel being placed at an effective cover of 0.10d. The materials used are M20 grade concrete and HYSD steel of grade Fe415.

T.A.12 A doubly reinforced concrete beam of rectangular section of size 300 × 725 mm overall is reinforced by 4 × 25 ϕ bars each on tension and compression sides. The effective covers to compression and tension steels are 25 mm and 45 mm, respectively. Determine the maximum stresses induced in concrete and steel when the beam carries a uniformly distributed load of 40 kN/m inclusive of its dead weight over an effective span of 6.5 m. Take $m = 13$.

[**Ans.** $x = 199.14$ mm, $\sigma'_{cbc} = 5.398$ MPa, $\sigma'_{sc} = 92.05$ MPa and $\sigma'_{st} = 169.45$ MPa]

T.A.13 A simply supported reinforced concrete beam carries a uniformly distributed load of 20 kN/m inclusive of self-weight over an effective span of 8.25 m. The cross-section of the beam is restricted to 250 mm width and 600 mm depth up to the centre of the tension steel and the compression steel is placed at an effective cover of 50 mm from the top of the beam. Determine: (a) the areas of tension and compression reinforcements to be provided, if the materials used are M20 grade concrete and mild steel reinforcement, and (b) the area of reinforcement if the effective depth is reduced by 25 per cent.

T.A.14 Determine the moment of resistance of a reinforced concrete beam of rectangular cross-section of size $b \times d$ mm effective, for the following conditions. The materials are M20 grade concrete and mild steel reinforcement. The effective cover to compression steel is $0.10d$(i.e., $d'/d = 0.10$).

(i) The area of compression steel is one-third that of the tension steel

(ii) The area of compression steel is limited to 2 per cent of the cross-sectional area bd

(iii) The area of compression steel is limited to 1 per cent of the cross-sectional area.

[***Ans.*** (i) $M_r = 1.5702bd^2$ (ii) $M_r = 2.9374bd^2$ and (iii) $M_r = 2.0677bd^2$]

T.A.15 A reinforced concrete T-beam having a flange of effective width 1250 mm and thickness of 100 mm is reinforced with 4 × 22 mm ϕ tension bars provided at a depth of 500 mm below the top of the flange in a 250 mm wide rib. Determine the uniformly distributed load, inclusive of self-weight, that the beam can support safely over a simply supported effective span of 6.5 m, if the materials used are concrete mix of grade M20 and HYSD steel of grade Fe415.

[**Ans.** $M_r = 162.12$ kNm and $w = 30.697$ kN/m]

T.A.16 A reinforced concrete L-beam has a flange of width 1200 mm and thickness 100 mm. The width of the web of the beam is 300 mm. The beam is provided with mild steel tension reinforcement of area 2000 mm^2 placed at an effective depth of 600 mm from the top of the flange. The concrete mix used is of grade M20 with a modular ratio of 13. Determine: (a) allowable moment of resistance of the section and (b) stresses in concrete and steel when the beam supports a uniformly distributed load of 12 kN/m over an effective simply supported span of 10 m.

[***Ans.*** (a) $M_r = 256.9$ kNm, (b) $\sigma'_{cbc} = 3.38$ MPa and $\sigma'_{st} = 134.83$ MPa]

T.A.17 Analyze a flanged beam spanning an effective distance of 9.5 m, and reinforced with 6 × 25 mm ϕ bars on the tension side. The flange is of size 1500 × 120 mm and the rib of size 300 × 425 mm effective (projected below the flange). The beam carries a uniformly distributed load of 25 kN/m inclusive of self-weight. Determine the stresses induced in concrete and steel at the centre of the span. The concrete used is grade M20.

[***Ans.*** σ'_{cbc} = 5.35 MPa and σ'_{st} = 191.86 MPa]

T.A.18 A reinforced concrete T-beam section consists of a flange of size 1250 × 100 mm and a rib of size 250 × 350 mm below the flange. The tension and compression steels consist of 5 × 25 mm ϕ and 3 × 20 mm ϕ, respectively, at effective covers of 40 mm. The grades of concrete and HYSD steel used are M20 and Fe415, respectively. Determine the moment of resistance and stresses induced in the top compression fibre of concrete, and in tension and compression steels.

[***Ans.*** z = 372.84 mm, M_r = 210.77 kNm, σ'_{cbc} = 6.89 MPa, σ'_{sc} = 87.54 MPa and σ'_{st} = 230 MPa]

T.A.19 Analyze the end beam of a reinforced concrete floor system consisting of a 125 mm thick slab cast monolithically and connected with 300 mm wide beams spaced 3.25 m centre-to-centre. The effective length of the beam is 7.5 m. The beam is reinforced with 6 × 25 mm ϕ bars placed on the tension side at a depth of 550 mm from the top of the flange slab. Determine the maximum stresses developed in concrete and steel when the beam is subjected to a bending moment of 225 kNm. The concrete mix used is of grade M20 with m = 13.

[***Ans.*** σ'_{cbc} = 4.617 MPa, σ'_{st} = 152.335 MPa]

T.A.20 A 120 mm thick slab of clear span of 4.0 m is cast monolithically with two simply supported edge beams having an effective span of 6 m. The load on each of the two end beams is 20.011 kN/m inclusive of self-weight. The width of the rib is 250 mm. The materials to be used are M20 grade concrete and HYSD reinforcement of grade Fe415. Ignoring torsion, design the beam for flexure.

T.A.21 (a) Explain the terms shear stress, diagonal tension, bond stress and development length with reference to reinforced concrete beams.

(b) Discuss the utility of bent-up bars in reinforced concrete beams in resisting the shear.

(c) Derive the expressions for computation of bond stress and shear stress in case of reinforced concrete beams of rectangular cross-section with tension reinforcement of diameter ϕ. Also obtain the relationship between shear stress and bond stress.

T.A.22 A simply supported reinforced concrete beam of rectangular cross-section of size 300 × 700 mm overall carries a concentrated load of 125 kN at 1 m from one support, in addition to a uniformly distributed load of 20 kN/m over a clear span of 6 m. The section is reinforced with six 20 mm ϕ bars of HYSD steel of grade Fe415. The concrete used is of grade M20. Design shear reinforcement when: (a) only vertical stirrups are provided, (b) two of the six bars are bent up at 45° at the same cross-section, and (c) four bars are bent up at 45° in two groups, each consisting of 2 bars. The two groups are separated by 500 mm.

T.A.23 A reinforced concrete cantilever beam projects 1.2 m from the face of the support with a bearing on the support of 1.75 m. The tension reinforcement consists of 3 bars of 20 mm ϕ provided straight on the top face of the beam. Check the adequacy of the anchorage. If inadequate, suggest the maximum size of the bar that can be used as reinforcement. The materials used are M20 grade concrete and Fe415 grade steel. The side cover is 50 mm.

T.A.24 A reinforced concrete beam of rectangular cross-section of size 300 × 600 mm effective is provided with 6 bars of 20 mm ϕ on its tension face. Two of the bars on the tension face are bent up at the support for shear. Determine the development length required for bars from the face of the supporting square column of size 300 × 300 mm in tension and compression zones. The materials used are M20 grade concrete and mild steel of grade I (σ_{st} = 190 MPa).

T.A.25 A 120 mm thick reinforced concrete cantilever slab projects 1.10 m beyond the face of lintel of cross-section 225 × 200 mm. The slab is to carry a uniformly distributed service load of 7 kN/m^2. It is reinforced with 10 mm ϕ bars @ 250 mm c/c. Determine the development length required from the face of lintel. The materials used are M20 grade concrete and HYSD steel of grade Fe500 (assume σ_{st} = 275 MPa).

T.A.26 A simply supported reinforced concrete beam of clear span of 7.5 m has a cross-section of size 200 × 500 mm effective depth. It is reinforced with 5 bars of 16 mm ϕ on the tension side. The beam carries a uniformly distributed service load of 20 kN/m including its own weight. Determine: (a) the length over which vertical stirrups are to be designed and their spacing and (b) spacing of stirrups when 3 bars are bent up at the same section.

T.A.27 In a residential building a reinforced concrete slab is to be designed for a room measuring 3 m × 6.75 m to the centre lines of the supporting 230 mm thick walls on all the four edges. On one of its long edges the slab projects out by 1.2 m beyond the face of the wall to form a balcony with a 75 × 100 mm high concrete parapet. The room and balcony slabs are to support superimposed loads of 2 kN/m^2 and 3 kN/m^2, respectively. The dead load due to floor finish may be taken as 1.0 kN/m^2. The materials are M20 grade concrete and mild steel of grade Fe250.

[***Hint:*** Design the slab as a beam of 1000 mm width with an overhang. For computing maximum moment in the room slab consider full load on it and dead load (without parapet) only on the balcony slab. For maximum negative moment in balcony slab consider full load on it. The slab thickness will be governed by the deflection criterion for a cantilever balcony slab.]

T.A.28 A reinforced concrete column of cross-section of size 350 × 450 mm is reinforced with 4 × 28 mm ϕ HYSD longitudinal bars of grade Fe415. The concrete mix used is of grade M25. Determine safe load the column can support if the effective length is (a) 3.5 m and (b) 6 m.

T.A.29 A reinforced concrete column of circular cross-section of 300 mm diameter is reinforced with 8 × 22 mm ϕ mild steel longitudinal bars at a clear cover of 25 mm, and is provided with a spiral reinforcement of 6 mm diameter bar at a pitch of 50 mm. Concrete used is of grade M25. Determine the safe load the column can support if the effective length is (a) 3 m and (b) 6 m.

T.A.30 A reinforced concrete tied column is to support an axial load of 2250 kN. Calculate the dimensions of the column for an effective length of 4 m so as to obtain: (a) the smallest square column and (b) the most economical square column. Also determine the area of compression steel required for a column of size 350 × 500 mm. The materials used are M25 grade concrete and mild steel reinforcement.

T.A.31 Design a square tied column of effective length of 6 m to support an axial load of 540 kN. The materials used are M25 grade concrete and mild steel reinforcement.

[***Hint:*** Adopt a square column of size 300 × 300 mm with 1 per cent longitudinal steel, i.e., 8 × 12 mm ϕ bars and 6 mm ϕ bar ties at about 190 mm c/c]

T.A.32 Design a circular column provided with helical reinforcement to support an axial load of 1250 kN. The ends of the column are restrained both in position and direction with an unsupported length of 6 m. The materials used are: M25 grade concrete and mild steel reinforcement.

T.A.33 A reinforced concrete tied square column of size 400 mm is reinforced with four bars of 25 mm ϕ placed at corners with an effective cover of 50 mm. It carries a direct load of 550 kN and a uniaxial moment of 44.0 kNm. Determine the stresses and check the adequacy of the section as an uncracked section. The materials used are M25 grade concrete and HYSD steel of grade Fe415. The modulus of rupture at 7 days is 2.4 MPa.

T.A.34 A reinforced concrete tied column of size 350 × 350 mm is reinforced with four bars of 25 mm ϕ placed at the corners with an effective cover of 55 mm. The materials used are M25 grade concrete and HYSD steel of grade Fe415. Calculate: (a) the maximum stresses in compression in concrete and steel and check the adequacy of the section when it is subjected to a direct load of 500 kN placed at an eccentricity of 20 mm, (b) the maximum eccentricity at which the load can be applied, and (c) the eccentricity when the resulting minimum stress due to P = 500 kN becomes zero.

T.A.35 A reinforced concrete tied column of size 300 × 400 mm is reinforced with 4 × 25 mm ϕ bars placed at corners at an effective cover of 50 mm. The materials used are M20 grade concrete and HYSD steel of grade Fe415. Check the suitability of the section if it is subjected to an axial load of 300 kN, and moments M_{xx} = 18 kNm and M_{yy} = 12 kNm. For M20 grade concrete, 7 days' modulus of rupture is 2.4 MPa.

[***Ans.*** $\sigma'_{cc,\max}$ = 4.835 MPa (compression); $\sigma'_{cc,\min}$ = 0.995 MPa (tension), section is adequate]

T.A.36 Design a square (uncracked) column section to support a direct compressive force of 500 kN and uniaxial moment of 50 kNm under the service state. Assume the effective cover d' such that d'/D = 0.125. Use M25 grade concrete with m = 13 and f_r (at 7 days) = 2.4 MPa, and Fe415 grade steel.

Appendix

B

Gravity Loads

Table B.1 Unit Weights of Common Building Materials

	Material	*Unit weight,* kN/m^3
1.	Bituminous substances	
	Asphalt	12.8
	Bitumen	10.2
2.	Bricks	
	Common burnt clay bricks	15.8–19.2
	Refractory bricks	17.3–19.6
	Brick ballast	11.8
	Brick dust (*surkhi*)	14.1
3.	*Brick masonry*	19.0–21.0
4.	*Cement*	14.4
5.	*Cement plaster*, 10 mm	226 N/m^2
6.	Concrete	
	Plain	24.0
	Reinforced	25.0
7.	Excavated earth	
	Earth (dry), loose	12.8
	compact	15.5
	Earth (moist), loose	14.1–15.7
	compact	17.8–18.4

(*Contd.*)

Table B.1 Unit Weights of Common Building Materials (*Contd.*)

	Material	*Unit weight,* kN/m^3
	Gravel, loose	15.7
	rammed	19.2–20.4
	Sand, dry, clean	15.4–16.0
	river sand	18.4
8.	*Floor finishes*, 22–24 mm	216–236 N/m^2
9.	*Glass sheets*	23–26.5
	Thickness, mm	Weight, N/m^2
	1.3	33.4
	2.0	52.0
	2.6	64.8
	3.2	78.5
10.	Granite	26.4–28.0
11.	*Liquids*	
	Water, fresh	9.81
	Salt	10.0
	Vegetable oil	9.2
12.	*Mortars*	
	Cement sand mortar	20.4
	Lime mortar	17.6
13.	*Stones*	
	Marble	26.7
	Sandstone, chalk	22.0–23.5
	Limestone	24.0–26.4
	Stone ballast	15.7–18.8
	Stone masonry	20.8–27.0
14.	*Steel*	78.0
15.	*Timber*	
	Fir	4.56
	Deodar	5.50
	Sissom	7.55
	Chir	5.65
	Sal	7.85
	Teak	6.13

Table B.2(a) Live Loads on Floors

Loading class number	*Types of floors*	*Minimum live load,* kN/m^2	*Alternative minimum live load*
2.0	Floors in dwelling houses, tenements, hospital wards, bedrooms and private sitting rooms in hostels and dormitories	2.0	Subject to a minimum total load of 2.5 times the values for any given slab panel and 6 times the values for any given beam. This total load shall be assumed uniformly distributed on the entire area of the slab panel or the entire length of the beam. (applies to classes 2.0 to 5.0)
2.5	Office floors other than entrance halls, floors of light workrooms:		
	(a) with separate storage	2.5	
	(b) without storage	4.0	
3.0	Floors of banking halls, office entrance halls and reading rooms	3.0	
4.0	Shop floors used for the display and sale of merchandise, floors of workrooms, floors of classrooms in schools, floors of places of assembly with fixed seating, restaurants, circulation spaces in machinery halls, power stations, etc., which are not occupied by plant or equipment.	4.0	
5.0	Floors of warehouses, workshops, factories and other buildings or parts of buildings of a similar category for light-weight loads, office floors for storage and filing purposes, floors of places of assembly without fixed seating, e.g. public rooms in hotels, dance halls, waiting halls.	5.0	
7.5	Floors of warehouses, workshops, factories and other buildings or parts of buildings of a similar category for medium weight loads.	7.5	
10.0	Floors of warehouses, workshops, factories and other buildings or parts of buildings of a similar category for heavy-weight loads, floors of book stores and libraries, roofs and pavement lights over basements projecting under public footpath	10.0	
Garage (light)	Floors used for garages for vehicles not exceeding 25 kN gross weight:		or the worst combination of actual wheel loads, whichever is greater.
	(a) Slabs	4.0	
	(b) Beam	2.5	

(Contd.)

Table B.2(a) Live Loads on Floors (*Contd.*)

Loading class number	*Types of floors*	*Minimum live load,* kN/m^2	*Alternative minimum live load*
Garage (heavy)	Floors used for garages for vehicles not exceeding 40 kN gross weight.	7.5	Subject to a minimum of one-and-a half times the maximum wheel load but not less than 9.0 kN considered to be distributed over 750 mm square area.
Stairs	Stairs, landings and corridors for class 2.0 loading:		Subject to a minimum of 1.3 kN concentrated load at unsupported end of each step for stairs constructed out of structurally independent cantilever steps.
	(a) Not liable to overcrowding	3.0	
	(b) Liable to overcrowding and for all other classes	5.0	
Balcony	Balconies		
	For class 2.0 loading and classes not liable to overcrowding	3.0	
	For all other classes liable to overcrowding	5.0	
Parapets	Light access stairway landing and balconies (domestic)	0.35 kN/m	Vertical and horizontal
	All other stairways, parapets	0.75 kN/m	Vertical and horizontal
	Theaters and schools (assembly)	2.25 kN/m	–do–

Notes:

1. In the preceding table a reference to a 'floor' includes a reference to any part of that floor, and a reference to 'slabs' includes boarding and beams or ribs spaced not further apart than 1 m between centres, and a reference to 'beams' means all other beams and rib.
2. Under loading class No. 2.5, the reference to 'light workrooms' envisages rooms in which some light machines (for example, sewing machines used by milliners or tailors) are operated without a central power driven unit, that is, the machines are independently operated, either by hand or by small motors. Under loading class no. 4.0, the reference to 'workrooms' generally envisages the installation of machines operated by a central power-driven unit, with the individual machines being belt-driven.
3. 'Fixed seating' implies that the removal of the seating and the use of the space for other purposes is improbable. The maximum likely load in this case, is, therefore, closely controlled.
4. The loading in workshops, warehouses and factories varies considerably and so three loading categories under the terms 'light', 'medium' and 'heavy' are introduced in order to allow for more economical designs but the terms have no special meaning in themselves other than the live load for which the relevant floor is designed. It is, however, important, particularly in the case of heavy weight loads, to assess the actual loads to ensure that they are not in excess of 10.0 kN/m^2, in cases where they are excess, the design shall be based on the actual loading.
5. The load classification for stairs, corridors, balconies and landings, provides for the fact that these often serve several occupancies and are used for transporting furniture and goods.
6. The horizontal loads on parapet and balustrades act at handrail or coping level. These loads should not be considered to act simultaneously with vertical forces on these components.

Table B.2(b) Live Loads on Roofs

Sr. No	*Types of roofs*	*Minimum live load measured on plan, kN/m²*	*Alternative minimum live load*
(i)	Flat, sloping or curved roof with slopes up to and including 10 degrees:		
	(a) Access provided	1.5	3.75 kN uniformly distributed over any span of 1.0 m width of the roof slab and 9.0 kN uniformly distributed over the span in the case of all beams
	(b) Access not provided, except for maintenance	0.75	1.90 kN uniformly distributed over any span of 1.0 m width of the roof slab and 4.50 kN uniformly distributed over the span in the case of beams.
(ii)	Sloping roof with a slope greater than 10 degrees:		
	(a) For roof membrane, sheets or purlins	0.75 kN/m^2 less 0.02 kN/m^2 for every degree increase in slope over 10 degrees.	
	(b) For members supporting the roof membrane and roof purlins such as trusses, beams, and girders	2/3 of loads in (a).	
(iii)	Curved roofs with a slope at springing greater than 10 degrees	$0.75-3.45(h/l)^2$ kN/m^2, where h is the height of the highest point of the structure measured from its springing and l is the chord width of the roof if singly curved, and the shorter of the two sides if doubly curved.	Subject to a minimum of 0.40 kN/m^2.

Note: For special types of roofs with highly permeable and absorbent material, the contingency of roof material increasing in weight due to absorption of moisture shall be provided for.

Appendix C

Design Forces

Table C.1 Maximum Positive and Negative Bending Moments, and Reactions coefficients in Multispan Continuous-beams of Equal Spans Carrying Dead and Live Loads

Uniformly distributed loads

All spans carry dead load simultaneously

0.125
0.070 0.070
0.38 0.62 0.62 0.38

0.10 0.10
0.080 0.025 0.080
0.40 0.60 0.50 0.50 0.60 0.40

0.107 0.071 0.107
0.077 0.036 0.036 0.077
0.39 0.61 0.54 0.46 0.46 0.54 0.61 0.39

0.105 0.08 0.08 0.105
0.078 0.053 0.046 0.053 0.078
0.40 0.60 0.53 0.47 0.50 0.50 0.47 0.53 0.60 0.40

Pattern live load causing the worst effect

0.125
0.096 0.096
0.44 0.62 0.62 0.44

0.117 0.117
0.101 0.075 0.101
0.45 0.62 0.58 0.58 0.62 0.45

0.121 0.107 0.121
0.099 0.081 0.081 0.099
0.45 0.62 0.60 0.57 0.57 0.60 0.62 0.45

0.120 0.111 0.111 0.120
0.100 0.080 0.086 0.080 0.100
0.45 0.62 0.60 0.58 0.59 0.59 0.58 0.60 0.62 0.45

Concentrated loads

Centre point dead loads on all spans | *Pattern centre point live loads for worst effect*

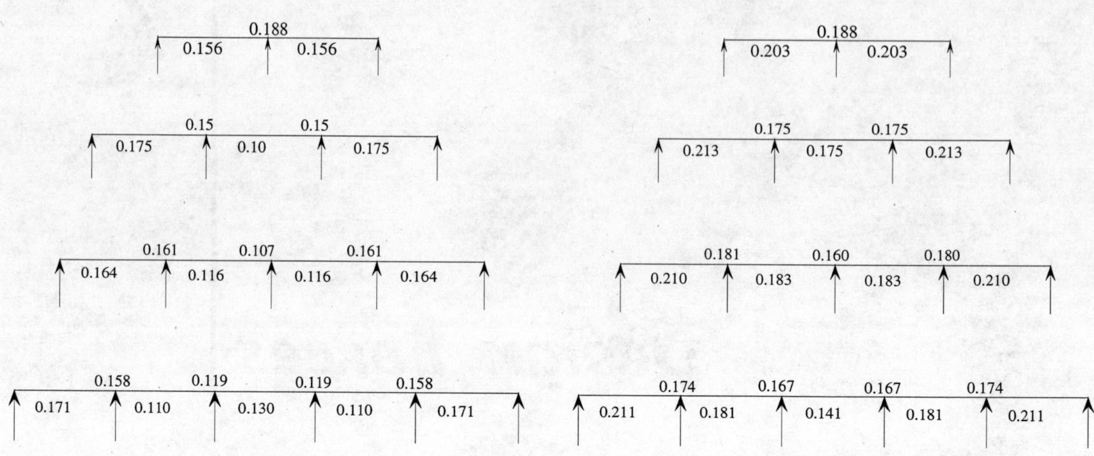

Point dead loads at 1/3rd span on all spans | *Pattern point loads at 1/3rd span for worst effect*

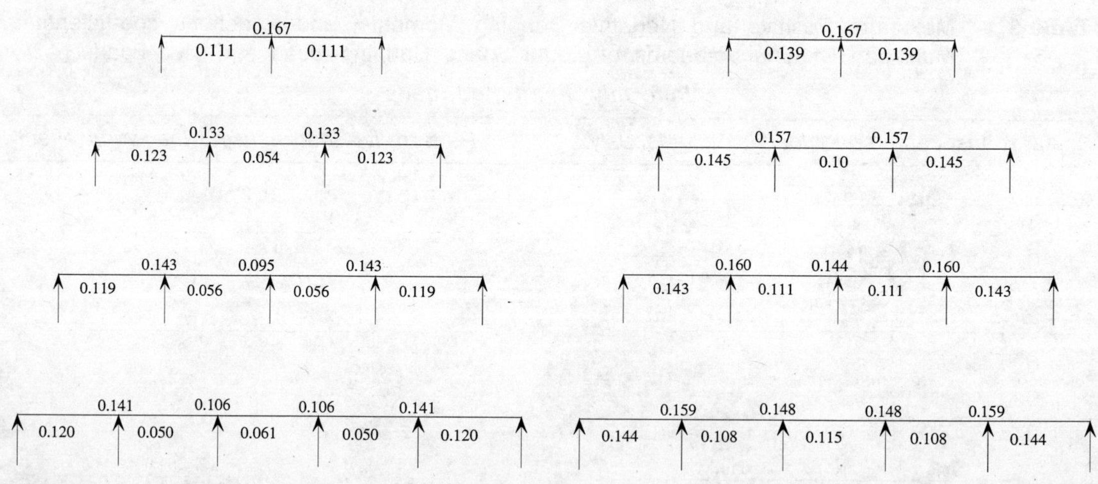

Bending moment = Coefficient × W × l
Reaction = Coefficient × W
where W = Total uniformly distributed or concentrated load on one span only
l = Effective span

Appendix

D

Design Aids

TABLE D.1 Moment Resisting Capacity of a Singly Reinforced Rectangular Beam with a Given Amount of Tension Reinforcement

Reinforcement, p_t, per cent	$R_u = (M_u/bd^2)$ MPa								
	Fe 250			Fe 415			Fe 500		
	M20	M25	M30	M20	M25	M30	M20	M25	M30
0.12	0.257	0.258	0.258	0.422	0.424	0.426	0.506	0.509	0.511
0.13	0.278	0.279	0.280	0.456	0.459	0.461	0.547	0.551	0.553
0.14	0.299	0.300	0.301	0.491	0.493	0.495	0.587	0.592	0.594
0.15	0.320	0.321	0.322	0.524	0.528	0.530	0.628	0.633	0.636
0.16	0.341	0.342	0.343	0.558	0.562	0.565	0.668	0.673	0.677
0.18	0.383	0.384	0.385	0.625	0.630	0.633	0.747	0.754	0.759
0.20	0.424	0.426	0.428	0.692	0.698	0.702	0.826	0.835	0.841
0.22	0.465	0.468	0.469	0.758	0.765	0.770	0.904	0.914	0.921
0.24	0.506	0.509	0.511	0.823	0.832	0.837	0.981	0.993	1.002
0.26	0.547	0.551	0.553	0.888	0.898	0.905	1.057	1.072	1.081
0.28	0.587	0.592	0.594	0.952	0.963	0.971	1.132	1.149	1.161
0.30	0.628	0.633	0.636	1.015	1.029	1.038	1.207	1.226	1.239
0.32	0.668	0.673	0.677	1.078	1.093	1.104	1.280	1.302	1.317
0.34	0.708	0.714	0.718	1.140	1.158	1.169	1.353	1.378	1.394

(Contd.)

TABLE D.1 Moment Resisting Capacity of a Singly Reinforced Rectangular Beam with a Given Amount of Tension Reinforcement (*Contd.*)

Reinforcement, p_t, per cent	$R_u = (M_u/bd^2)$ MPa								
	Fe250			*Fe415*			*Fe500*		
	M20	*M25*	*M30*	*M20*	*M25*	*M30*	*M20*	*M25*	*M30*
0.36	0.747	0.754	0.759	1.202	1.221	1.234	1.424	1.453	1.471
0.38	0.787	0.795	0.800	1.263	1.285	1.299	1.495	1.527	1.548
0.40	0.826	0.835	0.841	1.324	1.348	1.364	1.565	1.600	1.623
0.42	0.865	0.875	0.881	1.384	1.410	1.428	1.634	1.673	1.698
0.44	0.904	0.914	0.921	1.443	1.472	1.491	1.703	1.745	1.773
0.46	0.943	0.954	0.962	1.502	1.533	1.554	1.770	1.816	1.847
0.48	0.981	0.993	1.002	1.560	1.594	1.617	1.837	1.887	1.920
0.50	1.019	1.033	1.042	1.617	1.655	1.680	1.902	1.957	1.993
0.52	1.057	1.072	1.081	1.674	1.715	1.742	1.967	2.026	2.065
0.54	1.095	1.111	1.121	1.730	1.774	1.803	2.031	2.094	2.137
0.56	1.132	1.149	1.161	1.786	1.833	1.864	2.094	2.162	2.208
0.58	1.169	1.188	1.200	1.841	1.892	1.925	2.156	2.229	2.278
0.60	1.207	1.226	1.239	1.896	1.950	1.986	2.217	2.296	2.348
0.62	1.243	1.264	1.278	1.950	2.007	2.045	2.278	2.361	2.417
0.64	1.280	1.302	1.317	2.003	2.064	2.105	2.337	2.426	2.486
0.66	1.316	1.340	1.356	2.056	2.121	2.164	2.396	2.491	2.554
0.68	1.353	1.378	1.394	2.108	2.177	2.223	2.454	2.554	2.621
0.70	1.389	1.415	1.433	2.159	2.233	2.281	2.511	2.617	2.688
0.72	1.424	1.453	1.471	2.210	2.288	2.339	2.567	2.680	2.755
0.74	1.460	1.490	1.509	2.260	2.342	2.397	2.622	2.741	2.821
0.76	1.495	1.527	1.548	2.310	2.397	2.454	2.677	2.802	2.886
0.78	1.530	1.563	1.585	2.359	2.450	2.511	2.730	2.862	2.950
0.80	1.565	1.600	1.623	2.408	2.504	2.567	2.783	2.922	3.014
0.82	1.600	1.636	1.661	2.456	2.556	2.623	2.834	2.981	3.078
0.84	1.634	1.673	1.698	2.503	2.609	2.679	2.885	3.039	3.141
0.86	1.669	1.709	1.736	2.550	2.660	2.734	2.935	3.096	3.203
0.88	1.703	1.745	1.773	2.596	2.712	2.789	2.984	3.153	3.265
0.90	1.736	1.780	1.810	2.641	2.763	2.843	3.033	3.209	3.326
0.92	1.770	1.816	1.847	2.686	2.813	2.897	3.080	3.264	3.387
0.94	1.803	1.851	1.883	2.731	2.863	2.951	3.127	3.319	3.447

(*Contd.*)

TABLE D.1 Moment Resisting Capacity of a Singly Reinforced Rectangular Beam with a Given Amount of Tension Reinforcement (*Contd.*)

Reinforcement, p_t, per cent	$R_u = (M_u/bd^2)$ MPa								
	Fe250			*Fe415*			*Fe500*		
	M20	*M25*	*M30*	*M20*	*M25*	*M30*	*M20*	*M25*	*M30*
0.96	1.837	1.887	1.920	2.774	2.912	3.004	3.172	3.373	3.506
0.98	1.869	1.922	1.956	2.817	2.961	3.057	3.217	3.426	3.565
1.00	1.902	1.957	1.993	2.860	3.010	3.109	3.261	3.478	3.623
1.02	1.935	1.991	2.029	2.902	3.058	3.161	3.304	3.530	3.681
1.04	1.967	2.026	2.065	2.943	3.105	3.213	3.346	3.581	3.738
1.06	1.999	2.060	2.101	2.984	3.152	3.264	3.387	3.632	3.794
1.08	2.031	2.094	2.137	3.024	3.199	3.315	3.428	3.681	3.850
1.10	2.063	2.128	2.172	3.064	3.245	3.366	3.467	3.730	3.906
1.12	2.094	2.162	2.208	3.102	3.290	3.416	3.506	3.779	3.961
1.14	2.125	2.196	2.243	3.141	3.335	3.465	3.544	3.826	4.015
1.16	2.156	2.229	2.278	3.178	3.380	3.514	3.581	3.873	4.068
1.18	2.187	2.263	2.313	3.216	3.424	3.563	3.617	3.920	4.121
1.20	2.217	2.296	2.348	3.252	3.468	3.612	3.652	3.965	4.174
1.22	2.248	2.329	2.383	3.288	3.511	3.660	3.687	4.010	4.226
1.24	2.278	2.361	2.417	3.323	3.554	3.707	3.720	4.054	4.277
1.26	2.308	2.394	2.452	3.358	3.596	3.754	3.753	4.098	4.328
1.28	2.337	2.426	2.486	3.392	3.638	3.801	3.784	4.141	4.378
1.30	2.367	2.459	2.520	3.426	3.679	3.848	3.815	4.183	4.428
1.32	2.396	2.491	2.554	3.459	3.720	3.894	3.845	4.224	4.477
1.34	2.425	2.523	2.588	3.491	3.760	3.939	3.874	4.265	4.525
1.36	2.454	2.554	2.621	3.523	3.800	3.985	3.903	4.305	4.573
1.38	2.483	2.586	2.655	3.554	3.839	4.029	3.930	4.344	4.620
1.40	2.511	2.617	2.688	3.585	3.878	4.074	3.957	4.383	4.667
1.42	2.539	2.649	2.722	3.614	3.916	4.118	3.982	4.421	4.713
1.44	2.567	2.680	2.755	3.644	3.954	4.161	4.007	4.458	4.758
1.46	2.595	2.711	2.788	3.673	3.992	4.205	4.031	4.494	4.803
1.48	2.622	2.741	2.821	3.701	4.029	4.247	4.054	4.530	4.848
1.50	2.649	2.772	2.853	3.728	4.065	4.290	4.076	4.565	4.891
1.52	2.677	2.802	2.886	3.755	4.101	4.332	4.097	4.600	4.934
1.54	2.703	2.832	2.918	3.782	4.137	4.373	4.118	4.633	4.977

(*Contd.*)

TABLE D.1 Moment Resisting Capacity of a Singly Reinforced Rectangular Beam with a Given Amount of Tension Reinforcement (*Contd.*)

Reinforcement, p_t, per cent	$R_u = (M_u/bd^2)$ MPa								
	Fe250			*Fe415*			*Fe500*		
	M20	*M25*	*M30*	*M20*	*M25*	*M30*	*M20*	*M25*	*M30*
1.56	2.730	2.862	2.950	3.807	4.172	4.415	4.137	4.666	5.019
1.58	2.756	2.892	2.983	3.832	4.206	4.456	4.156	4.699	5.061
1.60	2.783	2.922	3.014	3.857	4.240	4.496	4.174	4.730	5.101
1.62	2.809	2.951	3.046	3.881	4.274	4.536	4.191	4.761	5.142
1.64	2.834	2.981	3.078	3.904	4.307	4.576	4.207	4.792	5.181
1.66	2.860	3.010	3.109	3.927	4.340	4.615	4.222	4.821	5.221
1.68	2.885	3.039	3.141	3.949	4.372	4.654	4.237	4.850	5.259
1.70	2.910	3.067	3.172	3.971	4.404	4.692	4.250	4.878	5.297
1.72	2.935	3.096	3.203	3.992	4.435	4.730	4.263	4.906	5.334
1.74	2.960	3.124	3.234	4.012	4.465	4.768	4.274	4.933	5.371
1.76	2.984	3.153	3.265	4.032	4.496	4.805	4.285	4.959	5.408
1.78	3.009	3.181	3.296	4.051	4.525	4.842	4.295	4.984	5.443
1.80	3.033	3.209	3.326	4.070	4.555	4.878	4.304	5.009	5.478
1.82	3.056	3.236	3.356	4.087	4.584	4.914	4.313	5.033	5.513
1.84	3.080	3.264	3.387	4.105	4.612	4.950	4.320	5.056	5.547
1.86	3.103	3.291	3.417	4.122	4.640	4.985	4.327	5.079	5.580
1.88	3.127	3.319	3.447	4.138	4.667	5.020	4.332	5.101	5.613
1.90	3.149	3.346	3.476	4.153	4.694	5.054	4.337	5.122	5.645
1.92	3.172	3.373	3.506	4.168	4.720	5.088	4.341	5.142	5.677
1.94	3.195	3.399	3.536	4.183	4.746	5.122	4.344	5.162	5.708
1.96	3.217	3.426	3.565	4.196	4.772	5.155	4.346	5.181	5.738
1.98	3.239	3.452	3.594	4.210	4.797	5.188	4.347	5.200	5.768
2.00	3.261	3.478	3.623	4.222	4.821	5.221	4.348	5.217	5.797
2.02	3.283	3.504	3.652	4.234	4.845	5.253	4.347	5.234	5.826
2.04	3.304	3.530	3.681	4.246	4.869	5.284	4.346	5.251	5.854
2.06	3.325	3.556	3.709	4.256	4.892	5.315	4.344	5.266	5.881
2.08	3.346	3.581	3.738	4.266	4.914	5.346	4.341	5.281	5.908
2.10	3.367	3.607	3.766	4.276	4.936	5.377	4.337	5.296	5.935
2.12	3.387	3.632	3.794	4.285	4.958	5.407	4.332	5.309	5.961
2.14	3.408	3.657	3.823	4.293	4.979	5.436	4.327	5.322	5.986

(*Contd.*)

TABLE D.1 Moment Resisting Capacity of a Singly Reinforced Rectangular Beam with a Given Amount of Tension Reinforcement (*Contd.*)

Reinforce-ment, p_t, per cent	$R_u = (M_u/bd^2)$ MPa								
	Fe250			*Fe415*			*Fe500*		
	M20	*M25*	*M30*	*M20*	*M25*	*M30*	*M20*	*M25*	*M30*
2.16	3.428	3.681	3.850	4.301	5.000	5.466	4.320	5.334	6.010
2.18	3.448	3.706	3.878	4.308	5.020	5.495	4.313	5.346	6.034
2.20	3.467	3.730	3.906	4.315	5.040	5.523	4.304	5.357	6.058
2.22	3.487	3.755	3.933	4.321	5.059	5.551	4.295	5.367	6.081
2.24	3.506	3.779	3.961	4.326	5.078	5.579	4.285	5.376	6.103
2.26	3.525	3.803	3.988	4.331	5.096	5.606	4.274	5.385	6.125
2.28	3.544	3.826	4.015	4.335	5.114	5.633	4.263	5.393	6.146
2.30	3.563	3.850	4.042	4.339	5.131	5.659	4.250	5.400	6.167
2.32	3.581	3.873	4.068	4.342	5.148	5.685	4.237	5.407	6.187
2.34	3.599	3.897	4.095	4.344	5.164	5.711	4.222	5.413	6.206
2.36	3.617	3.920	4.121	4.346	5.180	5.736	4.207	5.418	6.225
2.38	3.635	3.943	4.148	4.347	5.195	5.761	4.191	5.422	6.243
2.40	3.652	3.965	4.174	4.348	5.210	5.785	4.174	5.426	6.261
2.42	3.669	3.988	4.200	4.348	5.225	5.810	4.156	5.429	6.278
2.44	3.687	4.010	4.226	4.347	5.239	5.833	4.137	5.432	6.294
2.46	3.703	4.032	4.252	4.346	5.252	5.856	4.118	5.433	6.310
2.48	3.720	4.054	4.277	4.344	5.265	5.879	4.097	5.434	6.326
2.50	3.736	4.076	4.303	4.342	5.278	5.902	4.076	5.435	6.341
2.52	3.753	4.098	4.328	4.339	5.290	5.924	4.054	5.434	6.355
2.54	3.769	4.119	4.353	4.335	5.301	5.945	4.031	5.433	6.368
2.56	3.784	4.141	4.378	4.331	5.312	5.967	4.007	5.432	6.381
2.58	3.800	4.162	4.403	4.326	5.323	5.988	3.982	5.429	6.394
2.60	3.815	4.183	4.428	4.321	5.333	6.008	3.957	5.426	6.406
2.62	3.830	4.203	4.452	4.315	5.343	6.028	3.930	5.422	6.417

TABLE D.2 Design of Singly Reinforced Rectangular Beam for the Given Ultimate Moment

(M_u/bd^2) MPa	Reinforcement, p_t, per cent								
	Fe250			Fe415			Fe500		
	M20	M25	M30	M20	M25	M30	M20	M25	M30
0.32	0.150	0.149	0.149	-	-	-	-	-	-
0.34	0.160	0.159	0.158	-	-	-	-	-	-
0.36	0.169	0.168	0.168	-	-	-	-	-	-
0.38	0.179	0.178	0.177	-	-	-	-	-	-
0.40	0.188	0.188	0.187	-	-	-	-	-	-
0.42	0.198	0.197	0.196	0.119	0.119	0.118	-	-	-
0.44	0.208	0.207	0.206	0.125	0.125	0.124	-	-	-
0.46	0.218	0.216	0.215	0.131	0.130	0.130	-	-	-
0.48	0.227	0.226	0.225	0.137	0.136	0.136	-	-	-
0.50	0.237	0.236	0.235	0.143	0.142	0.141	0.119	0.118	0.117
0.52	0.247	0.245	0.244	0.149	0.148	0.147	0.123	0.123	0.122
0.54	0.257	0.255	0.254	0.155	0.154	0.153	0.128	0.127	0.127
0.56	0.266	0.265	0.263	0.161	0.159	0.159	0.133	0.132	0.132
0.58	0.276	0.274	0.273	0.166	0.165	0.164	0.138	0.137	0.137
0.60	0.286	0.284	0.283	0.172	0.171	0.170	0.143	0.142	0.141
0.62	0.296	0.294	0.292	0.178	0.177	0.176	0.148	0.147	0.146
0.64	0.306	0.304	0.302	0.184	0.183	0.182	0.153	0.152	0.151
0.66	0.316	0.313	0.312	0.190	0.189	0.188	0.158	0.157	0.156
0.68	0.326	0.323	0.321	0.196	0.195	0.194	0.163	0.162	0.161
0.70	0.336	0.333	0.331	0.202	0.201	0.199	0.168	0.167	0.166
0.72	0.346	0.343	0.341	0.209	0.207	0.205	0.173	0.171	0.170
0.74	0.356	0.353	0.351	0.215	0.213	0.211	0.178	0.176	0.175
0.76	0.366	0.363	0.360	0.221	0.219	0.217	0.183	0.181	0.180
0.78	0.377	0.373	0.370	0.227	0.225	0.223	0.188	0.186	0.185
0.80	0.387	0.383	0.380	0.233	0.231	0.229	0.193	0.191	0.190
0.82	0.397	0.393	0.390	0.239	0.237	0.235	0.198	0.196	0.195
0.84	0.407	0.403	0.400	0.245	0.243	0.241	0.204	0.201	0.200
0.86	0.417	0.413	0.410	0.251	0.249	0.247	0.209	0.206	0.205
0.88	0.428	0.423	0.419	0.258	0.255	0.253	0.214	0.211	0.210
0.90	0.438	0.433	0.429	0.264	0.261	0.259	0.219	0.216	0.215

(Contd.)

TABLE D.2 Design of Singly Reinforced Rectangular Beam for the Given Ultimate Moment

(M_u/bd^2) MPa	Reinforcement, p_t, per cent								
	Fe250			*Fe415*			*Fe500*		
	M20	*M25*	*M30*	*M20*	*M25*	*M30*	*M20*	*M25*	*M30*
0.92	0.448	0.443	0.439	0.270	0.267	0.265	0.224	0.221	0.220
0.94	0.459	0.453	0.449	0.276	0.273	0.271	0.229	0.226	0.225
0.96	0.469	0.463	0.459	0.283	0.279	0.277	0.235	0.232	0.230
0.98	0.480	0.473	0.469	0.289	0.285	0.283	0.240	0.237	0.235
1.00	0.490	0.483	0.479	0.295	0.291	0.289	0.245	0.242	0.240
1.02	0.501	0.494	0.489	0.302	0.297	0.295	0.250	0.247	0.245
1.04	0.511	0.504	0.499	0.308	0.303	0.301	0.256	0.252	0.250
1.06	0.522	0.514	0.509	0.314	0.310	0.307	0.261	0.257	0.255
1.08	0.532	0.524	0.519	0.321	0.316	0.313	0.266	0.262	0.260
1.10	0.543	0.535	0.529	0.327	0.322	0.319	0.271	0.267	0.265
1.12	0.553	0.545	0.539	0.333	0.328	0.325	0.277	0.272	0.270
1.14	0.564	0.555	0.550	0.340	0.334	0.331	0.282	0.278	0.275
1.16	0.575	0.566	0.560	0.346	0.341	0.337	0.287	0.283	0.280
1.18	0.586	0.576	0.570	0.353	0.347	0.343	0.293	0.288	0.285
1.20	0.596	0.586	0.580	0.359	0.353	0.349	0.298	0.293	0.290
1.22	0.607	0.597	0.590	0.366	0.360	0.356	0.304	0.298	0.295
1.24	0.618	0.607	0.600	0.372	0.366	0.362	0.309	0.304	0.300
1.26	0.629	0.618	0.611	0.379	0.372	0.368	0.315	0.309	0.305
1.28	0.640	0.628	0.621	0.386	0.378	0.374	0.320	0.314	0.310
1.30	0.651	0.639	0.631	0.392	0.385	0.380	0.325	0.319	0.316
1.32	0.662	0.649	0.641	0.399	0.391	0.386	0.331	0.325	0.321
1.34	0.673	0.660	0.652	0.405	0.398	0.393	0.337	0.330	0.326
1.36	0.684	0.671	0.662	0.412	0.404	0.399	0.342	0.335	0.331
1.38	0.695	0.681	0.672	0.419	0.410	0.405	0.348	0.341	0.336
1.40	0.706	0.692	0.683	0.426	0.417	0.411	0.353	0.346	0.341
1.42	0.718	0.703	0.693	0.432	0.423	0.418	0.359	0.351	0.347
1.44	0.729	0.713	0.704	0.439	0.430	0.424	0.364	0.357	0.352
1.46	0.740	0.724	0.714	0.446	0.436	0.430	0.370	0.362	0.357
1.48	0.751	0.735	0.725	0.453	0.443	0.436	0.376	0.367	0.362
1.50	0.763	0.746	0.735	0.459	0.449	0.443	0.381	0.373	0.368

(Contd.)

TABLE D.2 Design of Singly Reinforced Rectangular Beam for the Given Ultimate Moment (*Contd.*)

(M_u/bd^2) MPa	Reinforcement, p_t, per cent								
	Fe250			*Fe415*			*Fe500*		
	M20	*M25*	*M30*	*M20*	*M25*	*M30*	*M20*	*M25*	*M30*
1.52	0.774	0.756	0.746	0.466	0.456	0.449	0.387	0.378	0.373
1.54	0.786	0.767	0.756	0.473	0.462	0.455	0.393	0.384	0.378
1.56	0.797	0.778	0.767	0.480	0.469	0.462	0.399	0.389	0.383
1.58	0.809	0.789	0.777	0.487	0.475	0.468	0.404	0.395	0.389
1.60	0.820	0.800	0.788	0.494	0.482	0.475	0.410	0.400	0.394
1.62	0.832	0.811	0.798	0.501	0.489	0.481	0.416	0.405	0.399
1.64	0.843	0.822	0.809	0.508	0.495	0.487	0.422	0.411	0.404
1.66	0.855	0.833	0.820	0.515	0.502	0.494	0.427	0.416	0.410
1.68	0.867	0.844	0.830	0.522	0.508	0.500	0.433	0.422	0.415
1.70	0.878	0.855	0.841	0.529	0.515	0.507	0.439	0.428	0.420
1.72	0.890	0.866	0.852	0.536	0.522	0.513	0.445	0.433	0.426
1.74	0.902	0.877	0.862	0.543	0.529	0.520	0.451	0.439	0.431
1.76	0.914	0.889	0.873	0.551	0.535	0.526	0.457	0.444	0.437
1.78	0.926	0.900	0.884	0.558	0.542	0.532	0.463	0.450	0.442
1.80	0.938	0.911	0.895	0.565	0.549	0.539	0.469	0.455	0.447
1.82	0.950	0.922	0.906	0.572	0.556	0.546	0.475	0.461	0.453
1.84	0.962	0.934	0.916	0.580	0.562	0.552	0.481	0.467	0.458
1.86	0.974	0.945	0.927	0.587	0.569	0.559	0.487	0.472	0.464
1.88	0.986	0.956	0.938	0.594	0.576	0.565	0.493	0.478	0.469
1.90	0.999	0.968	0.949	0.602	0.583	0.572	0.499	0.484	0.475
1.92	1.011	0.979	0.960	0.609	0.590	0.578	0.505	0.490	0.480
1.94	1.023	0.991	0.971	0.616	0.597	0.585	0.512	0.495	0.485
1.96	1.036	1.002	0.982	0.624	0.604	0.592	0.518	0.501	0.491
1.98	1.048	1.014	0.993	0.631	0.611	0.598	0.524	0.507	0.496
2.00	1.061	1.025	1.004	0.639	0.618	0.605	0.530	0.513	0.502
2.02	1.073	1.037	1.015	0.646	0.624	0.611	0.537	0.518	0.508
2.04	1.086	1.048	1.026	0.654	0.632	0.618	0.543	0.524	0.513
2.06	1.098	1.060	1.037	0.662	0.639	0.625	0.549	0.530	0.519
2.08	1.111	1.072	1.048	0.669	0.646	0.632	0.556	0.536	0.524
2.10	1.124	1.083	1.060	0.677	0.653	0.638	0.562	0.542	0.530

(*Contd.*)

TABLE D.2 Design of Singly Reinforced Rectangular Beam for the Given Ultimate Moment

(M_u/bd^2) MPa	Reinforcement, p_t, per cent								
	Fe250			Fe415			Fe500		
	M20	M25	M30	M20	M25	M30	M20	M25	M30
2.12	1.137	1.095	1.071	0.685	0.660	0.645	0.568	0.548	0.535
2.14	1.150	1.107	1.082	0.693	0.667	0.652	0.575	0.553	0.541
2.16	1.163	1.119	1.093	0.700	0.674	0.659	0.581	0.559	0.547
2.18	1.176	1.131	1.104	0.708	0.681	0.665	0.588	0.565	0.552
2.20	1.189	1.143	1.116	0.716	0.688	0.672	0.594	0.571	0.558
2.22	1.202	1.154	1.127	0.724	0.695	0.679	0.601	0.577	0.564
2.24	1.215	1.166	1.138	0.732	0.703	0.686	0.607	0.583	0.569
2.26	1.228	1.178	1.150	0.740	0.710	0.693	0.614	0.589	0.575
2.28	1.241	1.191	1.161	0.748	0.717	0.699	0.621	0.595	0.581
2.30	1.255	1.203	1.173	0.756	0.724	0.706	0.627	0.601	0.586
2.32	1.268	1.215	1.184	0.764	0.732	0.713	0.634	0.607	0.592
2.34	1.282	1.227	1.196	0.772	0.739	0.720	0.641	0.613	0.598
2.36	1.295	1.239	1.207	0.780	0.746	0.727	0.648	0.620	0.604
2.38	1.309	1.251	1.219	0.789	0.754	0.734	0.654	0.626	0.609
2.40	1.323	1.264	1.230	0.797	0.761	0.741	0.661	0.632	0.615
2.42	1.336	1.276	1.242	0.805	0.769	0.748	0.668	0.638	0.621
2.44	1.350	1.288	1.253	0.813	0.776	0.755	0.675	0.644	0.627
2.46	1.364	1.301	1.265	0.822	0.784	0.762	0.682	0.650	0.632
2.48	1.378	1.313	1.277	0.830	0.791	0.769	0.689	0.657	0.638
2.50	1.392	1.326	1.288	0.839	0.799	0.776	0.696	0.663	0.644
2.52	1.406	1.338	1.300	0.847	0.806	0.783	0.703	0.669	0.650
2.54	1.421	1.351	1.312	0.856	0.814	0.790	0.710	0.675	0.656
2.56	1.435	1.364	1.324	0.864	0.821	0.797	0.718	0.682	0.662
2.58	1.449	1.376	1.335	0.873	0.829	0.804	0.725	0.688	0.668
2.60	1.464	1.389	1.347	0.882	0.837	0.812	0.732	0.694	0.674
2.62	1.478	1.402	1.359	0.891	0.844	0.819	0.739	0.701	0.680
2.64	1.493	1.414	1.371	0.899	0.852	0.826	0.747	0.707	0.686
2.66	1.508	1.427	1.383	0.908	0.860	0.833	0.754	0.714	0.691
2.68	1.523	1.440	1.395	0.917	0.868	0.840	0.761	0.720	0.697
2.70	1.537	1.453	1.407	0.926	0.875	0.848	-	0.727	0.703

(*Contd.*)

TABLE D.2 Design of Singly Reinforced Rectangular Beam for the Given Ultimate Moment (*Contd.*)

(M_u/bd^2) MPa	Reinforcement, p_t, per cent								
	Fe250			*Fe415*			*Fe500*		
	M20	*M25*	*M30*	*M20*	*M25*	*M30*	*M20*	*M25*	*M30*
2.72	1.552	1.466	1.419	0.935	0.883	0.855	-	0.733	0.709
2.74	1.568	1.479	1.431	0.944	0.891	0.862	-	0.740	0.716
2.76	1.583	1.492	1.443	0.953	0.899	0.869	-	0.746	0.722
2.78	1.598	1.505	1.455	0.963	0.907	0.877	-	0.753	0.728
2.80	1.613	1.519	1.467	-	0.915	0.884	-	0.759	0.734
2.82	1.629	1.532	1.480	-	0.923	0.891	-	0.766	0.740
2.84	1.644	1.545	1.492	-	0.931	0.899	-	0.773	0.746
2.86	1.660	1.558	1.504	-	0.939	0.906	-	0.779	0.752
2.88	1.676	1.572	1.516	-	0.947	0.914	-	0.786	0.758
2.90	1.692	1.585	1.529	-	0.955	0.921	-	0.793	0.764
2.92	1.708	1.599	1.541	-	0.963	0.928	-	0.799	0.771
2.94	1.724	1.612	1.554	-	0.971	0.936	-	0.806	0.777
2.96	1.740	1.626	1.566	-	0.980	0.943	-	0.813	0.783
2.98	1.756	1.640	1.578	-	0.988	0.951	-	0.820	0.789
3.00	-	1.653	1.591	-	0.996	0.958	-	0.827	0.795
3.02	-	1.667	1.603	-	1.004	0.966	-	0.834	0.802
3.04	-	1.681	1.616	-	1.013	0.974	-	0.840	0.808
3.06	-	1.695	1.629	-	1.021	0.981	-	0.847	0.814
3.08	-	1.709	1.641	-	1.029	0.989	-	0.854	0.821
3.10	-	1.723	1.654	-	1.038	0.996	-	0.861	0.827
3.12	-	1.737	1.667	-	1.046	1.004	-	0.868	0.833
3.14	-	1.751	1.679	-	1.055	1.012	-	0.876	0.840
3.16	-	1.765	1.692	-	1.063	1.019	-	0.883	0.846
3.18	-	1.779	1.705	-	1.072	1.027	-	0.890	0.853
3.20	-	1.794	1.718	-	1.081	1.035	-	0.897	0.859
3.22	-	1.808	1.731	-	1.089	1.043	-	0.904	0.865
3.24	-	1.823	1.744	-	1.098	1.050	-	0.911	0.872
3.26	-	1.837	1.757	-	1.107	1.058	-	0.919	0.878
3.28	-	1.852	1.770	-	1.115	1.066	-	0.926	0.885
3.30	-	1.866	1.783	-	1.124	1.074	-	0.933	0.891

(*Contd.*)

TABLE D.2 Design of Singly Reinforced Rectangular Beam for the Given Ultimate Moment

(M_u/bd^2) MPa	Reinforcement, p_t, per cent								
	Fe250			*Fe415*			*Fe500*		
	M20	*M25*	*M30*	*M20*	*M25*	*M30*	*M20*	*M25*	*M30*
3.32	-	1.881	1.796	-	1.133	1.082	-	0.941	0.898
3.34	-	1.896	1.809	-	1.142	1.090	-	-	0.905
3.36	-	1.911	1.822	-	1.151	1.098	-	-	0.911
3.38	-	1.926	1.836	-	1.160	1.106	-	-	0.918
3.40	-	1.941	1.849	-	1.169	1.114	-	-	0.924
3.42	-	1.956	1.862	-	1.178	1.122	-	-	0.931
3.44	-	1.971	1.876	-	-	1.130	-	-	0.938
3.46	-	1.986	1.889	-	-	1.138	-	-	0.944
3.48	-	2.001	1.902	-	-	1.146	-	-	0.951
3.50	-	2.017	1.916	-	-	1.154	-	-	0.958
3.52	-	2.032	1.929	-	-	1.162	-	-	0.965
3.54	-	2.048	1.943	-	-	1.170	-	-	0.972
3.56	-	2.063	1.957	-	-	1.179	-	-	0.978
3.58	-	2.079	1.970	-	-	1.187	-	-	0.985
3.60	-	2.095	1.984	-	-	1.195	-	-	0.992
3.62	-	2.111	1.998	-	-	1.203	-	-	0.999
3.64	-	2.127	2.012	-	-	1.212	-	-	1.006
3.66	-	2.143	2.025	-	-	1.220	-	-	1.013
3.68	-	2.159	2.039	-	-	1.229	-	-	1.020
3.70	-	2.175	2.053	-	-	1.237	-	-	1.027
3.72	-	2.191	2.067	-	-	1.245	-	-	1.034
3.74	-	2.208	2.081	-	-	1.254	-	-	1.041
3.76	-	-	2.096	-	-	1.262	-	-	1.048
3.78	-	-	2.110	-	-	1.271	-	-	1.055
3.80	-	-	2.124	-	-	1.279	-	-	1.062
3.82	-	-	2.138	-	-	1.288	-	-	1.069
3.84	-	-	2.153	-	-	1.297	-	-	1.076
3.86	-	-	2.167	-	-	1.305	-	-	1.083
3.88	-	-	2.181	-	-	1.314	-	-	1.091
3.90	-	-	2.196	-	-	1.323	-	-	1.098

(Contd.)

TABLE D.2 Design of Singly Reinforced Rectangular Beam for the Given Ultimate Moment (*Contd.*)

(M_u/bd^2) MPa	Reinforcement, p_t, per cent								
	Fe250			*Fe415*			*Fe500*		
	M20	*M25*	*M30*	*M20*	*M25*	*M30*	*M20*	*M25*	*M30*
3.92	-	-	2.210	-	-	1.332	-	-	1.105
3.94	-	-	2.225	-	-	1.340	-	-	1.112
3.96	-	-	2.240	-	-	1.349	-	-	1.120
3.98	-	-	2.254	-	-	1.358	-	-	-
4.00	-	-	2.269	-	-	1.367	-	-	-
4.02	-	-	2.284	-	-	1.376	-	-	-
4.04	-	-	2.299	-	-	1.385	-	-	-
4.06	-	-	2.314	-	-	1.394	-	-	-
4.08	-	-	2.329	-	-	1.403	-	-	-
4.10	-	-	2.344	-	-	1.412	-	-	-
4.12	-	-	2.359	-	-	1.421	-	-	-
4.14	-	-	2.374	-	-	-	-	-	-
4.16	-	-	2.389	-	-	-	-	-	-
4.18	-	-	2.405	-	-	-	-	-	-
4.20	-	-	2.420	-	-	-	-	-	-
4.22	-	-	2.436	-	-	-	-	-	-
4.24	-	-	2.451	-	-	-	-	-	-
4.26	-	-	2.467	-	-	-	-	-	-
4.28	-	-	2.482	-	-	-	-	-	-
4.30	-	-	2.498	-	-	-	-	-	-
4.32	-	-	2.514	-	-	-	-	-	-
4.34	-	-	2.530	-	-	-	-	-	-
4.36	-	-	2.546	-	-	-	-	-	-
4.38	-	-	2.562	-	-	-	-	-	-
4.40	-	-	2.578	-	-	-	-	-	-

TABLE D.3 Spacing of Two-legged Stirrups for Given Value of Shear Per Unit Depth

V_{us}/d	Spacing of stirrups, S_v, mm								V_{us}/d	Spacing of stirrups, S_v, mm							
	Fe250				Fe415					Fe250				Fe415			
N/mm	Size of stirrup bar, mm				Size of stirrup bar, mm				N/mm	Size of stirrup bar, mm				Size of stirrup bar, mm			
	6	8	10	12	6	8	10	12		6	8	10	12	6	8	10	12
40	307.2	-	-	-	-	-	-	-	440	-	49.9	77.6	111.7	-	82.9	128.8	185.4
60	204.8	366.1	-	-	340.0	-	-	-	460	-	-	74.2	106.9	-	79.3	123.2	177.4
80	153.6	274.6	426.8	-	255.0	-	-	-	480	-	-	71.1	102.4	-	76.0	118.1	170.0
100	122.9	219.7	341.5	-	204.0	364.7	-	-	500	-	-	68.3	98.3	-	72.9	113.4	163.2
120	102.4	183.1	284.6	409.6	170.0	303.9	-	-	520	-	-	65.7	94.5	-	70.1	109.0	156.9
140	87.8	156.9	243.9	351.1	145.7	260.5	404.9	-	540	-	-	63.2	91.0	-	67.5	105.0	151.1
160	76.8	137.3	213.4	307.2	127.5	227.9	354.3	-	560	-	-	61.0	87.8	-	65.1	101.2	145.7
180	68.3	122.0	189.7	273.1	113.3	202.6	314.9	453.3	580	-	-	58.9	84.8	-	62.9	97.7	140.7
200	61.4	109.8	170.7	245.8	102.0	182.3	283.4	408.0	600	-	-	56.9	81.9	-	60.8	94.5	136.0
220	55.9	99.9	155.2	223.4	92.7	165.8	257.7	370.9	620	-	-	55.1	79.3	-	58.8	91.4	131.6
240	51.2	91.5	142.3	204.8	85.0	151.9	236.2	340.0	640	-	-	53.4	76.8	-	57.0	88.6	127.5
260	-	84.5	131.3	189.1	78.5	140.3	218.0	313.8	660	-	-	51.7	74.5	-	55.3	85.9	123.6
280	-	78.5	122.0	175.6	72.9	130.2	202.4	291.4	680	-	-	50.2	72.3	-	53.6	83.4	120.0
300	-	73.2	113.8	163.9	68.0	121.6	188.9	272.0	700	-	-	-	70.2	-	52.1	81.0	116.6
320	-	68.6	106.7	153.6	63.7	114.0	177.1	255.0	720	-	-	-	68.3	-	50.6	78.7	113.3
340	-	64.6	100.4	144.6	60.0	107.3	166.7	240.0	740	-	-	-	66.4	-	-	76.6	110.3
360	-	61.0	94.9	136.5	56.7	101.3	157.5	226.7	760	-	-	-	64.7	-	-	74.6	107.4
380	-	57.8	89.9	129.4	53.7	96.0	149.2	214.7	780	-	-	-	63.0	-	-	72.7	104.6
400	-	54.9	85.4	122.9	51.0	91.2	141.7	204.0	800	-	-	-	61.4	-	-	70.9	102.0
420	-	52.3	81.3	117.0	-	86.8	135.0	194.3									

(Contd.)

TABLE D.3 Spacing of Two-legged Stirrups for Given Value of Shear Per Unit Depth (*Contd.*)

V_{us}/d	*Spacing of stirrups, S_v, mm*								V_{us}/d	*Spacing of stirrups, S_v, mm*							
	Fe250				*Fe415*					*Fe250*				*Fe415*			
N/	*Size of stirrup bar, mm*				*Size of stirrup bar, mm*				N/	*Size of stirrup bar, mm*				*Size of stirrup bar, mm*			
mm	*6*	*8*	*10*	*12*	*6*	*8*	*10*	*12*	mm	*6*	*8*	*10*	*12*	*6*	*8*	*10*	*12*
820	-	-	-	59.9	-	-	69.1	99.5	1220	-	-	-	-	-	-	-	66.9
840	-	-	-	58.5	-	-	67.5	97.1	1240	-	-	-	-	-	-	-	65.8
860	-	-	-	57.2	-	-	65.9	94.9	1260	-	-	-	-	-	-	-	64.8
880	-	-	-	55.9	-	-	64.4	92.7	1280	-	-	-	-	-	-	-	63.7
900	-	-	-	54.6	-	-	63.0	90.7	1300	-	-	-	-	-	-	-	62.8
920	-	-	-	53.4	-	-	61.6	88.7	1320	-	-	-	-	-	-	-	61.8
940	-	-	-	52.3	-	-	60.3	86.8	1340	-	-	-	-	-	-	-	60.9
960	-	-	-	51.2	-	-	59.0	85.0	1360	-	-	-	-	-	-	-	60.0
980	-	-	-	50.2	-	-	57.8	83.3	1380	-	-	-	-	-	-	-	59.1
1000	-	-	-	-	-	-	56.7	81.6	1400	-	-	-	-	-	-	-	58.3
1020	-	-	-	-	-	-	55.6	80.0	1420	-	-	-	-	-	-	-	57.5
1040	-	-	-	-	-	-	54.5	78.5	1440	-	-	-	-	-	-	-	56.7
1060	-	-	-	-	-	-	53.5	77.0	1460	-	-	-	-	-	-	-	55.9
1080	-	-	-	-	-	-	52.5	75.6	1480	-	-	-	-	-	-	-	55.1
1100	-	-	-	-	-	-	51.5	74.2	1500	-	-	-	-	-	-	-	54.4
1120	-	-	-	-	-	-	50.6	72.9	1520	-	-	-	-	-	-	-	53.7
1140	-	-	-	-	-	-	49.7	71.6	1540	-	-	-	-	-	-	-	53.0
1160	-	-	-	-	-	-	-	70.3	1560	-	-	-	-	-	-	-	52.3
1180	-	-	-	-	-	-	-	69.2	1580	-	-	-	-	-	-	-	51.6
1200	-	-	-	-	-	-	-	68.0	1600	-	-	-	-	-	-	-	51.0

TABLE D.4 Values of p_t and p_c for the Doubly Reinforced Rectangular Beam Sections for the Given M_u/bd^2.

$\frac{M_u}{bd^2}$	*M20*								*M25*							
	d'/d = 0.05		*d'/d = 0.10*		*d'/d = 0.15*		*d'/d = 0.20*		*d'/d = 0.05*		*d'/d = 0.10*		*d'/d = 0.15*		*d'/d = 0.20*	
MPa	p_t	p_c	p_t	p_c	p_t	p_c	p_t	p_c	p_t	p_c	p_t	p_c	p_t	p_c	p_t	p_c
2.76	Singly Reinforced Beam: Balanced Design								-	-	-	-	-	-	-	-
2.78	0.962	0.001	0.962	0.001	0.962	0.001	0.962	0.001	-	-	-	-	-	-	-	-
2.80	0.968	0.007	0.968	0.007	0.968	0.008	0.969	0.009	-	-	-	-	-	-	-	-
2.82	0.974	0.013	0.974	0.014	0.975	0.015	0.976	0.017	-	-	-	-	-	-	-	-
2.84	0.979	0.019	0.980	0.020	0.982	0.022	0.983	0.025	-	-	-	-	-	-	-	-
2.86	0.985	0.025	0.987	0.027	0.988	0.029	0.990	0.032	-	-	-	-	-	-	-	-
2.88	0.991	0.031	0.993	0.033	0.995	0.036	0.997	0.040	-	-	-	-	-	-	-	-
2.90	0.997	0.037	0.999	0.040	1.001	0.043	1.004	0.048	-	-	-	-	-	-	-	-
2.92	1.003	0.043	1.005	0.046	1.008	0.051	1.011	0.056	-	-	-	-	-	-	-	-
2.94	1.009	0.050	1.011	0.053	1.014	0.058	1.017	0.064	-	-	-	-	-	-	-	-
2.96	1.014	0.056	1.017	0.059	1.021	0.065	1.024	0.071	-	-	-	-	-	-	-	-
2.98	1.020	0.062	1.023	0.066	1.027	0.072	1.031	0.079	-	-	-	-	-	-	-	-
3.00	1.026	0.068	1.030	0.072	1.034	0.079	1.038	0.087	-	-	-	-	-	-	-	-
3.02	1.032	0.074	1.036	0.079	1.040	0.086	1.045	0.095	-	-	-	-	-	-	-	-
3.04	1.038	0.080	1.042	0.085	1.047	0.093	1.052	0.103	-	-	-	-	-	-	-	-
3.06	1.044	0.086	1.048	0.092	1.053	0.100	1.059	0.111	-	-	-	-	-	-	-	-
3.08	1.049	0.092	1.054	0.098	1.060	0.107	1.066	0.118	-	-	-	-	-	-	-	-
3.10	1.055	0.098	1.060	0.105	1.066	0.114	1.073	0.126	-	-	-	-	-	-	-	-
3.12	1.061	0.104	1.067	0.111	1.073	0.121	1.080	0.134	-	-	-	-	-	-	-	-
3.14	1.067	0.110	1.073	0.118	1.079	0.128	1.087	0.142	-	-	-	-	-	-	-	-
3.16	1.073	0.116	1.079	0.124	1.086	0.135	1.094	0.150	-	-	-	-	-	-	-	-
3.18	1.078	0.123	1.085	0.131	1.092	0.142	1.101	0.157	-	-	-	-	-	-	-	-
3.20	1.084	0.129	1.091	0.137	1.099	0.149	1.107	0.165	-	-	-	-	-	-	-	-
3.22	1.090	0.135	1.097	0.143	1.105	0.156	1.114	0.173	-	-	-	-	-	-	-	-
3.24	1.096	0.141	1.103	0.150	1.112	0.164	1.121	0.181	-	-	-	-	-	-	-	-
3.26	1.102	0.147	1.110	0.156	1.118	0.171	1.128	0.189	-	-	-	-	-	-	-	-
3.28	1.108	0.153	1.116	0.163	1.125	0.178	1.135	0.196	-	-	-	-	-	-	-	-
3.30	1.113	0.159	1.122	0.169	1.131	0.185	1.142	0.204	-	-	-	-	-	-	-	-
3.32	1.119	0.165	1.128	0.176	1.138	0.192	1.149	0.212	-	-	-	-	-	-	-	-
3.34	1.125	0.171	1.134	0.182	1.144	0.199	1.156	0.220	-	-	-	-	-	-	-	-
3.36	1.131	0.177	1.140	0.189	1.151	0.206	1.163	0.228	-	-	-	-	-	-	-	-
3.38	1.137	0.183	1.147	0.195	1.157	0.213	1.170	0.236	-	-	-	-	-	-	-	-
3.40	1.143	0.190	1.153	0.202	1.164	0.220	1.177	0.243	-	-	-	-	-	-	-	-
3.42	1.148	0.196	1.159	0.208	1.171	0.227	1.184	0.251	-	-	-	-	-	-	-	-
3.44	1.154	0.202	1.165	0.215	1.177	0.234	1.191	0.259	M_u/bd^2 = 3.450 is for a Singly Reinforced Balanced Beam:							
3.46	1.160	0.208	1.171	0.221	1.184	0.241	1.197	0.267	1.198	-0.004	1.197	0.004	1.197	-0.004	1.197	-0.005
3.48	1.166	0.214	1.177	0.228	1.190	0.248	1.204	0.275	1.203	0.002	1.203	0.003	1.204	0.003	1.204	0.003
3.50	1.172	0.220	1.183	0.234	1.197	0.255	1.211	0.282	1.209	0.009	1.210	0.009	1.210	0.010	1.211	0.011

(*Contd.*)

TABLE D.4 Values of p_t and p_c for the Doubly Reinforced Rectangular Beam Sections for the Given M_u/bd^2. *(Contd.)*

$\frac{M_u}{bd^2}$	M20								M25							
	d'/d = 0.05		d'/d = 0.10		d'/d = 0.15		d'/d = 0.20		d'/d = 0.05		d'/d = 0.10		d'/d = 0.15		d'/d = 0.20	
MPa	p_t	p_c	p_t	p_c	p_t	p_c	p_t	p_c	p_t	p_c	p_t	p_c	p_t	p_c	p_t	p_c
3.52	1.178	0.226	1.190	0.241	1.203	0.262	1.218	0.290	1.215	0.015	1.216	0.016	1.217	0.017	1.218	0.019
3.54	1.183	0.232	1.196	0.247	1.210	0.270	1.225	0.298	1.221	0.021	1.222	0.022	1.223	0.024	1.225	0.027
3.56	1.189	0.238	1.202	0.254	1.216	0.277	1.232	0.306	1.227	0.027	1.228	0.029	1.230	0.031	1.231	0.035
3.58	1.195	0.244	1.208	0.260	1.223	0.284	1.239	0.314	1.232	0.033	1.234	0.035	1.236	0.038	1.238	0.042
3.60	1.201	0.250	1.214	0.267	1.229	0.291	1.246	0.321	1.238	0.039	1.240	0.042	1.243	0.046	1.245	0.050
3.62	1.207	0.256	1.220	0.273	1.236	0.298	1.253	0.329	1.244	0.045	1.247	0.048	1.249	0.053	1.252	0.058
3.64	1.213	0.263	1.227	0.280	1.242	0.305	1.260	0.337	1.250	0.051	1.253	0.055	1.256	0.060	1.259	0.066
3.66	1.218	0.269	1.233	0.286	1.249	0.312	1.267	0.345	1.256	0.058	1.259	0.061	1.262	0.067	1.266	0.074
3.68	1.224	0.275	1.239	0.292	1.255	0.319	1.274	0.353	1.262	0.064	1.265	0.068	1.269	0.074	1.273	0.082
3.70	1.230	0.281	1.245	0.299	1.262	0.326	1.281	0.360	1.267	0.070	1.271	0.074	1.275	0.081	1.280	0.090
3.72	1.236	0.287	1.251	0.305	1.268	0.333	1.287	0.368	1.273	0.076	1.277	0.081	1.282	0.088	1.287	0.098
3.74	1.242	0.293	1.257	0.312	1.275	0.340	1.294	0.376	1.279	0.082	1.283	0.087	1.288	0.095	1.294	0.105
3.76	1.248	0.299	1.264	0.318	1.281	0.347	1.301	0.384	1.285	0.088	1.290	0.094	1.295	0.102	1.301	0.113
3.78	1.253	0.305	1.270	0.325	1.288	0.354	1.308	0.392	1.291	0.094	1.296	0.100	1.301	0.110	1.308	0.121
3.80	1.259	0.311	1.276	0.331	1.294	0.361	1.315	0.400	1.297	0.100	1.302	0.107	1.308	0.117	1.315	0.129
3.82	1.265	0.317	1.282	0.338	1.301	0.368	1.322	0.407	1.302	0.107	1.308	0.113	1.314	0.124	1.321	0.137
3.84	1.271	0.323	1.288	0.344	1.307	0.375	1.329	0.415	1.308	0.113	1.314	0.120	1.321	0.131	1.328	0.145
3.86	1.277	0.329	1.294	0.351	1.314	0.383	1.336	0.423	1.314	0.119	1.320	0.126	1.327	0.138	1.335	0.153
3.88	1.283	0.336	1.300	0.357	1.320	0.390	1.343	0.431	1.320	0.125	1.327	0.133	1.334	0.145	1.342	0.160
3.90	1.288	0.342	1.307	0.364	1.327	0.397	1.350	0.439	1.326	0.131	1.333	0.140	1.340	0.152	1.349	0.168
3.92	1.294	0.348	1.313	0.370	1.333	0.404	1.357	0.446	1.332	0.137	1.339	0.146	1.347	0.159	1.356	0.176
3.94	1.300	0.354	1.319	0.377	1.340	0.411	1.364	0.454	1.337	0.143	1.345	0.153	1.353	0.166	1.363	0.184
3.96	1.306	0.360	1.325	0.383	1.346	0.418	1.371	0.462	1.343	0.149	1.351	0.159	1.360	0.174	1.370	0.192
3.98	1.312	0.366	1.331	0.390	1.353	0.425	1.377	0.470	1.349	0.156	1.357	0.166	1.367	0.181	1.377	0.200
4.00	1.318	0.372	1.337	0.396	1.360	0.432	1.384	0.478	1.355	0.162	1.363	0.172	1.373	0.188	1.384	0.208
4.02	1.323	0.378	1.344	0.403	1.366	0.439	1.391	0.485	1.361	0.168	1.370	0.179	1.380	0.195	1.391	0.216
4.04	1.329	0.384	1.350	0.409	1.373	0.446	1.398	0.493	1.367	0.174	1.376	0.185	1.386	0.202	1.398	0.223
4.06	1.335	0.390	1.356	0.416	1.379	0.453	1.405	0.501	1.372	0.180	1.382	0.192	1.393	0.209	1.405	0.231
4.08	1.341	0.396	1.362	0.422	1.386	0.460	1.412	0.509	1.378	0.186	1.388	0.198	1.399	0.216	1.411	0.239
4.10	1.347	0.402	1.368	0.428	1.392	0.467	1.419	0.517	1.384	0.192	1.394	0.205	1.406	0.223	1.418	0.247
4.12	1.353	0.409	1.374	0.435	1.399	0.474	1.426	0.525	1.390	0.198	1.400	0.211	1.412	0.230	1.425	0.255
4.14	1.358	0.415	1.380	0.441	1.405	0.481	1.433	0.532	1.396	0.205	1.407	0.218	1.419	0.238	1.432	0.263
4.16	1.364	0.421	1.387	0.448	1.412	0.489	1.440	0.540	1.402	0.211	1.413	0.224	1.425	0.245	1.439	0.271
4.18	1.370	0.427	1.393	0.454	1.418	0.496	1.447	0.548	1.407	0.217	1.419	0.231	1.432	0.252	1.446	0.278
4.20	1.376	0.433	1.399	0.461	1.425	0.503	1.454	0.556	1.413	0.223	1.425	0.237	1.438	0.259	1.453	0.286
4.22	1.382	0.439	1.405	0.467	1.431	0.510	1.461	0.564	1.419	0.229	1.431	0.244	1.445	0.266	1.460	0.294
4.24	1.388	0.445	1.411	0.474	1.438	0.517	1.468	0.571	1.425	0.235	1.437	0.250	1.451	0.273	1.467	0.302

(Contd.)

TABLE D.4 Values of p_t and p_c for the Doubly Reinforced Rectangular Beam Sections for the Given M_u/bd^2. (Contd.)

$\frac{M_u}{bd^2}$	M20								M25							
	d'/d = 0.05		d'/d = 0.10		d'/d = 0.15		d'/d = 0.20		d'/d = 0.05		d'/d = 0.10		d'/d = 0.15		d'/d = 0.20	
MPa	p_t	p_c	p_t	p_c	p_t	p_c	p_t	p_c	p_t	p_c	p_t	p_c	p_t	p_c	p_t	p_c
4.26	1.393	0.451	1.417	0.480	1.444	0.524	1.474	0.579	1.431	0.241	1.444	0.257	1.458	0.280	1.474	0.310
4.28	1.399	0.457	1.424	0.487	1.451	0.531	1.481	0.587	1.437	0.247	1.450	0.263	1.464	0.287	1.481	0.318
4.30	1.405	0.463	1.430	0.493	1.457	0.538	1.488	0.595	1.442	0.253	1.456	0.270	1.471	0.294	1.488	0.326
4.32	1.411	0.469	1.436	0.500	1.464	0.545	1.495	0.603	1.448	0.260	1.462	0.276	1.477	0.302	1.495	0.334
4.34	1.417	0.475	1.442	0.506	1.470	0.552	1.502	0.610	1.454	0.266	1.468	0.283	1.484	0.309	1.502	0.341
4.36	1.423	0.482	1.448	0.513	1.477	0.559	1.509	0.618	1.460	0.272	1.474	0.289	1.490	0.316	1.508	0.349
4.38	1.428	0.488	1.454	0.519	1.483	0.566	1.516	0.626	1.466	0.278	1.480	0.296	1.497	0.323	1.515	0.357
4.40	1.434	0.494	1.460	0.526	1.490	0.573	1.523	0.634	1.472	0.284	1.487	0.303	1.503	0.330	1.522	0.365
4.42	1.440	0.500	1.467	0.532	1.496	0.580	1.530	0.642	1.477	0.290	1.493	0.309	1.510	0.337	1.529	0.373
4.44	1.446	0.506	1.473	0.539	1.503	0.587	1.537	0.649	1.483	0.296	1.499	0.316	1.516	0.344	1.536	0.381
4.46	1.452	0.512	1.479	0.545	1.509	0.594	1.544	0.657	1.489	0.302	1.505	0.322	1.523	0.351	1.543	0.389
4.48	1.458	0.518	1.485	0.552	1.516	0.602	1.551	0.665	1.495	0.309	1.511	0.329	1.529	0.358	1.550	0.396
4.50	1.463	0.524	1.491	0.558	1.522	0.609	1.558	0.673	1.501	0.315	1.517	0.335	1.536	0.366	1.557	0.404
4.52	1.469	0.530	1.497	0.565	1.529	0.616	1.564	0.681	1.507	0.321	1.524	0.342	1.542	0.373	1.564	0.412
4.54	1.475	0.536	1.504	0.571	1.535	0.623	1.571	0.689	1.512	0.327	1.530	0.348	1.549	0.380	1.571	0.420
4.56	1.481	0.542	1.510	0.577	1.542	0.630	1.578	0.696	1.518	0.333	1.536	0.355	1.556	0.387	1.578	0.428
4.58	1.487	0.548	1.516	0.584	1.549	0.637	1.585	0.704	1.524	0.339	1.542	0.361	1.562	0.394	1.585	0.436
4.60	1.492	0.555	1.522	0.590	1.555	0.644	1.592	0.712	1.530	0.345	1.548	0.368	1.569	0.401	1.592	0.444
4.62	1.498	0.561	1.528	0.597	1.562	0.651	1.599	0.720	1.536	0.351	1.554	0.374	1.575	0.408	1.598	0.452
4.64	1.504	0.567	1.534	0.603	1.568	0.658	1.606	0.728	1.542	0.358	1.560	0.381	1.582	0.415	1.605	0.459
4.66	1.510	0.573	1.540	0.610	1.575	0.665	1.613	0.735	1.547	0.364	1.567	0.387	1.588	0.422	1.612	0.467
4.68	1.516	0.579	1.547	0.616	1.581	0.672	1.620	0.743	1.553	0.370	1.573	0.394	1.595	0.430	1.619	0.475
4.70	1.522	0.585	1.553	0.623	1.588	0.679	1.627	0.751	1.559	0.376	1.579	0.400	1.601	0.437	1.626	0.483
4.72	1.527	0.591	1.559	0.629	1.594	0.686	1.634	0.759	1.565	0.382	1.585	0.407	1.608	0.444	1.633	0.491
4.74	1.533	0.597	1.565	0.636	1.601	0.693	1.641	0.767	1.571	0.388	1.591	0.413	1.614	0.451	1.640	0.499
4.76	1.539	0.603	1.571	0.642	1.607	0.700	1.648	0.774	1.577	0.394	1.597	0.420	1.621	0.458	1.647	0.507
4.78	1.545	0.609	1.577	0.649	1.614	0.708	1.654	0.782	1.582	0.400	1.604	0.426	1.627	0.465	1.654	0.514
4.80	1.551	0.615	1.584	0.655	1.620	0.715	1.661	0.790	1.588	0.407	1.610	0.433	1.634	0.472	1.661	0.522
4.82	1.557	0.621	1.590	0.662	1.627	0.722	1.668	0.798	1.594	0.413	1.616	0.439	1.640	0.479	1.668	0.530
4.84	1.562	0.628	1.596	0.668	1.633	0.729	1.675	0.806	1.600	0.419	1.622	0.446	1.647	0.486	1.675	0.538
4.86	1.568	0.634	1.602	0.675	1.640	0.736	1.682	0.814	1.606	0.425	1.628	0.452	1.653	0.494	1.682	0.546
4.88	1.574	0.640	1.608	0.681	1.646	0.743	1.689	0.821	1.611	0.431	1.634	0.459	1.660	0.501	1.688	0.554
4.90	1.580	0.646	1.614	0.688	1.653	0.750	1.696	0.829	1.617	0.437	1.640	0.466	1.666	0.508	1.695	0.562
4.92	1.586	0.652	1.620	0.694	1.659	0.757	1.703	0.837	1.623	0.443	1.647	0.472	1.673	0.515	1.702	0.569
4.94	1.592	0.658	1.627	0.701	1.666	0.764	1.710	0.845	1.629	0.449	1.653	0.479	1.679	0.522	1.709	0.577
4.96	1.597	0.664	1.633	0.707	1.672	0.771	1.717	0.853	1.635	0.456	1.659	0.485	1.686	0.529	1.716	0.585
4.98	1.603	0.670	1.639	0.714	1.679	0.778	1.724	0.860	1.641	0.462	1.665	0.492	1.692	0.536	1.723	0.593
5.00	1.609	0.676	1.645	0.720	1.685	0.785	1.731	0.868	1.646	0.468	1.671	0.498	1.699	0.543	1.730	0.601

(Contd.)

TABLE D.4 Values of p_t and p_c for the Doubly Reinforced Rectangular Beam Sections for the Given M_u/bd^2. *(Contd.)*

$\frac{M_u}{bd^2}$	M20								M25							
	d'/d = 0.05		d'/d = 0.10		d'/d = 0.15		d'/d = 0.20		d'/d = 0.05		d'/d = 0.10		d'/d = 0.15		d'/d = 0.20	
MPa	p_t	p_c	p_t	p_c	p_t	p_c	p_t	p_c	p_t	p_c	p_t	p_c	p_t	p_c	p_t	p_c
5.02	1.615	0.682	1.651	0.726	1.692	0.792	1.738	0.876	1.652	0.474	1.677	0.505	1.705	0.550	1.737	0.609
5.04	1.621	0.688	1.657	0.733	1.698	0.799	1.744	0.884	1.658	0.480	1.684	0.511	1.712	0.558	1.744	0.617
5.06	1.627	0.694	1.664	0.739	1.705	0.806	1.751	0.892	1.664	0.486	1.690	0.518	1.718	0.565	1.751	0.625
5.08	1.632	0.701	1.670	0.746	1.711	0.813	1.758	0.899	1.670	0.492	1.696	0.524	1.725	0.572	1.758	0.632
5.10	1.638	0.707	1.676	0.752	1.718	0.821	1.765	0.907	1.676	0.498	1.702	0.531	1.731	0.579	1.765	0.640
5.12	1.644	0.713	1.682	0.759	1.724	0.828	1.772	0.915	1.681	0.505	1.708	0.537	1.738	0.586	1.772	0.648
5.14	1.650	0.719	1.688	0.765	1.731	0.835	1.779	0.923	1.687	0.511	1.714	0.544	1.745	0.593	1.778	0.656
5.16	1.656	0.725	1.694	0.772	1.737	0.842	1.786	0.931	1.693	0.517	1.720	0.550	1.751	0.600	1.785	0.664
5.18	1.662	0.731	1.701	0.778	1.744	0.849	1.793	0.938	1.699	0.523	1.727	0.557	1.758	0.607	1.792	0.672
5.20	1.667	0.737	1.707	0.785	1.751	0.856	1.800	0.946	1.705	0.529	1.733	0.563	1.764	0.615	1.799	0.680
5.22	1.673	0.743	1.713	0.791	1.757	0.863	1.807	0.954	1.711	0.535	1.739	0.570	1.771	0.622	1.806	0.687
5.24	1.679	0.749	1.719	0.798	1.764	0.870	1.814	0.962	1.716	0.541	1.745	0.576	1.777	0.629	1.813	0.695
5.26	1.685	0.755	1.725	0.804	1.770	0.877	1.821	0.970	1.722	0.547	1.751	0.583	1.784	0.636	1.820	0.703
5.28	1.691	0.761	1.731	0.811	1.777	0.884	1.828	0.978	1.728	0.554	1.757	0.589	1.790	0.643	1.827	0.711
5.30	1.697	0.767	1.737	0.817	1.783	0.891	1.834	0.985	1.734	0.560	1.764	0.596	1.797	0.650	1.834	0.719
5.32	1.702	0.774	1.744	0.824	1.790	0.898	1.841	0.993	1.740	0.566	1.770	0.602	1.803	0.657	1.841	0.727
5.34	1.708	0.780	1.750	0.830	1.796	0.905	1.848	1.001	1.746	0.572	1.776	0.609	1.810	0.664	1.848	0.735
5.36	1.714	0.786	1.756	0.837	1.803	0.912	1.855	1.009	1.751	0.578	1.782	0.615	1.816	0.671	1.855	0.743
5.38	1.720	0.792	1.762	0.843	1.809	0.919	1.862	1.017	1.757	0.584	1.788	0.622	1.823	0.679	1.862	0.750
5.40	1.726	0.798	1.768	0.850	1.816	0.927	1.869	1.024	1.763	0.590	1.794	0.629	1.829	0.686	1.868	0.758
5.42	1.732	0.804	1.774	0.856	1.822	0.934	1.876	1.032	1.769	0.596	1.800	0.635	1.836	0.693	1.875	0.766
5.44	1.737	0.810	1.781	0.862	1.829	0.941	1.883	1.040	1.775	0.603	1.807	0.642	1.842	0.700	1.882	0.774
5.46	1.743	0.816	1.787	0.869	1.835	0.948	1.890	1.048	1.781	0.609	1.813	0.648	1.849	0.707	1.889	0.782
5.48	1.749	0.822	1.793	0.875	1.842	0.955	1.897	1.056	1.786	0.615	1.819	0.655	1.855	0.714	1.896	0.790
5.50	1.755	0.828	1.799	0.882	1.848	0.962	1.904	1.063	1.792	0.621	1.825	0.661	1.862	0.721	1.903	0.798
5.52	1.761	0.834	1.805	0.888	1.855	0.969	1.911	1.071	1.798	0.627	1.831	0.668	1.868	0.728	1.910	0.805
5.54	1.767	0.840	1.811	0.895	1.861	0.976	1.918	1.079	1.804	0.633	1.837	0.674	1.875	0.735	1.917	0.813
5.56	1.772	0.847	1.817	0.901	1.868	0.983	1.925	1.087	1.810	0.639	1.844	0.681	1.881	0.743	1.924	0.821
5.58	1.778	0.853	1.824	0.908	1.874	0.990	1.931	1.095	1.816	0.645	1.850	0.687	1.888	0.750	1.931	0.829
5.60	1.784	0.859	1.830	0.914	1.881	0.997	1.938	1.103	1.821	0.651	1.856	0.694	1.894	0.757	1.938	0.837
5.62	1.790	0.865	1.836	0.921	1.887	1.004	1.945	1.110	1.827	0.658	1.862	0.700	1.901	0.764	1.945	0.845
5.64	1.796	0.871	1.842	0.927	1.894	1.011	1.952	1.118	1.833	0.664	1.868	0.707	1.907	0.771	1.952	0.853
5.66	1.802	0.877	1.848	0.934	1.900	1.018	1.959	1.126	1.839	0.670	1.874	0.713	1.914	0.778	1.959	0.861
5.68	1.807	0.883	1.854	0.940	1.907	1.025	1.966	1.134	1.845	0.676	1.880	0.720	1.920	0.785	1.965	0.868
5.70	1.813	0.889	1.861	0.947	1.913	1.032	1.973	1.142	1.851	0.682	1.887	0.726	1.927	0.792	1.972	0.876
5.72	1.819	0.895	1.867	0.953	1.920	1.040	1.980	1.149	1.856	0.688	1.893	0.733	1.934	0.799	1.979	0.884
5.74	1.825	0.901	1.873	0.960	1.926	1.047	1.987	1.157	1.862	0.694	1.899	0.739	1.940	0.807	1.986	0.892
5.76	1.831	0.907	1.879	0.966	1.933	1.054	1.994	1.165	1.868	0.700	1.905	0.746	1.947	0.814	1.993	0.900

(Contd.)

TABLE D.4 Values of p_t and p_c for the Doubly Reinforced Rectangular Beam Sections for the Given M_u/bd^2. (*Contd.*)

$\frac{M_u}{bd^2}$	M20								M25							
	d′/d = 0.05		d′/d = 0.10		d′/d = 0.15		d′/d = 0.20		d′/d = 0.05		d′/d = 0.10		d′/d = 0.15		d′/d = 0.20	
MPa	p_t	p_c	p_t	p_c	p_t	p_c	p_t	p_c	p_t	p_c	p_t	p_c	p_t	p_c	p_t	p_c
5.78	1.837	0.913	1.885	0.973	1.940	1.061	2.001	1.173	1.874	0.707	1.911	0.752	1.953	0.821	2.000	0.908
5.80	1.842	0.920	1.891	0.979	1.946	1.068	2.008	1.181	1.880	0.713	1.917	0.759	1.960	0.828	2.007	0.916
5.82	1.848	0.926	1.897	0.986	1.953	1.075	2.015	1.188	1.886	0.719	1.924	0.765	1.966	0.835	2.014	0.923
5.84	1.854	0.932	1.904	0.992	1.959	1.082	2.021	1.196	1.891	0.725	1.930	0.772	1.973	0.842	2.021	0.931
5.86	1.860	0.938	1.910	0.999	1.966	1.089	2.028	1.204	1.897	0.731	1.936	0.779	1.979	0.849	2.028	0.939
5.88	1.866	0.944	1.916	1.005	1.972	1.096	2.035	1.212	1.903	0.737	1.942	0.785	1.986	0.856	2.035	0.947
5.90	1.872	0.950	1.922	1.011	1.979	1.103	2.042	1.220	1.909	0.743	1.948	0.792	1.992	0.863	2.042	0.955
5.92	1.877	0.956	1.928	1.018	1.985	1.110	2.049	1.228	1.915	0.749	1.954	0.798	1.999	0.871	2.049	0.963
5.94	1.883	0.962	1.934	1.024	1.992	1.117	2.056	1.235	1.921	0.756	1.961	0.805	2.005	0.878	2.055	0.971
5.96	1.889	0.968	1.941	1.031	1.998	1.124	2.063	1.243	1.926	0.762	1.967	0.811	2.012	0.885	2.062	0.979
5.98	1.895	0.974	1.947	1.037	2.005	1.131	2.070	1.251	1.932	0.768	1.973	0.818	2.018	0.892	2.069	0.986
6.00	1.901	0.980	1.953	1.044	2.011	1.138	2.077	1.259	1.938	0.774	1.979	0.824	2.025	0.899	2.076	0.994
6.02	1.906	0.986	1.959	1.050	2.018	1.146	2.084	1.267	1.944	0.780	1.985	0.831	2.031	0.906	2.083	1.002
6.04	1.912	0.993	1.965	1.057	2.024	1.153	2.091	1.274	1.950	0.786	1.991	0.837	2.038	0.913	2.090	1.010
6.06	1.918	0.999	1.971	1.063	2.031	1.160	2.098	1.282	1.956	0.792	1.997	0.844	2.044	0.920	2.097	1.018
6.08	1.924	1.005	1.977	1.070	2.037	1.167	2.105	1.290	1.961	0.798	2.004	0.850	2.051	0.927	2.104	1.026
6.10	1.930	1.011	1.984	1.076	2.044	1.174	2.111	1.298	1.967	0.805	2.010	0.857	2.057	0.935	2.111	1.034
6.12	1.936	1.017	1.990	1.083	2.050	1.181	2.118	1.306	1.973	0.811	2.016	0.863	2.064	0.942	2.118	1.041
6.14	1.941	1.023	1.996	1.089	2.057	1.188	2.125	1.313	1.979	0.817	2.022	0.870	2.070	0.949	2.125	1.049
6.16	1.947	1.029	2.002	1.096	2.063	1.195	2.132	1.321	1.985	0.823	2.028	0.876	2.077	0.956	2.132	1.057
6.18	1.953	1.035	2.008	1.102	2.070	1.202	2.139	1.329	1.991	0.829	2.034	0.883	2.083	0.963	2.139	1.065
6.20	1.959	1.041	2.014	1.109	2.076	1.209	2.146	1.337	1.996	0.835	2.041	0.889	2.090	0.970	2.145	1.073
6.22	1.965	1.047	2.021	1.115	2.083	1.216	2.153	1.345	2.002	0.841	2.047	0.896	2.096	0.977	2.152	1.081
6.24	1.971	1.053	2.027	1.122	2.089	1.223	2.160	1.352	2.008	0.847	2.053	0.902	2.103	0.984	2.159	1.089
6.26	1.976	1.059	2.033	1.128	2.096	1.230	2.167	1.360	2.014	0.854	2.059	0.909	2.109	0.991	2.166	1.097
6.28	1.982	1.066	2.039	1.135	2.102	1.237	2.174	1.368	2.020	0.860	2.065	0.915	2.116	0.999	2.173	1.104
6.30	1.988	1.072	2.045	1.141	2.109	1.244	2.181	1.376	2.025	0.866	2.071	0.922	2.122	1.006	2.180	1.112
6.32	1.994	1.078	2.051	1.148	2.115	1.251	2.188	1.384	2.031	0.872	2.077	0.928	2.129	1.013	2.187	1.120
6.34	2.000	1.084	2.057	1.154	2.122	1.259	2.195	1.392	2.037	0.878	2.084	0.935	2.136	1.020	2.194	1.128
6.36	2.006	1.090	2.064	1.160	2.129	1.266	2.201	1.399	2.043	0.884	2.090	0.942	2.142	1.027	2.201	1.136
6.38	2.011	1.096	2.070	1.167	2.135	1.273	2.208	1.407	2.049	0.890	2.096	0.948	2.149	1.034	2.208	1.144
6.40	2.017	1.102	2.076	1.173	2.142	1.280	2.215	1.415	2.055	0.896	2.102	0.955	2.155	1.041	2.215	1.152
6.42	2.023	1.108	2.082	1.180	2.148	1.287	2.222	1.423	2.060	0.903	2.108	0.961	2.162	1.048	2.222	1.159
6.44	2.029	1.114	2.088	1.186	2.155	1.294	2.229	1.431	2.066	0.909	2.114	0.968	2.168	1.055	2.229	1.167
6.46	2.035	1.120	2.094	1.193	2.161	1.301	2.236	1.438	2.072	0.915	2.121	0.974	2.175	1.063	2.235	1.175
6.48	2.041	1.126	2.101	1.199	2.168	1.308	2.243	1.446	2.078	0.921	2.127	0.981	2.181	1.070	2.242	1.183
6.50	2.046	1.132	2.107	1.206	2.174	1.315	2.250	1.454	2.084	0.927	2.133	0.987	2.188	1.077	2.249	1.191

(*Contd.*)

TABLE D.4 Values of p_t and p_c for the Doubly Reinforced Rectangular Beam Sections for the Given M_u/bd^2. (*Contd.*)

$\frac{M_u}{bd^2}$	M20								M25							
	d'/d = 0.05		d'/d = 0.10		d'/d = 0.15		d'/d = 0.20		d'/d = 0.05		d'/d = 0.10		d'/d = 0.15		d'/d = 0.20	
MPa	p_t	p_c	p_t	p_c	p_t	p_c	p_t	p_c	p_t	p_c	p_t	p_c	p_t	p_c	p_t	p_c
6.52	2.052	1.139	2.113	1.212	2.181	1.322	2.257	1.462	2.090	0.933	2.139	0.994	2.194	1.084	2.256	1.199
6.54	2.058	1.145	2.119	1.219	2.187	1.329	2.264	1.470	2.095	0.939	2.145	1.000	2.201	1.091	2.263	1.207
6.56	2.064	1.151	2.125	1.225	2.194	1.336	2.271	1.477	2.101	0.945	2.151	1.007	2.207	1.098	2.270	1.215
6.58	2.070	1.157	2.131	1.232	2.200	1.343	2.278	1.485	2.107	0.952	2.157	1.013	2.214	1.105	2.277	1.222
6.60	2.076	1.163	2.138	1.238	2.207	1.350	2.285	1.493	2.113	0.958	2.164	1.020	2.220	1.112	2.284	1.230
6.62	2.081	1.169	2.144	1.245	2.213	1.357	2.291	1.501	2.119	0.964	2.170	1.026	2.227	1.119	2.291	1.238
6.64	2.087	1.175	2.150	1.251	2.220	1.365	2.298	1.509	2.125	0.970	2.176	1.033	2.233	1.127	2.298	1.246
6.66	2.093	1.181	2.156	1.258	2.226	1.372	2.305	1.517	2.130	0.976	2.182	1.039	2.240	1.134	2.305	1.254
6.68	2.099	1.187	2.162	1.264	2.233	1.379	2.312	1.524	2.136	0.982	2.188	1.046	2.246	1.141	2.312	1.262
6.70	2.105	1.193	2.168	1.271	2.239	1.386	2.319	1.532	2.142	0.988	2.194	1.052	2.253	1.148	2.319	1.270
6.72	2.111	1.199	2.174	1.277	2.246	1.393	2.326	1.540	2.148	0.994	2.201	1.059	2.259	1.155	2.325	1.277
6.74	2.116	1.205	2.181	1.284	2.252	1.400	2.333	1.548	2.154	1.001	2.207	1.065	2.266	1.162	2.332	1.285
6.76	2.122	1.212	2.187	1.290	2.259	1.407	2.340	1.556	2.160	1.007	2.213	1.072	2.272	1.169	2.339	1.293
6.78	2.128	1.218	2.193	1.297	2.265	1.414	2.347	1.563	2.165	1.013	2.219	1.078	2.279	1.176	2.346	1.301
6.80	2.134	1.224	2.199	1.303	2.272	1.421	2.354	1.571	2.171	1.019	2.225	1.085	2.285	1.183	2.353	1.309
6.82	2.140	1.230	2.205	1.309	2.278	1.428	2.361	1.579	2.177	1.025	2.231	1.091	2.292	1.191	2.360	1.317
6.84	2.146	1.236	2.211	1.316	2.285	1.435	2.368	1.587	2.183	1.031	2.237	1.098	2.298	1.198	2.367	1.325
6.86	2.151	1.242	2.218	1.322	2.291	1.442	2.375	1.595	2.189	1.037	2.244	1.105	2.305	1.205	2.374	1.332
6.88	2.157	1.248	2.224	1.329	2.298	1.449	2.382	1.602	2.195	1.043	2.250	1.111	2.311	1.212	2.381	1.340
6.90	2.163	1.254	2.230	1.335	2.304	1.456	2.388	1.610	2.200	1.049	2.256	1.118	2.318	1.219	2.388	1.348
6.92	2.169	1.260	2.236	1.342	2.311	1.463	2.395	1.618	2.206	1.056	2.262	1.124	2.325	1.226	2.395	1.356
6.94	2.175	1.266	2.242	1.348	2.318	1.470	2.402	1.626	2.212	1.062	2.268	1.131	2.331	1.233	2.402	1.364
6.96	2.181	1.272	2.248	1.355	2.324	1.478	2.409	1.634	2.218	1.068	2.274	1.137	2.338	1.240	2.409	1.372
6.98	2.186	1.278	2.254	1.361	2.331	1.485	2.416	1.641	2.224	1.074	2.281	1.144	2.344	1.248	2.416	1.380
7.00	2.192	1.285	2.261	1.368	2.337	1.492	2.423	1.649	2.230	1.080	2.287	1.150	2.351	1.255	2.422	1.388
7.02	2.198	1.291	2.267	1.374	2.344	1.499	-	-	2.235	1.086	2.293	1.157	2.357	1.262	2.429	1.395
7.04	2.204	1.297	2.273	1.381	2.350	1.506	-	-	2.241	1.092	2.299	1.163	2.364	1.269	2.436	1.403
7.06	2.210	1.303	2.279	1.387	2.357	1.513	-	-	2.247	1.098	2.305	1.170	2.370	1.276	2.443	1.411
7.08	2.216	1.309	2.285	1.394	2.363	1.520	-	-	2.253	1.105	2.311	1.176	2.377	1.283	2.450	1.419
7.10	2.221	1.315	2.291	1.400	2.370	1.527	-	-	2.259	1.111	2.317	1.183	2.383	1.290	2.457	1.427
7.12	-	-	-	-	-	-	-	-	2.265	1.117	2.324	1.189	2.390	1.297	2.464	1.435
7.14	-	-	-	-	-	-	-	-	2.270	1.123	2.330	1.196	2.396	1.304	2.471	1.443
7.16	-	-	-	-	-	-	-	-	2.276	1.129	2.336	1.202	2.403	1.312	2.478	1.450
7.18	-	-	-	-	-	-	-	-	2.282	1.135	2.342	1.209	2.409	1.319	2.485	1.458
7.20	-	-	-	-	-	-	-	-	2.288	1.141	2.348	1.215	2.416	1.326	2.492	1.466
7.22	-	-	-	-	-	-	-	-	2.294	1.147	2.354	1.222	2.422	1.333	2.499	1.474
7.24	-	-	-	-	-	-	-	-	2.300	1.154	2.361	1.228	2.429	1.340	2.506	1.482
7.26	-	-	-	-	-	-	-	-	2.305	1.160	2.367	1.235	2.435	1.347	2.512	1.490

(*Contd.*)

TABLE D.4 Values of p_t and p_c for the Doubly Reinforced Rectangular Beam Sections for the Given M_u/bd^2. *(Contd.)*

$\frac{M_u}{bd^2}$	M20								M25							
	d'/d = 0.05		d'/d = 0.10		d'/d = 0.15		d'/d = 0.20		d'/d = 0.05		d'/d = 0.10		d'/d = 0.15		d'/d = 0.20	
MPa	p_t	p_c	p_t	p_c	p_t	p_c	p_t	p_c	p_t	p_c	p_t	p_c	p_t	p_c	p_t	p_c
7.28	-	-	-	-	-	-	-	-	2.311	1.166	2.373	1.241	2.442	1.354	2.519	1.498
7.30	-	-	-	-	-	-	-	-	2.317	1.172	2.379	1.248	2.448	1.361	2.526	1.506
7.32	-	-	-	-	-	-	-	-	2.323	1.178	2.385	1.254	2.455	1.368	-	-
7.34	-	-	-	-	-	-	-	-	2.329	1.184	2.391	1.261	2.461	1.376	-	-
7.36	-	-	-	-	-	-	-	-	2.335	1.190	2.398	1.268	2.468	1.383	-	-
7.38	-	-	-	-	-	-	-	-	2.340	1.196	2.404	1.274	2.474	1.390	-	-
7.40	-	-	-	-	-	-	-	-	2.346	1.203	2.410	1.281	2.481	1.397	-	-
7.42	-	-	-	-	-	-	-	-	2.352	1.209	2.416	1.287	2.487	1.404	-	-
7.44	-	-	-	-	-	-	-	-	2.358	1.215	2.422	1.294	2.494	1.411	-	-
7.46	-	-	-	-	-	-	-	-	2.364	1.221	2.428	1.300	2.500	1.418	-	-
7.48	-	-	-	-	-	-	-	-	2.370	1.227	2.434	1.307	2.507	1.425	-	-
7.50	-	-	-	-	-	-	-	-	2.375	1.233	2.441	1.313	2.514	1.432	-	-
7.52	-	-	-	-	-	-	-	-	2.381	1.239	2.447	1.320	2.520	1.440	-	-
7.54	-	-	-	-	-	-	-	-	2.387	1.245	2.453	1.326	2.527	1.447	-	-
7.56	-	-	-	-	-	-	-	-	2.393	1.252	2.459	1.333	2.533	1.454	-	-
7.58	-	-	-	-	-	-	-	-	2.399	1.258	2.465	1.339	2.540	1.461	-	-
7.60	-	-	-	-	-	-	-	-	2.405	1.264	2.471	1.346	2.546	1.468	-	-
7.62	-	-	-	-	-	-	-	-	2.410	1.270	2.478	1.352	2.553	1.475	-	-
7.64	-	-	-	-	-	-	-	-	2.416	1.276	2.484	1.359	2.559	1.482	-	-
7.66	-	-	-	-	-	-	-	-	2.422	1.282	2.490	1.365	2.566	1.489	-	-
7.68	-	-	-	-	-	-	-	-	2.428	1.288	2.496	1.372	-	-	-	-
7.70	-	-	-	-	-	-	-	-	2.434	1.294	2.502	1.378	-	-	-	-
7.72	-	-	-	-	-	-	-	-	2.439	1.301	2.508	1.385	-	-	-	-
7.74	-	-	-	-	-	-	-	-	2.445	1.307	2.514	1.391	-	-	-	-
7.76	-	-	-	-	-	-	-	-	2.451	1.313	2.521	1.398	-	-	-	-
7.78	-	-	-	-	-	-	-	-	2.457	1.319	2.527	1.404	-	-	-	-
7.80	-	-	-	-	-	-	-	-	2.463	1.325	2.533	1.411	-	-	-	-

Inadmissible zones

$\frac{M_u}{bd^2} < \frac{M_{u,\text{lim}}}{bd^2}$

$p_t + p_c > 4.0$

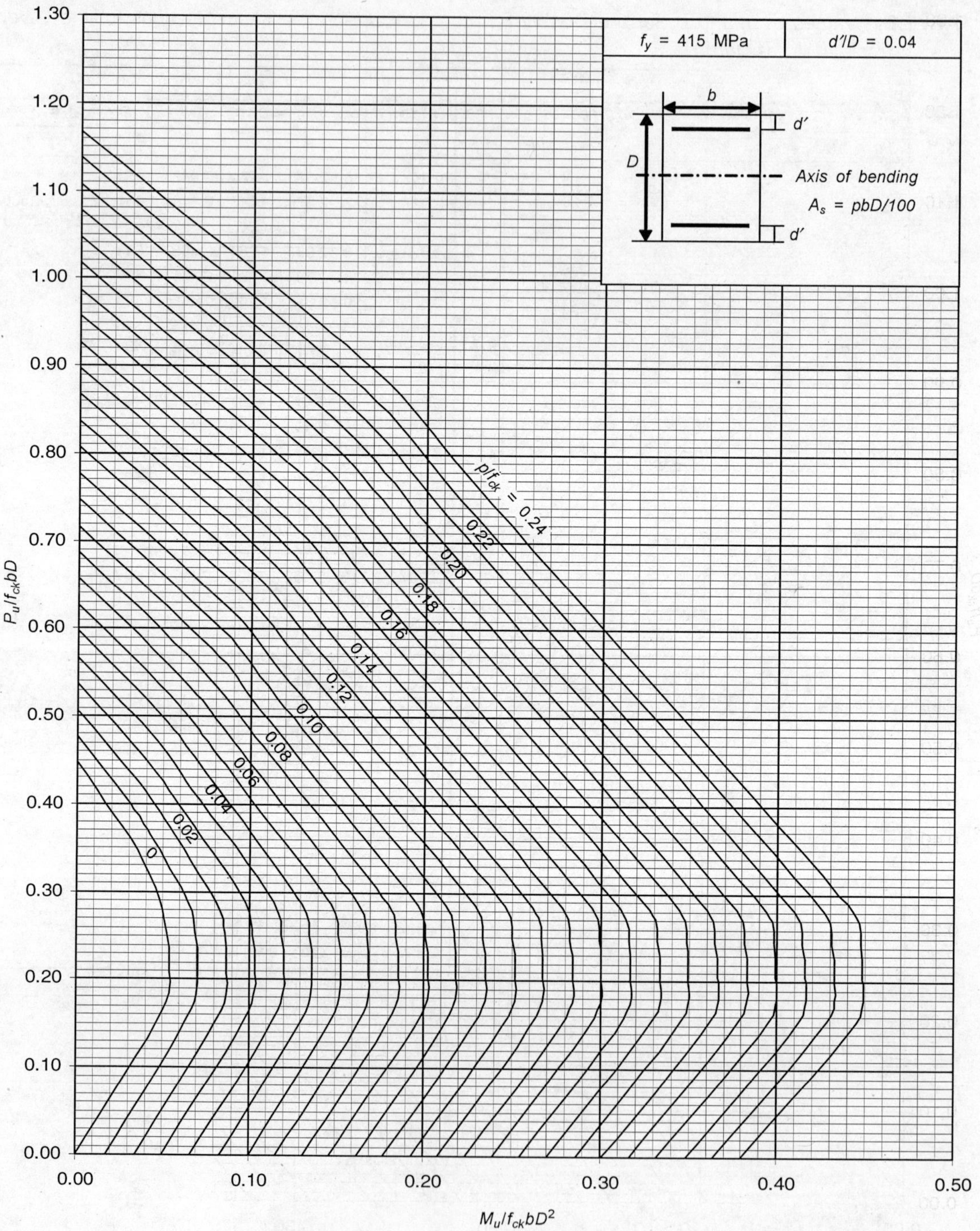

Fig. D.1 *Rectangular section subjected to compression and bending with reinforcement distributed equally on two faces.*

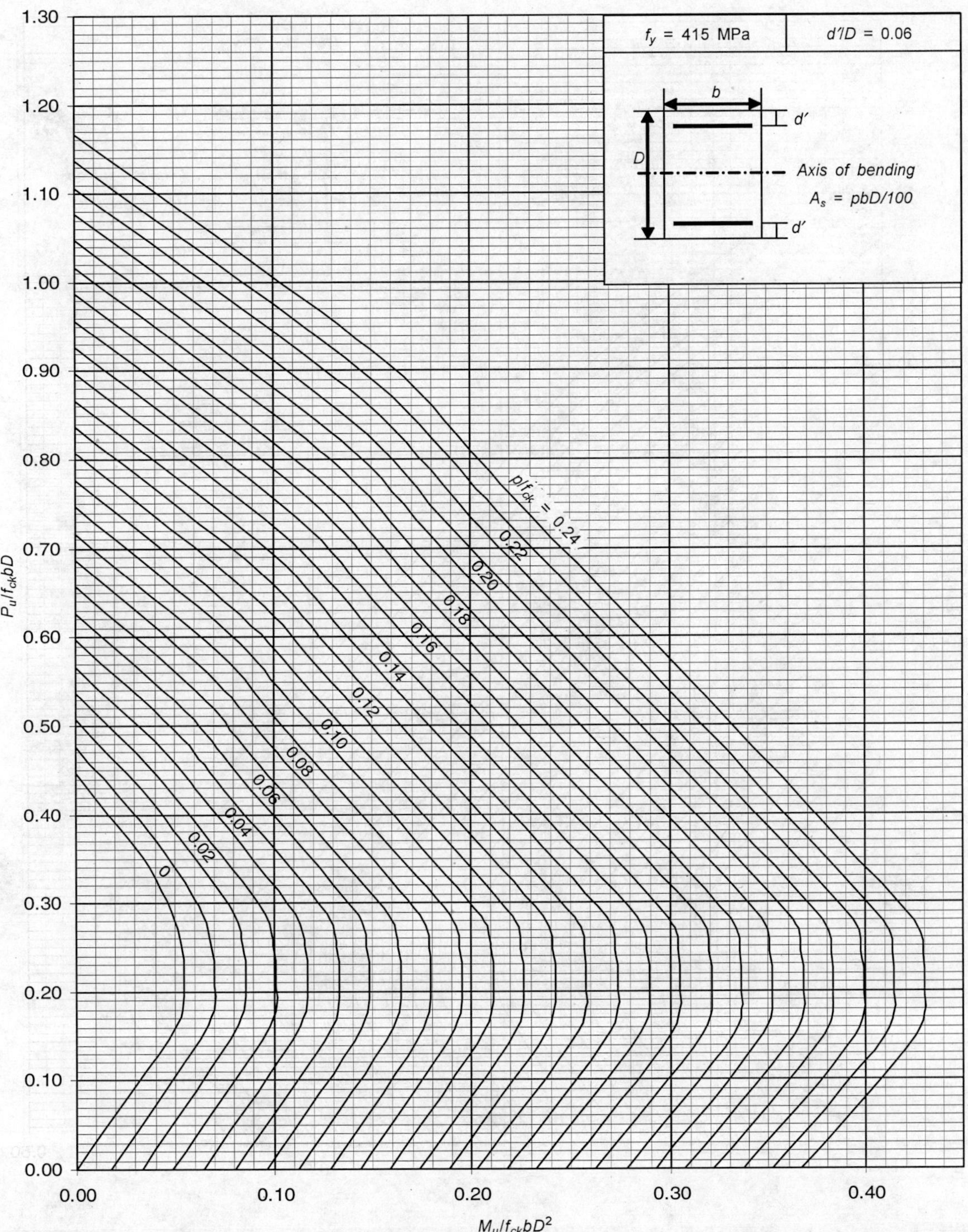

Fig. D.2 *Rectangular section subjected to compression and bending with reinforcement distributed equally on two faces.*

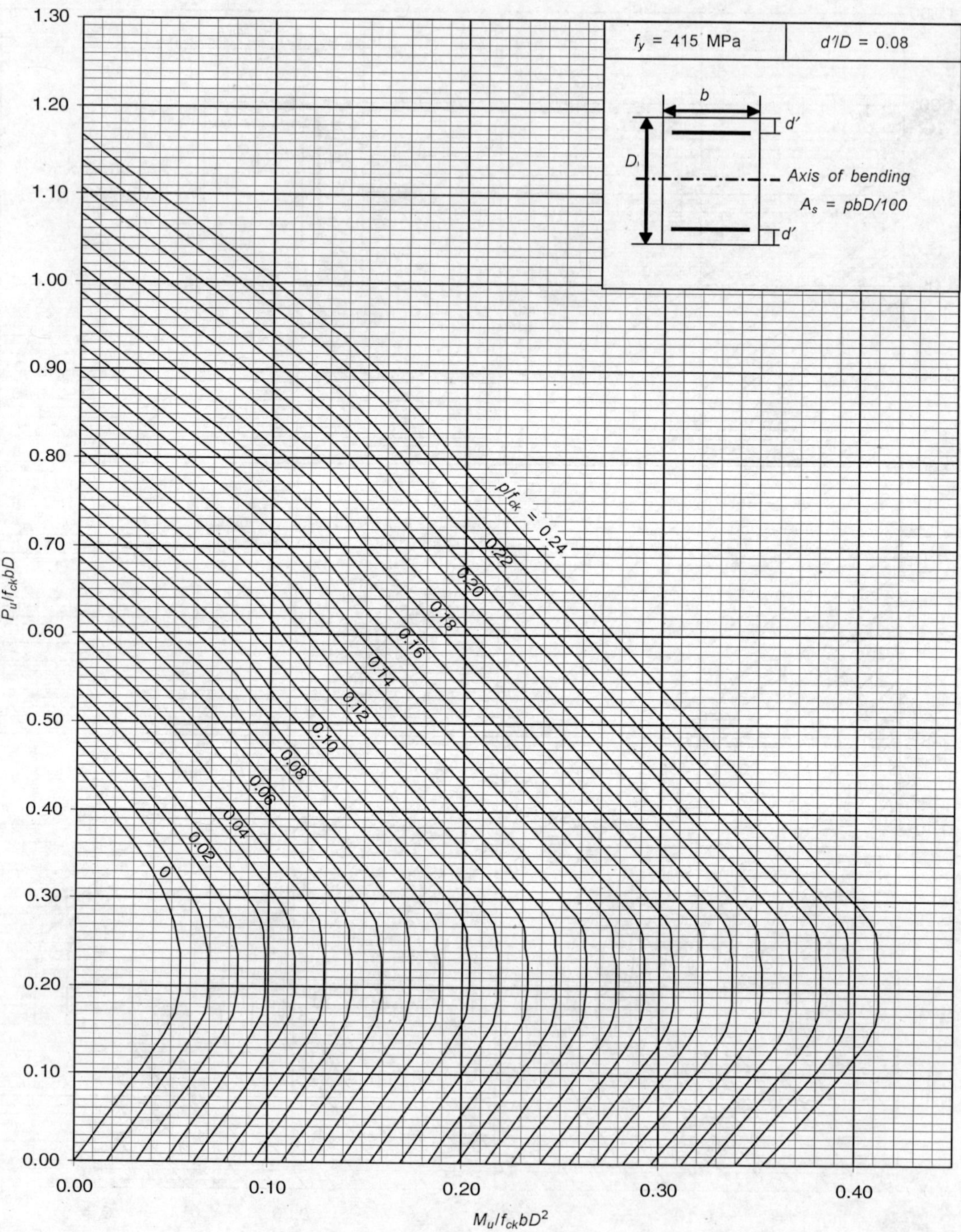

Fig. D.3 Rectangular section subjected to compression and bending with reinforcement distributed equally on two faces.

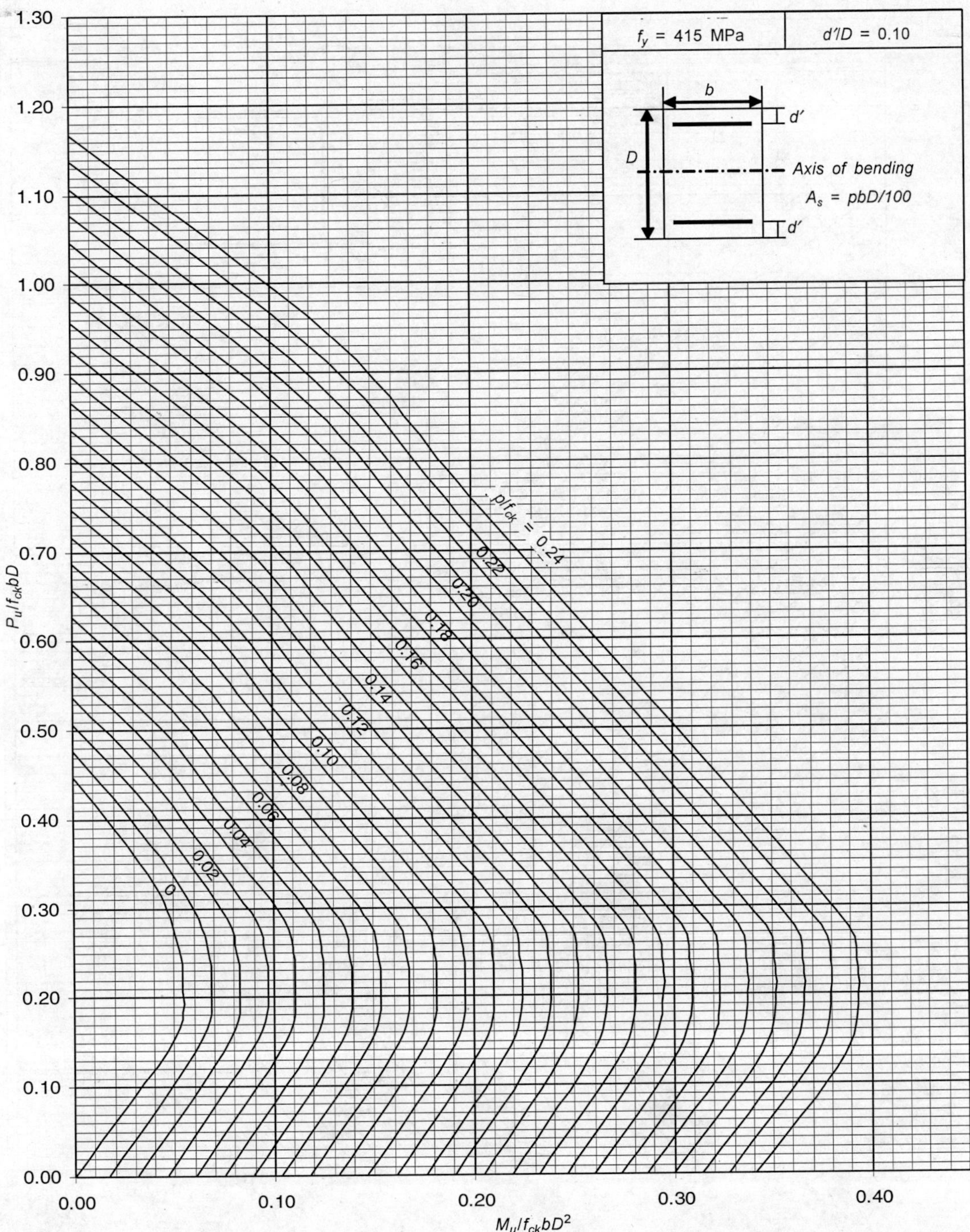

Fig. D.4 *Rectangular section subjected to compression and bending with reinforcement distributed equally on two faces.*

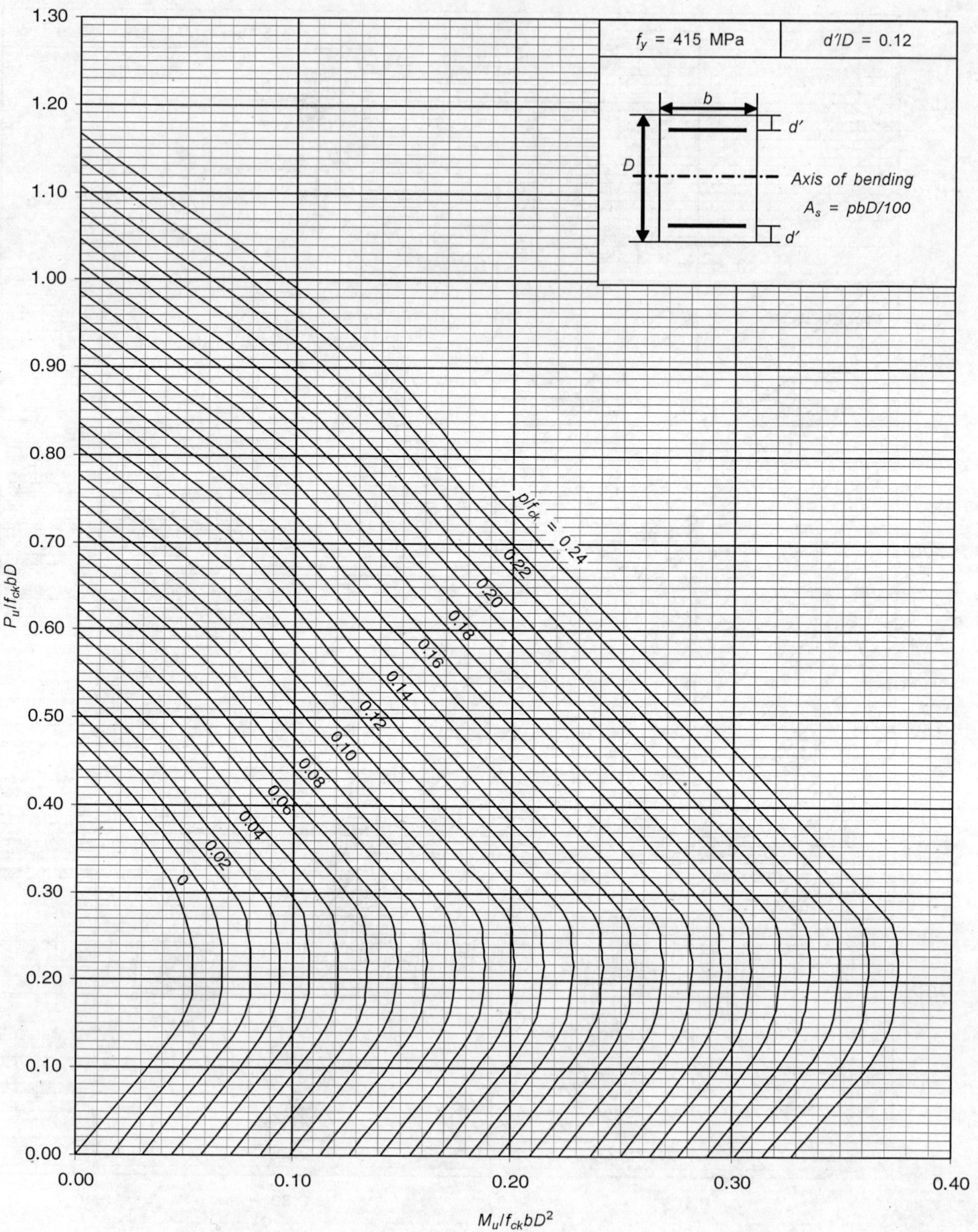

Fig. D.5 *Rectangular section subjected to compression and bending with reinforcement distributed equally on two faces.*

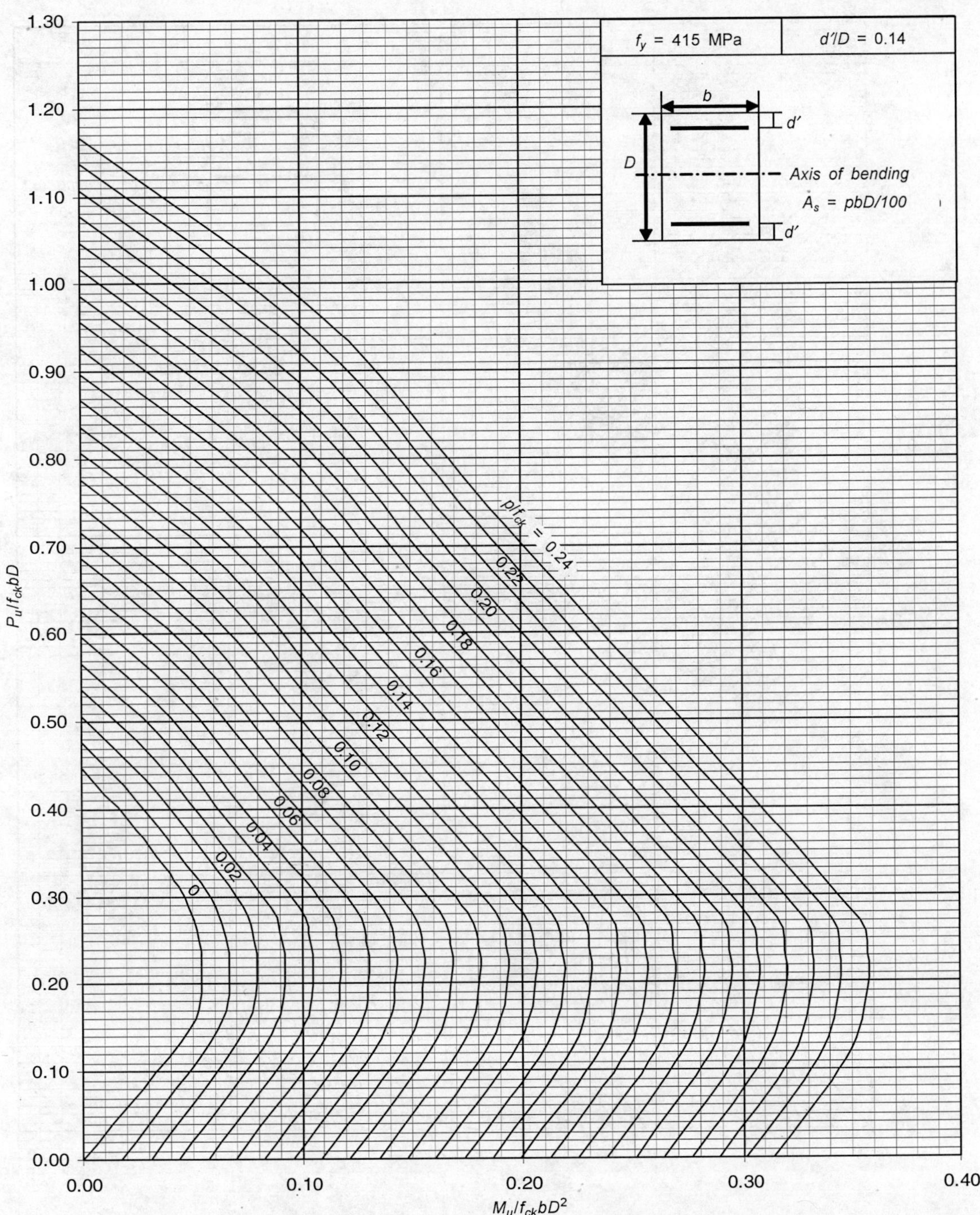

Fig. D.6 *Rectangular section subjected to compression and bending with reinforcement distributed equally on two faces.*

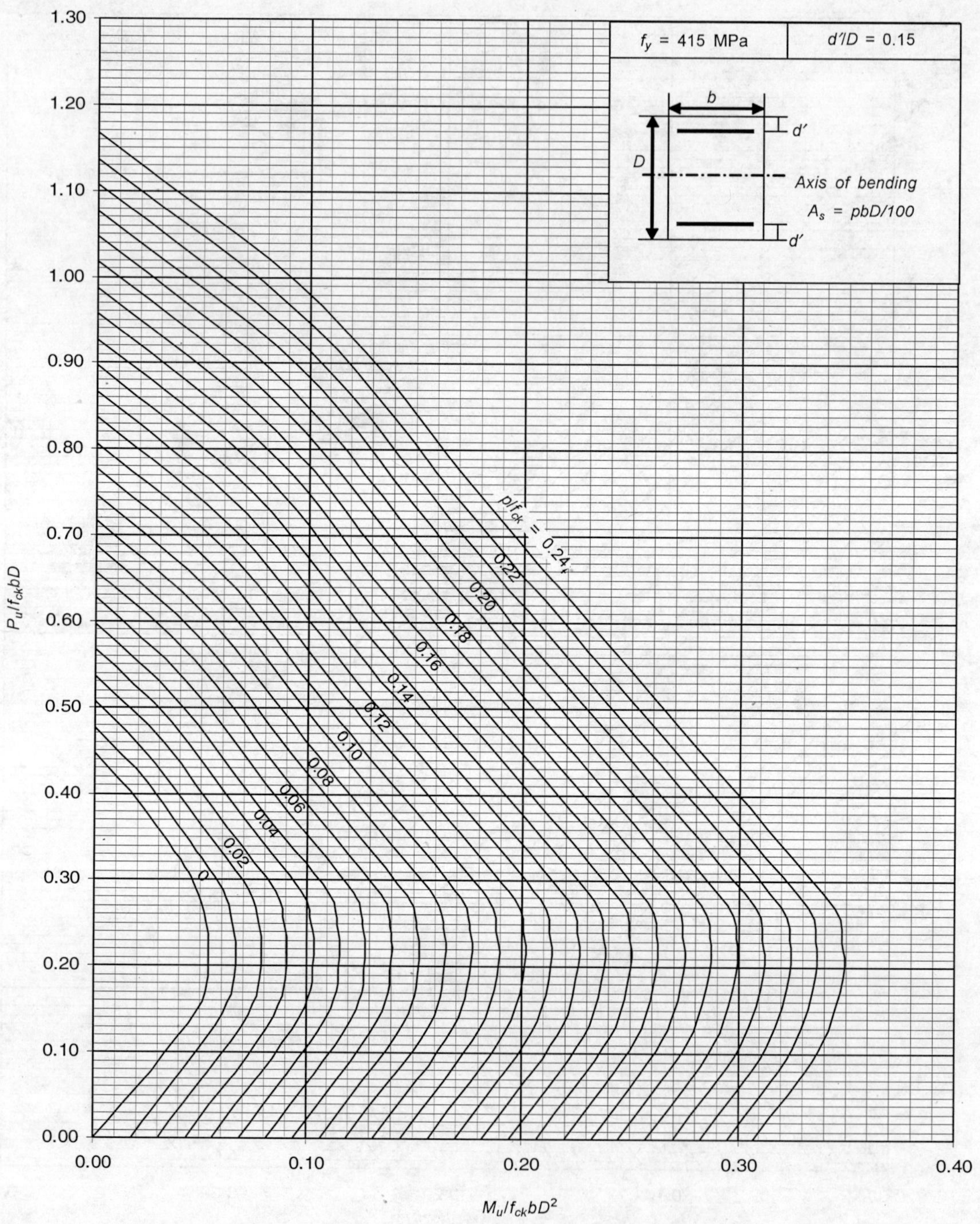

Fig. D.7 *Rectangular section subjected to compression and bending with reinforcement distributed equally on two faces.*

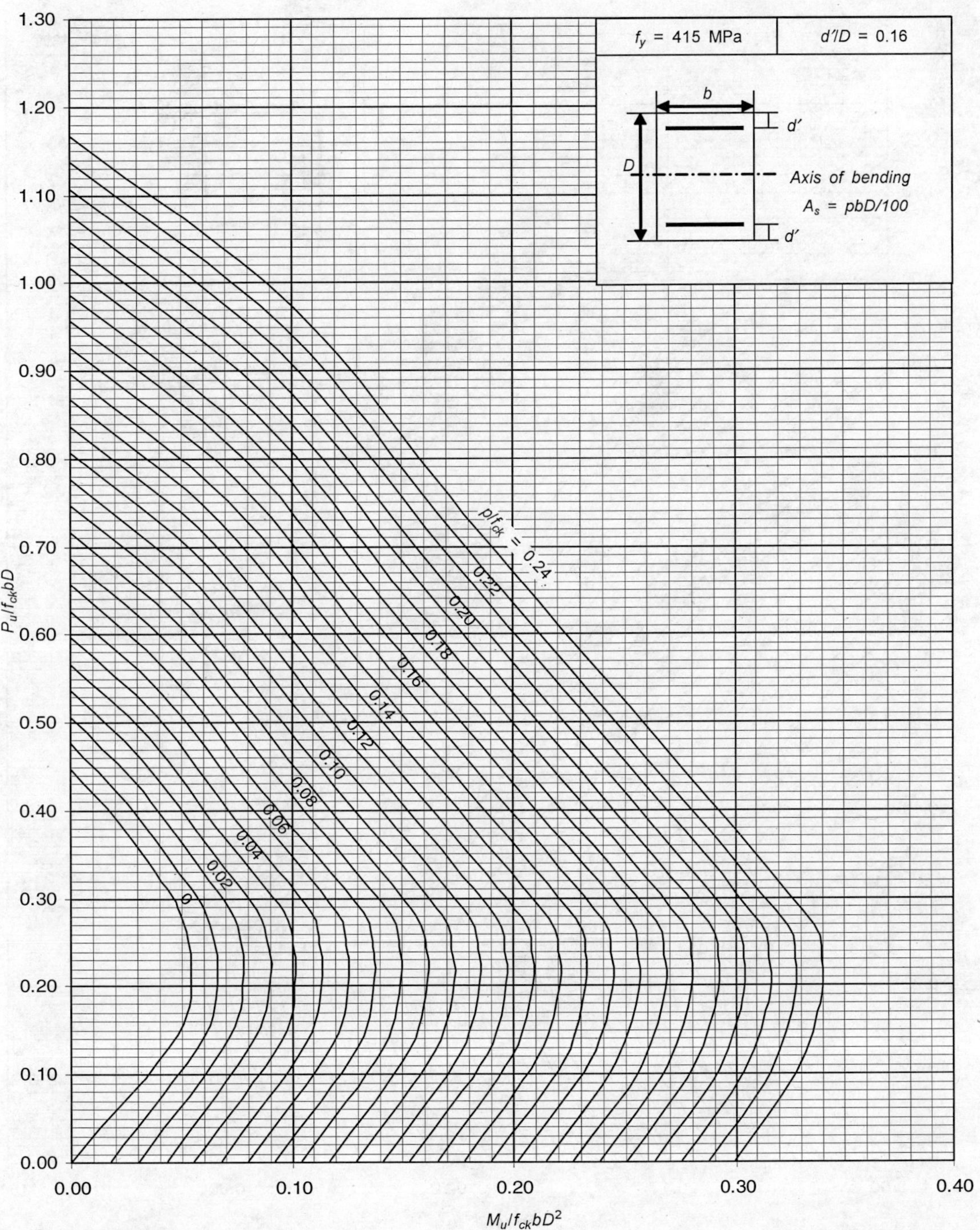

Fig. D.8 *Rectangular section subjected to compression and bending with reinforcement distributed equally on two faces.*

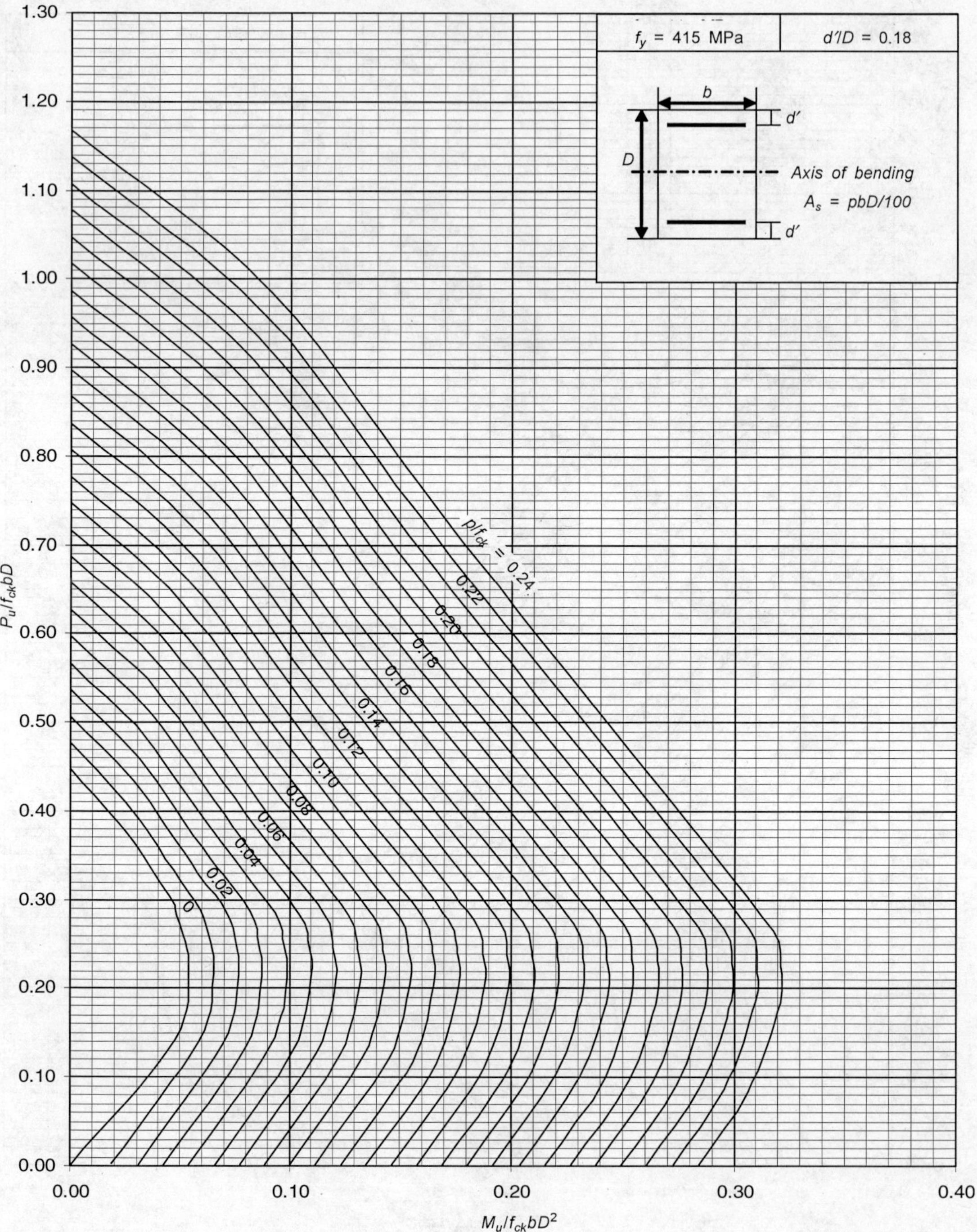

Fig. D.9 *Rectangular section subjected to compression and bending with reinforcement distributed equally on two faces.*

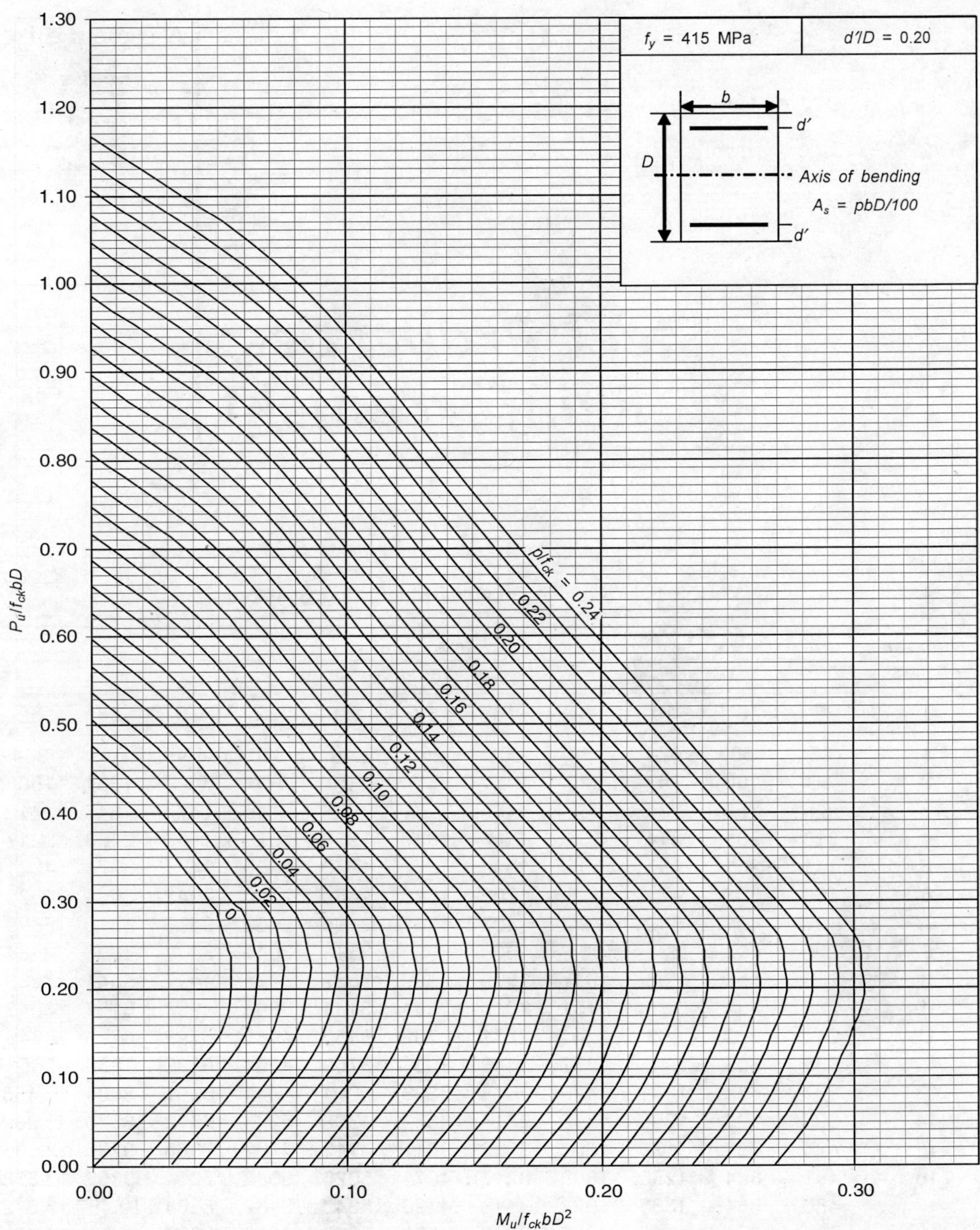

Fig. D.10 *Rectangular section subjected to compression and bending with reinforcement distributed equally on two faces.*

Appendix

E

Steel Properties: Reinforcement

TABLE E.1 Area of Group of Reinforcing Standard Bars

Number of bars	*Area,* mm²										
	Bar diameter, mm										
	6	8	10	12	16	18	20	22	25	28	32
1	28	50	078	113	201	254	314	380	490	615	804
2	56	100	157	226	402	508	628	760	981	1231	1608
3	84	150	235	339	603	763	942	1140	1472	1847	2412
4	113	201	314	452	804	1017	1256	1520	1963	2463	3217
5	141	251	392	565	1005	1272	1570	1900	2454	3078	4021
6	169	301	471	678	1206	1526	1885	2280	2945	3694	4825
7	197	351	549	791	1407	1781	2199	2660	3436	4310	5629
8	226	402	628	904	1608	2035	2513	3041	3927	4926	6434
9	254	452	706	1017	1809	2290	2827	3421	4417	5541	7238
10	282	502	785	1131	2010	2544	3141	3801	4908	6157	8042
11	311	552	863	1244	2211	2799	3455	4181	5399	6773	8846
12	339	603	942	1357	2412	3053	3769	4561	5890	7389	9651
13	367	653	1021	1470	2613	3308	4084	4941	6381	8004	10,455
14	395	703	1099	1583	2814	3562	4398	5321	6872	8620	11,259
15	424	754	1178	1696	3015	3817	4712	5702	7363	9236	12,063
16	452	804	1256	1809	3217	4071	5026	6082	7854	9852	12,868
17	480	854	1335	1922	3418	4326	5340	6462	8384	10,467	13,672
18	508	904	1413	2035	3619	4580	5654	6842	8835	11,083	14,476
19	537	955	1492	2148	3820	4834	5969	7222	9326	11,699	15,280
20	565	1005	1570	2262	4021	5089	6283	7602	9817	12,315	16,085

TABLE E.2 Number of Standard Reinforcing Bars for the Given Area

Area,	*Diameter of bar, mm*								*Area,*	*Diameter of bar, mm*							
mm²	*12*	*16*	*18*	*20*	*22*	*25*	*28*	*32*	mm²	*12*	*16*	*18*	*20*	*22*	*25*	*28*	*32*
200	2	-	-	-	-	-	-	-	2050	-	10	8	7	5	4	3	3
250	2	-	-	-	-	-	-	-	2100	-	10	8	7	6	4	3	3
300	3	-	-	-	-	-	-	-	2150	-	11	8	7	6	4	3	3
350	3	2	-	-	-	-	-	-	2200	-	11	9	**7**	6	4	4	3
400	4	**2**	-	-	-	-	-	-	2250	-	11	9	7	6	5	4	3
450	**4**	2	2	-	-	-	-	-	2300	-	11	9	7	6	5	4	3
500	4	2	**2**	-	-	-	-	-	2350	-	12	9	7	6	5	4	3
550	5	3	2	2	-	-	-	-	2400	-	12	9	8	6	5	4	**3**
600	5	**3**	2	2	-	-	-	-	2450	-	-	10	8	6	**5**	**4**	3
650	6	3	3	2	2	-	-	-	2500	-	-	10	**8**	7	5	4	3
700	6	3	3	2	2	-	-	-	2550	-	-	**10**	8	7	5	4	3
750	7	4	3	2	**2**	-	-	-	2600	-	-	10	8	7	5	4	3
800	7	**4**	3	3	2	-	-	-	2650	-	-	10	8	**7**	5	4	3
850	8	4	3	3	2	2	-	-	2700	-	-	11	9	7	6	4	3
900	**8**	4	4	3	2	2	-	-	2750	-	-	11	9	7	6	4	3
950	8	5	4	**3**	2	2	-	-	2800	-	-	**11**	9	7	6	5	3
1000	9	**5**	4	3	3	**2**	-	-	2850	-	-	11	9	7	6	5	4
1050	9	5	4	3	3	2	2	-	2900	-	-	11	9	8	6	5	4
1100	10	5	4	4	3	2	2	-	2950	-	-	12	9	8	**6**	5	4
1150	10	6	5	4	**3**	2	2	-	3000	-	-	12	10	8	6	5	4
1200	11	**6**	5	4	3	2	2	-	3050	-	-	**12**	10	**8**	6	**5**	4
1250	11	6	5	**4**	3	3	**2**	-	3100	-	-	-	10	8	6	5	4
1300	11	6	5	4	3	3	2	-	3150	-	-	-	**10**	8	6	5	4
1350	12	7	5	4	4	3	2	-	3200	-	-	-	10	8	7	5	**4**
1400	-	7	6	4	4	3	2	2	3250	-	-	-	10	9	7	5	4
1450	-	7	6	5	4	**3**	2	2	3300	-	-	-	11	9	7	5	4
1500	-	7	6	5	4	3	2	2	3350	-	-	-	11	9	7	5	4
1550	-	8	6	5	4	3	3	2	3400	-	-	-	11	9	7	6	4
1600	-	8	6	5	4	3	3	**2**	3450	-	-	-	**11**	9	**7**	6	4
1650	-	8	6	5	4	3	3	2	3500	-	-	-	11	9	7	6	4
1700	-	8	7	5	4	3	3	2	3550	-	-	-	11	9	7	6	4
1750	-	9	7	6	5	4	3	2	3600	-	-	-	11	9	7	6	4
1800	-	9	7	6	5	4	3	2	3650	-	-	-	12	10	7	6	5
1850	-	9	7	6	5	4	**3**	2	3700	-	-	-	12	10	8	**6**	5
1900	-	9	7	**6**	**5**	4	3	2	3750	-	-	-	12	10	8	6	5
1950	-	10	8	6	5	**4**	3	2	3800	-	-	-	-	**10**	8	6	5
2000	-	10	8	6	5	4	3	2	3850	-	-	-	-	10	8	6	5

(Contd.)

TABLE E.2 Number of Standard Reinforcing Bars for the Given Area (*Contd.*)

Area,	*Diameter of bar, mm*								*Area,*	*Diameter of bar, mm*							
mm^2	*12*	*16*	*18*	*20*	*22*	*25*	*28*	*32*	mm^2	*12*	*16*	*18*	*20*	*22*	*25*	*28*	*32*
3900	-	-	-	-	10	8	6	5	5750	-	-	-	-	-	12	9	7
3950	-	-	-	-	10	**8**	6	5	5800	-	-	-	-	-	12	9	7
4000	-	-	-	-	11	8	6	**5**	5850	-	-	-	-	-	12	10	7
4050	-	-	-	-	11	8	7	5	5900	-	-	-	-	-	-	10	7
4100	-	-	-	-	11	8	7	5	5950	-	-	-	-	-	-	10	7
4150	-	-	-	-	11	8	7	5	6000	-	-	-	-	-	-	10	7
4200	-	-	-	-	**11**	9	7	5	6050	-	-	-	-	-	-	10	8
4250	-	-	-	-	11	9	7	5	6100	-	-	-	-	-	-	10	8
4300	-	-	-	-	11	9	**7**	5	6150	-	-	-	-	-	-	**10**	8
4350	-	-	-	-	11	9	7	5	6200	-	-	-	-	-	-	10	**8**
4400	-	-	-	-	12	**9**	7	5	6250	-	-	-	-	-	-	10	8
4450	-	-	-	-	12	9	7	6	6300	-	-	-	-	-	-	**10**	8
4500	-	-	-	-	12	9	7	6	6350	-	-	-	-	-	-	10	8
4550	-	-	-	-	**12**	9	7	6	6400	-	-	-	-	-	-	10	8
4600	-	-	-	-	-	9	7	6	6450	-	-	-	-	-	-	10	**8**
4650	-	-	-	-	-	9	8	6	6500	-	-	-	-	-	-	11	8
4700	-	-	-	-	-	10	8	6	6550	-	-	-	-	-	-	11	8
4750	-	-	-	-	-	10	8	6	6600	-	-	-	-	-	-	11	8
4800	-	-	-	-	-	10	8	**6**	6650	-	-	-	-	-	-	11	8
4850	-	-	-	-	-	10	8	6	6700	-	-	-	-	-	-	11	8
4900	-	-	-	-	-	**10**	**8**	6	6750	-	-	-	-	-	-	**11**	8
4950	-	-	-	-	-	10	**8**	6	6800	-	-	-	-	-	-	11	8
5000	-	-	-	-	-	10	8	6	6850	-	-	-	-	-	-	11	9
5050	-	-	-	-	-	10	8	6	6900	-	-	-	-	-	-	11	9
5100	-	-	-	-	-	10	8	6	6950	-	-	-	-	-	-	11	9
5150	-	-	-	-	-	10	8	6	7000	-	-	-	-	-	-	11	**9**
5200	-	-	-	-	-	11	8	6	7050	-	-	-	-	-	-	11	9
5250	-	-	-	-	-	11	9	7	7100	-	-	-	-	-	-	12	9
5300	-	-	-	-	-	11	9	7	7150	-	-	-	-	-	-	12	9
5350	-	-	-	-	-	11	9	7	7200	-	-	-	-	-	-	12	9
5400	-	-	-	-	-	**11**	9	7	7250	-	-	-	-	-	-	12	**9**
5450	-	-	-	-	-	11	9	7	7300	-	-	-	-	-	-	12	9
5500	-	-	-	-	-	11	9	7	7350	-	-	-	-	-	-	**12**	9
5550	-	-	-	-	-	11	**9**	7	7400	-	-	-	-	-	-	-	9
5600	-	-	-	-	-	11	9	7	7450	-	-	-	-	-	-	-	9
5650	-	-	-	-	-	12	9	**7**	7500	-	-	-	-	-	-	-	9
5700	-	-	-	-	-	12	9	7	7550	-	-	-	-	-	-	-	9

(*Contd.*)

TABLE E.2 Number of Standard Reinforcing Bars for the Given Area (*Contd.*)

Area,	*Diameter of bar,* mm								*Area,*	*Diameter of bar,* mm							
mm^2	*12*	*16*	*18*	*20*	*22*	*25*	*28*	*32*	mm^2	*12*	*16*	*18*	*20*	*22*	*25*	*28*	*32*
7600	-	-	-	-	-	-	-	9	8650	-	-	-	-	-	-	-	11
7650	-	-	-	-	-	-	-	10	8700	-	-	-	-	-	-	-	11
7700	-	-	-	-	-	-	-	10	8750	-	-	-	-	-	-	-	11
7750	-	-	-	-	-	-	-	10	8800	-	-	-	-	-	-	-	11
7800	-	-	-	-	-	-	-	10	8850	-	-	-	-	-	-	-	**11**
7850	-	-	-	-	-	-	-	10	8900	-	-	-	-	-	-	-	11
7900	-	-	-	-	-	-	-	10	8950	-	-	-	-	-	-	-	11
7950	-	-	-	-	-	-	-	10	9000	-	-	-	-	-	-	-	11
8000	-	-	-	-	-	-	-	10	9050	-	-	-	-	-	-	-	11
8050	-	-	-	-	-	-	-	**10**	9100	-	-	-	-	-	-	-	11
8100	-	-	-	-	-	-	-	10	9150	-	-	-	-	-	-	-	11
8150	-	-	-	-	-	-	-	10	9200	-	-	-	-	-	-	-	11
8200	-	-	-	-	-	-	-	10	9250	-	-	-	-	-	-	-	12
8250	-	-	-	-	-	-	-	10	9300	-	-	-	-	-	-	-	12
8300	-	-	-	-	-	-	-	10	9350	-	-	-	-	-	-	-	12
8350	-	-	-	-	-	-	-	10	9400	-	-	-	-	-	-	-	12
8400	-	-	-	-	-	-	-	10	9450	-	-	-	-	-	-	-	12
8450	-	-	-	-	-	-	-	11	9500	-	-	-	-	-	-	-	12
8500	-	-	-	-	-	-	-	11	9550	-	-	-	-	-	-	-	12
8550	-	-	-	-	-	-	-	11	9600	-	-	-	-	-	-	-	12
8600	-	-	-	-	-	-	-	11	9650	-	-	-	-	-	-	-	**12**

Note: Shaded figures give exact or very close area to that shown in column no. 1.

TABLE E.3 Areas of Bars in Reinforcement Mesh, e.g., Slab Reinforcement

Spacing mm	*Area,* mm²/m									
	Bar diameter, mm									
	6	*8*	*10*	*12*	*14*	*16*	*18*	*20*	*22*	*25*
50	565	1005	1571	2262	3079	4021	5089	6283	7603	9817
60	471	838	1309	1885	2566	3351	4241	5236	6336	8181
70	404	718	1122	1616	2199	2872	3635	4448	5430	7012
80	353	628	982	1414	1924	2513	3181	3927	4752	6136
90	314	558	873	1257	1710	2234	2827	3491	4224	5454
100	283	503	785	1131	1539	2011	2545	3142	3801	4909
110	257	457	714	1028	1399	1828	2313	2856	3456	4462
120	236	419	654	942	1283	1675	2121	2618	3168	4091
130	217	387	604	870	1184	1547	1957	2417	2924	3776
140	202	359	561	808	1100	1436	1818	2244	2715	3506
150	188	335	524	754	1026	1340	1696	2094	2534	3272
160	177	314	491	707	962	1257	1590	1963	2376	3068
170	166	296	462	665	905	1183	1497	1848	2236	2887
180	157	279	436	628	855	1117	1444	1745	2112	2727
190	149	265	413	595	810	1058	1339	1653	2001	2584
200	141	251	393	565	770	1005	1272	1571	1901	2454
210	135	239	374	539	733	957	1212	1496	1810	2337
220	128	228	357	514	700	914	1157	1428	1728	2231
230	123	218	341	492	669	874	1106	1366	1653	2134
240	118	209	327	471	641	838	1060	1309	1584	2045
250	113	201	314	452	616	804	1018	1257	1520	1963
260	109	193	302	435	592	773	979	1208	1462	1888
270	105	186	291	419	570	745	942	1164	1408	1818
280	101	179	280	404	550	718	909	1122	1358	1753
290	97	173	271	390	531	693	977	1083	1311	1693
300	94	168	262	377	513	670	848	1047	1267	1636
320	88	157	245	353	481	628	795	982	1188	1534
340	83	148	231	333	453	591	748	924	1418	1444
360	78	140	218	314	428	558	707	873	1056	1363
380	74	132	207	298	405	529	670	827	1000	1292
400	71	126	196	283	385	503	636	785	950	1227

TABLE E.4 Perimeter of Group of Reinforcing Bars

Number of bars	Perimeter, mm Bar diameter, mm										
	6	8	10	12	16	18	20	22	25	28	32
1	19	25	31	38	50	57	63	69	79	88	100
2	37	50	62	75	100	113	125	138	157	175	201
3	56	75	94	113	150	169	188	207	235	263	301
4	75	100	125	150	201	226	251	276	314	351	402
5	94	125	157	188	251	282	314	345	392	439	502
6	113	150	188	226	301	339	376	414	471	527	603
7	131	175	219	263	351	395	439	483	549	615	703
8	150	201	251	301	402	452	502	552	628	703	804
9	169	226	282	339	452	508	565	622	706	791	904
10	188	251	314	376	502	565	628	691	785	879	1005
11	207	276	345	414	552	622	691	760	863	967	1105
12	226	301	376	452	603	678	753	829	942	1055	1206
13	245	326	408	490	653	735	816	898	1021	1143	1306
14	263	351	439	527	703	791	879	967	1099	1231	1407
15	282	376	471	565	753	848	942	1036	1178	1319	1507
16	301	402	502	603	804	904	1005	1105	1256	1407	1608
17	320	427	534	640	854	961	1068	1174	1335	1495	1709
18	339	452	565	678	904	1017	1130	1244	1413	1583	1809
19	358	477	596	716	955	1074	1193	1313	1492	1671	1910
20	376	502	628	753	1005	1130	1256	1382	1570	1759	2010

TABLE E.5 Area, Perimeter, Mass m, and Mass of Steel in a mesh for Specified Spacing of Bars

Dia ϕ, mm	*Area*, mm^2	*Peri-meter*, mm	*Mass*, kg/m	*Length /tonne*, m	*Mass of steel per sq.m*, kg — *Spacing of bars*, mm										
					50	*75*	*100*	*125*	*150*	*175*	*200*	*225*	*250*	*275*	*300*
6	28.3	18.8	0.222	4505.7	4.44	2.96	2.22	1.78	1.48	1.27	1.11	0.99	0.89	0.81	0.74
8	50.3	25.1	0.395	2534.5	7.89	5.26	3.95	3.16	2.63	2.25	1.97	1.75	1.58	1.43	1.32
10	78.5	31.4	0.617	1622.1	12.33	8.22	6.17	4.93	4.11	3.52	3.08	2.74	2.47	2.24	2.06
12	113.1	37.7	0.888	1126.4	17.76	11.84	8.88	7.10	5.92	5.07	4.44	3.95	3.55	3.23	2.96
16	201.1	50.3	1.578	633.6	31.57	21.04	15.78	12.63	10.52	9.02	7.89	7.01	6.31	5.74	5.26
18	254.5	56.5	1.997	500.6	39.95	26.63	19.98	15.98	13.32	11.41	9.99	8.88	7.99	7.26	6.66
20	314.2	62.8	2.466	405.5	49.32	32.88	24.66	19.73	16.44	14.09	12.33	10.96	9.86	8.97	8.22
22	380.1	69.1	2.984	335.1	59.68	39.79	29.84	23.87	19.89	17.05	14.92	13.26	11.94	10.85	9.95
25	490.9	78.5	3.853	259.5	77.07	51.38	38.53	30.83	25.69	22.02	19.27	17.13	15.41	14.01	12.84
28	615.8	88.0	4.833	206.9	96.67	64.45	48.34	38.67	32.22	27.62	24.17	21.48	19.33	17.58	16.11
32	804.2	100.5	6.313	158.4	126.27	84.18	63.13	50.51	42.09	36.08	31.57	28.06	25.25	22.96	21.04
36	1017.9	113.1	7.990	125.2	159.81	106.54	79.90	63.92	53.27	45.66	39.95	35.51	31.96	29.06	26.63

Note: The information in the table is based on the following:

Basic weight	= 0.00785 kg/mm^2/m
Weight per metre	= 0.0061654 ϕ^2 (kg)
Weight per mm^2 at spacings (mm)	= 6.1654 ϕ^2/s (kg)
ϕ	= Diameter of bar in mm.
S	= Spacing of bars in mm

Appendix

F

References

Selected Codes of Practice

ACI:318–1986, *Building Code Requirements for Reinforced Concrete*, The American Concrete Institute, Detroit, 1986.

ACI:318R–95, *Commentary on Building Code Requirements for Structural Concrete,* The American Concrete Institute, Detroit, Michigan, 1995.

BS:8007–1987, *Design of Concrete Structures for Retaining Aqueous Liquids,* The British Standards Institution, London, 1987.

BS:8110–1985, *Structural Use of Concrete*: Part I—*Code of Practice for Design and Construction;* Part II—*Code of Practice for Special Circumstances*, The British Standards Institution, London,1989.

CP:110–1972, *The Structural Use of Concrete,* Part-I—*Design, Materials and Workmanship*, The British Standards Institution, London, 1972.

CSA-CAN:A23.3–M84, *Design of Concrete Structures for Buildings,* The Canadian Standards Association, Rexdale, Ontario, 1994.

IS:432 (Parts I and II)-1982, *Code of Practice for Mild Steel and Medium Tensile Steel Bars and Hard Drawn Steel Wire for Concrete Reinforcement:* Part I—*Mild Steel and Medium Tensile Steel Bars*, and Part II—*Hard Drawn Steel Wire*, Bureau of Indian Standards, New Delhi, 1982.

IS:456–2000, *Code of Practice for Plain and Reinforced Concrete,* Bureau of Indian Standards, New Delhi, 2000.

IS:875 (Parts I–V)–1987, *Code of Practice for Design Loads for Buildings and Structures*: Part I—*Dead Loads;* Part II—*Imposed (Live)* Loads; Part III—*Wind Loads*; Part IV—*Snow Loads;* and Part V—*Special Loads and Load Combinations,* Bureau of Indian Standards, New Delhi, 1989.

IS:1139–1966, *Code of Practice for Deformed Bars for Concrete Reinforcement, Hot-rolled Mild Steel and Medium Tensile Steel,* Bureau of Indian Standards, New Delhi, 1966.

IS:1566–1982, *Code of Practice for Fabric for Concrete Reinforcement, Hard-dawn Steel Wires,* Bureau of Indian Standards, New Delhi, 1982.

IS:1786–1979, *Code of Practice for Cold-worked High Strength Deformed Bars for Concrete Reinforcement,* Bureau of Indian Standards, New Delhi, 1979.

IS:1893–2002, *Criteria for Earthquake Resistant Design of Structures,* Bureau of Indian Standards, New Delhi, 2002.

IS:1904–1986, *Code of Practice for Structural Safety of Buildings: Shallow Foundations,* Bureau of Indian Standards, New Delhi, 1986.

IS:2131–1981, *Method of Standard Penetration Test for Soils,* Bureau of Indian Standards, New Delhi, 1981.

IS:2502–1963, *Code of Practice for Bending and Fixing of Bars for Concrete Reinforcement,* Bureau of Indian Standards, New Delhi, 1963.

IS:3370 (Part IV)–1967, *Code of Practice for Concrete Structures for Storage of Liquids:* Part IV—*Design Tables*, Bureau of Indian Standards, New Delhi, 1967.

IS:4326–1993 (reaffirmed 1998), *Code of Practice for Earthquake Resistant Design and Construction of Buildings,* Bureau of Indian Standards, New Delhi, 1993.

IS:5525–1969, *Recommendations for Detailing of Reinforcement in Reinforced Concrete Works*, Bureau of Indian Standards, New Delhi, 1969.

IS:10262–1982, *Recommended Guidelines for Concrete Mix Design*, Bureau of Indian Standards, New Delhi, 1982.

IS:13920–1993, Ductile Detailing of Reinforced Concrete Structures Subjected to Seismic Forces, Bureau of Indian Standards, New Delhi, 1993.

Selected Special Publications

SP:16–1980, *Design Aids for Reinforced Concrete* to IS:456–1978, Bureau of Indian Standards, New Delhi, 1980.

SP:22–1982, *Explanatory Handbook on Codes for Earthquake Engineering,* Bureau of Indian Standards, New Delhi, 1982.

SP:24–1983, *Explanatory Handbook* on IS: 456–1978, Bureau of Indian Standards, New Delhi, 1983.

SP:34–1987, *Handbook on Concrete Reinforcement and Detailing,* Bureau of Indian Standards, New Delhi, 1987.

Selected Books

Dayaratnam, P., *Design of Reinforced Concrete Structures,* Oxford & IBH, New Delhi, 1984.

Ferguson, P.M., *Reinforced Concrete Fundamentals,* John Wiley & Sons Inc., New York, 1981.

Gambhir, M.L., *Concrete Technology,* Tata McGraw-Hill, 3rd ed., New Delhi, 2004.

Gambhir, M.L., *Stability Analysis and Design of Structures*, Springer-Verlag, Berlin, Heidelberg, 2004.

Gambhir, M.L., *Concrete Manual: Laboratory Testing for Quality Control of Concrete*, 3rd ed., Dhanpat Rai & Sons, Delhi, 1987.

Hughes, B.P., *Limit State Theory Reinforced Concrete*, Pitman Publishing Co., London, 1971.

Jain, A.K., *Reinforced Concrete,* Nem Chand & Bros, Roorkee, 2002.

Krishna, J. and Jain, O.P., *Plain and Reinforced Concrete,* Vol. 1, 8th ed., Nem Chand & Bros., Roorkee, 1981.

Mallick, S.K. and Gupta, A.P., *Reinforced Concrete,* Oxford & IBH, Calcutta, 1993.

Park, P. and Paulay, T., *Reinforced Concrete Structures,* John Wiley & Sons Inc., New York, 1981.

Pillai, S.U. and Menon, D., *Reinforced Concrete Design,* 1st ed., Tata McGraw-Hill, New Delhi, 1998.

Purushothaman, P., *Reinforced Concrete Structural Elements—Behaviour, Analysis and Design,* Tata McGraw-Hill, New Delhi, 1984.

Ranganathan, R., *Reliability Analysis and Design of Structures,* Tata McGraw-Hill, New Delhi, 1990.

Raynolds, C.E. and Steadman, J.C., *Reinforced Concrete Designer's Handbook,* Cement and Concrete Association, London, 1981.

Scordelis, A.C., *Internal Forces in Uniformly Loaded Helicoidal Girders,* ACI Journal, Vol. 56, pp. 1013–1026, April 1960.

Sinha, S.N., *Reinforced Concrete Design,* Tata McGraw-Hill, New Delhi, 2002.

UBC–1994, *Uniform Building Code*, International Conference of Building Officials, Whittier, California, USA, 1994.

Varghese, P.C., *Limit States Design of Reinforced Concrete,* Prentice-Hall of India, New Delhi, 1994.

Index